Lecture Notes in Mathematics 1599

Editors:
A. Dold, Heidelberg
F. Takens, Groningen

T0211428

Klaus Johannson

Topology
and Combinatorics
of 3-Manifolds

 Springer

Author

Klaus Johannson
Department of Mathematics
University of Tennessee
Knoxville, TN 37996, USA
E-mail: johann@math.utk.edu

Mathematics Subject Classification (1991): 57M99

ISBN 3-540-59063-3 Springer-Verlag Berlin Heidelberg New York

CIP-Data applied for

This work is subject to copyright. All rights are reserved, whether the whole or part of the material is concerned, specifically the rights of translation, reprinting, re-use of illustrations, recitation, broadcasting, reproduction on microfilms or in any other way, and storage in data banks. Duplication of this publication or parts thereof is permitted only under the provisions of the German Copyright Law of September 9, 1965, in its current version, and permission for use must always be obtained from Springer-Verlag. Violations are liable for prosecution under the German Copyright Law.

© Springer-Verlag Berlin Heidelberg 1995
Printed in Germany

Typesetting: Camera-ready TeX output by the author
SPIN: 10130263   46/3142-543210 - Printed on acid-free paper

INTRODUCTION

This book is concerned with the combinatorial rigidity of 3-manifolds. Indeed, the main result of this book says that combinatorial structures of all Haken 3-manifolds without non-trivial Stallings fibrations are virtually rigid, i.e., rigid up to finitely many choices. It follows in particular that the combinatorial structures of all hyperbolic Haken 3-manifolds with infinite volume are virtually rigid. Recall that a Haken 3-manifold is a compact piecewise linear 3-manifold which is irreducible, in the sense that every PL-embedded disc or 2-sphere separates a 3-ball, and which is sufficiently large, in the sense that it contains an orientable surface whose fundamental group is non-trivial and whose embedding induces an injection on the fundamental groups. For instance, all non-trivial knot spaces are Haken 3-manifolds.

At the very basis of 3-manifold theory there are the classical theorems that every compact 3-manifold has a piecewise linear structure, i.e., an atlas of charts whose transition maps are piecewise linear, and that any topological homeomorphism between piecewise linear 3-manifolds can be isotoped into a piecewise linear homeomorphism. In short every compact 3-manifold has a rigid piecewise linear structure. This theorem has been proved by Moise [Moi 1, 2] (for other proofs see [Bin], [Sh] and [Ham]). It is common to realize the piecewise linear structure of a 3-manifold by a combinatorial structure, e.g., by a triangulation. But note that in doing so one loses information which is reflected by the fact that there are various different triangulations which realize the underlying piecewise linear structure of a 3-manifold. Thus the combinatorics of a 3-manifold is no longer rigid. From the rigidity of piecewise linear structures one can only deduce that any two triangulations of a 3-manifold have a common sub-division. Thus any two triangulations are stably equivalent but in general not equivalent. This deficiency has proved to be an obstacle. For instance, it is the reason for the well-known difficulties in approaching the classification problem for 3-manifolds from a combinatorial point of view (i.e, via Heegaard diagrams, etc.). Virtually all results in the present theory of 3-manifolds are ultimately based on the *existence* of triangulations while, in contrast, stable equivalence has received much less attention.

It appears that stable equivalence is an equivalence relation which is too strong for most applications. It therefore suggests itself to look for other, better suited equivalence relations on the set of all combinatorial structures. From now on (and for convenience) we assume that all 3-manifolds are irreducible. Observe that a triangulation of an irreducible 3-manifold is completely determined (modulo isotopy) by its 1-skeleton alone, and so we can concentrate on graphs. More precisely, we can concentrate on Heegaard graphs. By definition, a graph, Γ, in a 3-manifold, N, is a Heegaard graph if $(N - U(\Gamma))^-$ is a handlebody (here $U(..)$ denotes regular neighborhoods). We say two Heegaard graphs $\Gamma_1, \Gamma_2 \subset N$ are slide-equivalent if $U(\Gamma_1 \cup \partial N)$ and $U(\Gamma_2 \cup \partial N)$ are ambient isotopic in N. We will see that, equivalently, slide-equivalent Heegaard graphs can be obtained from each other by sliding their edges. A surface $F \subset N$ is called a Heegaard surface, provided there is a Heegaard graph $\Gamma \subset N$ such that F is isotopic to $(\partial U(\Gamma \cup \partial N) - \partial N)^-$. Note that slide-equivalence classes of Heegaard graphs are geometrically represented by their associated Heegaard surfaces and vice versa. It turns out that the extra information encoded in Heegaard graphs is useful for a study of Heegaard surfaces.

By the very definition, Heegaard surfaces are totally compressible from both sides by disc-compressions. Thus Heegaard surfaces and incompressible surfaces are two extreme cases with respect to their compression properties. This book deals with both cases. It turns out that there is a strong interaction between them which can be exploited.

One of the intriguing features of 3-manifolds is the fact that they are finite combinatorial objects, and yet they still have enough room for very complicated and globally very different embeddings of codim 1 submanifolds. For instance, it is known that 3-manifolds with boundaries (such as knot spaces) have in general infinitely many non-isotopic incompressible surfaces. Moreover, Casson and Gordon (unpublished) have constructed 3-manifolds which contain (irreducible) Heegaard surfaces of arbitrarily large genus. In fact, it is now known that the set of all Heegaard surfaces in a 3-manifold is usually quite rich (see section 26). On the other hand, Haken [Ha 1] has demonstrated that every simple 3-manifold (i.e., every Haken 3-manifold without essential annuli and tori) contains only finitely many non-isotopic incompressible surfaces of given Euler characteristic, and that furthermore every incompressible surface in an irreducible 3-manifold can be obtained from a finite and constructable set of fundamental surfaces by using cut-and-paste alone. Haken's approach to incompressible surfaces has been extended by various authors and is the basis of many deep theorems in 3-manifold theory. Far less is known about Heegaard surfaces and in [Wa 5] Waldhausen raised the problem whether similar results are conceivable for Heegaard surfaces in 3-manifolds as well. Our main result concerning Heegaard surfaces (see thm. 32.17) gives an answer to this question for torus-free Haken 3-manifolds with non-empty boundary. More generally, we will show:

Rigidity Theorem. *Let N be a Haken 3-manifold (with or without boundary) which contains no non-trivial, essential Stallings fibration. Then the set of all isotopy classes of Heegaard surfaces in N of any given genus is finite and constructable. In particular, the Heegaard genus of N can be determined.*

Remark. This theorem has been proved by Pitts and Rubinstein for all closed 3-manifolds using different methods (see [Ru]).

(The term "constructable" has to be taken here with a grain of salt, its precise meaning is given in remark 32.21). Recall that a Stallings fibration is the mapping torus of a homeomorphism of some orientable surface. A Stallings fibration $X \subset N$ is essential and non-trivial if ∂X is incompressible in N and if X cannot be isotoped into $U(\partial N)$.

As an immediate corollary we get that, modulo Nielsen-equivalence, every fundamental group, $\pi_1 N$, of a 3-manifold, N, as in the theorem has only finitely many "geometric" presentations of given rank. In other words the "geometric presentation space" of these groups is compact. At this point recall that, according to Thurston [Th 2], a hyperbolic Haken 3-manifold, N, may be characterized as an atoroidal Haken 3-manifold (except for one small exception) and that hyperbolic structures on N are given by "geometric" representations $\pi_1(N) \to \mathrm{PSL}_2\mathbf{C}$. As a fundamental extension of Mostow's rigidity theorem it has been shown by Thurston [Th 4] that the "geometric representation space" of $\pi_1 N$ is compact, provided N contains no essential annuli or tori. Thurston proves this theorem by estimating the hyperbolic lengths of fixed curves for sequences of representations (hyperbolic structures). We prove the above theorem by introducing a combinatorial length for geometric presentations and estimating the combinatorial lengths for sequences of geometric presentations. See below for more details.

The above Rigidity Theorem reflects only one side of the available combinatorial rigidity for Haken 3-manifolds. Indeed, a second side is provided by the finiteness of the set of all great hierarchies in a Haken 3-manifold (see e.g. [Joh 1]). Both these rigidity theorems complement each other and they are due to the rigid way incompressible surfaces and Heegaard surfaces interact. The study of this interaction is the centerpiece of this book.

A first result concerning this interaction of essential surfaces (in the sense of [Joh 1]) with Heegaard surfaces has been established by Haken's theorem [Ha 3] (for a new proof see prop. 23.21), that the existence of an essential 2-sphere in a 3-manifold with fixed Heegaard surface implies the existence of such a 2-sphere intersecting the Heegaard surface in exactly one curve. Another formulation of this statement is that the Heegaard genus is additive under connected sums. This result of Haken has been extended to projective planes [Och] and to non-separating, essential tori in Heegaard genus two 3-manifolds [Ko 1] (see also [Joh 5]). Our starting point is a generalization of Haken's result to all surfaces as expressed by the following theorem (see cor. 23.34):

General Handle-Addition Theorem. *To every compact and irreducible 3-manifold, N, there is associated a polynomial function $\varphi : \mathbf{N} \times \mathbf{N} \to \mathbf{N}$ so that, for every essential surface $S \subset N$ and every Heegaard surface $F \subset N$, we have*

$$\#(\alpha_1 S \cap hF) \leq \varphi(\chi S, \chi F),$$

where α_t, $t \in [0,1]$, is some ambient isotopy in N and where h is some finite product of Dehn twist along essential tori in N.

This general form of Haken's theorem is due to a hidden connection of Haken's 2-sphere theorem on the one hand and Haken's powerful theory of normal surfaces [Ha 1] on the other. To make this connection apparent, we generalize Haken's approach. While Haken's normal surface theory is based on a fixed chosen handlebody-structure of the underlying 3-manifold, we now allow the handlebody-structures to be changed. A convenient working context for this approach is provided by the concept of n-relator 3-manifolds. Recall from [Joh 3] that an n-relator 3-manifold, $M^+(k)$, is given by a handlebody, M, with n 2-handles attached along a system, k, of n simple closed curves in ∂M. Now, consider an essential surface, S^+, in such an n-relator 3-manifold. In general, we cannot expect any relationship between S^+ and the intersection $S := S^+ \cap M$. The reason being that the Euler characteristic of S does not only depend on that of S^+, but also on the intersection of S^+ with the attached 2-handles of $M^+(k)$; and the latter can be increased arbitrarily by sliding these 2-handles around. Of course, none of these slides changes the homeomorphism type of the manifold $M^+(k)$ or the embedding type of the surface S^+. Thus we can replace $M^+(k)$ by any other n-relator 3-manifold obtained from it by handle-slides. Taking this extra flexibility into account we develop a normal surface theory for sets of n-relator structures. Normal surfaces for sets of n-relator structures will be called strictly normal. It turns out that, as in Haken's theory, every incompressible surface in an (irreducible) 3-manifold can be isotoped into strictly normal form. The main advantage of strictly normal surfaces is that their intersection with ∂M can be estimated. Thus, replacing $M^+(k)$ by an n-relator structure for which S^+ is strictly normal, we now find that S^+ and $S = S^+ \cap M$ are closely related. From this point of view, it suggests itself to study surfaces in $M^+(k)$ (and so surfaces in Haken 3-manifolds in general) by studying surfaces in the relative handlebody (M, k) which do not meet k. It turns out that a variety of properties of essential surfaces in 3-manifolds can be obtained by studying essential surfaces in relative handlebodies. This is the starting point of this book.

In special cases, the estimates from the General Handle-Addition Theorem take an especially simple form. For instance, for 1-relator 3-manifolds we know (see [Joh 3] or prop. 15.11) that there are essential, non-separating surfaces which do not meet the attached 2-handle at all. Moreover, the intersection of a strictly normal 2-sphere with 2-handles is connected. In this way, the two former seemingly unrelated theorems of Haken, concerning the finiteness of incompressible surfaces on the one hand and the summability of the Heegaard genus on the other, now appear as special consequences of the virtual rigid interaction of Heegaard surfaces and incompressible surfaces.

So far the General Handle-Addition Theorem has been viewed as a statement concerning essential surfaces. But our point is that it can also be taken as a statement about Heegaard surfaces, or sets of Heegaard surfaces. Indeed, it tells us that *any* Heegaard surface can meet a given essential surface in some limited way only. As a first expression of this viewpoint we prove a sum-formula for Heegaard genera concerning splittings along essential surfaces other than 2-spheres (see prop. 23.40). We further derive Frohmann's criterion for the reducibility of Heegaard surfaces (see prop. 23.22) and give, following an observation of Frohmann and Scharlemann, a quick proof of Waldhausen's theorem that all Heegaard surfaces of S^3 are standard (see thm. 23.23).

To prove the rigidity theorem concerning Heegaard surfaces it is certainly not enough to consider the intersection of Heegaard surfaces with a single incompressible surface alone; we rather have to study the intersection of Heegaard surfaces, or Heegaard graphs, with a complete hierarchy of surfaces. Now hierarchies are a familiar tool in the theory of Haken 3-manifolds. They are designed to set up an induction by splitting a given Haken 3-manifold into small pieces - usually 3-balls. For our purpose, however, it turns out that we are not allowed to split along hierarchies. We therefore use hierarchies to construct Haken 2-complexes. Basically, a Haken 2-complex is the union of all incompressible surfaces of a hierarchy. So conceptually a Haken 2-complex is not much different from a hierarchy. Nevertheless, one can formulate certain additional features which make them useful for us and preferable over hierarchies. Unfortunately, they are technically also rather hard to handle. To get a quicker idea for the special flexibility of this concept the reader may wish to consult [Joh 9]. There it is studied how singular discs and 2-spheres may intersect a Haken 2-complex, and as a result it is shown that a new proof of the loop-theorem can be given, using Haken 2-complexes instead of the tower-construction. To study the intersection of Heegaard graphs with Haken 2-complexes, we first associate a Heegaard 2-complex to the Heegaard graph, by attaching to it a complete system of meridian-discs of its complement. Then we study the intersection of Heegaard 2-complexes with Haken 2-complexes. Thus we investigate the intersection of two CW-complexes. Of course, in general such a study is hopeless but in our case the CW-complexes come with some nice internal structures which make them more accessible.

Here is a more detailed outline of this book in form of a Leitfaden.

We begin with handlebodies (chapt. 1), we turn to relative handlebodies (chapt. 2), we study 1-relator manifolds (chapt. 3), we expand our investigation to cover n-relator manifolds (chapt. 4) and finally we are ready to analyze the set of all n-relator structures of a given n-relator 3-manifold (chapt. 5).

Chapter I sets the tone. In section 1 we discuss slides and give a proof that all Heegaard graphs in handlebodies are slide-equivalent. We then use the same technique to give a quick proof for Dehn's theorem [De][Li 1] that the mapping class group of orientable surfaces is generated by Dehn twists. We take this as an indication that a good knowledge of Heegaard graphs can produce useful results which go beyond Heegaard graphs. In section 2 and 3 we recall that the mapping class group of a handlebody is finitely presented and generated by Dehn twists along discs and annuli [McC 2]. We further discuss some decision problems connected with handlebody homeomorphisms. In section 4 we give a simple construction for pseudo Anosov homeomorphisms of handlebodies. In section 5, we introduce, in the spirit of Thurston's train tracks, branching complexes for generating curve systems on boundaries of handlebodies.

Chapter II is concerned with relative handlebodies or, equivalently, with equivalence classes of Heegaard diagrams. A relative handlebody is a handlebody with boundary-pattern whose boundary-pattern consists of annuli alone. We study essential surfaces in relative handlebodies. First we study essential discs and annuli. Specifically, we give in section 7 a new proof of Whitehead's theorem concerning minimal discs in relative handlebodies, and we use this to classify relative handlebodies. In section 8 we calculate the mapping class group for relative handlebodies. In section 9 we introduce the piping process and discuss its relevance for Poincaré duality in the context of relative handlebodies. In particular, a computational form of the Poincaré map and an elementary proof for the existence of incompressible surfaces in 3-manifolds with boundary will be given. Section 10 gives a convenient presentation for surfaces in relative handlebodies. Section 11 establishes some estimates for the Seifert characteristic and the Thurston norm for relative handlebodies. In section 12 we introduce circle-patterns in Heegaard diagrams, and we show their relevance for essential surfaces in relative handlebodies. In particular, we use them to give fast algorithms for the isotopy and incompressibility problem of surfaces in relative handlebodies. In section 13 we introduce the Haken matrix associated to a relative handlebody. We show that Haken matrices have a very simple structure. In the spirit of Haken, essential surfaces occur as positive vectors of the null-spaces of these Haken matrices. We describe the relationship between our Haken matrices and Haken's original surface theory. Moreover, we describe the decision procedures of Jaco-Oertel in the context of relative handlebodies. As a result we get a collection of tools for searching and classifying essential surfaces in relative handlebodies (in the appendix these tools are illustrated by means of some concrete examples). Finally, we introduce the Haken spectrum for relative handlebodies, an arithmetic function counting essential surfaces, and use Haken matrices to establish its polynomial growth.

Chapter III   deals with (generalized) one-relator 3-manifolds. One-relator 3-manifolds are genuine 3-manifolds, but still very close to relative handlebodies. By their very definition, we obtain a one-relator 3-manifold by attaching a 2-handle along a curve, k, in the boundary of a handlebody, M. Dually, a 3-manifold, N, is a one-relator 3-manifold if it contains a proper arc ℓ, $\ell \cap \partial N = \partial \ell$, such that $(N - U(\ell))^-$ is a handlebody. Such an arc will be called a Heegaard string for N. We show that minimal essential surfaces avoid Heegaard strings, at least in the case when ℓ joins different components of ∂N. In the other case the same is true modulo certain modifications controlled by the characteristic submanifold. This is our basic observation. It is proved in prop. 15.11. Throughout this book we extend this observation to more and more general situations. In section 16, we generalize this proposition to 3-manifolds with boundary-patterns (see thm. 16.23). It turns out that this generalization can be used to study the interaction of Heegaard strings and their homotopies with hierarchies of surfaces. As a result we obtain (cor. 17.32):

Homotopy Theorem. *Heegaard strings in simple 3-manifolds are homotopic if and only if they are ambient isotopic.*

As an immediate application of this theorem we get (see thm. 26.27) a solution to the isotopy problem for Heegaard strings (the homeomorphism problem is solved by Whitehead's theorem mentioned above). Indeed, doubling the underlying simple 3-manifold, we reduce the problem of deciding whether two given Heegaard strings are ambient isotopic to the problem of deciding whether two curves in a closed, simple 3-manifold are homotopic. But the latter problem is the conjugacy problem for 3-manifold groups. Now closed, simple 3-manifolds are hyperbolic and, according to a result of Cannon [Can], the conjugacy problem for finite volume hyperbolic 3-manifolds is solvable.

As a further consequence of our investigation and the properties of the characteristic submanifold, we obtain in section 18 a structure theorem for the mapping class group of one-relator 3-manifolds (see prop. 18.5). In section 19 and 20 we establish a close relationship between incompressible surfaces in one-relator 3-manifolds and essential surfaces in their associated relative handlebody. In section 21 we show how to extend results from chapter II for a calculation of the Haken spectrum of one-relator 3-manifolds. Many of the results discussed in this section do not really use the fact that $(N - U(\ell))^-$ is a handlebody. We often only need to know that $(N - U(\ell))^-$ is ∂-reducible and we therefore include these "generalized" one-relator 3-manifolds in our discussion.

In *Chapter IV* we consider n-relator 3-manifolds in general. The relevance of this chapter rests in the fact that all 3-manifolds are (completed) n-relator 3-manifolds. Moreover, an n-relator structure is almost a handlebody decomposition for the underlying 3-manifold (it only remains to specify a meridian-system in the handlebody). In particular, there is a canonical Heegaard surface (or rather a homeomorphism class of them) associated to any n-relator 3-manifold. We show how to generalize the results of the previous chapter from one-relator to arbitrary n-relator 3-manifolds. Specifically, we proof in section 23 the General Handle-Addition Theorem mentioned before. This theorem has various applications. Perhaps the most striking one is the observation (due to Frohmann and Scharlemann) that it can be combined with Gabai's thin-position concept to give a simple proof of Waldhausen's uniqueness theorem for Heegaard decompositions of S^3. As mentioned before the General Handle-Addition Theorem contains as a special case Haken's 2-sphere theorem. But, moreover, it can also be used to prove a generalization of Haken's 2-sphere theorem from essential 2-spheres to inessential 2-spheres. As a consequence we can show that local knots can be removed from Heegaard graphs. Specifically, we show (thm. 24.5):

Unknotting Theorem. *Let N be an irreducible 3-manifold different from the 3-ball or the 3-sphere. Let $E \subset N$ be a system of 3-balls and let $\Gamma \subset N$ be an irreducible Heegaard graph such that $(\Gamma - E)^-$ consists of arcs. Then Γ is slide-equivalent to a Heegaard graph, Γ', such that $\Gamma' \cap (N - E)^- \subset \Gamma \cap (N - E)^-$ and that $\Gamma' \cap E$ is the cone over $\Gamma' \cap \partial E$.*

In section 25 we use this theorem in order to give a new proof of the Reidemeister-Singer theorem. Moreover, we relate the number of stabilizations in the Reidemeister-Singer theorem to the "winding number" of Heegaard graphs.

Chapter V is the culmination of the book. It is devoted to the study of Heegaard graphs in Haken 3-manifolds. Section 26 describes various methods for constructing Heegaard graphs. As a result we will see that the set of Heegaard graphs is usually quite rich. We show that there is already an abundance of non-homeomorphic Heegaard strings and so an abundance of counter-examples to the Magnus conjecture. In section 27 we introduce the concept of Haken 2-complexes and Heegaard 2-complexes and we establish some of their relevant properties. The intersection of a Heegaard 2-complex with a Haken 2-complex gives rise to Haken graphs in discs. In section 28 we study Haken graphs per se. We introduce various complexities for them and we concentrate on Haken graphs whose complexities are bounded from above. We establish the existence of "short companions" for such Haken graphs. The existence of these short companions is a crucial rigidity result for Haken graphs and one of the two main ingredients for the rigidity of Heegaard graphs. In section 29 we prove a technical result concerning sequences of non-proper arcs in surfaces. This result is needed for spotting Stallings fibrations in 3-manifolds. In section 30 we show that the complexities of Haken graphs, coming from intersections of Heegaard 2-complexes with a Haken 2-complex, have a universal upper bound. This is the second crucial ingredient for the rigidity of Heegaard graphs. To prove it we need the technical result from section 29. In section 31 and 32 we are finally ready to turn to the rigidity of Heegaard graphs. First, we show that the universal bound for the complexities of Haken graphs as established in section 30 can be enhanced to a universal bound for the combinatorial length of Heegaard graphs. More precisely, using the existence of short companions in a crucial way, we prove:

Finiteness Theorem. *Let N be a Haken 3-manifold (with or without boundary) which contains no non-trivial, essential Stallings fibration and let $\Psi \subset N$ be a (great and useful) Haken 2-complex. Then there is a polynomial $p_N(x) \in \mathbf{Z}[x]$ with the property that every Heegaard graph $\Gamma \subset N$ can be changed, using isotopies and slides, so that afterwards $\#(\Gamma \cap \Psi) \leq p_N(g\Gamma)$, where $g\Gamma := |\chi\Gamma - \chi\partial\Gamma|$.*

Combining this theorem with our version of the Reidemeister-Singer theorem we get the following result on the number of necessary stabilizations for Heegaard graphs in Haken 3-manifolds:

Theorem. *For any Haken 3-manifold N (with or without boundary) which contains no non-trivial, essential Stallings fibration, there is a two-variable polynomial $q_N(x, y) \in \mathbf{Z}[x, y]$ with the property that one can pass from a Heegaard graph $\Gamma_1 \in \mathbf{N}$ to a Heegaard graph $\Gamma_2 \in N$ in less than $q_N(g\Gamma_1, g\Gamma_2)$ stabilizations.*

Remark. Rubinstein and Scharlemann have recently shown that, for all non-Haken 3-manifolds, the function q_N can be taken to be linear [RS].

Moreover, it turns out that the Finiteness Theorem puts a severe limitation on the possible embeddings for Heegaard graphs. Analyzing these limitations we eventually get the Rigidity Theorem mentioned in the beginning. To deduce it we need the Finiteness Theorem, the General Handle-Addition Theorem and the Unknotting Theorem. We conclude chapter V with a brief discussion of the Reidemeister-Zieschang problem concerning homotopy classes of curves on handlebodies (section 33).

In the Appendix we illustrate our algorithms for essential surfaces in 3-manifolds. Specifically, we show that there are essential surfaces in the relative handlebody whose associated n-relator 3-manifold is Poincare's homology sphere (which in turn is known to contain no incompressible surface whatsoever).

Apart from some exceptions, the results in this book are new and have never been published before. The main results have been announced in [Joh 8]. An early version of this book has been circulated in a small number under the title "Computations in 3-manifolds".

The research reported in this book has been carried out in the course of several years. In this period I held positions or visiting positions at the Universität Bielefeld, Germany, at Rice University, Houston, at the University of Oklahoma, Norman and at the Universität Frankfurt, Germany. At all those places I enjoyed a stimulating atmosphere and I like to take the opportunity to thank the Departments of Mathematics at those institutions for their kind hospitality. I would also like to take the opportunity to acknowledge gratefully the support I have received through NSF-grants DMS-8803256, DMS-8923061 and DMS-9210044, the Deutsche Forschung Gemeinschaft and several grants from the Science Alliance, a Center of Excellence at The University of Tennessee, Knoxville.

Knoxville, October 1994

Contents

Chapter IV. N-Relator 3-Manifolds

I. HANDLEBODIES

Throughout this book we work in the PL-category. We refer to [Hem] for standard results and terminology concerning 3-manifolds. All 3-manifolds will be compact, orientable and irreducible (unless stated otherwise). As indicated in the introduction, we approach 3-manifolds via a study of handlebodies. Handlebodies form one of the simplest classes of 3-manifolds, in the same way as free groups form one of the simplest classes of groups. Nevertheless, their internal structure is quite involved. Fortunately, most of their more intricate properties are confined to handlebodies alone and disappear for more general 3-manifolds. In particular, it will be sufficient for our purpose to discuss the basic elementary features concerning handlebodies and their homeomorphisms. This is the content of the first chapter.

§1. Standard Graphs and Meridian-Systems.

A *handlebody* is defined to be a 3-manifold M which is homeomorphic to the regular neighborhood of some finite graph Γ in the 3-sphere, S^3. The solid depicted in the following picture is the standard model for a handlebody (of genus three):

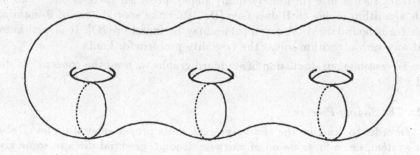

(Figure 1.1.)

Γ is a "standard graph" in the handlebody M. Here a finite graph Γ in a 3-manifold M, $\Gamma \cap \partial M = \emptyset$, is called a *standard graph* in M if $(M - U(\Gamma))^-$ is homeomorphic to $\partial M \times I$ (as always throughout this book $U(..)$ denotes the regular neighborhood). Notice that the standard graph Γ is *not* part of the definition of the handlebody M. In fact, M has usually many different standard graphs. The choice of any one of them selects a combinatorial structure for M. We are interested in the set of all these combinatorial structures, i.e. in the set $\mathcal{G}(M)$ of all standard graphs in M modulo isotopy; notably because the mapping class group of M acts on this set. To study $\mathcal{G}(M)$ we utilize the duality of standard graphs and "meridian-systems" in M.

1.1. Definition. *A system* $\mathcal{D}$ *of pairwise disjoint discs in a 3-manifold* M *with* $\mathcal{D} \cap \partial M = \partial \mathcal{D}$ *is called a* <u>*meridian-system*</u> *in* M *if* $(M - U(\mathcal{D}))^-$ *is homeomorphic to a 3-ball.*

To any standard graph Γ in the handlebody M there is associated (the isotopy class of) a meridian-system $\mathcal{D}_\Gamma$ in M. We obtain $\mathcal{D}_\Gamma$ by first fixing a spanning tree $T_\Gamma \subset \Gamma$ and then extending the disc-system $\mathcal{B} \subset U(\Gamma)$, associated to the mid-points of the edges of $(\Gamma - T_\Gamma)^-$, by the system of vertical annuli in $(M - U(\Gamma))^-$ above $\mathcal{B} \cap \partial U(\Gamma)$. Vice versa, any meridian-system $\mathcal{D}$ in M gives rise to a standard graph $\Gamma_{\mathcal{D}}$ defined as the cone over the mid-points of all discs from $\mathcal{D}$ with cone-point the mid-point of $M - \mathcal{D}$. Clearly, $\mathcal{D}(\Gamma_{\mathcal{D}})$ is isotopic to $\mathcal{D}$ and $\Gamma_{\mathcal{D}(\Gamma)}$ is isotopic to Γ. Thus the assignment $\Gamma \mapsto \mathcal{D}_\Gamma$ defines a one-to-one correspondence between the set $\mathcal{G}(M)$ and the set of all meridian-systems of M modulo isotopy.

By what we have just seen the study of standard graphs can be turned into the study of meridian-systems. This is an improvement since codim 1 objects are usually easier to handle than codim 2 objects. For instance it follows immediately from the above correspondence that the property of a 3-manifold to contain a standard graph characterizes handlebodies. Moreover, it can be decided whether or not a given 3-manifold is a handlebody. To do this we have to search for meridian-systems. But there are well-known algorithms for deciding whether a given 3-manifold contains an essential disc. These algorithms actually construct such a disc (in finitely many steps), provided there is one. The first such algorithm is due to Haken (see [Ha 1] for the special case of 3-manifolds with torus-boundaries). A second one, may be found in [JO]. It is well-known that any such algorithm solves the triviality-problem for knots.

To continue our discussion of standard graphs we need the concept of "disc-slides".

1.2. *The Sliding-Process.*

In order to describe the *sliding-process* in its proper generality, let $\mathcal{D}$ be a disc-system, i.e. a finite union of pairwise disjoint, essential discs in some given handlebody M with $\mathcal{D} \cap \partial M = \partial \mathcal{D}$ (a disc is *essential* if it is proper and not boundary-parallel).

Now, given $\mathcal{D}$, choose a simple arc k in the boundary ∂M of M with $k \cap \mathcal{D} = \partial k$ and meeting two different discs from $(\partial U(\mathcal{D}) - \partial M)^-$. Then the end-points of k lie in discs from $\mathcal{D}$, say D_1 and D_2, which may or may not be equal. In any case, consider

$$\mathcal{C} := (\partial U(D_1 \cup D_2 \cup k) - \partial M)^-.$$

It is easily checked that the system $\mathcal{C}$ consists either of one or three discs according whether or not the arc k joins two different discs.

Let D_0 denote the component of C which is neither parallel to D_1 nor to D_2. Then

$$\mathcal{D}' := (\mathcal{D} - D_1) \cup D_0$$

is a new disc-system in M. Intuitively, we obtain $\mathcal{D}'$ from $\mathcal{D}$ by "sliding" the disc D_1 along the sliding arc k and across the disc D_2. Therefore we say that $\mathcal{D}'$ is obtained from $\mathcal{D}$ by a *disc-slide* of D_1 *along* k.

1.3. Example.

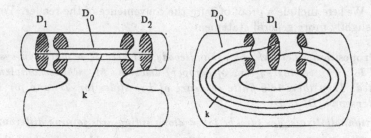

(Figure 1.2.)

Note that the choice of the sliding arc k is important for the outcome of the sliding-process. The sliding process turns meridian-systems into meridian-systems, provided the sliding arc joins different discs. In fact, these sliding processes generate an equivalence relation on the set of all meridian-systems (mod isotopy) of any given handlebody M. Dually, the sliding process for meridian-systems has an equivalent for standard graphs in M. To describe this let $\mathcal{D}$ and $\mathcal{D}'$ be two slide-equivalent meridian-systems in M. Then the corresponding standard graphs $\Gamma_{\mathcal{D}'}$, $\Gamma_{\mathcal{D}} \subset M$ are Whitehead-equivalent. Here *Whitehead-equivalence* is the equivalence relation on the set of standard graphs generated by the processes of collapsing an edge to a point and expanding a vertex to an edge.

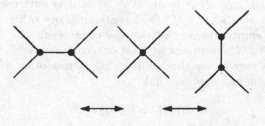

(Figure 1.3.)

Note that the above picture is slightly misleading. It only describes the effect of
Whitehead moves on the combinatorics of graphs. But Whitehead moves applied
to graphs in 3-manifolds also have an additional effect on their embedding types.

1.4. *Whitehead Equivalence Classes of Standard Graphs.*

We already know that the set of Whitehead equivalence classes of $\mathcal{G}(M)$ is
in one-to-one correspondence to the set of slide-equivalent classes of meridian-
systems of the handlebody M. Now any two meridian-systems of M are
slide-equivalent (using sliding arcs which join different discs). This fact is well-
known. We here include a proof of it for the convenience of the reader. We begin
with a slightly more general statement.

1.5. Proposition. *Let M be a handlebody, let $\mathcal{D}$ be a meridian-system of
M and let $\mathcal{D}^*$ be any system of (proper) discs in M whose complement is
connected. Then there is a finite sequence of disc-slides for $\mathcal{D}$ and for $\mathcal{D}^*$ so
that afterwards $\mathcal{D}^* \subset \mathcal{D}$.*

In addition, all disc-slides may be taken along siding-arcs joining different discs.

Proof. The proof uses the classical theorem due to Alexander [Al] that any
(piecewise-linear) embedding of a 2-sphere separates the 3-sphere into two 3-
balls.

It is convenient to break the proof into the following two cases.

Case 1. The systems $\mathcal{D}$ and $\mathcal{D}^*$ are disjoint.

Since $\mathcal{D}$ is supposed to be a meridian-system, we have that $M^* = (M - U(\mathcal{D}))^-$ is a 3-ball. Every disc from $\mathcal{D}^*$ is contained in this 3-ball. Moreover,
as a consequence of Alexander's theorem, any one of them splits the 3-ball into
two balls. If one of the discs is not outermost (in the sense that it does not
separate a ball from M^* not containing any other disc from $\mathcal{D}^*$), then we
easily find a sequence of disc-slides across discs from $\mathcal{D}^*$ which increases the
number of outermost discs. Thus w.l.o.g. we may suppose that *all* discs from
$\mathcal{D}^*$ are in fact outermost.

Consider a disc, say D_0^*, from $\mathcal{D}^*$. Since it is outermost, it separates
a ball M_0^* from the 3-ball M^* not containing any other disc from $\mathcal{D}^*$.
Since $\mathcal{D} \cap \mathcal{D}^*$ is empty, there is at least one component, D_0, from $\mathcal{D}$ such
that $C := M_0^* \cap \partial M^*$ contains at least one component of $U(D_0) \cap \partial M^*$.
Furthermore, it is easy to see that, sliding D_0 across other discs from $\mathcal{D}$ if
necessary, we may in fact suppose that

$$U(\mathcal{D}) \cap C = U(D_0) \cap C.$$

Observe that then, in addition, the intersection $U(D_0) \cap C$ has to be connected,
for otherwise the disc D_0^* would be separating in the handlebody M contradicting the hypothesis that the complement of $\mathcal{D}^*$ is connected (a property
which remains unchanged under disc-slides along siding-arcs which join different
discs). Thus D_0^* is parallel to D_0. So D_0^* can be isotoped into D_0.

Splitting M along D_0 and repeating the previous process, the proposition follows inductively.

Case 2. The systems $\mathcal{D}$ and $\mathcal{D}^*$ are not disjoint.

It is our aim to reduce Case 2 to Case 1. For this it remains to show, how to remove the intersection of $\mathcal{D}$ with $\mathcal{D}^*$, by using disc-slides (and ambient isotopies) alone.

To begin with, let us first isotope the system $\mathcal{D}$ so that it intersects $\mathcal{D}^*$ in a finite number of arcs and closed curves (general position) and that, in addition, the number of these intersection-curves is as small as possible.

Then $\mathcal{D} \cap \mathcal{D}^*$ consists of arcs. To see this claim, we apply the well-known cut-and-paste argument. Indeed, assume for a moment that one of the intersection curves of $\mathcal{D} \cap \mathcal{D}^*$, say t, is a closed curve. Then recall the fact that a closed curve in a disc separates a unique disc from the original disc (Schönflies-theorem). Thus we may suppose that t is chosen to be innermost, i.e. in such a way that it bounds a disc D_0^* in $\mathcal{D}^*$ with $D_0^* \cap \mathcal{D} = t$. But, by the same argument, k separates a unique disc D_0 from $\mathcal{D}$. It follows from our choice of t that the union $D_0 \cup D_0^*$ is a 2-sphere. Since a handlebody is contained in S^3 we conclude that this 2-sphere separates S^3 into two 3-balls, one of them not containing the boundary of the handlebody M. This latter ball is therefore contained in M and meets the meridian-system $\mathcal{D}$ in D_0 alone. Thus we may isotope D_0 across this ball and extend this isotopy to an isotopy of $\mathcal{D}$ which reduces the intersection $\mathcal{D} \cap \mathcal{D}^*$. This, however, is a contradiction to our minimal choice of $\mathcal{D}$, proving our claim.

By what we have seen so far, the intersection $\mathcal{D} \cap \mathcal{D}^*$ consists of arcs. We now claim this intersection is in fact empty. To see this claim note that any intersection-arc splits $\mathcal{D}^*$ into discs. We therefore may pick an outermost arc, i.e. an arc which separates a disc from $\mathcal{D}^*$, say D_0^*, containing no other intersection-arc. Then let D_0 denote the disc from $\mathcal{D}$ containing this specific arc. At this point, recall that $(M - U(\mathcal{D}))^-$ is a 3-ball. The disc D_0^* separates this 3-ball into two balls (a consequence of Alexander's theorem). Exactly one of these balls contains a component of $(\partial U(D_0) - \partial M)^-$. Let M_0 be the other one. Then no component of $(\partial U(D_0) - \partial M)^-$ is contained in M_0. It is then easy to check that an isotopy of the disc D_0 across the ball M_0, can actually be realized as a sequence of disc-slides (across the discs corresponding to the components of $(\partial U(\mathcal{D}) - \partial M)^-$ contained in M_0). Finally, observe that this sequence of disc-slides, reduces the intersection $\mathcal{D} \cap \mathcal{D}^*$. Therefore and by our minimal condition on the intersection $\mathcal{D} \cap \mathcal{D}^*$, it follows that this intersection has to be empty. So we are in Case 1. $\Diamond$

As a corollary we obtain the above mentioned fact that all meridian-systems are slide-equivalent (using disc-slides whose sliding-arcs join different discs).

1.6. Corollary. *Let $\mathcal{D}$ and $\mathcal{D}^*$ be two meridian-systems of the handlebody M. Then $\mathcal{D}^*$ can be obtained from $\mathcal{D}$ by some finite sequence of disc-slides of $\mathcal{D}$.*

Proof. By the previous proposition, we know that $\mathcal{D}^*$ is contained in $\mathcal{D}$, modulo disc-slides of $\mathcal{D}$ *and* $\mathcal{D}^*$. The inverse of a disc-slide may be realized by some disc-slide again. So the previous statement remains even true, if we allow only disc-slides for $\mathcal{D}$. We further claim that a meridian-system $\mathcal{D}$ containing another meridian-system $\mathcal{D}^*$ is actually equal to $\mathcal{D}^*$. Indeed, otherwise at least one component of $\mathcal{D}$ lies in the complement of $\mathcal{D}^*$ and so separates that complement into two 3-balls. It then follows that the complement of $\mathcal{D}$ cannot be connected. This, however, contradicts the hypothesis that $\mathcal{D}$ is a meridian-system.

But even if (as we have shown) $\mathcal{D} = \mathcal{D}^*$, it still need not be the case that the individual discs coincide. However, the following sequence of disc-slides show how to interchange two components of a meridian-system and so the missing property can easily be obtained.

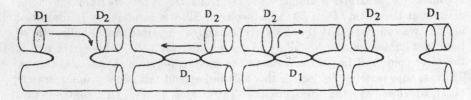

(Figure 1.4.)

This completes the proof of cor. 1.6. ◊

As an immediate geometric consequence of the previous result note that the number of components of meridian-systems is a homeomorphy-invariant for handlebodies, i.e. two handlebodies are homeomorphic iff they admit meridian-systems with the same number of components. In particular, a handlebody is determined by the genus of its boundary. Note further that any non-separating disc in a handlebody can be completed to a meridian-system of that handlebody. In particular, every non-separating disc splits a handlebody into a handlebody. All these geometric observations illustrate the special relevance of standard graphs or meridian-systems for a classification of handlebodies.

In this context it is worth recalling that any presentation of the fundamental group of M can be realized by a standard graph (unique modulo homotopy). To see this, let $< a_1, a_2, ..., a_n >$ be a presentation of $\pi_1 M$ associated to some fixed standard graph $\Gamma \subset M$, and let $< b_1, b_2, ..., b_n >$ be some other presentation. (In the following it is to be understood that the edges of Γ are labeled by the corresponding a_i's.) According to Nielsen (see e.g. [ZVC] or [LS]), the presentation $< b_1, b_2, ..., b_n >$ can be obtained from $< a_1, a_2, ..., a_n >$ by a finite sequence of Nielsen transformations. But, modulo re-labeling of edges, there is only one transformation, namely replacing a_i by $a_i a_j$, $i \neq j$, while leaving all other generators unchanged. This Nielsen transformation in turn gives rise to an edge-slide of the edge a_i over the edge a_j. This slide is the result of an isotopy in $M - (\Gamma - a_i)$ and so the complement of the resulting graph is homeomorphic to $M - \Gamma$. The claim therefore follows inductively.

To illustrate the potential of graphs for a study of 3-manifolds, let us lower the dimension by one and let us apply the above observations to 2-manifolds.

1.7. *Standard Graphs and Dehn Twists.*

In analogy to handlebodies let us call a (finite) graph Γ in an orientable surface S a *standard graph* if it has exactly one vertex and its complement, $S - \Gamma$, is either an open disc or homeomorphic to $\partial S \times [0, 1)$. Any basis of $\pi_1 S$ is given by a map $\Gamma \rightarrow S$ which induces an injection of fundamental groups, but, in contrast to handlebodies, this map can generally not taken to be a standard graph (this is the case when self-singularities are unavoidable). At this point, however, recall Nielsen's theorem (see e.g. [ZVC]) that every isomorphism between closed orientable surfaces is induced by a homeomorphism. In other words, the outer automorphism group of a closed orientable surface acts on the set of all basis given by standard graphs. Thus for questions concerning this group if suffices to consider those special basis alone. (A similar fact is true for Haken 3-manifolds and is one of the basic motivations for this book). In this context we note the following analogy of the previous considerations.

1.8. Lemma. *Any two standard graphs* Γ_1, Γ_2 *in an orientable surface* S *are slide equivalent.*

Proof. If S is closed, let D be a disc in S and consider the surface $S' = (S - D)^-$. Since Γ_1 and Γ_2 are standard graphs, their regular neighborhoods $U(\Gamma_1), U(\Gamma_2)$ are surfaces with non-empty boundaries whose complement is a disc, and so they are ambient isotopic to S'. Replacing S by S' if necessary, we may therefore suppose w.l.o.g. that ∂S is non-empty.

If S is non-empty, then we may associate a dual arc-system B_i to the standard graph Γ_i (in the same way we previously associated meridian-systems to standard graphs in handlebodies). Similar to prop. 1.5 (but in fact much easier) it can be shown that B_1 can be obtained from B_2 by some finite sequence of arc-slides (similar to disc-slides). Arc-slides for B_i are dual to edge-slides for Γ_i, and the lemma follows. $\diamond$

As a consequence of this lemma we obtain the following theorem due to Dehn [De] (see also [Li 1]).

1.9. Theorem. *The orientation-preserving mapping class group of any closed orientable surface is generated by Dehn twists along curves.*

Remark. Recall a *Dehn twist along a curve* k of a surface S is defined to be a homeomorphism with support in the regular neighborhood of the simple closed curve k in S (i.e. a homeomorphism which is the identity on $(S - U(k))^-$).

Proof. We begin with an observation relating slides and Dehn twists. For this let Γ be a standard graph in S. Suppose Γ' is a graph obtained from Γ by a single edge-slide, i.e. by a slide of some edge a along some other edge b of Γ. Denote by a' the result of a under this slide. Then a' is isotopic to a in $S - (\Gamma - a)$. Moreover, we claim there is a product $g : S \to S$ of Dehn twists with $g(a) = a'$ (but *not* necessarily $g(\Gamma) = \Gamma'$).

To prove the claim note that the edges of a standard graph are simple closed curves in S since standard graphs have only one vertex. The simple closed curves a and b intersect in exactly one point, but this intersection may or may not be transversal. If a and b happen to intersect transversally, then a' is clearly the result of a Dehn twist g along b. If not, then they touch in exactly one point. In particular, the regular neighborhood $U(a \cup b)$ is then a disc with two holes. Two components of $\partial U(a \cup b)$ are homotopic to a or b (we denote them again by a and b). The remaining component is homotopic to a' (and will be denoted by a'). As an edge of a standard graph in a closed surface, the curve a is not only non-separating in S but also in $S - (\Gamma - a)$. Thus there is a simple closed curve in $S - (\Gamma - a)$ which intersects Γ in one point contained in a. This curve does not meet b and it has to intersect a'. In fact, we may suppose it is chosen so that, in addition, it intersects a' in exactly one point. The claim now follows from the simple fact that every simple closed curve, which intersects another curve transversally in exactly one point, is the image of that curve under some product of Dehn twists along these two curves.

Now, let $h : S \to S$ be any orientation-preserving homeomorphism. We have to show that h is a product of Dehn twists. Let a be an edge of the standard graph Γ. $h(\Gamma)$ is a standard graph again and so, by lemma 1.8, slide-equivalent to Γ. Thus, applying the above procedure over and over again, we find a product g of Dehn twists such that $g(a) = h(a)$. Thus, multiplying h with Dehn twists if necessary, we may suppose $h(a) = a$. Moreover, there is a simple closed curve l which intersects a in exactly one point since a is a non-separating, simple closed curve in S. Multiplying h with some appropriate Dehn twists along a and l if necessary, we may further suppose that $h|a = \mathrm{id}|a$. In particular, h does not interchange the sides of a since h is supposed to be orientation-preserving.

Splitting S along a and contracting the resulting boundary components to points z_1, z_2 say, we obtain from S a closed surface S' and from h a homeomorphism $h' : S' \to S'$ with $h'|z_1 \cup z_2 = \mathrm{id}|z_1 \cup z_2$. Every orientation-preserving homeomorphism of the 2-sphere is isotopic to the identity. Thus, by induction on the genus of S, h' is isotopic to some product g' of Dehn-twists. So the induction step would be complete if the isotopy would fix $z_1 \cup z_2$. This however is not necessarily the case. But the restriction of an isotopy α_t, $t \in [0,1]$, to z_1, say, is a path and any path is a composition of simple closed paths, say w_i, $i \geq 1$. An isotopy of the identity along a simple closed curve w_i is the same as the product of Dehn twists along $\partial U(w_i)$. The same with the point z_2. Thus, multiplying g' with all these Dehn twists (or better their inverses), we finally obtain a product of Dehn twists which is isotopic to h', using an isotopy which fixes $z_1 \cup z_2$. ◊

Standard graphs are very special indeed. All 2-manifolds admit them, but, as mentioned before, only those 3-manifolds which are handlebodies. However, any standard graph can be turned into a Heegaard graph - by simply joining it, using some fiber-like edge, with the boundary of the underlying manifold. In this way standard graphs are closely related to Heegaard graphs. But the latter is a much more general notion in that *every* 3-manifold contains at least one Heegaard graph (triangulation theorem). In later chapters we will study Heegaard graphs extensively, and then the above results on standard graphs will become relevant again.

§2. The Disc-Complex.

As we have seen in the previous section, the various meridian-systems of a given handlebody are interrelated: any two of them can be joined by disc-slides. It is interesting to note that this property can be expressed in a more organized way by constructing a CW-complex (due to Kramer). Observe that (the isotopy class of) any meridian-system may be considered as a point and any disc-slide as an edge. In this way the set of meridian-systems becomes a graph which is connected (by cor. 1.6). This graph, may be called the *graph of disc-slides*. It is not a finite graph, not even a locally finite graph, but it lies in a bigger simplicial complex which is contractible. This bigger simplicial complex is the *disc-complex* which we are going to describe next.

Let us define the *disc-complex* of the handlebody M to be the simplicial complex $\Delta(M)$ whose m-simplices are (isotopy-classes of) $(m+1)$-tuples $(D_0, D_1, ..., D_m)$ of essential and pairwise non-isotopic discs in M. Note that two m-simplices $(D_1, D_2, ..., D_m)$ and $(D_1^*, D_2^*, ..., D_m^*)$ are equal iff D_i is (properly) isotopic to D_i^*, for all $1 \leq i \leq m$. Note further that the disc-complex has dimension $3n - 3$, where n is the genus of M.

Now, in order to get a first feeling for this complex, let us consider the proof of the following interesting property of Δ (taken from [McC 2], but replacing the geometric argument given there by a purely combinatorial proof).

2.1. Lemma. *Every complete subgraph in $\Delta(M)$ is contained in a simplex.*

Remark. By definition, a graph is *complete* if every pair of different vertices is joined by exactly one edge. Note that any two vertices of the complex $\Delta(M)$ are joined by at most one edge.

Proof. A complete graph Γ in $\Delta = \Delta(M)$ is given by a finite collection of meridian-discs $D_1, D_2, ..., D_m$ with the property that each pair of them can be isotoped to be disjoint. The lemma claims that all these discs can be isotoped so that afterwards the whole set of discs is pairwise disjoint.

A first step in showing this property, consists in proving that any system of pairwise disjoint, simple closed curves in ∂M bounds a system of pairwise disjoint discs in M, provided every individual curve bounds a disc at all. This, however, can be seen inductively as follows. Suppose $D_1, D_2, ..., D_m$ is the set of discs bound by the given curves and suppose the discs $D_1, ..., D_q$, $q \leq m$, are already pairwise disjoint. Then isotope the disc D_{q+1}, using an isotopy which is constant on the boundary, so that afterwards the number of intersection-curves of $(D_1 \cup ... \cup D_q) \cap D_{q+1}$ is as small as possible. It is easily checked that in this case, D_{q+1} is actually disjoint to all discs with smaller indices. This fact proves our claim inductively.

To finish the proof, set $k_i := \partial D_i$, $1 \leq i \leq m$. It remains to prove that all these curves can be isotoped in ∂M so that afterwards they form a system of pairwise disjoint curves.

Let us suppose that the k_i's are chosen within their respective isotopy classes so that the number

$$\sum \#(k_i \cap k_j),$$

is minimal, where the sum is taken over all pairs of different curves. Suppose the set $k_1, ..., k_m$ of curves is not pairwise disjoint. Then w.l.o.g. there is a deformation of k_1 in ∂M which reduces the number of intersection points with k_2. Considering the pre-image of k_2 under this deformation, we find a 2-faced disc D embedded in ∂M such that $k_i \cap \partial D$, $i = 1, 2$, are two arcs in ∂D whose union forms ∂D. Choosing different discs, if necessary, we find that w.l.o.g. the intersection of D with any other curve from our original curve-system has to consist of arcs joining the two faces of D. Thus, pushing k_1 across D, we may reduce the intersection $k_1 \cap k_2$ without increasing the number of intersection points with any other curve. This contradicts our minimal choice of our original set of curves, and this contradiction concludes the proof. $\diamond$

We next show the important property that $\Delta(M)$ is not only connected, but even simply connected. This is the crucial observation in showing that the mapping class group of handlebodies is finitely presented (see next section).

2.2. Proposition. *The simplicial complex* $\Delta(M)$ *is connected and simply connected, i.e.* $\pi_1(\Delta(M)) = 1$.

Remark. This result has been proved in [McC 2] for a closely related complex. It is shown there that that complex is contractible. For our purpose the above weaker form is sufficient and we prove it by a slightly different argument.

Proof. By definition of the complex Δ any meridian-disc D_i in M may in fact be considered as a point of Δ. Fix a base-point D_0 of Δ. We have to show that any point in Δ can be joined by a path in Δ with D_0 and that furthermore any based loop in Δ can be contracted. We first show the latter.

For this let w be a based edge-path in Δ and let $D_0, D_1, ..., D_m$ be the set of all vertices met by the path w. Let these discs be isotoped so that they pairwise intersect in arcs. Let the *energy* $E(w)$ of the path w defined by setting

$$E(w) := \sum \#(D_i \cap D_0),$$

where the sum is taken over all vertices from w.

If the energy of the path w is zero, then all vertices of it are adjacent to D_0 and the path is clearly contractible. So it remains to show how to reduce the energy using deformations alone.

To this end note that, by our choice of the discs involved, every disc D_i intersects the base-disc D_0 in a system of arcs. Every such arc separates a 2-faced disc from D_0 (as a matter of fact any arc splits D_0 into two such discs). Thus we find an outermost arc. Here an outermost arc is an intersection arc from some intersection $D_0 \cap D_i$ with the property that it separates a 2-faced disc E from D_0 which does not contain an entire intersection arc from any intersection $D_0 \cap D_j$. Thus, for every $j \neq i$, the intersection $E \cap D_j$ consists of arcs and none of them having both its end-points in $E \cap \partial D_0$.

2.3. Example.

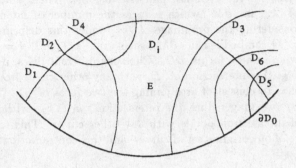

(Figure 2.1.)

Observe that at least one component from

$$(\partial U(D_i \cup E) - \partial M)^-$$

is a meridian-disc not parallel to D_i. Denote this disc by D_i^*.

It also follows from our choice of E that D_i^* is disjoint to any disc D_j for which $D_i \cap D_j = \emptyset$. Furthermore, D_i^* is disjoint to D_i. Thus D_i and D_i^* are adjacent vertices in $\triangle$. In fact, they are joined by an edge which in turn is a common face of two 2-simplices containing the two edges in w incident to D_i. Thus, replacing D_i by D_i^*, we obtain an edge-path w^* homotopic to w with strictly smaller energy (recall the way the discs D_j, $j \neq i$, intersect E).

The previous argument shows also that any vertex can be joined by an edge-path to some vertex adjacent to D_0 and so $\triangle$ is connected. $\Diamond$

We end this section with the following remarks.

2.4. Remark. First note that the dual complex $\triangle^d$ of $\triangle$ is closely related to the disc-slides considered in the previous section. Indeed, recall from the definition of $\triangle$ that the end-points of an edge in $\triangle^d$ correspond to two meridian-systems $\mathcal{D}_1$ and $\mathcal{D}_2$ which have a common face. This common face is given by a system $\mathcal{D}_0$ of $n-1$ pairwise disjoint discs such that $M^* := (M - U(\mathcal{D}_0))^-$ is connected. Observe that M^* has to be a solid torus and $U(\mathcal{D}_0) \cap \partial M$ is a system of pairwise disjoint discs in the boundary of that solid torus. Moreover, there are meridian-discs D_1 and D_2 in M^* disjoint to $U(\mathcal{D}_0)$ such that $\mathcal{D}_i = \mathcal{D}_0 \cup D_i$, $i = 1, 2$. We see that in particular $\mathcal{D}_1$ and $\mathcal{D}_2$ have a common face in $\triangle$, if they differ by a disc-slide. Thus the graph of disc-slides is a sub-graph of the 1-dimensional skeleton of the dual complex $\triangle^d$.

For later use, we now may state the following result which in turn is an immediate consequence of the previous proposition. Note that it also yields a different proof of cor. 1.6.

2.5. Corollary. *The dual complex Δ^d is simply connected. In particular, the 2-skeleton of Δ^d is simply connected. The 1-skeleton of Δ^d contains the graph of disc-slides as a sub-complex.* $\Diamond$

2.6. Remark. We further remark that many 3-manifolds, including all 3-manifolds with non-empty boundary, can be cut into a 3-ball. The handlebodies, however, are the only 3-manifolds for which this can be achieved by cutting along discs which are pairwise disjoint. For more general 3-manifolds, we will have to use more complicated surfaces which furthermore also may meet each other. A special object emerges known as a hierarchy. The set of hierarchies, although not a simplicial complex in a natural way, will then replace the set of all meridian-systems. The complex $\Delta(M)$ considered so far will then appear as a special, although interestingly coherent, object specific for handlebodies.

§3. Homeomorphisms.

The topic of this section is the discussion of (isotopy classes of) handlebody homeomorphisms and their groups. We describe generators for the mapping class group of handlebodies and we discuss a few decision problems concerning handlebody homeomorphisms.

3.1. *The Mapping Class Group.*

Throughout this section let M denote a handlebody. There are three different types of locally defined homeomorphisms of M, classified according to the shape of their supporting regions:

A *Dehn twist along a disc* is a homeomorphism $M \to M$ with support in the regular neighborhood of some essential disc in M.

A *Dehn twist along an annulus* is a homeomorphism $M \to M$ with support in the regular neighborhood of some incompressible annulus in M.

A *flip of a handle* is a homeomorphism $M \to M$ with support in some "handle" $U := U(D \cup k)$, where D is some non-separating disc in M and where k is some simple closed curve in ∂M which intersects $\partial D \subset \partial M$ transversally in a single point.

All these locally defined homeomorphisms give rise to subgroups in the mapping class group $\pi_0 \mathrm{Diff}(M)$ of the handlebody M. Note that the subgroup given by all Dehn twists along a given disc, or annulus, is free cyclic. Moreover, the subgroup of all flips of a given handle is isomorphic to $\mathbf{Z}_2$, i.e. every handle supports exactly one non-trivial flip. Now, the special significance of the above homeomorphisms is due to the following well-known theorem (notice the similarity to thm. 1.9).

3.2. Theorem. *The mapping class group $\pi_0 \mathrm{Diff}^+(M)$ of all orientation preserving homeomorphisms of M is generated by flips and by Dehn-twists along discs and annuli.*

Proof. Let $\mathcal{D}$ be a meridian-system of M. Set $E := (M - U(\mathcal{D}))^-$.

Let $h : M \to M$ be a homeomorphism. We have to show that h is a finite product of flips and of Dehn twists along discs and annuli.

Note first that $h(\mathcal{D})$ is also a meridian-system of the handlebody M. Thus, by cor. 1.6, $h(\mathcal{D})$ is slide-equivalent to $\mathcal{D}$. But, as illustrated in figure 3.1, every disc-slide (whose sliding-arc joins different discs) can be realized by a Dehn twist along some incompressible annulus. Thus, multiplying h with Dehn twists along appropriate annuli if necessary, we may suppose h maps every component of $\mathcal{D}$ to itself.

Let D_1 be a disc from $\mathcal{D}$. Then D_1 is a non-separating disc in the solid torus $(M - U(\mathcal{D} - D_1))^-$. Thus there is a simple closed curve $k_1 \subset \partial M$ with $k_1 \cap \partial \mathcal{D} \subset \partial D_1$ and which intersects ∂D_1 transversally in exactly one point. Moreover, there is a flip, say g_1, of the handle $U(D_1 \cup k_1)$ with $g_1 U(D_1) =$

$U(D_1)$ and which interchanges the two components of $(\partial U(D_1) - \partial M)^-$. Thus, multiplying h with such flips if necessary, we may suppose that $h(U(\mathcal{D}) = U(\mathcal{D})$ and that, moreover, h maps every component of $(\partial U(\mathcal{D}) - \partial M)^-$ to itself. Thus $h(E) = E$.

In particular, we have that h restricts to a homeomorphism of the planar surface $(\partial E - U(\mathcal{D}))^-$ to itself which also maps each boundary component of that surface to itself. Now, such a homeomorphism is easily seen to be a finite product of Dehn twists along simple closed curves (see proof of thm. 1.9). But every Dehn twist along a simple closed curve in $(\partial E - U(\mathcal{D}))^-$ extends to a Dehn twist of M along a disc. Thus, multiplying h with Dehn twists along appropriate discs if necessary, we may suppose h is isotoped so that $h|E = \mathrm{id}|E$. In this case, h is a homeomorphism with support in $U(\mathcal{D})$. So it is a product of Dehn twists along discs.

This proves the theorem. $\Diamond$

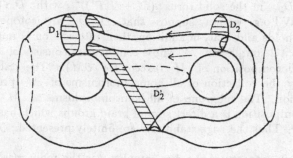

(Figure 3.1.)

Actually, a more careful choice of the local homeomorphisms in the previous proof would reveal that $\pi_0 \mathrm{Diff}(M)$ is generated by a *finite* collection of flips of handles and Dehn twists along discs and annuli. Moreover, the properties of the disc-complex Δ from section 2 allow us to deduce that $\pi_0 \mathrm{Diff}(M)$ is not only finitely generated, but also finitely presented. For this we need the following observation.

3.3. Theorem. *The mapping class group of a handlebody M acts simplicially on the disc-complex $\Delta = \Delta(M)$ with finite quotient complex, and so also on the dual complex Δ^d. The point- and edge-stabilizers for the action on Δ^d are finitely presented groups.*

Proof. The image of a disc-system in M under a homeomorphism $M \to M$ is again a disc-system in M. This defines a natural action of $\pi_0 \mathrm{Diff}(M)$ on the disc-complex Δ which is clearly simplicial.

Now, fix a meridian-system $\mathcal{D}$ of M. By prop. 1.5, every system $\mathcal{D}^*$ of pairwise disjoint discs with $M - \mathcal{D}^*$ connected, can be pushed into $\mathcal{D}$, using disc-slides alone. Furthermore, by what we have seen above, we know that any of these disc-slides can be realized by a Dehn-twist along an annulus. In other words, a fundamental-region of the action is given by all disc-systems contained in $\mathcal{D}$, i.e. by the n-simplex given by $\mathcal{D}$, where $n + 1 = \#\mathcal{D}$. Thus $\triangle / \pi_0 \mathrm{Diff}(M)$ is finite.

It remains to show the statement on the stabilizers.

A vertex of the dual complex $\triangle^d$ is given by a meridian-system of M, say $\mathcal{D}$. The stabilizer of that vertex is given by the collection of all homeomorphisms fixing every disc from $\mathcal{D}$. Set $E := (M - U(\mathcal{D}))^-$. The subgroup of all homeomorphisms $h : M \to M$ with $h(E) = E$ and fixing every disc from $(\partial E - \partial M)^-$ is a braid group which has finite index in the stabilizer (see the argument from 3.2). Thus the point-stabilizers are finitely presented since braid groups are known to be finitely presented.

An edge in $\triangle^d$ is dual to an $(n - 1)$-simplex incident to two n-simplices from $\triangle$. Thus an edge is given by some disc-system in M, say $\mathcal{D}'$, with $n - 1$ discs (and connected complement) plus a choice of two non-separating discs, say D_1, D_2, in the solid torus $(M - U(\mathcal{D}'))^-$ with $D_i \cap U(\mathcal{D}') = \emptyset$, for $i = 1, 2$. W.l.o.g. we may suppose that D_1, D_2 are isotoped in M so that the above holds and that, in addition, their intersection is minimal. Then $\mathcal{D}' \cup D_1 \cup D_2$ is a 2-complex in M whose complement consists of 3-balls. Thus we have a cell-decomposition of M (except for ∂M). The stabilizer of the edge is given by the collection of all homeomorphisms of M preserving this cell-decomposition. The subgroup of all homeomorphisms of M fixing every cell of this decomposition is a direct sum of braid groups which has finite index in the stabilizer. Thus the edge-stabilizers are finitely presented. $\Diamond$

The previous property of the disc-complex was the main reason for its invention in [Kr]. It is the key for the following well-known result (see e.g. [McC]).

3.4. Corollary. *The mapping class group of a handlebody is finitely presented. In fact, a finite presentation of this group may explicitly be given.*

Proof. Because of thm. 3.3 and prop. 2.2, the procedure given in [Br] for producing finite presentations for groups acting simplicially on simply connected complexes applies. $\Diamond$

3.5. Remark. We here point out that the situation for the mapping class group of Haken 3-manifolds is formally quite similar. Indeed, while in the case of handlebodies, a finite presentation for the mapping class group can be deduced from its action on the set of all meridian-systems, the same can be deduced, in the case of Haken 3-manifolds, from its action on the set of all (great) hierarchies [Joh 1].

Let us now turn to some *decision problems* concerning handlebody homeomorphisms.

3.6. *The Conjugacy Problem.*

Given a handlebody M note that the restriction $h \to h \mid \partial M$ defines a homomorphism between the mapping class groups of M and its boundary. This homomorphism is easily seen to be injective. However, it is far from being surjective. As a matter of fact the cokernel of the above map is still unknown. The reason for this being that the structure of the mapping class group of surfaces gives no direct access to that of the subgroup of handlebody homeomorphisms. In some particular instances, however, it really suffices to know the action of a handlebody homeomorphism on the boundary. As an example take the conjugacy problem for handlebody homeomorphisms. Indeed, let f and g be two homeomorphisms of the handlebody M. According to a theorem of Hemion [He], there is a finite, constructable set of homeomorphisms on ∂M which contains (the isotopy class of) any homeomorphism $h : \partial M \to \partial M$ with $h \circ (f \mid \partial M) \simeq (g \mid \partial M) \circ h$ in ∂M. Thus in order to check whether f and g are conjugate in the mapping class group of M we only have to check whether one of the homeomorphisms h in the previous finite set actually extends to a handlebody homeomorphism. Now, in order to check whether a homeomorphism $h : \partial M \to \partial M$ extends to all of M, it suffices to fix a meridian-system $\mathcal{D}$ in M and to check whether the image $h(\partial \mathcal{D})$ bounds discs in M. By the loop-theorem [Pa 2][Joh 9], this follows e.g. from the solution of the word-problem for free groups.

3.7. *The Torsion Problem.*

The mapping class group of a handlebody is not torsion-free and torsion elements are easy to construct. Given a homeomorphism $h : M \to M$ of a handlebody, the question arises whether we can decide, if this homeomorphism represents a torsion element. The following result answers this question:

3.8. Proposition. *There is an algorithm which decides, in a finite number of steps, whether or not a given homeomorphism of a handlebody is finite, up to isotopy.*

Proof. This is clear for the solid torus since every homeomorphism of the solid torus preserves the meridian-disc modulo isotopy. Thus suppose M is a handlebody of genus at least two. Now, a homeomorphism h of the handlebody M defines an automorphism $\varphi_h := (h|\partial M)_*$ of the surface group $\pi_1 \partial M$ (preserving the kernel of the canonical homomorphism $\pi_1 \partial M \to \pi_1 M$). The assignment $h \mapsto \varphi_h$ defines a homomorphism $\pi_0 \mathrm{Diff}(M) \to \mathrm{Out}(\pi_1 \partial M)$. In fact, an application of Nielsen's theorem shows that this homomorphism must be an injection. It follows in particular that h is finite mod isotopy if and only if there is a positive integer m such that φ_h^m is an inner automorphism. But if there is any such integer m at all there must be one with $m \leq 84(\mathrm{genus}(M) - 1)$. This is because any surface-homeomorphism which is finite mod isotopy is isotopic to a finite homeomorphism [Nie 1][Zie 4][Ke]. Now, every finite homeomorphism of a surface (of genus ≥ 2) is an isometry for some appropriate hyperbolic metric of that surface and, by a classical result due to Hurwitz, the order of the isometry group of a closed Riemannian surface of genus n is at most $84(n-1)$ [BS, p.580] (see also [MMZ,thm. 7.2] for a somewhat sharper bound). This proves the claim. To finish the proof we need an algorithm to decide whether or not a given automorphism φ of a surface group is inner. But this is the case if and only if $\varphi(a)$ is conjugate to a, for all generators a of the group. The proposition therefore follows from Dehn's solution of the conjugacy problem for surface groups (see [LS]). ◊

3.9. *The Fixpoint Problem.*

We know that the mapping class group of a handlebody acts simplicially on its associated disc-complex. This action is certainly not fixpoint-free since e.g. any Dehn-twist around a meridian-disc has at least one fixpoint. Vice versa, the problem arises, whether we are able to check if a given homeomorphism acts fixpoint-free. This is a special case of a more general decision-problem concerning pseudo-Anosov homeomorphisms which we are going to discuss next.

3.10. *Pseudo-Anosov homeomorphisms.* It is not entirely clear what may be meant by a "pseudo-Anosov" homeomorphism of a handlebody M. For lack of a better definition we here simply define it to be a homeomorphism whose restriction to the boundary ∂M is pseudo-Anosov. At this point recall that, due to Thurston, pseudo-Anosov homeomorphisms of surfaces can be character-ized by the property that they are neither finite, nor reducible. (Recall that a surface homeomorphism is *reducible*, if it leaves invariant (modulo isotopy) an essential codim 1 submanifold). Observe that *this* characterization of pseudo-Anosov homeomorphisms does *not* carry over to handlebodies. Indeed, there are homeomorphisms of handlebodies M which do not leave invariant any non-trivial codim 1 submanifold of M, but whose restriction to ∂M is nevertheless reducible.

Here is a concrete example. Let $p : M \to F$ be the I-bundle over some (possibly non-orientable) surface F with boundary, and let $g : F \to F$ be a pseudo-Anosov homeomorphism of F (see [FLP] or section 4 for the existence of such homeomorphisms). Then M is certainly a handlebody. Moreover, g can be lifted to a homeomorphism $h : M \to M$. The restriction $h \mid \partial M$ is reducible since it leaves invariant the submanifold $p^{-1}(\partial F)$, but it has no non-trivial invariant codim 1 submanifold in M, for every essential surface in an I-bundle is either vertical or horizontal (see e.g. [Joh 1]).

The previous example illustrates the possibility of essential curves in ∂M, which are not sufficiently complicated. Here an essential, simple closed curve k in ∂M is called *sufficiently complicated*, if (1) or (2) holds:

(1) there is no meridian-disc in M which intersects k in exactly one point, or

(2) there is no projection $p : M \to F$ onto a non-orientable surface which turns M into an I-bundle such that $p^{-1}(\partial F)$ is a regular neighborhood of k in ∂M.

A curve-system is called *sufficiently complicated*, if one of its curves is. The next result tells us that the above example is somehow typical.

3.11. Proposition. *Let* $h : M \to M$ *be a homeomorphism of some handlebody* M. *Then* h *is periodic if and only if its restriction to* ∂M *is. If* h *is not periodic, then* h *leaves invariant an incompressible, not boundary-parallel codim 1 submanifold, provided the restriction* $h \mid \partial M$ *leaves invariant a sufficiently complicated curve-system.*

Remark. More precisely, the proof will show that the invariant codim 1 submanifold may be chosen to consist of at most two components, provided it is not a system of annuli.

Proof. It remains to prove the second part of the proposition, for we have already verified the first part. We suppose M is not the solid torus, for otherwise the proposition is obvious.

Suppose $h \mid \partial M$ leaves invariant a sufficiently complicated codim 1 submanifold, i.e. a sufficiently complicated system of essential, simple closed curves. Then, passing to some power of h if necessary, we may suppose that there is a sufficiently complicated curve k in ∂M with $h(k) = k$. Given this curve, define

$$F := (\partial M - U(k))^-.$$

where $U(k)$ denotes the regular neighborhood in ∂M. Then F consists of two or one surface according whether the curve k is separating in ∂M or not. We say that k is *compressible* if one of these surfaces is compressible in M.

Case 1. The curve k *is compressible.*

If k bounds a disc in M, then this disc is invariant under h (modulo isotopy), and we are done. Thus we may assume the converse.

But we are in Case 1, and so there is at least one essential disc D in M with $D \cap (\partial M - k) = \partial D$ and such that ∂D is a non-contractible curve in F. Let $\mathcal{D}$ be a maximal system of pairwise disjoint and pairwise non-parallel of such discs. Take the union N of $U(\mathcal{D})$ with all those components of $(M - U(\mathcal{D}))^-$ which are 3-balls. Then N is a disjoint union of handlebodies which do not meet k since $U(\mathcal{D})$ does not meet k and since k cannot be contained in one of the 3-balls of $(M - U(\mathcal{D}))^-$ (k does not bound a disc in M). But we do not claim that N is invariant under the homeomorphism h.

Consider the submanifold N^+, defined by setting

$$N^+ := U(N \cup F).$$

Note first that no component of $(M - N^+)^-$ is a 3-ball since, by our choice of N, no component of $(M - N)^-$ can possibly be a 3-ball.

Note further that the frontier $A := (\partial N^+ - \partial M)^-$ is incompressible in M. To see this, take a disc D_0 contained in N^+ or in $(M - N^+)^-$ with $D_0 \cap A = \partial D_0$. If D_0 is contained in $(M - N^+)^-$, then it is contained in a component M_0 of $(M - N)^-$, and observe that $(M_0 \cap N^+)^-$ consists of product I-bundles. In particular, D_0 can be extended to a disc D_0^+ in M_0 with $D_0^+ \cap \partial M_0 = \partial D_0^+$. Since $M_0 \cap N$ consists of discs, the disc D_0^+ can be isotoped in M_0, fixing D_0, so that afterwards it misses N. Then it is a proper disc in M which is disjoint to N and so, by our maximal choice of $\mathcal{D}$, it follows that D_0^+ separates a ball from M_0 which does not meet k. This ball intersects the frontier of N^+ in discs and so it follows that ∂D_0 bounds a disc in A, provided $D_0 \subset (M - N^+)^-$. If D_0 is contained in N^+, observe that D_0 can be isotoped (fixing ∂D_0) in N^+ so that afterwards $D_0 \cap \mathcal{D} = \emptyset$. But $\mathcal{D}$ splits N^+ into a system of sub-manifolds and those of them containing components of A are product I-bundles. It therefore follows again that ∂D_0 separates a disc in A. Altogether, we thus conclude that A is incompressible indeed.

We next claim that h can be isotoped so that afterwards

$$h(N^+) = N^+.$$

To prove this claim, let us suppose the homeomorphism h is isotoped so that the disc-system $\mathcal{D}' := h(\mathcal{D})$ intersects $\mathcal{D}$ in arcs alone. Then $\mathcal{D}' \cap (M - N)^-$ is a system of discs. Let D_0' be one of them. Then D_0' is a proper disc in $(M - N)^-$. But recall $U(\mathcal{D})$ intersects $\partial(M - N)^-$ in a system of discs too. Thus D_0' can be isotoped in $(M - N)^-$ so that afterwards it does not meet $U(\mathcal{D})$ anymore. It therefore follows from our maximal choice of the disc-system $\mathcal{D}$ that D_0' must be boundary-parallel in $(M - N)^-$. Moreover, $D_0' \cap k = \emptyset$ since $\mathcal{D} \cap k = \emptyset$ and $h(k) = k$. Thus $\partial D_0' \subset F$ and so w.l.o.g. $D_0' \subset U(N \cup F)$. But D_0' has been chosen arbitrarily. Therefore it follows that $h(U(\mathcal{D})) \subset U(N \cup F)$.

Moreover, we may conclude that $h(N) \subset U(N \cup F)$. To see this, recall $(N - U(\mathcal{D}))^-$ is a system of 3-balls. Let E be one of them and set $E' = h(E)$. Now, by what has been verified above, we have $(\partial E' - \partial M)^- \subset U(N \cup F) = N^+$. W.lo.g. $(\partial E' - \partial M)^- \cap (\partial N^+ - \partial M)^- = \emptyset$. Set $A' := A \cap E' = (\partial N^+ - \partial M)^- \cap E'$. Then A' consists of components of A. Moreover, A' consists of discs since A is incompressible (see above) and since discs are the only incompressible surfaces in a 3-ball. Now, discs split a 3-ball into a system of 3-balls, and so $(E' - N^+)^-$ consists of components of $(M - N^+)^-$ which are 3-balls. But no component of $(M - N^+)^-$ is a 3-ball (see above) and so E' is must be contained in N^+. Since E has been chosen arbitrarily, it follows that $h(N) \subset N^+$. Therefore $h(N \cup F) = h(N) \cup h(F) \subset N^+$ since $h(F) = F \subset N^+$, and so $h(N^+) \subset N^+$. Now, h^{-1} is a homeomorphism with the same property as h and so the previous reasoning also demonstrates $h^{-1}(N^+) \subset N^+$, i.e. $N^+ = h(h^{-1}(N^+)) \subset h(N^+)$ modulo isotopy. It follows that h can be isotoped so that afterwards $N^+ \subset h(N^+)$ and $h(N^+ \cap \partial M) = N^+ \cap \partial M$. Moreover, $\partial h(N^+)$ can be deformed (fixing boundary) into N^+. As in [Wa 2], it follows that $\partial h(N^+)$ and ∂N^+ are homotopic and so isotopic relative boundary. Thus h can indeed be isotoped so that afterwards $h(N^+) = N^+$.

By what we have seen so far, the frontier $A = (\partial N^+ - \partial M)^-$ is a codim 1 submanifold in M which is invariant under h. Moreover, $\partial A = \partial U(F)$, where the regular neighborhood $U(F)$ is taken in ∂M. In particular, ∂A and so A has at most two components.

It remains to show that at least one component from A is not boundary-parallel in M. As a matter of fact it will turn out that no component of A can be boundary-parallel. To see this, assume the converse. Then there is a component A_0 of A which is parallel to some surface B_0 in ∂M. Now, w.l.o.g. B_0 is either a component of $N^+ \cap \partial M$ or a component of $(\partial M - N^+)^-$. But B_0 cannot be a component of $N^+ \cap \partial M$. Otherwise the parallelity region between A_0 and B_0 is a component of N^+. It follows that every proper disc in M contained in this component is boundary-parallel. But, by construction of N^+, every component of N^+ contains at least one disc from $\mathcal{D}$. So at least one disc from $\mathcal{D}$ would be boundary-parallel in M. This however is impossible. Thus B_0 must be a component of $(\partial M - N^+)^-$. So, by definition of N^+, B_0 is an annulus in $U(F)$. By construction of N^+ and since A_0 is parallel to B_0, it follows that that component W from $(M - U(\mathcal{D}))^-$ containing the curve k is actually a solid torus. Moreover, the winding number of k with respect to the solid torus W is one since A_0 is parallel to B_0. Therefore there is a meridian-disc in the solid torus W which intersects k in exactly one point and which misses $W \cap U(\mathcal{D})$ altogether. By construction, it follows that this disc is actually a disc in M. This contradicts the hypothesis that k is a sufficiently large curve-system. Thus no component from A is boundary-parallel in M.

This completes the proof of prop. 3.11 in Case 1.

Case 2. The curve k is incompressible.

In this Case, let $U(k)$ be again a regular neighborhood of k in ∂M. Then the components of $(\partial M - U(k))^-$, together with the surface $U(k)$, form a boundary-pattern $\underline{m}$ of M in the sense of [Joh 1, def. 1.1] . Since we are in Case 2 and since k is sufficiently complicated, it follows that this boundary-pattern is useful (in the sense of [Joh 1, def. 2.2]). Thus, by [Joh 1, prop. 9.4], it follows that $(M, \underline{m})$ contains a characteristic submanifold V. Since the homeomorphism h preserves the curve k it is an admissible homeomorphism of $(M, \underline{m})$, and so, by [Joh 1, cor. 10.9], it can be admissibly isotoped so that afterwards $h(V) = V$. Now, by hypothesis, we may suppose that h is not periodic (modulo admissible isotopy). It follows that the characteristic submanifold V is non-trivial since, by [Joh 1, prop. 27.1], the mapping class group of simple 3-manifolds is finite.

We further claim the frontier $(\partial V - \partial M)^-$ is non-empty. If not, then $(M, \underline{m})$ is a component of the characteristic submanifold V and so either an admissible I-bundle or an admissible Seifert fiber space. But it cannot be a Seifert fiber space, for otherwise ∂M is a torus and so M a solid torus, which is excluded (a homeomorphism of a solid torus preserving an incompressible curve is periodic mod isotopy). But it also cannot be an admissible I-bundle.

Otherwise $(\partial M - U(k))^-$ is connected since k is connected. So $(M, \underline{m})$ is the (twisted) I-bundle over some non-orientable surface. This is excluded, for k is supposed to be sufficiently complicated.

Now, the frontier $(\partial V - \partial M)^-$ is an invariant codim 1 submanifold of h, and none of its components is boundary parallel in M. To see the latter, assume for a moment that some component, say A_0, is boundary-parallel. Then A_0 is parallel to some annulus B_0 in ∂M (the frontier of V consists of annuli). Now, this annulus B_0 is either a regular neighborhood of k, or entirely contained in $(\partial M - U(k))^-$. But it cannot be contained in $(\partial M - U(k))^-$, for otherwise A_0 is inessential in $(M, \underline{m})$ which contradicts the fact that V is essential in $(M, \underline{m})$ [Joh 1, def. 8.2]. But it also cannot be a regular neighborhood of k, for otherwise that component of $(M - V)^-$ containing k is the product I-bundle over the annulus, and this contradicts the fact that the characteristic submanifold is complete (see [Joh 1, cor. 10.10]).

Altogether, this proves prop. 3.11 in Case 2. $\Diamond$

§4. Pseudo-Anosov Homeomorphisms.

To complete our discussion from the last section, we here give an explicit example of a pseudo-Anosov handlebody homeomorphism. Many more explicit examples may be found by invoking an algorithm due to Bestvina and Handel [BH 1, 2] for deciding whether a given surface homeomorphism is pseudo-Anosov.

In order to describe this example, let M be a handlebody of genus two and let D_1, D_2, D_3 be three pairwise disjoint, pairwise non-parallel and non-separating discs in M. The basis for the envisioned example is the existence of a disc D in the handlebody M such that, for no $1 \leq i \leq 3$, there is an isotopy of D which makes D disjoint to D_i. In [FL] the *existence* of such a disc has been shown, using Thurston's theory of curves on surfaces (as presented in [FLP]). Here we rather wish to give an *explicit example* of such a disc.

4.1. Example.

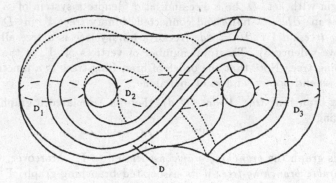

(Figure 4.1.)

Given the above disc D, a pseudo-Anosov homeomorphism $h : M \to M$ can now be constructed as in [FL].

First, observe that Dehn-twists around the meridian-discs D_1, D_2 and D_3 map D to another disc D^* with the crucial property that $(\partial M - \partial D \cup \partial D^*)^-$ consists of discs. Thus, by a criterion of Thurston (see [FLP, exposé 13], the product $T_{\partial D} \circ T_{\partial D^*}$ of Dehn-twists along the curves ∂D and ∂D^* is actually a pseudo-Anosov homeomorphism $g : \partial M \to \partial M$. But, by construction, these Dehn-twists can be extended to Dehn-twists around the discs D and D^*. Thus g can be extended to a homeomorphism $h : M \to M$ which is a pseudo-Anosov homeomorphism.

§5. Branching Complexes.

If one wishes to parametrize simple closed curves on the boundary of a handlebody, one needs some sort of reference-system first. For instance one could choose a generating set of loops for the fundamental group of the boundary-surface and then express curves as words in these generators. In this way, it is very easy to generate the set of all curves. However, this approach has the obvious shortcomings that it does not distinguish between singular and non-singular curves. To fix this problem, we may, in addition, choose one of the available algorithms (notably [Chi]) to find out which one of the words generated in this way is actually represented by a *simple* closed curve. Of course the overwhelming majority of loops in a surface is singular, and so this "algebraic" approach is neither very instructive nor efficient. In fact, it may take a long time before one ever finds a single non-singular curve by randomly picking words. Therefore one looks for a more geometrically defined approach to the present problem. This motivates the present section.

5.1. *Branching-Trees.* The notion of "branching-trees" is closely related to Thurston's concept of "train-tracks". Indeed, we will see in this section that, in the spirit of Thurston's concept, branching-trees can be used to represent systems of simple closed curves on handlebodies.

To begin with, let D be a disc and let d denote a system of n pairwise disjoint arcs in ∂D. A finite and connected, binary tree Γ in D is called a *branching-tree*, if $\Gamma \cap \partial D = \partial d$ (recall a binary tree is a tree all of whose vertices have valence 3). The total number of vertices of Γ is the *order* of the branching-tree. Note that the order of branching-trees is a function of the number of its end-points (and so of n) alone.

To any branching-tree Γ in (D, d) there is associated a graph Γ^+, by simply setting

$$\Gamma^+ := \Gamma \cup d.$$

We call this graph the *branching-graph* associated to Γ. Moreover, we call Γ a *non-separable branching-tree* if its associated branching-graph Γ^+ is non-separable in the sense that the complement of every single point in Γ^+ is connected.

By taking first the regular neighborhood U (in D) of the union of all vertices of Γ^+ (of valence 3), and then contracting all arcs from $(\Gamma^+ - U)^-$ to points, we obtain from U a 2-complex, say $\Delta := \Delta(\Gamma)$. In fact, Δ may be considered as a collection of triangles whose vertices are pairwise identified. This 2-complex Δ is called the *branching-complex* associated to the tree Γ. The number of triangles is the *order of* Δ and of course this number equals the order of the underlying branching-tree. $\Delta(\Gamma)$ is non-separable if Γ is.

5.2. *Branching-Matrices.* To any branching-complex Δ as given before, we associate a *branching-matrix* $A(\Delta)$ in the following way.

First, we distinguish between *outer* and *inner* vertices of Δ according whether they are incident to one or two triangles of Δ (note that the branching-complexes associated to non-separable branching-trees have inner vertices only). Then we associate to Δ a (homogeneous) system of linear equations. For this, we first label all those edges from Δ, which have at least one inner point as end-point, with symbols x_i, $1 \leq i \leq 6g-6$ (where g denotes the order of Δ). Choose these labels in such a way that two edges of a triangle from Δ carry the same label, if they both join the same inner point with outer points. Observe that to any inner point we may associate a linear equation $a_1 + a_2 = b_1 + b_2$ in the symbols x_i above. Indeed, recall that any inner point is incident to exactly two triangles σ_1, σ_2 from Δ_g and so z is end-point of exactly two edges from σ_1, and the same with σ_2. We then obtain an equation as above by setting equal the sum of the labels of each of these two pairs of edges. In this way, there is a unique homogeneous system of linear equations associated to any branching-complex Δ.

We define the *branching matrix* $A(\Delta)$ to be the coefficient matrix of this system.

Note that the number of columns of the branching-matrix $A(\Delta)$ equals the number of edges, while the number of rows equals the number of inner points of Δ. Thus, from a computational point of view, it is worth mentioning that the complexity of solving the equation

$$A(\Delta)x = 0$$

depends heavily on the number of inner points (or outer points for that matter). Observe further that, for a given order g, the number of inner points may differ substantially for the various branching complexes Δ of order g (especially when g is large). In any case, however, the solution-space of $A(\Delta)x = 0$ has a finite basis. In particular, there are only finitely many linear independent solution-vectors whose entries are all positive. We will see in a moment that these special solutions are closely related to curves. But first an example.

5.3. Example. The following branching complex of order 4 has four inner points.

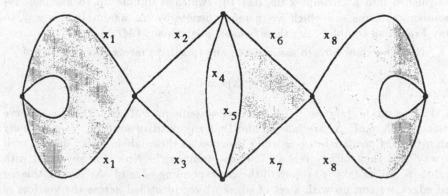

(Figure 5.1)

The following system of four equations is associated to the previous branching complex:

$$
\begin{aligned}
2x_1 \quad -x_2 \quad -x_3 \qquad\qquad\qquad\qquad\qquad &= 0 \\
x_2 \qquad\qquad +x_4 \quad -x_5 \quad -x_6 \qquad\qquad &= 0 \\
x_3 \quad +x_4 \quad -x_5 \qquad\qquad -x_7 \qquad &= 0 \\
x_6 \quad +x_7 \quad -2x_8 &= 0
\end{aligned}
$$

and the following elements from $\mathbf{Z}^8$ form a basis for the solution-space:

$$
\begin{aligned}
m_1 &= (0,0,0,1,1,0,0,0), \\
m_2 &= (1,1,1,0,1,0,0,0), \\
m_3 &= (1,1,1,0,0,1,1,1), \\
m_4 &= (1,2,0,0,0,2,0,1),
\end{aligned}
$$

5.4. *Curves and branching-matrices.* Let us now discuss the relation of positive solutions and curve-systems.

For this, let Γ be some branching-tree in the disc (D,d), embedded in Euclidean space $\mathbf{R}^3$. Then the regular neighborhood (in $\mathbf{R}^3$) of the associated branching-graph Γ^+ is a handlebody, denoted by $M(\Gamma)$ or $M(\Delta)$ in order to indicate the connection with the branching-tree Γ resp. the branching-complex Δ. The homeomorphism-type of this handlebody is certainly independent of the actual choice of the branching-tree Γ. In fact, for every binary graph without end-points whose regular neighborhood is a handlebody of genus $\#d$, there is a branching-tree Γ in (D,d) whose associated branching-graph is homeomorphic to the original graph. In particular, we may note the close relationship between the total number of branching trees (which can be calculated to be a Catalan number [Knu]) and the total number of maximal disc-systems in the associated handlebody, modulo homeomorphisms.

Now, observe that the branching-complex Δ associated to Γ^+ can be pushed into a 2-complex in $\partial M(\Gamma)$, which is unique up to handlebody-homeomorphisms and which we therefore denote by Δ again. In this way, to any branching complex Δ, there is associated a pair $(M(\Delta), \Delta)$.

Next, we show how to associate to any (positive) integer solution z of

$$
A(\Delta)x = 0
$$

a class, notation $[z]_\Delta$ or $[z]_\Gamma$, of curve-systems in $\partial M(\Delta)$. For this observe that the edges of Δ are labeled and that any positive solution z is actually an n-tuple of positive integers which provides all these labels with values in such a way that they satisfy certain "switch-conditions". Now, for any label with value p, let us take p copies of the corresponding edge of Δ. Doing this for all edges, we end up with a set of edges which, identified across the vertices of Δ, produces a curve-system in $\partial M(\Delta)$. Still more curves can be obtained by "shifting". Specifically, observe that, for any vertex z of Δ, we may choose a disc D in $M(\Delta)$ which intersects Δ in z alone. Cutting the handlebody open along D and re-attaching the copies of D, via a rotation of some appropriate angle, possibly less than 2π, produces other curve-systems. Moreover, values for the edges joining outer points produce new curve-systems too. All these new curve-systems are *shift-equivalent* to the original curve-system.

Thus, indeed, to any positive integer solution z of $A(\Delta)x = 0$, there is associated a shift-equivalence class $[z]_\Delta$ of (infinitely many) curve-systems in $\partial M(\Delta)$. Since the branching-matrix $A(\Delta) = A(\Gamma)$ is determined by a choice of a branching-complex Δ (i.e. by the choice of some branching-tree Γ), we also say that the curves from $[z]_\Delta$ are *carried by* Δ (or Γ for that matter).

The relevance of the previous construction lies in the following result:

5.5. Proposition. *Let M be a handlebody. Let $\mathcal{K}$ be a system of simple closed curves in ∂M such that $\partial M - \mathcal{K}$ is incompressible in M.*

Then $\mathcal{K}$ is carried by some non-separable branching-complex Δ.

In particular, there is a solution z of $A(\Delta)x = 0$ such that $\mathcal{K}$ is contained in $[z]_\Delta$.

The proof of this proposition is based on an observation concerning "tight" planar surfaces in ∂M. Here a planar surface $F \subset \partial M$ is called *tight* if the following holds:

(1) Every curve in F is contractible in M.

(2) Every component from ∂F is non-separating in ∂M.

(2) Every component from $\mathcal{K} \cap F$ is a non-recurrent arc in F.

Recall a proper arc in a surface is *recurrent*, if both its end-points lie in one boundary component of the surface.

5.6. Lemma. *Let M and $\mathcal{K}$ be given as in 5.5. Let $F \subset \partial M$ be a tight planar surface with $\#\partial F > 3$. Then there is a simple closed, not boundary-parallel curve in F which is non-separating in ∂M and which intersects every arc from $\mathcal{K} \cap F$ in at most one point.*

Proof. For lack of a better name we say that the union $r_1 \cup r_2$ of two boundary curves r_1, r_2 of ∂F is *separating* if it lies in a single component of $(\partial M - F)^-$; otherwise it is *non-separating*. The reason for this name is that if $r_1 \cup r_2$ is non-separating then, for every arc $k \subset F$ joining r_1 and r_2, the closed curve $\partial U(r_1 \cup k \cup r_2) - \partial F$ is non-separating in ∂M in the usual sense (the regular neighborhood is taken in F). To see this fact let $r_1 \cup r_2$ be non-separating and let A_i, $i = 1, 2$, denote the component of $(\partial M - F)^-$ containing r_i. Note that r_1 cannot be the only boundary component of A_1 since r_1 is non-separating in ∂M. So A_1 must meet $(F - U(r_1 \cup k \cup r_2))^-$ since $\partial A_1 \subset F$

and $A_1 \neq A_2$. The same with A_2. Hence the claim follows from the fact that every non-recurrent arc (such as k) in a planar surface is non-separating.

We next claim that there is at least one arc k from $\mathcal{K}' := \mathcal{K} \cap F$ which joins two different components r_1, r_2 of ∂F such that $\partial U(r_1 \cup k \cup r_2) - \partial F$ is a non-separating curve in ∂M. To see this claim note that $\mathcal{K}'$ is non-empty since F is tight and since $\partial M - \mathcal{K}$ is incompressible. Let k_1 be one of them. Then k_1 joins different components of ∂F, say r_1, r_2, since F is tight. W.l.o.g. $\ell_1 := \partial U(r_1 \cup k_1 \cup r_2) - \partial F$ is separating, for otherwise we are done. So, in particular, $r_1 \cup r_2$ is separating (see above). Now ℓ_1 separates F into two planar surfaces F_1, F_2 where the indices are chosen so that $r_1 \cup r_2 \subset F_1$. The curve ℓ_1 is contractible in M since F is tight. So it has to intersect $\mathcal{K}'$ since $\partial M - \mathcal{K}$ is incompressible. Moreover, no arc from $\mathcal{K}'$ is recurrent since F is tight. So there is at least one arc k_2 from $\mathcal{K}'$ which joins r_1 (or r_2) with a component r_3 of $F_2 \cap \partial F$. Set $\ell_2 := \partial U(r_1 \cup k_2 \cup r_3) - \partial F$. Note that r_1 and r_2 lie in the same component of $\partial M - \ell_2$ but in different components of $F - \ell_2$. Hence it follows that ℓ_2 is non-separating in ∂M since every component of $(\partial M - F)^-$ meets F. The claim is therefore established.

Let k_1 be an arc from $\mathcal{K}'$ joining boundary curves r_1, r_2 from ∂F such that $\partial U(r_1 \cup k_1 \cup r_2) - \partial F$ is a non-separating curve in ∂M (the regular neighborhood is taken in F). Such an arc k_1 exists by what we have just seen.

Suppose *every* arc from $\mathcal{K}'$ which joins r_1 with r_2 is parallel in F to k_1. Consider

$$\ell := \partial U(r_1 \cup k_1 \cup r_2) - \partial F.$$

Recall ℓ is a curve which is non-separating in ∂M. Moreover, it is not boundary-parallel in F since $\#\partial F > 3$. Finally, ℓ intersects every arc from $\mathcal{K}'$ in at most one point since no arc from $\mathcal{K}'$ is recurrent (F is tight) and since every arc from $\mathcal{K}'$ joining r_1 and r_2 is parallel to k_1 (and may be supposed to lie in $U(k_1)$). Hence, altogether, ℓ is the desired curve in F.

Now suppose there is at least one arc, say k_2, from $\mathcal{K}'$ which joins r_1 with r_2 and which is not parallel in F to k_1. Note that $(F - U(k_1 \cup k_2))^-$ consists of two different components since F is planar. Let F_1 be one of them. W.l.o.g. we may suppose that no arc k from $F_1 \cap \mathcal{K}$ joins two components r_i, r_j from ∂F such that $r_i \cup r_j$ is non-separating, for otherwise either $\partial U(r_i \cup k \cup r_j) - \partial F$ is the desired curve or we get a reduction. Let $\mathcal{G}$ be the graph whose vertices are the boundary curves of ∂F meeting F_1 and whose edges are the arcs from $F_1 \cap \mathcal{K}$. Then this graph is disconnected, by what we have just seen. In fact, this graph consists of exactly two components since F is tight and since $\partial M - \mathcal{K}$ is incompressible. Let $\mathcal{G}_i$, $i = 1, 2$, be that component of $\mathcal{G}$ containing r_i. It follows that there is an arc s in F_1 which joins a vertex r_i from $\mathcal{G}_1$ with a vertex r_j from $\mathcal{G}_2$ without meeting $\mathcal{K}$. Moreover, we may suppose that $r_i \cup r_j \neq r_1 \cup r_2$, for F_1 is not a disc and $\mathcal{G} = \mathcal{G}_1 \cup \mathcal{G}_2$. Thus none of the arcs from $\mathcal{K}$ join r_i with r_j (recall r_i and r_j lie in different components of $\mathcal{G}$). Therefore

$$\ell := \partial U(r_i \cup s \cup r_j) - \partial F$$

is a closed curve which intersects every arc from $\mathcal{K}$ in at most one point (recall no arc from $\mathcal{K}$ is recurrent). It also is non-separating in ∂M. Indeed, $r_i \cup r_j$ is non-separating, for any two vertices of $\mathcal{G}$ lie in one component of $\mathcal{G}$ if and only if they lie (as boundary components) in one component of $\partial M - F$ (see our assumption above). Thus again we have found the desired curve.

This proves the lemma. $\Diamond$

5.7. Lemma. *Let* M *and* $\mathcal{K}$ *be given as in 5.5. Then there is a system* $\mathcal{D}$ *of essential discs in* M, *containing a meridian-system* $\mathcal{D}'$, *with the following properties:*

(1) $\mathcal{D}$ *splits* ∂M *into a system of three-punctured 2-spheres,*

(2) *any disc from* $\mathcal{D}$ *is non-separating in* M, *and*

(2) *any sub-system of* $\mathcal{D}$ *(possibly equal to* $\mathcal{D}$*), containing* $\mathcal{D}'$, *splits* ∂M *into a system of surfaces whose intersection with* $\mathcal{K}$ *consists of non-recurrent arcs.*

Remark. It will be apparent from the proof that *any* minimal meridian-system in M can be extended to a system of discs satisfying the conclusion of the previous lemma.

Proof. Suppose M is not a solid torus (otherwise the lemma follows immediately). Then we construct $\mathcal{D}$ from an appropriate meridian-system of M. For this choose a meridian-system $\mathcal{D}'$ of M intersecting $\mathcal{K}$ in a minimal number of points (such a meridian-system not only exists but can be constructed in finitely many steps; see section 7). Set

$$F_0 := (\partial M - U(\mathcal{D}'))^- \text{ and } \mathcal{K}_0 := \mathcal{K} \cap F_0,$$

where the regular neighborhood is taken in M.

We claim F_0 is a tight planar surface. Indeed, it is is a planar surface in ∂M and all curves in F_0 are contractible in M. Moreover, $\mathcal{K}_0$ is a system of arcs in F_0 since $\partial M - \mathcal{K}$ is supposed to be incompressible in M. Assume for a moment there is an arc from $\mathcal{K}_0$, say k, which is recurrent in F_0. Then there is a disc B in $(M - U(\mathcal{D}'))^-$ such that $B \cap F_0 = k$. In particular, the arc $(\partial B - k)^-$ lies in $U(D_0)$, for some disc D_0 from $\mathcal{D}'$. Thus $(\partial U(B \cup U(D_0)) - \partial M)^-$ consists of three discs. Replacing D_0 by some appropriate component from $(\partial U(B \cup U(D_0)) - \partial M)^-$, we get another meridian-system from $\mathcal{D}'$ which intersects $\mathcal{K}$ in strictly fewer points than $\mathcal{D}'$. But this contradicts our minimal choice of $\mathcal{D}'$. This proves the claim.

Since $\#\partial F_0 > 3$, there is a simple closed curve l in F as given by lemma 5.6. Set $F_1 := (F_0 - U(l))^-$. Then F_1 is again a tight surface. Repeating this process we get a tight planar 2-manifold F_2 and so on. But the process must stop after finitely many steps. So eventually we obtain a system of tight thrice punctured 2-spheres. The desired system $\mathcal{D} \subset M$ is then a disc-system bounded by the boundary of these tight planar surfaces. $\Diamond$

Proof of Proposition 5.5. We are asked to construct a non-separating branching-complex Δ which carries the curve-system $\mathcal{K}$.

In order to construct Γ, let $\mathcal{D}$ be a system of non-separating, essential discs in the handlebody M which (1) splits ∂M into a system of thrice punctured 2-spheres and which (2) splits the curve-system $\mathcal{K}$ into non-recurrent arcs. Such a disc-system exists because of the previous lemma.

Now, fix a system $\mathcal{L}$ of arcs contained in $\mathcal{D}$ so that every disc from $\mathcal{D}$ contains exactly one arc from $\mathcal{L}$. Note further that $\mathcal{D}$ splits the handlebody M into a set of 3-balls. In any one of these 3-balls E_i choose a disc B_i with $\mathcal{L} \cap E_i \subset B_i$ and $B_i \cap \mathcal{D} \subset L$. It is easy to see that, collapsing all arcs from $\mathcal{L}$ to points, the union $\bigcup B_i$ becomes a 2-complex Δ of triangles. Δ can be pushed into a branching complex in ∂M which carries $\mathcal{K}$. To see the latter recall $\mathcal{D}$ cuts the curve-system $\mathcal{K}$ into non-recurrent arcs (see our choice of $\mathcal{D}$). Moreover, $\mathcal{D}$ cuts ∂M into thrice punctured 2-spheres. Any 2-sphere with three holes has a unique maximal system R_i of pairwise non-parallel and non-recurrent arcs, which consists of three arcs. W.l.o.g. we may suppose that the discs B_i from above are chosen so that, in addition $B_i \cap \partial M = R_i$. It follows from our choice of $\mathcal{D}$ that (modulo shifts) the curve-system $\mathcal{K}$ can be isotoped in ∂M into $\bigcup B_i$, and so Δ carries $\mathcal{K}$ indeed.

Note that Δ is in fact associated to the dual-graph of $\mathcal{D}$. The spanning-tree of this dual-graph is a branching-tree Γ with $\Delta = \Delta(\Gamma)$. It remains to verify that Γ is in fact non-separable. But this follows immediately from the fact that $\mathcal{D}$ consists of non-separating discs (see above). $\Diamond$

5.8. By what has been seen earlier, every (positive integer) solution of $A(\Delta)x = 0$ is a curve-solution, i.e. represents a system of simple closed curves $\mathcal{K}$ on the boundary $\partial M(\Delta)$. (In fact, it represents an infinite set of shift-equivalent curve-systems). *All* these curves are essential in ∂M. To see this, recall that to the set of vertices of Δ, there is associated a system of discs which splits ∂M into a set of 3-punctured 2-spheres which in turn contain no recurrent sub-arc of $\mathcal{K}$. So no curve from $\mathcal{K}$ can possibly be contracted in ∂M to a point.

Under a slightly stronger hypothesis, we have the following result:

5.9. Proposition. *Let Δ be a branching-complex and let z be an n-tuple of non-zero, positive integers which solves the equation $A(\Delta)x = 0$. Then any curve-system in $[z]_\Delta$ is a curve-system in $\partial M(\Delta)$ such that $\partial M(\Delta) - \mathcal{K}$ is incompressible in $M(\Delta)$.*

Proof. Recall from its definition that $M(\Delta)$ is actually the regular neighborhood of the triangle-complex Δ embedded in $\mathbf{R}^3$. Let $\mathcal{D}$ be the disc-system in $M(\Delta)$ associated to the set of vertices of Δ. Then $\mathcal{D}$ splits $\partial M(\Delta)$ into a set of thrice punctured 2-spheres, say S_i, $1 \le i \le n$ (pants).

Now, let $\mathcal{K}$ be any curve-system in $[z]_\Delta$. By our hypothesis on z, it follows that, for all $1 \le i \le n$, any two boundary curves of S_i are in S_i joined by a sub-arc from $\mathcal{K}$. In particular, no pant S_i contains a recurrent arc which is disjoint from $\mathcal{K}$.

Assume $\partial M(\Delta) - \mathcal{K}$ is compressible in $M(\Delta)$. Then there is a disc A in $M(\Delta)$ with $A \cap \partial M(\Delta) \subset \partial M(\Delta) - \mathcal{K}$ and such that ∂A is not contractible in $\partial M(\Delta) - \mathcal{K}$. Let A be isotoped so that the above holds and that, in addition, the number of components of the intersection $A \cap \mathcal{D}$ is as small as possible. Then $A \cap \mathcal{D}$ is a system of arcs, and so it splits the disc A into a system of discs. There is at least one of these discs, say A_0, such that $A_0 \cap \mathcal{D}$ is connected. It follows that $A_0 \cap \partial M(\Delta)$ is a recurrent arc in some pant S_i. But this arc does not meet $\mathcal{K}$ since $\partial A \subset \partial M(\Delta) - \mathcal{K}$, by our choice of A. This, however, contradictions the above property of $\mathcal{K}$. So the proposition is proved. $\Diamond$

5.10. We have seen that the set of all systems $\mathcal{K}$ of simple closed curves (with incompressible complement) in a handlebody boundary is parametrized by the solutions of the system of homogeneous linear equations associated to branching complexes. Now, the solution-set of a finite homogeneous system of linear equations has a finite basis. As a matter of fact, given the system $A(\Delta)x = 0$, we may even construct a finite basis of curve-solutions so that *any* curve-solution of $A(\Delta)x = 0$ is a finite *sum* of this basis. Guided by Haken [Ha 1], we call such a basis a basis of *fundamental curve-solutions*. To see that a basis of fundamental solutions exists, recall that the solution-space of a homogeneous system is a sub-space and the curve solutions are given by those points from the integer-lattice which are contained in the intersection of the solution-space with the positive quadrant. Moreover, observe that the solution space intersects every coordinate-plane (spanned by two coordinate-axis) in the origin or in a straight line with rational slope (provided it does not contain this coordinate-plane). By picking appropriate vectors in those lines the required basis can be constructed inductively (see [Schu] and the appendix of this book).

5.11. Example. As an illustration, we here show the complete set of fundamental curves for the handlebody of genus 3 carried by the branching-complex given in example 5.3 (and corresponding to the basis given there). Note that these curves are *not* pairwise disjoint.

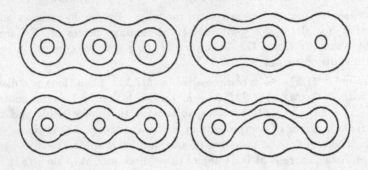

(Figure 5.2.)

5.12. The importance of the observation from 5.10 rests on the fact that *sums* (but not necessarily differences) *of curve-solutions* have a simple geometric interpretation. To describe it, let $[z_1]$ and $[z_2]$ be two curve-solutions of $A(\Delta)x = 0$, and let $\mathcal{K}_1, \mathcal{K}_2$ be curve-systems in $[z_1], [z_2]$, respectively. Then $\mathcal{K}_1, \mathcal{K}_2$ are curve-systems in $\partial M(\Delta)$. Observe that, for every intersection-point from $\mathcal{K}_1 \cap \mathcal{K}_2$, there is a unique cut-and-paste such that in the end we obtain a curve-system contained in $[z_1 + z_2]$. Thus the sum of curve-solutions is always geometrically realized by a certain cut-and-paste operation.

5.13. *Curves in shift-equivalence classes.* As we have seen, the concept of branching complexes presents a convenient way to *generate* the set of all (interesting) curve-systems on handlebodies. The *classification* of those curve-systems, however, appears to be a different, more complicated matter. As an example take the classification modulo handlebody-homeomorphisms. Although, it can be decided whether two given curve-systems are equivalent under a handlebody-homeomorphism (see section 8), it seems to be difficult to generate the set of all such *equivalence-classes* of curve-systems. E.g. the question arises, when some given set of curve-systems $\mathcal{K}$ on a handlebody M, with $\partial M - \mathcal{K}$ incompressible, is really infinite, modulo handlebody homeomorphisms, and how to construct such infinite sequences of equivalence-classes of curve-systems. This question will be addressed in section 11, after an appropriate complexity for curve-systems on M has been introduced.

At this point, a shortcomings of the above described representation of curve-systems via branching-complexes has to be pointed out, though. On the one hand, *every* essential curve-system on a handlebody can be described by an appropriate, labeled branching-complex. It is further not difficult to see that two curve-systems, carried by the same branching-complex, are isotopic if and only if the corresponding labels are the same. On the other hand, these curve-systems are often disconnected. Later however we will be interested in generating sequences of closed curves (not curve-systems), and the question arises whether we perhaps always find a closed curve in $[z]_\Delta$. In other words, given a curve-system $\mathcal{K}$ carried by some branching-complex Δ, is it possible to change $\mathcal{K}$ into a single closed curve by shifts along the meridian-discs associated to the vertices of Δ? Unfortunately, the answer to this question is negative, as can be shown by the following counter-example.

5.14. Example. Consider the following labeled branching-complex Δ, where the shift-factors are all set to be zero.

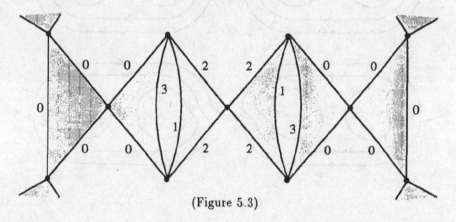

(Figure 5.3)

The curve-system $\mathcal{K}$ carried by this complex may schematically be drawn as follows.

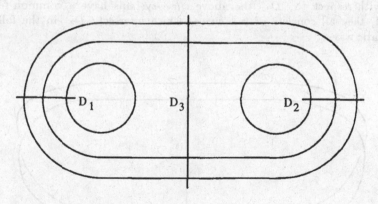

(Figure 5.4.)

Note that this curve-system crosses five vertices of the complex Δ, but observe that only the shift factors of three of these vertices are relevant. As explained above, these three vertices correspond to three meridian-discs and the positions of these meridian-discs are indicated in the figure above by the lines

D_1, D_2, D_3 (which actually represent *closed* curves). In the following, we list the result of all relevant combinations of shifts along the discs D_1 and D_2 (modulo handlebody homeomorphisms).

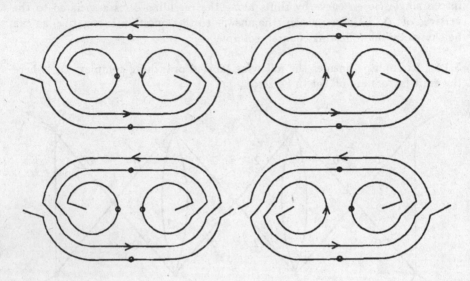

(Figure 5.5.)

As a result we find that none of the above shifts turns the curve-system $\mathcal{K}$ into a closed curve. It remains to consider the result of a shift along D_3. But observe that with respect to D_3 the above curve-systems have a common feature. Indeed, they all consists of two curves which intersect D_3 in the following schematic way.

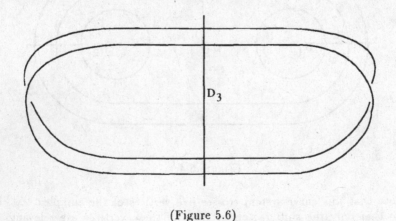

(Figure 5.6)

There are four different shifts along D_3 possible (modulo handlebody-homeo-morphisms) and they yield the following curve-systems.

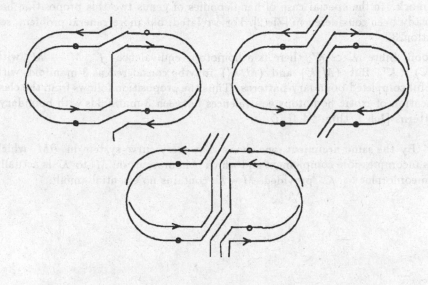

(Figure 5.7)

Observe that none of the resulting curve-systems are connected. ◊

Thus the following problem still remains:

5.15. Problem. *Find a systematic way to represent all (connected) simple closed curves.*

As has been indicated before, the more algebraic way of representing curves on handlebodies (briefly discussed in the beginning of this section) produces indeed only closed curves, but allows *singular* curves as well, while the more geometric way based on branching-complexes produces simple closed curve-systems, but this time allows *disconnected* curve-systems in an essential way. A procedure is asked for which combines the good features of both the approaches above. (It is certainly possible to write a computer-program which mechanically produces all curve-systems and sorts out the disconnected ones by brute force, but such a procedure would not be very instructive).

5.16. We conclude this section with an interesting feature concerning the relation of curve-systems with incompressible complement and handlebody homeomorphisms.

5.17. Proposition. *Let M be a handlebody. Let K be a curve-systems in ∂M with incompressible complement, i.e., $\partial M - K$ is incompressible in M. Then,* *modulo handlebody homeomorphisms, there are only finitely many curve-systems with incompressible complement which are homotopic in M to $\mathcal{K}$.*

Remark. In the special case of handlebodies of genus two this proposition has already been considered in [Zie 2]. For a related, but more general problem, see section 32.

Proof. Since $\mathcal{K} \simeq \mathcal{K}'$, there is a homotopy equivalence $f : M \to M$ with $f(\mathcal{K}) = \mathcal{K}'$. But $(M, \mathcal{K})$ and $(M, \mathcal{K}')$ may be considered as 3-manifolds with useful completed boundary-patterns. Thus the proposition follows from the classification of exotic homotopy equivalences between 3-manifolds with boundary-patterns [Joh 1, thm. 24.4]. $\Diamond$

By the same argument, one shows that every curve-system in ∂M which has incompressible complement and which is homotopic in M to $\mathcal{K}$ is actually homeomorphic to $\mathcal{K}$, provided $M - \mathcal{K}$ contains no essential annuli.

II. Relative Handlebodies

In the last chapter, handlebodies and curves on handlebody boundaries have been considered separately. A new object emerges by forming the pair (M, k) of a handlebody M and a system k of simple closed (pairwise non-parallel) curves in ∂M. Such pairs will be called *relative handlebodies*. Relative handlebodies are the object of the present chapter. The very name may be slightly misleading since, as we will see, relative handlebodies behave differently from handlebodies. Indeed, while handlebodies have some rather peculiar properties, relative handlebodies share many important features with Haken 3-manifolds; notably their rigidity.

Relative handlebodies are also special cases of 3-manifolds with boundary-patterns. So the theory of 3-manifolds with boundary-patterns, as developed in [Joh 1], is readily available for the study of relative handlebodies. In fact, the impact of the boundary-pattern is overwhelming in this case since the internal topology of the underlying manifold is so primitive. It is the boundary-pattern which makes the combinatorics of relative handlebodies surprisingly complex. In our approach, a good understanding of these combinatorics is important though, for it will ultimately yield a better understanding of the topology of 3-manifolds in general.

In this chapter, we will be mainly concerned with essential surfaces in relative handlebodies. Specifically, we will show how to find them. For *non-separating* surfaces, we will first introduce the Heegaard-matrix for relative handlebodies. We will use the Heegaard-matrix to obtain not only a constructive version of the Poincaré duality map, but also a certain algebraic invariant which estimates the Seifert characteristic of relative handlebodies. For *separating* surfaces, we will introduce the Haken matrix for relative handlebodies. We will use the Haken matrix to investigate the Haken spectrum; a certain counting function for essential surfaces.

§6. Heegaard-Diagrams.

Recall that, by its very definition, a relative handlebody is given by some handlebody, together with a system of simple closed curves on its boundary. Thus, increasing the genus of the underlying handlebody, is one way to make a relative handlebody more complicated. Increasing the complexity of the curve system, however, is much more interesting. In fact, understanding the curve-systems on a given handlebody is known to be hard. For some aspects of this problem, branching complexes, as discussed in the previous section, are helpful. For other aspects, we use the classical presentation of curve-systems on handlebodies by means of Heegaard-diagrams.

In this section, we introduce Heegaard-diagrams and the notion of circle-patterns in Heegaard-diagrams in a rather formal way. We adopt this more abstract point of view since it appears to be better adapted for our later study of surfaces in 3-manifolds.

6.1. We begin with *Heegaard-diagrams*. Our definition differs slightly from what is usually known under this term. The difference lies mainly in the fact that we allow diagrams to be disconnected and the number of relator curves to be smaller than the genus.

6.2. Definition. *A Heegaard-diagram (or simply: a diagram) is given by a triple $\underline{D} = (D, h, \underline{K})$ such that*

(1) *D is a finite system of (pairwise disjoint) discs in a finite system $F = F(\underline{D})$ of 2-spheres.*

(2) *h is a fixpoint-free and orientation-reversing involution on D.*

(3) *$\underline{K}$ is a finite system of essential arcs in $(F - D)^-$ with $\underline{K} \cap \partial D = \partial \underline{K}$ and such that $h(\partial \underline{K}) = \partial \underline{K}$.*

(It is to be understood that an orientation of the underlying surface F has been fixed once and for all and that the system D of discs in F carries the induced orientation). The curves from ∂D are called the *meridian-curves* and the components from D are the *meridian-discs* (of the diagram $\underline{D}$). $F(\underline{D})$ is the *underlying 2-manifold* of $\underline{D}$. Any union $\underline{K}_i$, $i \geq 0$, of components from $\underline{K}$ is a *relator-cycle*, provided the quotient $\underline{K}_i/h$ is a circle. Moreover, the following invariants are naturally associated to (the homeomorphism type) of any diagram $\underline{D} = (D, h, \underline{K})$:

$$\text{genus of } \underline{D} = g(\underline{D}) := \tfrac{1}{2}\#(D),$$
$$\text{rank of } \underline{D} = r(\underline{D}) := \#(\underline{K}/h),$$
$$\text{complexity of } \underline{D} = c(\underline{D}) := \tfrac{1}{2}\#(\partial \underline{K}).$$

As always throughout this book $\#A$ denotes the total number of components of the topological space A. Furthermore, $\underline{D}$ is an *n-relator diagram*, if $n = r(D)$.

There are diagrams with especially nice properties. We single them out because of their later importance.

6.3. Definition. *Let $\underline{D} = (D, h, \underline{K})$ be a Heegaard-diagram and let F denote its underlying 2-manifold. Then we say:*

(1) *$\underline{D}$ is useful, if, for no disc D_0 from D, there is an arc in $(F - D)^-$ which is disjoint to $\underline{K}$ and which joins two different components of $\partial D_0 - \underline{K}$.*

(2) *$\underline{D}$ is irreducible, if $F_0 \cap (D \cup \underline{K})$ is connected, for every component F_0 from F.*

(3) *$\underline{D}$ is full, if any two components of $F_0 \cap D$ are joined by at least one arc from $\underline{K}$, for every component F_0 from F.*

Remark. A useful diagram is a Heegaard-diagram without "waves", in the terminology of [VKF]. We prefer to call these diagrams "useful" since they are closely related to 3-manifolds with useful boundary-patterns as introduced in [Joh 1].

Notice that a diagram $\underline{D} = (D, h, \underline{K})$ is reducible or not useful if an arc from $\underline{K}$ is recurrent (in the sense that it has both end-points in one meridian-disc). Observe further that it is easy to decide whether a diagram is useful or irreducible.

6.4. Definition. *A* pant-*diagram is a Heegaard-diagram* $\underline{D} = (D, h, \underline{K})$ *with the following two properties:*

(1) every arc from $\underline{K}$ *joins different components from* D, *and*

(2) the underlying 2-manifold consists of 2-spheres each of which containing exactly three discs from D.

Let us next turn to *circle-patterns*. Here is the formal definition.

6.5. Definition. *Let* $\underline{D} = (D, h, \underline{K})$ *be any diagram and let* F *denote its underlying 2-manifold. Then a system* C *of (pairwise disjoint) simple closed curves in* $F - \underline{K}$ *is called a* circle-system *in* $\underline{D}$ *if*

(1) $C \cap D$ *consists of proper arcs with* $C \cap D = h(C \cap D)$,

and it is called a circle-pattern *in* $\underline{D}$ *if, in addition,*

(2) no component of $C \cap D$ *separates a disc from* D *disjoint to* $\partial \underline{K}$, *and*

(3) no component of $(C - D)^-$ *separates a disc from* $(F - D)^-$.

A circle-pattern in $\underline{D}$ is *orientable* if it has a normal orientation which is compatible under the attaching map h. Two circle-patterns in $\underline{D}$ are *pattern-isotopic*, if they can be joined by a continuous family of circle-patterns. $\mathcal{C}(\underline{D})$ denotes the set of all pattern-isotopy classes of circle-patterns in $\underline{D}$, and $\mathcal{C}^+(\underline{D}) \subset \mathcal{C}(\underline{D})$ is the subset of all orientable circle-patterns. To any $C \in \mathcal{C}(\underline{D})$, we associate its *Euler characteristic* defined by

$$\chi(C) := \#C - \tfrac{1}{2}\#(C \cap D).$$

From a mere formal point of view, it may be somewhat surprising that Heegaard diagrams $\underline{D} = (D, h, \underline{K})$ really do contain circle-patterns. Of course, a trivial example is given by the boundary components of the regular neighborhood of $\underline{K}$ in $F(\underline{D})$. But other examples are usually hard to find in an ad hoc way. The reason being that the abundance of circles satisfying condition (1) makes it a very time-consuming task to filter out those which satisfy condition (2) and (3) as well. From this point of view circle-patterns are like paths in a maze. Indeed, one might even wonder whether there are *any* non-trivial examples at all. But we will see later that there is actually not only an abundance of circles, but of circle-patterns as well. The problem will be to construct and classify them all. This will be made easier by the simple observation that circle-patterns in Heegaard-diagrams correspond to surfaces in 3-manifolds (an observation which is also the starting point in the work of Kneser and Haken; see [Kne] and [Ha 1]).

6.6. Heegaard-diagrams have long been known to provide concrete models for 3-manifolds (see e.g. [ST]). The fact that they can be drawn on a piece of paper and yet store all the relevant information of the 3-manifold they represent is still very appealing. This property is unique. It remains to make good use of it for a study of 3-manifolds.

To be more specific, note that, to any n-relator diagram $\underline{D} = (D, h, \underline{K})$, there is associated a unique relative handlebody $(M, k) = \underline{M(\underline{D})}$ given by setting

(1) M to be the quotient under h of the union of all 3-balls, bound by the underlying 2-manifold F, and

(2) k to be the union of all components of $\underline{K}/h$.

Of course, for any relative handlebody (M, k) there is at least one Heegaard diagram $\underline{D}$ with $(M, k) = M(\underline{D})$. $M(\underline{D})$ and $M(\underline{D}')$ are called *admissibly homeomorphic* if there is a homeomorphism $M(\underline{D}) \to M(\underline{D}')$ mapping relator curves to relator curves. Many different Heegaard diagrams can yield admissibly homeomorphic relative handlebodies.

Notice that, under the above quotient map, $D \subset F$ becomes a system of discs in the relative handlebody $M(\underline{D})$. This is the *disc-system associated to* $\underline{D}$. The disc-system associated to $\underline{D}$ contains at least one meridian-system for $M(\underline{D})$. Any such meridian-system is called a *meridian-system associated to the diagram* $\underline{D}$.

Vice versa, given a relative handlebody (M, k), together with a meridian-system $\mathcal{D}$ in M, we obtain a Heegaard-diagram $\underline{D}$ by splitting (M, k) along $\mathcal{D}$. The above attaching-process is dual to the previous splitting-process in that $M(\underline{D})$ is (admissibly) homeomorphic to (M, k). Thus we may consider $\underline{D}$ as a Heegaard-diagram for (M, k). However, we have to keep in mind that this Heegaard-diagram depends strongly on our choice of the meridian-system $\mathcal{D}$. Since we know that there are many different meridian-systems in (M, k), there are many different Heegaard-diagrams describing the same relative handlebody.

Using the previous terminology, the following result is an immediate consequence of prop. 5.5.

6.7. Proposition. *For every relative handlebody (M, k) for which $\partial M - k$ is incompressible in M, there is an irreducible pant-diagram $\underline{D} = (D, h, \underline{K})$*

such that $M(\underline{D})$ *is homeomorphic to* (M, k) *(in particular, no arc from* $\underline{K}$ *is recurrent).* $\Diamond$

Now, observe further that, by attaching a 2-handle to every component of the relator-system k, we obtain from the relative handlebody $M(\underline{D}) = (M, k)$ a 3-manifold, denoted by $M^+(\underline{D}) = M^+(k)$. This 3-manifold is called an *n-relator 3-manifold*, if k has n components (note that its fundamental group has then a canonical presentation as an n-relator group). We call $M^*(\underline{D}) = M^*(k)$ the *completed n-relator 3-manifold*, if $M^*(\underline{D})$ is obtained from $M^+(\underline{D})$ by attaching 3-balls to all 2-sphere components of $\partial M^+(\underline{D})$. Recall every (orientable) 3-manifold is homeomorphic to some completed n-relator 3-manifold (triangulation theorem). But note also that the notion of a (completed) n-relator 3-manifold describes more than just the topological type of the underlying 3-manifold; it actually fixes a handlebody-structure (or: "relator-structure") for that manifold.

Turning finally to circle-patterns again, observe that, in a canonical way, every circle-pattern C in the Heegaard diagram $\underline{D} = (D, h, \underline{K})$ gives rise to an (admissible) surface $S(C)$ in the relative handlebody $M(\underline{D})$. This surface is obtained simply as the quotient under h of the system of proper 2-discs bound by C in the system of 3-balls bound by the 2-manifold $F(\underline{D})$. Furthermore, attaching 2-discs, contained in $M^+(\underline{D}) - M(\underline{D})$, to all those components of $\partial S(C)$ contained in $\partial M(\underline{D}) - \partial M^+(\underline{D})$, we obtain from $S(C)$ the surface $S^+(C)$ in the n-relator 3-manifold $M^+(\underline{D})$, and so in $M^*(\underline{D})$ for that matter. Note that $S(C)$ and $S^+(C)$ are orientable if and only if $C \in \mathcal{C}^+(\underline{D})$.

§7. Surfaces.

Throughout this book we will study various aspects of surfaces in relative handlebodies (M, k). In this section we establish some of their most basic properties. First we give a new proof of Whitehead's algorithm for finding minimal, non-separating discs in relative handlebodies. Then we give a direct proof (and estimates) for the finiteness of essential surfaces in relative handlebodies (avoiding Haken's machinery in this special situation). More precisely, we will show that the existence of an infinite sequence of essential surfaces of fixed genus in (M, k) implies the existence of an essential annulus in (M, k). Thus annuli appear here as obstruction to rigidity; a phenomenon which is familiar, e.g., for homotopy equivalences [Joh 1] and hyperbolic structures [Th 4] and which will be encountered in several other forms throughout this book.

7.1. Definition. *Let (M, k) be a relative handlebody. Then a surface S in (M, k) is <u>essential</u>, if for every disc $D \subset M$ with $D \cap S = (\partial D - \partial M)^-$ connected and $D \cap k = \emptyset$, there is a disc $D' \subset S$ with $(\partial D' - \partial S)^- = D \cap S$ and $D' \cap k = \emptyset$.*

Remark. Recall that the notion of an essential surface has been introduced in [Joh 1] for 3-manifolds with boundary patterns. In fact, the previous definition appears as a special case of that notion, provided we take as boundary-pattern for M the set of all components of $U(k)$ and $(\partial M - U(k))^-$. Thus, by Prop. 2.1. of [Joh 1], an incompressible surface S is essential in (M, k), if and only if every proper (possibly singular) arc in S, which can be deformed (fixing end-points) into $\partial M - k$, can also be deformed (fixing end-points) in S into $\partial S - k$. In particular, essential surfaces differ from incompressible ones in that the latter may still have ∂-compressions.

Essential surfaces are known to be important for a study of 3-manifolds. In our approach, a good understanding of essential surfaces in relative handlebodies is crucial. Here we concentrate on three aspects of these surfaces, and we will study them under the headlines *Simplicity, Minimality* and *Finiteness*.

7.2. *Simplicity of Relative Handlebodies.*

Before we enter a general study of essential surfaces in (M, k), let us first single out discs and annuli in (M, k) which do *not* meet k, for they are of special importance. In fact, as we will see, relative handlebodies (and ultimately 3-manifolds in general) are much more rigid, if these surfaces are not present. In this respect, it is interesting to note the following simple observation:

7.3. Proposition. *Let (M, k) be any given relative handlebody. Then it can be decided whether or not (M, k) contains an essential disc or annulus, respectively (which does not meet k at all).*

Proof. A (finite) algorithm has to be given which decides this question. The desired algorithm can be described as follows:

First pick some meridian-system $\mathcal{D}$ for M, in general position to k. Certainly, (M, k) contains an essential disc, if a closed curve from k is contained in $F := (\partial M - U(\mathcal{D}))^-$. So we suppose the converse. Then, in particular, $k \cap F$ consists of arcs.

Next search for recurrent arcs in $F - k$. Here an arc l in $F - k$ is called *recurrent* if ∂l is contained in one component, say b, of ∂F and if every component of $b - \partial l$ contains at least one intersection point from $k \cap b$. Observe that the existence of recurrent arcs in $F - k$ can be found out by inspection. Given such a recurrent arc l consider the disc D_0 from $\mathcal{D}$ with $\partial l \subset U(D_0)$. Observe that l, together with some arc in $\partial(U(D_0) \cap \partial M)$, bounds a non-separating disc in M disjoint to D_0. Replacing D_0 by this new disc, we get from $\mathcal{D}$ a meridian-system which intersects k in strictly fewer points. Thus, after at most $\#(k \cap \partial \mathcal{D})$ such steps, we eventually end up with a meridian-system having no recurrent arcs whatsoever. The Heegaard-diagram associated to this new meridian-system is useful (see 6.3. for definition). Thus, in particular, useful diagrams can be constructed. So prop. 7.3. is a consequence of the following lemma.

7.4. Lemma. *Let $\underline{\mathcal{D}} = (D, h, \underline{K})$ be a useful Heegaard-diagram, and let (M, k) be the relative handlebody $M(\underline{\mathcal{D}})$ (see section 6). Then the following holds:*

(1) There is an essential disc in the relative handlebody $M(\underline{\mathcal{D}})$ not meeting k, if and only if the diagram $\underline{\mathcal{D}}$ is reducible.

(2) Suppose the diagram $\underline{\mathcal{D}}$ is irreducible. Then there is an essential annulus or Möbius band in $M(\underline{\mathcal{D}})$ which is not parallel to $U(k)$ if and only if there is a circle-pattern C in $\underline{\mathcal{D}}$ which satisfies the following conditions:

(a) C consists of 4-cycles alone,

(b) no two circles from C are parallel in $\underline{\mathcal{D}}$, and

(c) C is not the boundary of the regular neighborhood in $F(\underline{\mathcal{D}})$ of some relator-cycle from $\underline{K}$.

Remarks. (1) A circle c from a circle-pattern in a diagram $\underline{\mathcal{D}} = (D, h, \underline{K})$ is a *4-cycle*, if $c \cap D$ consists of exactly two arcs, each one joining two different components of $\partial D - \underline{K}$. Two 4-cycles are *parallel*, if they can be joined by a path of 4-cycles.

(2) Note that the above conditions are easy to verify in a finite number of steps.

(3) If A is an essential Möbius band, then the frontier $(\partial U(A) - \partial M)^-$ is an essential annulus.

Proof of Lemma 7.4. Let $\mathcal{D}$ be a meridian-system in (M, k) associated to the diagram $\underline{\mathcal{D}}$.

Let us first verify (1). One direction is obvious. For the other one, let A be an essential disc in (M, k) which does not meet k at all. We suppose A is isotoped in (M, k) (i.e. without meeting k) so that the intersection $A \cap \mathcal{D}$

consists of curves and that, in addition, the number of these curves is as small as possible. Then $A \cap \mathcal{D}$ is a system of arcs. Let s be one of them which is outermost in A, in the sense that it separates a disc A_0 from A with $A_0 \cap \mathcal{D} = s$. By our minimal choice of A, $(\partial A_0 - s)^-$ is an essential arc in the planar surface $F := (\partial M - U(\mathcal{D}))^-$. This arc does not meet k and both its end-points are certainly contained in one component b of ∂F. But, by hypothesis, $\mathcal{D}$ is useful. Thus ∂s bounds an arc s' in b which does not meet k. Then $s \cup s'$ has to be a non-contractible and simple closed curve in $F - \underline{K}$. It follows that $\underline{D}$ is reducible. This shows (1) of lemma 7.4.

Let us now verify (2). If C is a circle-pattern of 4-cycles, then the surface $S(C)$ in (M, k) is either an annulus or a Möbius band. Since $\underline{D}$ is useful, the intersection of every 4-cycle with every disc from D is connected. It therefore follows that $S(C)$ cannot have any compression-disc. But it also cannot have any ∂-compression disc, for otherwise the end-points of at least one arc from $C \cap D$ would lie in the same component of $\partial D - \underline{K}$ which is impossible since C is a circle-pattern. Thus $S(C)$ is the required essential surface in (M, k).

To show the other direction, let A be an essential annulus or Möbius band in (M, k), not parallel to the regular neighborhood of k in ∂M. We suppose A is chosen so that, in addition, the number of curves from $A \cap \mathcal{D}$ is as small as possible. Then $A \cap \mathcal{D}$ is a system of arcs since A is incompressible in M. Moreover, every arc from $A \cap \mathcal{D}$ is essential in A. Otherwise at least one arc from $A \cap \mathcal{D}$ would separate a disc from A. This would imply that $\mathcal{D}$ is reducible since $\underline{D}$ is useful, which in turn would contradict the hypothesis. Thus the meridian-system $\mathcal{D}$ splits A into a system of essential squares. Hence there is a circle-pattern C in the diagram $\underline{D}$ with $A = S(C)$ and satisfying properties (a) and (c) of the lemma. We are done if no two circles from C are parallel in $\underline{D}$. But if there are two parallel circles from C, then there is a square B in $M - U(\mathcal{D})$ with $B \cap (A \cup \partial M)$ and such that $B \cap A$ consists of two essential arcs in A. $B \cap A$ splits A into two squares, say A_1 and A_2. So $B \cup A_1$ as well as $B \cup A_2$ is an annulus or Möbius band. At least one of them, say $B \cup A_1$, must be incompressible in M since A is. Thus $B \cup A_1$ is an essential annulus or Möbius band in (M, k) intersecting $\mathcal{D}$ in strictly fewer arcs than A. This, however, contradicts our choice of A. $\Diamond$

7.5. Definition. *A relative handlebody (M, k) is called $\underline{\partial\text{-irreducible}}$, if $\partial M - k$ is incompressible in M. A ∂-irreducible relative handlebody (M, k) is called $\underline{simple}$, if every essential annulus in (M, k) (not meeting k) is parallel to $U(k)$.*

7.6. Definition. *An irreducible and useful Heegaard-diagram $\underline{D}$ is called $\underline{simple}$, if it has no circle-pattern with properties 2(a,b,c) from lemma 7.4.*

Using these definitions, the above results may now be reformulated as follows.

7.7. Proposition. *Let $\underline{D}$ be any Heegaard-diagram whose underlying surface is connected. Then, starting with $\underline{D}$, a Heegaard-diagram $\underline{D}^*$ can be constructed in finitely many steps with the property that the relative handlebody $M(\underline{D})$ is simple if and only if the diagram $\underline{D}^*$ is simple.* ◊

Remarks. (1) It is easy to check whether a given diagram is simple. Moreover, it is easy to translate disc-slides into an operation for Heegaard-diagrams (see section 26), and, using this operation, the relevant algorithm (based on 7.4) becomes a finite algorithm involving only diagrams.

(2) By a variation of the above argument, we can slightly generalize the method in order to discover meridian-systems $\mathcal{D}$ whose discs intersect k in at most one point (rather than no point at all as before). This follows again from the observation that such a meridian-system has to lie in the complement of any meridian-system which separates no recurrent arc from k.

7.8. *Minimal Meridean-Systems.*

To any proper 2-manifold S in a relative handlebody (M, k) we associate a complexity by setting

$$c(S) := \#(S \cap k) - \chi(S).$$

Given a class $\mathcal{S}$ of (proper) 2-manifolds in (M, k), we say $S \in \mathcal{S}$ is *(geometrically) minimal* (w.r.t. $\mathcal{S}$) if $c(S) \leq c(S')$, for all $S' \in \mathcal{S}$. Of course, minima *exist* for all non-empty classes $\mathcal{S}$, but it is usually a challenging problem to *construct* them. For later use (and to illustrate the point) we will consider two extreme cases, namely the class of all surfaces of Euler characteristic one and the class of all surfaces not meeting k. In this section we consider the first case, i.e., discs, or disc-systems. The other case will be treated later (see sections 9-13).

We are interested in minimal meridian-systems. More precisely, we are interested in *constructing* meridian-systems $\mathcal{D}$ in (M, k) whose geometric intersection number with k is as small as possible. To put our approach taken below in perspective, let us indicate its relation to the classical approach of Whitehead. Note first that one may consider *algebraic* instead of *geometric* intersection numbers. In other words one may search for algebraic minima, i.e., for meridian-systems whose algebraic intersection-number with k is as small as possible. Now, according to a result of Zieschang [Zie 3] (see also [Wa 5]), an algebraic minimum is also a geometric minimum. So it would suffice to construct the former. In [Wh 2] this problem has been solved even for finite sets k of *singular* curves in ∂M. But in this general setting, even the special case $\min\{\#(k \cap \partial\mathcal{D})\} \leq \#k$, requires a clever argument (see [Wh 1]), while it is trivial for simple curve-systems (see remark to prop. 7.7). Thus it may be expected that the situation becomes conceptually easier, if k is non-singular. Next, we verify this intuition, by showing (without referring to [Wh 1,2]) how the concept of algebraic minima may be avoided altogether, when dealing with simple curve-systems k.

7.9. Theorem. [Wh 1,2] *Let (M, k) be a ∂-irreducible, relative handlebody. Then a meridian-system for M can be constructed which intersects k in a minimal number of points.*

Remark. Once again the mere *existence* of such a minimal meridian-system is no problem.

Proof. An algorithm has to be given which constructs minimal meridian-systems. Here is an algorithm which constructs "pseudo-minimal" meridian-systems. We call a meridian-system $\mathcal{D}$ in (M, k) *pseudo-minimal*, if there is no simple closed curve t in $F := (\partial M - U(\mathcal{D}))^-$ and no component D_0 from $\mathcal{D}$ such that (1) and (2) holds:

(1) $\#(k \cap t) < \#(k \cap \partial D_0)$.

(2) t separates the two components of ∂F contained in $U(D_0)$.

The Algorithm. First pick some meridian-system $\mathcal{D} \subset M$, in general position to k. Since (M, k) is supposed to be ∂-irreducible, it follows that k splits $F := (\partial M - U(\mathcal{D}))^-$ into discs. Thus, given any integer α, there are (modulo isotopy in F) only finitely many simple closed curves t in F with $\#(t \cap k) \leq \alpha$. In particular, it can be decided whether $\mathcal{D}$ is "pseudo-minimal". If it is not, then there is a closed curve $t \subset F$ and a disc D_0 from $\mathcal{D}$ with (1) and (2) above. Replacing D_0 by a disc in $M - (\mathcal{D} - D_0)$ bound by t, we get from $\mathcal{D}$ a meridian-system which intersects k in strictly fewer points and so on. The process terminates after a limited number of steps, and the result is a pseudo-minimal meridian-system.

It remains to show that pseudo-minimal meridian-systems are actually minimal. This will be shown in lemma 7.13, but first we need some preparation. The idea is to temporarily permit *singular* disc-systems, and to use a certain crucial observation concerning singular compression discs for pseudo-minimal meridian-systems.

To formulate this observation, let $\mathcal{D}$ be any meridian-system in (M, k). Let t be an arc in $\mathcal{D}$ with $\partial t \subset \partial \mathcal{D} - k$ (notice every singular arc in $\mathcal{D}$ can be straightened out to a simple arc). Let D be the component of $\mathcal{D}$ containing t and denote by d_1, d_2 the two arcs in which ∂D is split by ∂t. Finally, let t' be any, possibly singular, arc in ∂M with $\partial t = \partial t'$ and homotopic in M to t, by a homotopy fixing $\partial t'$.

7.10. Lemma. *If, in the previous setting, $\mathcal{D}$ is pseudo-minimal, then*

$$\#(t' \cap k) \geq \#(d_i \cap k), \quad \text{for } i = 1 \text{ or } 2.$$

Proof. By hypothesis, there is a homotopy $f : t \times I \to M$ with $f \mid t \times 0 = t$ and $f \mid (\partial(t \times I) - (t \times 0))^- = t'$. For simplicity, let us denote the arc

$(\partial(t \times I) - (t \times 0))^-$ by t' too. Let us further suppose that f is deformed, using a homotopy fixing $\partial(t \times I)$, so that $f^{-1}\mathcal{D}$ is a system of curves whose number is as small as possible. By asphericity of the handlebody M, it then follows that the pre-image $f^{-1}\mathcal{D}$ consists of arcs whose end-points are all contained in t'. Let α (possibly $\alpha = t$) be an outermost of these arcs. Then, by definition, α separates a disc A from $t \times I$ with $A \cap t'$ connected and $A \cap f^{-1}\mathcal{D} = \alpha$. Set

$$\alpha' := A \cap \partial(t \times I).$$

For simplicity, we again denote the singular arcs $f \mid \alpha$ and $f \mid \alpha'$ by α and α', respectively. Now, let D_1 be that component from $\mathcal{D}$ which contains the singular arc α and let β_1 and β_2 denote the two arcs in which ∂D_1 is split by $\partial \alpha$.

We claim

$$\#(k \cap \alpha') \geq \#(k \cap \beta_i), \quad \text{for } i = 1 \text{ or } 2.$$

Assume the converse. Let l be any simple closed curve in $f(\partial A)$ containing $\partial \alpha$, and set $l_1 := l \cap D_1$ and $l_2 := (l - l_1)^-$. Then, by our assumption and by our choice of l, we have $\#(k \cap l_2) < \#(k \cap \beta_i)$, for $i = 1$ and 2. After a small isotopy of the surface $F := (\partial M - U(\mathcal{D}))^-$ in ∂M, we may suppose that ∂D_1 lies in a component of F. Consider the regular neighborhood $U := U(l_2 \cup \partial D_1)$ in F. The boundary ∂U consists of three components. One of them is a copy of ∂D_1. Let b_1, b_2 denote the other two. By the above inequalities, it follows that b_1 as well as b_2 bounds a disc in M which intersects k in strictly less points than D_1. But F is planar and so b_i, $i = 1, 2$, separates F into two surfaces. Let B_i be that one them which does not contain ∂D_1. Then $B_1 \cap B_2 = \emptyset$ and $U \cup B_1 \cup B_2 = F$. Thus w.l.o.g. $U \cup B_1$ contains both of those components of ∂F which are isotopic to ∂D_1. Therefore b_1 is a simple closed curve in F which (1) intersects k in strictly less points than the component D_1 of $\mathcal{D}$ and which (2) separates those two components of ∂F which are isotopic to ∂D_1. This contradicts the hypothesis that $\mathcal{D}$ is pseudo-minimal, and so our claim is verified.

By the previous claim, we may suppose w.l.o.g. that $\#(k \cap \alpha') \geq \#(k \cap \beta_1)$. But $\mathcal{D}$ is a system of discs and so α can be deformed in $\mathcal{D}$ into β_1, using a homotopy fixing the end-points of α. Replacing the map $f \mid A$ by the latter homotopy, we obtain a new map $f' : t \times I \to M$ which has the same properties than f, but for which $g^{-1}\mathcal{D}$ consists of strictly fewer arcs than $f^{-1}\mathcal{D}$. Thus, applying the above exchange-process over and over again if necessary, we may suppose that $f^{-1}\mathcal{D}$ equals $t \times 0$. But then the conclusion of lemma 7.10 follows, by an argument used to show the claim above, for $\mathcal{D}$ is pseudo-minimal. This proves the lemma.

We are next interested in the question whether every minimal disc-system $\mathcal{D}' \subset M$ with connected complement can be enlarged to a minimal meridian-system, i.e., whether there is a meridian-system $\mathcal{D} \subset M$ with $\mathcal{D}' \subset \mathcal{D}$ so that $\#(k \cap \mathcal{D})$ is minimal w.r.t. to all meridian-systems. The following lemma provides an affirmative answer for single discs.

7.11. Lemma. *Let $\mathcal{D}$ be a minimal meridian-system of the relative handlebody (M, k) of genus n, and let $D_1, D_2, ..., D_n$ be the discs from $\mathcal{D}$. Then we may suppose the indices are chosen so that $\#(k \cap D_i) \leq \#(k \cap D')$, for all i, $1 \leq i \leq n$, and all non-separating discs D' in $M - U(D_1 \cup ... \cup D_{i-1})$ with $D' \cap M = \partial D'$.*

Remark. As a result we obtain a minimal meridian-system by successively picking non-separating and minimal discs.

Proof. Given a minimal meridian-system $\mathcal{D}$ and a non-separating, minimal disc D in (M, k), we only show that there is some disc D_0 from $\mathcal{D}$ which intersects k in the same number of points as D. (After splitting M along D_0, it will be apparent how to modify the argument in order to obtain the full conclusion of the lemma by induction).

Assume the converse. Then there is a non-separating, minimal disc D in M such that

$$\#(k \cap \partial D) < \#(k \cap \partial D_i),$$

for every disc D_i, $1 \leq i \leq n$, from $\mathcal{D}$. Since we are only interested in the intersection number with k, we may suppose w.l.o.g. that $\mathcal{D}$ and D have been *chosen* so that, in addition, the number of components of $D \cap \mathcal{D}$ is as small as possible. Then, in particular, the intersection, $D \cap \mathcal{D}$ consists of arcs alone. We now may distinguish between two cases and we will show that both cases lead to a contradiction.

Case 1. The intersection $D \cap \mathcal{D}$ is empty.

Observe that in this Case, D is contained in the 3-ball $M^* := (M - U(\mathcal{D}))^-$. Then, ∂D is a simple closed curve in ∂M^* which does not meet the discs from $U(\mathcal{D}) \cap \partial M^*$. Now, this curve splits the 2-sphere ∂M^* into two discs. Let C be one of them. Since D is non-separating in M, it follows the existence of at least one disc D_0 from $\mathcal{D}$ with the property that $U(D_0) \cap C$ is connected. Then D, as well as D_0, is contained in the solid torus $(M - U(\mathcal{D} - D_0))^-$. So, replacing D_0 by D, transforms $\mathcal{D}$ into another meridian-system. But this new meridian-system intersects the curve-system k in strictly fewer points since, by assumption, $\#(k \cap \partial D) < \#(k \cap \partial D_i)$, for every disc D_i from $\mathcal{D}$. This, however, contradicts the hypothesis that $\mathcal{D}$ is a minimal meridian-system.

Case 2. The intersection $D \cap \mathcal{D}$ is non-empty.

In this Case, let s be an arc from $D \cap \mathcal{D}$ chosen to be outermost in D. By definition, s separates a disc A from D with $A \cap \mathcal{D} = s$. Let D_0 be the disc from $\mathcal{D}$ containing the intersection-arc s. Then ∂s splits ∂D_0 into two arcs s_1^* and s_2^*. Furthermore, let s_1 and s_2 be the two arcs in which ∂D is split by ∂s. Finally, let the indices be chosen so that $s_1 = A \cap \partial M$.

The situation at hand may schematically be illustrated by the following picture (keep in mind that D and D_0 may have intersection curves different from s).

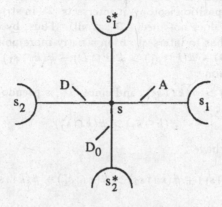

(Figure 7.1.)

We claim the indices may be chosen so that

$$\#(k \cap s_1) > \#(k \cap s_1^*)$$

Since $\mathcal{D}$ is supposed to be minimal, it is also pseudo-minimal, and so, by lemma 7.10, we have w.l.o.g. $\#(k \cap s_1) \geq \#(k \cap s_1^*)$. Assume $\#(k \cap s_1) = \#(k \cap s_1^*)$. If now $\#(k \cap s_1) > \#(k \cap s_2^*)$, we may replace s_1^* by s_2^*, and the claim follows. Thus we suppose $\#(k \cap s_1) \leq \#(k \cap s_2^*)$. Then consider the regular neighborhood $U := U(D_0 \cup A)$. The frontier $(\partial U - \partial M)^-$ consists of three discs since A is supposed to be a disc which lies in the complement of $\mathcal{D}$. One of these discs is a copy of D_0. But it follows that one of the other two discs, say D_0', is also parallel to D_0 since U is contained in the solid torus $(M - U(\mathcal{D} - D_0))^-$. Thus, replacing D_0 by D_0', we get from $\mathcal{D}$ another meridian-system for M. It is easily verified that this new meridian system intersects k in not more, but D in strictly fewer components. This, however, contradicts our minimal choice of $D \cap \mathcal{D}$. So the claim is established.

We next claim

$$\#(k \cap s_1^*) > \#(k \cap s_2).$$

To see this claim, note that the arc s separates the disc D_0 into discs E_1, E_2. Let the indices be chosen so that $E_i \cap D_0 = s_i^*$. Consider $D' := (D - A) \cup E_1$. Then D' is a, possibly singular, disc in M, which intersects k in strictly fewer points than D since $\#(k \cap s_1) > \#(k \cap s_1^*)$. But the singularities of D' are just double-arcs. Thus, by cut-and-paste along double arcs, we obtain a non-singular disc from D' which meets k in the same number of points as D'. Since D is minimal, it follows that the new disc must be separating in M. In particular, it represents the trivial element in the relative homology group $H_2(M, \partial M)$. Since cut-and-paste is an operation which does not change the $\mathbf{Z}_2$-homology class, it follows that also D' is trivial in $H_2(M, \partial M; \mathbf{Z}_2)$. Thus $\hat{D} := A \cup E_1$ cannot be trivial in $H_2(M, \partial M; \mathbf{Z}_2)$

since D is not. Thus $\hat{D}$ is a non-singular and non-separating disc in M. But, after a small general position isotopy, it intersects $\mathcal{D}$ in strictly fewer arcs than D (in fact, it then does not meet $\mathcal{D}$ at all). Thus, by our minimal choice of D, the disc $\hat{D}$ has to intersect k in strictly more points than D. Thus $\#(k \cap s_1^*) = \#(k \cap \hat{D}) - \#(k \cap s_1) > \#(k \cap D) - \#(k \cap s_1) = \#(k \cap s_2)$. This proves our second claim.

Since $\#(k \cap s_1^*) > \#(k \cap s_2)$ and since $\mathcal{D}$ is pseudo-minimal, it follows from lemma 7.10 that

$$\#(k \cap s_2) \geq \#(k \cap s_2^*).$$

Thus, altogether, we have

$$\#(k \cap D) = \#(k \cap s_1) + \#(k \cap s_2) > \#(k \cap s_1^*) + \#(k \cap s_2^*) = \#(k \cap D_0).$$

This, however, contradicts the fact that D is a minimal disc in (M, k).

Thus in any Case we get a contradiction. This proves the lemma.

There is an internal characterization of meridian-systems which applies to collections of singular discs as well. To formulate it let M be a handlebody of genus n. Let $\coprod D_i$ denote the disjoint union of n discs. By a *meridian-collection* in M we mean an immersion $\varphi : \coprod D_i \to M$ with the following properties:

(1) $\varphi^{-1} \partial M = \coprod \partial D_i$.
(2) $\#\varphi^{-1}(z) \leq 2$, for every point z in M.
(3) if $\sum_{1 \leq i \leq n} \epsilon_i \varphi(D_i)$, $\epsilon_i \in \mathbf{Z}_2$, represents the trivial element in $H_2(M, \partial M; \mathbf{Z}_2)$, then $\epsilon_i = 0$, for all $1 \leq i \leq n$.

In other words, a meridian-collection is a collection of n singular discs in M which are linearly independent mod 2 and whose intersections and self-intersections form a system of pairwise disjoint double-curves. Of course, a system of n discs in M (i.e., a collection of n non-singular and pairwise disjoint discs) is a meridian-system if and only if it is a meridian-collection. Thus "meridian-collection" is the desired generalization of "meridian-system". The next lemma tells us that the two notions are actually closely related.

7.12. Lemma. *Any meridian-collection for the handlebody M can be transformed into a meridian-system for M, by using cut-and-paste along double-curves.*

Proof. The proof is by induction on the number of double-curves. If this number is zero, we are done. Thus suppose the conclusion of the lemma holds for all meridian-collections whose number of double-curves is less than or equal to a fixed integer m, say. Let φ be a meridian-collection whose number $d(\varphi)$ of double curves equals $m + 1$.

Any self-intersection curve of a meridian-collection gives rise to a *unique* cut-and-paste operation which transforms the singular disc containing this curve into another singular disc with fewer double-curves. This operation does not change the Z_2-homology class. Thus, using cut-and-paste if necessary, we may suppose φ is a meridian-collection of non-singular discs, say D_i, $1 \leq i \leq n$. Then every double-curve d of φ is a component of the intersection of two different discs, say D_1, D_2, from φ. Moreover, d gives rise to two different cut-and-paste operations. Both of them reduce the number of double-curves. So it remains to show that one of them results in a meridian-collection again.

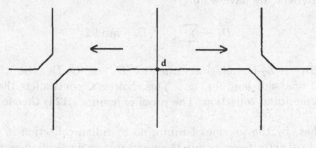

(Figure 7.2.)

If d is a closed curve, it bounds a unique disc E_i in D_i, $i = 1, 2$. W.l.o.g. $E_1 \cap D_2 = \partial E_1$. So $E_1 \cup E_2$ is a non-singular 2-sphere in the handlebody M, bounding a 3-ball in M. Thus one cut-and-paste procedures along d produces a 2-sphere. The other one transforms φ into a Z_2-homologous meridian-collection (which may contain singular discs again).

Thus we may suppose d is an arc. Then d splits D_1 into two discs, say D_1', D_1''. The same with D_2. Now, pick any one of the two possible cut-and-paste operations along the arc d. It changes the pair of discs D_1, D_2 into a new pair of (possibly singular) discs given by the unions $D_1^* := D_1' \cup D_2'$ and $D_2^* := D_1'' \cup D_2''$, say. All the other discs from the meridian-collection φ remain unchanged. In this way we obtain from φ a new disc-collection φ^* with $d(\varphi^*) < d(\varphi)$; whose (possibly singular) discs will be denoted by D_i^*.

If φ^* is a meridian-collection, we are done. Thus we suppose the converse. Then property (3) of the above definition of meridian-collection is violated. In this case there is a non-trivial linear combination

$$\sum_{1 \leq i \leq n} \epsilon_i D_i^*,$$

$\epsilon_i = 0, 1$, which is null-homologous in $(M, \partial M)$ mod 2. At least one of ϵ_1 or ϵ_2 has to be non-zero since φ is a meridian-collection. But observe that also at least one of these integers has to be zero. The latter follows from the fact that, by construction, $D_1^* + D_2^*$ is $\mathbf{Z}_2$-homologous to $D_1 + D_2$, that $D_i^* = D_i$, $i > 2$, and that φ_- is a meridian-collection. This shows that w.l.o.g.

$$D_1^* \sim \sum_{i \geq 3} \epsilon_i D_i \quad \text{mod } 2.$$

At this point carry out the other available cut-and-paste operation along d. As above, we obtain from the pair D_1, D_2 a new pair $\hat{D}_1, \hat{D}_2$ of (possibly singular) discs and from φ a new disc-collection $\hat{\varphi}$. If $\hat{\varphi}$ is a meridian-collection, we are done, and so we may suppose the converse. Then, by the previous argument, we have w.l.o.g.

$$\hat{D}_1 \sim \sum_{i \geq 3} \delta_i D_i \quad \text{mod } 2.$$

But $D_1^* + \hat{D}_1$ is $\mathbf{Z}_2$-homologous either to D_1 or to D_2. So it follows that property (3) does not hold for φ. This, however, contradicts the hypothesis that φ is a meridian-collection. The proof of lemma 7.12 is therefore complete.

Note that, by the previous lemma, no meridian-collection in (M, k) intersects k in strictly fewer points than any minimal meridian-system. Having established this property, we can now finish the proof of thm. 7.9.

7.13. Lemma. *A meridian-system in (M, k) is minimal if and only if it is pseudo-minimal.*

Proof. One direction is obvious. The proof of the other direction is similar to the proof of lemma 7.11. Let $\mathcal{D}^*$ be any pseudo-minimal meridian-system and let $\mathcal{D}$ be any minimal meridian-system in (M, k). Then, by definition, $\#(k \cap \mathcal{D}) = \min \#(k \cap \mathcal{D}')$, where the minimum is taken over all meridian-systems $\mathcal{D}'$ of (M, k). In particular,

$$\#(k \cap \mathcal{D}) \leq \#(k \cap \mathcal{D}^*).$$

It remains to show that equality holds in this formula. In order to prove this, we may suppose that *both* $\mathcal{D}$ as well as $\mathcal{D}^*$ are chosen, without enlarging their number of intersection points with k, so that, in addition, the number of curves from $\mathcal{D} \cap \mathcal{D}^*$ is as small as possible. Then, in particular, $\mathcal{D} \cap \mathcal{D}^*$ consists of arcs.

By lemma 7.11, we know that at least one of the components of $\mathcal{D}$, say D_0, is actually a minimal disc in (M, k).

We claim $D_0 \cap \mathcal{D}^*$ is empty. Assume the converse. Then there is at least one arc s from $D_0 \cap \mathcal{D}^*$ which is outermost in D_0. By definition, s separates a disc A from D_0 with $A \cap \mathcal{D}^* = s$. Let D_0^* be the disc from $\mathcal{D}^*$ containing the intersection-arc s. Now, ∂s splits ∂D_0 into two arcs s_1 and s_2 and let the indices be chosen so that $s_1 = A \cap \partial M$. Furthermore, let s_1^* and s_2^* be the two arcs in which ∂D_0^* is split by ∂s.

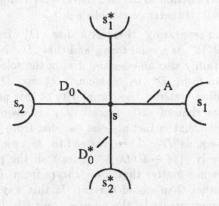

(Figure 7.3.)

$\mathcal{D}^*$ is pseudo-minimal. So, by lemma 7.10, we have w.l.o.g. $\#(k \cap s_1^*) \le \#(k \cap s_1)$. As in the proof of lemma 7.11, we even claim that w.l.o.g. $\#(k \cap s_1^*) < \#(k \cap s_1)$. If not, then $\#(k \cap s_1) = \#(k \cap s_1^*)$. If now $\#(k \cap s_2^*) < \#(k \cap s_1)$, we may replace s_1^* by s_2^*, and we are done. Thus we may suppose $\#(k \cap s_2^*) \ge \#(k \cap s_1)$. Then note that A is a disc which lies in the complement of $\mathcal{D}^*$ and so, by an argument from lemma 7.11, we may replace $\mathcal{D}^*$ by some other meridian-system which intersects k in no more but $\mathcal{D}$ in strictly fewer components. This in turn contradicts our minimal choice of $\mathcal{D} \cap \mathcal{D}^*$. So it follows indeed that w.l.o.g

$$\#(k \cap s_1^*) < \#(k \cap s_1).$$

Recall D_0 is a minimal disc in (M, k) (see above). By an argument from 7.11, it then follows that the union of $(D_0 - A)^-$ with the disc E_1 in $\mathcal{D}^*$ bound by $s \cup s_1^*$ is null-homologous in $(M, \partial M)$. Thus, replacing D_0 by the disc $E_1 \cup A$ and, applying a small general position isotopy, we obtain from $\mathcal{D}$ a meridian-collection which intersects $\mathcal{D}^*$ in strictly fewer components (all disjoint to the double curves of the meridian-collection) than $\mathcal{D}$. By lemma 7.12, this meridian-collection can be turned into a meridian-system, say $\mathcal{D}'$, with the same property. Thus it follows, from our choice of $\mathcal{D}$, that $\mathcal{D}'$ must intersect k in strictly more points than $\mathcal{D}$. This is only possible if

$$\#(k \cap s_2) < \#(k \cap s_1^*).$$

It then follows from lemma 7.10 that

$$\#(k \cap s_2^*) \le \#(k \cap s_2).$$

Thus, altogether, we have

$$\#(k \cap D_0^*) = \#(k \cap s_1^*) + \#(k \cap s_2^*) < \#(k \cap s_1) + \#(k \cap s_2) = \#(k \cap D_0)$$

This, however, is a contradiction to the fact that D_0 is a minimal disc in (M, k). The claim $D_0 \cap \mathcal{D}^* = \emptyset$ is therefore established.

Since D_0 is non-separating, there is a disc D_0^* from $\mathcal{D}^*$ such that $\hat{M} := (M - U(\mathcal{D}^* - D_0^*))^-$ is a solid torus, and that D_0 is an essential disc in $\hat{M}$. Since D_0^* is certainly also an essential disc in the solid torus $\hat{M}$, we get a new meridian-system from $\mathcal{D}^*$ by replacing D_0^* by D_0. By definition of pseudo-minimal meridian-systems, the complexity of this new meridian-system has to be the same as that of $\mathcal{D}^*$ (recall D_0 is disjoint to $\mathcal{D}^*$). Thus w.l.o.g. we may suppose that in fact at least one disc from the pseudo-minimal meridian-system $\mathcal{D}^*$ equals D_0. Then, instead to M, we apply the argument above to the handlebody $(M - U(D_0))^-$ (since all the processes have been defined locally, it does not matter that some curves from k intersect the new handlebody in arcs rather than closed curves). In this way we eventually turn $\mathcal{D}^*$ into $\mathcal{D}$ without ever increasing the intersection with k in the process. So $\#(k \cap \mathcal{D}^*) = \#(k \cap \mathcal{D})$.

The proof of thm. 7.9. is therefore complete. $\Diamond$

7.14. Example.

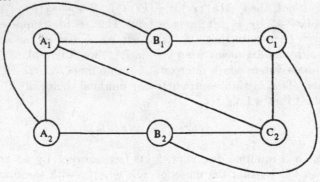

(Figure 7.4.)

Note that in the above Heegaard diagram $\mathcal{D}$ no essential curve, in the complement of the meridian-discs, meets the arcs in strictly fewer points than the meridian-discs. Thus the meridian-system in $M(\mathcal{D})$ associated to this diagram is pseudo-minimal. So, by lemma 7.13, $c(\mathcal{D}) \leq c(\mathcal{D}')$, for all diagrams $\mathcal{D}'$ with $M(\mathcal{D}') \cong M(\underline{\mathcal{D}})$ (see section 6 for notation).

7.15. *Finiteness for Surfaces.*

In general, minimal meridian-systems in relative handlebodies (M, k) are not unique, not even up to admissible isotopy in (M, k) (an isotopy in (M, k) is *admissible*, if it preserves k). To give a concrete example, let (M, k) be a relative handlebody which contains an essential annulus A (not meeting k). Given a minimal meridian-system in (M, k), we obtain an infinite sequence of pairwise not admissibly isotopic meridian-systems, by taking its iterates under a

Dehn-twist along A. (Recall from [Joh 1] that a *Dehn-twist along an annulus* A is defined to be a homeomorphism supported by $U(A)$, i.e. a homeomorphism which is the identity outside the regular neighborhood of A). This example, however, is typical. In particular, the set of minimal meridian-systems is finite in any simple relative handlebody. This fact is actually a consequence of a much more general finiteness result for surfaces in relative handlebodies (see remark of 7.18). To establish this result, we first prove a crude, but explicit estimate for the intersection of arcs in orientable surfaces. Afterwards we show how to generalize this estimate to one for surfaces in relative handlebodies. The envisioned finiteness result will then follow easily.

Here is the result concerning the intersection of arc-systems in surfaces.

7.16. Lemma. *Let F be an orientable surface. Let $\mathcal{K}$ and $\mathcal{L}$ be systems of essential arcs in F. Then there is a finite product $h : F \to F$ of Dehn twists in F such that*

$$\#(\mathcal{L} \cap h\mathcal{K}) \leq 2 \cdot \#\mathcal{L} \cdot 3^{\#\mathcal{K}},$$

modulo isotopy fixing the boundary.

Remark. Note that this estimate does not depend on the topological type of the underlying surface.

Proof. Let k be an arbitrary arc from $\mathcal{K}$. It suffices to show that there is an orientation-preserving homeomorphism $h_k : F \to F$ with $h_k | \partial F = \mathrm{id} | \partial F$ and $\#(\mathcal{L} \cap h_k k) \leq 2 \cdot \#\mathcal{L}$. Indeed, splitting F successively along the arcs from $\mathcal{K}$, we eventually get a homeomorphism $h : F \to F$ with $h | \partial F = \mathrm{id} | \partial F$ and $\#(\mathcal{L} \cap h\mathcal{K}) \leq 2 \cdot 3^0 \cdot \#\mathcal{L} + 2 \cdot 3^1 \cdot \#\mathcal{L} + ... + 2 \cdot 3^{n-1} \cdot \#\mathcal{L} \leq 2 \cdot 3^n \cdot \#\mathcal{L}$, where $n := \#\mathcal{K}$. So the lemma then follows from Dehn's theorem [De][Li 1] (see thm. 1.9) that every orientation-preserving homeomorphism $h : F \to F$ with $h | \partial F = \mathrm{id} | \partial F$ is a finite product of Dehn twists.

We next distinguish between two cases.

Case 1. k is separating in F.

In this Case there is a surface $G \subset F$ with $(\partial G - \partial F)^- = k$. W.l.o.g. we suppose there is no orientation-preserving homeomorphism $h : F \to F$ with $h | \partial F = \mathrm{id} | \partial F$ and so that $\mathcal{L} - h(G)$ has strictly fewer components than $\mathcal{L} - G$.

We claim $l - G$ is connected, for every component l from $\mathcal{L}$. Assume the converse. Then there must be a component β_1 from $\mathcal{L} \cap G$ and a component β_2 from $(\mathcal{L} - G)^-$ such that $\beta_1 \cap \beta_2 \neq \emptyset$ and $\partial \beta_i \subset k$, $i = 1, 2$. Let $\alpha \subset k$ be the arc joining the two points from $\partial \beta_2$. Let β_1' be that component of $(\partial U(\beta_1) - \partial G)^-$ meeting α, where the regular neighborhood is taken in G. Set $G' := (G - U(\beta_1'))^- \cup U(\beta_2)$. Then G' is a (connected) surface which is homeomorphic to G. More precisely, there is an orientation-preserving homeomorphism $g : G \to G'$ which is the identity on $(G - U(k \cup \beta_1))^-$. Both $U(\beta_1)$ as well as $U(\beta_2)$ join different components of $\partial(F - (G - U(\beta_1))^-)^-$. So they are both non-separating in $(F - (G - U(\beta_1))^-)^-$. It therefore follows that g extends to an orientation-preserving homeomorphism $h : F \to F$ with

$h|\partial F = \mathrm{id}|\partial F$. But $\mathcal{L} - hG = \mathcal{L} - G'$ has strictly fewer components than $\mathcal{L} - G$. This contradicts our hypothesis and the claim is established.

Thus there is an orientation-preserving homeomorphism $h_k : F \to F$ with $h_k|\partial F = \mathrm{id}|\partial F$ and so that $l - G$ is connected, for every arc l from $\mathcal{L}$. So $\#(\mathcal{L} \cap h_k k) \le 2 \cdot \#\mathcal{L}$.

Case 2. k is non-separating in F.

Note that $\mathcal{L}$ is contained in a single disc $D \subset F$ (which may be constructed by joining the components of $\mathcal{L}$ with arcs and taking the regular neighborhood of the ensuing 1-complex). Given such a disc, we next show how to construct a non-separating arc k' in F which intersects D in at most two components and for which $\partial k' = \partial k$ (mod isotopy of k). This will finish the proof since w.l.o.g. $\#(k' \cap \mathcal{L}) \le 2 \cdot \#\mathcal{L}$ and since there is a homeomorphism $h_k : F \to F$ with $h_k|\partial F = \mathrm{id}|\partial F$ and $h_k(k) = k'$ (recall k and k' are both non-separating).

To construct k' suppose first that ∂k lies in one component of ∂F, say r. Since k is non-separating, there must be at least one simple closed curve, say l, which is non-separating in F. If $l \cap D$ is disconnected, there is an arc α in D, $\alpha \cap l = \partial\alpha$, which joins two different components of $D \cap l$. The union of α with at least one of the two components of $l - \alpha$ is a non-separating curve intersecting D in fewer components than l. Thus we may suppose $l \cap D$ is connected. A similar short-cut argument yields an arc α in F, $\alpha \cap (l \cup \partial F) = \partial\alpha$, which joins r with l and for which $D \cap (\alpha \cup l)$ is connected. One component of $(\partial U(\alpha \cup l) - \partial F)^-$ is an arc. In fact, it is a non-separating arc in F which intersects D in at most two components (F is orientable). Since this arc as well as k have both end-points in r, we have found the desired arc. - If k joins two different boundary-curves from F, then every arc in F joining the end-points of k is non-separating. So, using short-cuts in D if necessary, we can always change k into an arc $k' \subset F$ with $\partial k' = \partial k$ and $k' \cap D$ connected. $\Diamond$

As a consequence of the previous lemma we get the following explicit estimate for the intersection of surfaces with squares:

7.17. Lemma. *Let (M, k) be a ∂-irreducible, relative handlebody. Let S be an essential surface in (M, k) with $S \cap k = \emptyset$. Suppose A is a square in M such that $A \cap \partial M$ consists of two arcs contained in $\partial M - k$ and that $(\partial A - \partial M)^- \subset S$ is an essential arc-system in S. Then there is product $h : M \to M$ of Dehn twists along essential annuli in (M, k) such that*

$$\#(A \cap hS) \le 2 \cdot (|\chi S| + 1) \cdot 3^{|\chi \partial M| + 1},$$

modulo isotopies constant on $(\partial A - \partial M)^-$.

Proof. Let S be isotoped, using an isotopy which is constant on $(\partial A - \partial M)^-$, so that afterwards the number of curves from $A \cap S$ is as small as possible. Then $S \cap A$ is a system of essential arcs in S since (M, k) is supposed to be

∂-irreducible. In particular, there is a system α of at most $|\chi S|+1$ essential arcs in S such that w.l.o.g. $A \cap S \subset U(\alpha)$. $U(A \cup U(\alpha))$ is an I-bundle whose lids lie in $\partial M - k$. Let X be the smallest essential I-bundle in (M, k), not meeting k, containing it. The orbit surface of X may or may not be orientable. But in any case $2|\chi F| = |\chi(X \cap \partial M)| \le |\chi \partial M|$ since $X \cap \partial M$ is an essential surface in ∂M. Thus there is a system $\beta \subset A$ of at most $|\chi F|+1 \le |\chi \partial M|+1$ arcs, contained in $A - S$ and joining the two components from $A \cap \partial M$, such that $U((A - U(\beta))^- \cup U(\alpha))$ is an I-bundle over an *orientable* surface. Let X' be the smallest essential I-bundle in (M, k), not meeting k, containing it. Let F' be the orbit surface of X'. Then both $S \cap X'$ and $A \cap X'$ are systems of *vertical*, essential squares in X'. Let $\mathcal{L}$ and $\mathcal{K}$ denote the fiber-projections of $S \cap X'$ resp. $A \cap X'$ to F'. Then $\mathcal{K}, \mathcal{L}$ are both systems of essential arcs in F'. By construction, $\#(\mathcal{K} \cap \mathcal{L}) = \#(S \cap A') = \#(S \cap A)$. Moreover, $\#\mathcal{L} = \#\alpha \le |\chi S|+1$ and $\#\mathcal{K} = \#\beta+1 \le |\chi \partial M|+2$.

Assume $\#(S \cap A) > 2 \cdot (|\chi S|+1) \cdot 3^{|\chi \partial M|+2}$. Then $\#(\mathcal{K} \cap \mathcal{L}) = \#(S \cap A') = \#(S \cap A) > 2 \cdot (|\chi S|+1) \cdot 3^{|\chi \partial M|+2} \ge 2 \cdot \#\mathcal{L} \cdot 3^{\#\mathcal{K}}$. Thus, according to lemma 7.16, the intersection $\mathcal{K} \cap \mathcal{L}$ can be reduced, using Dehn twists and isotopies of F' which are constant on $\partial F'$. But X' is a product I-bundle over F'. So any such Dehn twists and isotopies of F' can be lifted to X'. The lifted Dehn twists and isotopies reduce $\#(A \cap S)$.

The conclusion of lemma 7.17 therefore follows. $\Diamond$

We are now ready to prove the envisioned finiteness result for surfaces in relative handlebodies. Denote by $\mathcal{A}(M, k)$ the subgroup of the mapping class group $\pi_0 \mathrm{Diff}(M, k)$ generated by all Dehn twists along essential annuli in (M, k) (not meeting k).

7.18. Proposition. *Let (M, k) be a ∂-irreducible relative handlebody. Set $\mathcal{S}_n(M, k) = \{$admissible isotopy classes of essential surfaces S in (M, k) with $S \cap k = \emptyset$ and $|\chi S| = n\}$. Then $\mathcal{S}_n(M, k)/\mathcal{A}(M, k)$ is finite, and a complete set of representatives can be constructed.*

In particular, $\mathcal{S}_n(M, k)$ is a finite constructable set if (M, k) is simple.

Remarks. 1. This finiteness will be based on lemma 7.17, and so ultimately on Dehn's theorem [De][Li1] (see thm. 19) rather than Haken's normal surface theory (as presented in [Ha 1] (see also [Schu]). Modifications of this approach will later be used in other contexts as well.

2. In a same way a similar result can be shown for the set of all essential surfaces of given complexity $c(S)$ (see 7.8 for definition).

Proof. Fix an essential meridian-system $\mathcal{D}$ in (M, k) (not necessarily minimal). Denote $\tilde{M} := (M - U(\mathcal{D}))^-$ and $\tilde{k} := k \cup \partial(U(\mathcal{D}) \cap \partial M)$. Let S be any essential surface in (M, k). Suppose S is admissibly isotoped so that $\#(\mathcal{D} \cap S)$ is as small as possible. Then $S \cap \tilde{M}$ is a system of essential discs in $(\tilde{M}, \tilde{k})$. Note that there are only finitely many embedding types of essential discs in $(\tilde{M}, \tilde{k})$, modulo admissible isotopy (i.e., isotopy preserving $\tilde{k}$). Thus it follows that the set of all those essential surfaces in (M, k), which intersect $\mathcal{D}$ in a given (minimal) number of arcs, is finite. Furthermore, the latter set is certainly constructable since an essential meridian-system $\mathcal{D}$ can be constructed. Thus it suffices to establish a positive integer $\alpha_{\mathcal{D}}(n)$ with the property that, for every essential surface S in (M, k) with $|\chi(S)| = n$, there is a product h of Dehn twists (along essential annuli) such that $\#(\mathcal{D} \cap hS) \leq \alpha_{\mathcal{D}}(n)$, modulo admissible isotopy. But this is an easy consequence of lemma 7.17 since $\mathcal{D} \cap S$ falls into a limited number of (admissible) parallelity classes of arcs in $(\mathcal{D}, k \cap \partial \mathcal{D})$. $\Diamond$

§8. Homeomorphisms.

In this section we discuss briefly admissible homeomorphisms of relative handlebodies. In general, such a homeomorphism is a connected sum of an admissible homeomorphisms of a ∂-irreducible relative handlebody and some handlebody homeomorphism. Since homeomorphisms of handlebodies have already been considered in section 3, we here restrict our attention to ∂-irreducible relative handlebodies. Thus throughout this section (M, k) will always denote a ∂-irreducible relative handlebody. Given such a handlebody, recall that a homeomorphism $g : M \rightarrow M$ is an *admissible homeomorphism*, provided $g(k) = k$ (in the notion of [Joh 1], g is then admissible with respect to the boundary-pattern induced by the regular neighborhood of k in ∂M). Recall further that the mapping class group of a relative handlebody (M, k) is defined to be $\pi_0 \text{Diff}(M, k)$, i.e. it is the group of all admissible homeomorphisms modulo admissible isotopy (i.e., isotopy preserving k).

8.1. *Classification.*

As an application of the last section, we first solve the homeomorphism-problem for relative handlebodies. The result is well-known. In fact, it is known that the homeomorphism problem is solvable for all Haken 3-manifolds with useful boundary-patterns (see [Ha 2][He] and the exposition [Wa 4]).

8.2. Proposition. *Given two ∂-irreducible relative handlebodies (M, k) and (N, l), it can be decided whether or not (M, k) and (N, l) are admissibly homeomorphic.*

Remark. For example, these relative handlebodies may be "given" as (positive) integer solution-vectors of the homogeneous system of linear equations associated to some branching complex (see section 5).

Proof. Suppose (M, k) is a simple relative handlebody. Then construct an essential meridian-system $\mathcal{D}$ of (M, k) (not necessarily minimal; although, by thm. 7.9, the latter would also be constructable). Set $n := \#(k \cap \mathcal{D})$. The set $\mathcal{E}_n(N, l)$ of all (admissible isotopy classes of) meridian-systems $\mathcal{D}'$ with $\#(\mathcal{D}' \cap k) = n$ is finite and constructable (an actual construction can be carried out along the lines of the proof of prop. 7.18). But (M, k) is admissibly homeomorphic to (N, l) if and only if there is a meridian-system $\mathcal{D}' \in \mathcal{E}_n(N, l)$ and a homeomorphism $h : (M, k) \rightarrow (N, l)$ with $h(\mathcal{D}) = \mathcal{D}'$. Since this can be found out in finitely many steps, the proposition follows for simple relative handlebodies.

Using the characteristic submanifold, the above algorithm can be extended to all ∂-irreducible relative handlebodies. Indeed, the characteristic submanifold is unique (modulo admissible isotopy) and can be constructed for all given irreducible 3-manifolds with useful boundary-patterns; such as ∂-irreducible relative handlebodies [Joh 1]. The characteristic submanifold consists of I-bundles and Seifert fiber spaces and its complement is a simple 3-manifold. Moreover, recall any Seifert fiber space in a handlebody must be a solid torus since a handlebody contains no incompressible tori. Thus it can be decided whether characteristic submanifolds and their complements in ∂-irreducible relative handlebodies are admissibly homeomorphic. So the proposition follows. ◊

8.3. Remark. Incidentally, the classification-problem for ∂-irreducible relative handlebodies is also solvable. At this point recall that the *classification-problem* for a class of mathematical objects asks for an enumeration of these objects *without* repetition. But, using branching-complexes (see section 5), ∂-irreducible relative handlebodies can be enumerated and repetitions in that enumeration can be avoided by appealing to the previous solution of the homeomorphism-problem.

8.4. *Mapping Class Group.*

Let us now turn to the mapping class group $\pi_0\mathrm{Diff}(M, k)$ of relative handlebodies (M, k). In the following, we discuss how to compute the mapping class group for *simple* handlebodies. But before doing so, we should point out that the restriction to simple relative handlebodies is for convenience only. Indeed, using the existence of characteristic submanifolds again, our discussion below can easily be modified so as to yield a method for the calculation of

$$\pi_0\mathrm{Diff}(M, k)/\mathcal{A}(M, k)$$

for ∂-irreducible relative handlebodies (recall from the last section that $\mathcal{A}(M, k)$ denotes the subgroup of the mapping class group generated by all Dehn twists along essential annuli in (M, k) not meeting k).

Now, according to prop. 7.9, a minimal meridian-system $\mathcal{D}$ for (M, k) can be constructed. Thus $n := \min \#(k \cap \mathcal{D}')$ may be determined, where the minimum is taken over all meridian-systems for (M, k). Since (M, k) is supposed to be simple, it follows that the set $\mathcal{E}(M, k)$ of *all* (admissible isotopy classes of) minimal meridian-systems of (M, k) is finite and constructable (see remark of prop. 7.18). An actual construction may be derived from the proof of 7.18. The set $\mathcal{E}(M, k)$ can be used for an actual computation of the mapping class group based on the fact that $\pi_0\mathrm{Diff}(M, k)$ acts on the set $\mathcal{E}(M, k)$. We are therefore interested in the stabilizer and the orbits of this action.

8.5. *Stabilizer.* Let $\mathcal{D} \in \mathcal{E}(M, k)$. To calculate the stabilizer $\mathrm{stab}_\mathcal{D}\,\pi_0\mathrm{Diff}(M, k)$ of $\mathcal{D}$, recall from section 6 that we get a Heegaard-diagram $\underline{D} = (D, h, \underline{K})$ by splitting (M, k) along $\mathcal{D}$. The stabilizer above may therefore be characterized as the group of all (admissible isotopy classes of) homeomorphisms of the 2-sphere $S^+(\underline{D})$, underlieing the diagram $\underline{D}$, which map D to D and relator-cycles to relator-cycles and which commute with the attaching homeomorphism h (so, in particular, the stabilizer is finite).

Actually, it is often not hard to determine the stabilizer. To do this note that it acts w.l.o.g. as a finite group on the underlying 2-sphere $S^+(\underline{D})$. So every homeomorphism from the stabilizer is represented by some rotation of this 2-sphere (leaving $\underline{D}$ invariant and commuting with h). Any such rotation has a pair of fixpoints, and any two of them lieing in the same components of either D, or $\underline{K}$, or $F - (D \cup \underline{K})$ may be considered as equivalent. Thus we have only finitely many candidates for pairs of fixpoints. It remains to check them one by one to find out which one of them are fixpoints of a rotation with the above properties. In this way, the stabilizers of the action of the mapping class group $\pi_0\mathrm{Diff}(M, k)$ on the set $\mathcal{E}(M, k)$ of all minimal meridian-systems can be computed.

8.6. Example. Consider the three diagrams below. They are examples of Heegaard diagrams. The attaching homeomorphisms, h, for these Heegaard diagrams are specified according to the following convention: First orient the boundaries of the discs A, B, C clockwise and the boundaries of A', B', C' counter-clockwise. Then the homeomorphism h maps A to A' etc. in an orientation preserving fashion so that dots get mapped to dots.

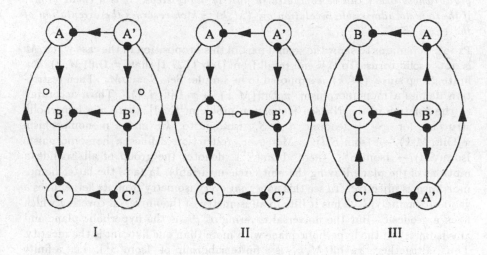

I   II   III

(Figure 8.1.)

The above Heegaard diagrams look similar, only their attaching homeomorphisms differ. It is easy to verify that these Heegaard diagrams are all one-relator diagrams. But note that their automorphism groups display different aspects: Heegaard diagram I has a rotational symmetry of order 3 (around the two little circles) which *preserves* the orientation of the relator-cycle. Heegaard diagram II has a rotational symmetry of order 2 (around the little circle and the point at infinity) which *reverses* the orientation of the relator-cycle. Heegaard diagram III has *no* non-trivial automorphims whatsoever. ◊

8.7. Orbits. By what has been seen so far, it remains to determine the orbits of the action of the mapping class group on $\mathcal{E}(M, k)$. Since we know from prop. 7.18 that $\mathcal{E}(M,k)$ is finite, we may conclude that all these orbits are finite, and since we know that the stabilizers are finite, we have that the *mapping class group of* (M,k) *is finite*. In fact, the *group-table for this finite group is computable* since the stabilizers are computable and since it can be decided, for any two elements $\mathcal{D}, \mathcal{D}'$ from $\mathcal{E}(M,k)$, whether or not there is an admissible homeomorphism h from (M,k) which maps $\mathcal{D}$ to $\mathcal{D}'$.

But still more can be said in the case when k is connected. As a matter of fact, there is then only one orbit and it is the calculation of the mere order of this orbit which is left for a calculation of the mapping class group. The following proposition is a formal way of stating this fact.

8.8.   Proposition. *Let* (M,k) *be a simple relative handlebody. Then* $\pi_0\mathrm{Diff}(M,k)$ *is a finite group whose group table can be determined in finitely many steps.*

If, in addition, k is connected, then $\pi_0\mathrm{Diff}(M,k)$ is a subgroup of a Dieder group whose order can be computed in finitely many steps. It is a cyclic group, if there is no admissible involution on (M,k) which reverses the orientation of the curve k.

Proof. It remains to show the second part of this proposition in the case when M is not a solid torus. To this end recall from [Joh 1, 27.1] that $\pi_0\mathrm{Diff}(M,k)$ is a finite group since (M,k) is supposed to be simple. Set $S := \partial M$. Then restriction defines a monomorphism $\pi_0\mathrm{Diff}(M,k) \to \pi_0\mathrm{Diff}(S,k)$. Thus, according to the solution of the Nielsen realization problem [Ke 1], there is a hyperbolic structure for S, a geodesic $l \subset S$, isotopic to k, and a monomorphism $\pi_0\mathrm{Diff}(M,k) \to \mathrm{Isom}(S,l)$. Moreover, restriction defines a homomorphism $\mathrm{Isom}(S,l) \to \mathrm{Isom}(S^1)$ (here $\mathrm{Isom}(S^1)$ denotes the group of all Euclidean motions of the plane leaving the unit circle invariant). In fact, the latter homomorphism is injective. To see this note that every isometry from its kernel fixes a geodesic, namely l. Thus it lifts to an isometry of the universal covering which fixes a geodesic. But the universal covering of S is the hyperbolic plane and any isometry of the hyperbolic plane with more than one fixpoint is the identity. Thus, altogether, $\pi_0\mathrm{Diff}(M,k)$ is a finite subgroup of $\mathrm{Isom}(S^1)$, i.e., a finite subgroup of a Dieder group. $\Diamond$

8.9.   Remarks. (1) A similar statement as in prop. 8.8 may be derived for the quotient-group $\pi_0\mathrm{Diff}(M,k)/\mathcal{A}(M,k)$ of arbitrary ∂-irreducible relative handlebodies, by using the theory of characteristic submanifolds (details are left to the reader).

(2) For a real calculation of $\pi_0\mathrm{Diff}(M,k)$ note that this group is often cyclic (provided k is connected). In this case it remains to calculate its order. But this order is the product of the orders of the stabilizer and the orbit of the action of $\pi_0\mathrm{Diff}(M,k)$ on $\mathcal{E}(M,k)$. Thus the size of $\mathcal{E}(M,k)$ provides a (crude) upper bound. In principal, the size of $\mathcal{E}(M,k)$ can be determined for all simple relative handlebodies (M,k) (but it may be tedious process).

(3) Obviously, finite subgroups of the mapping class group of handlebodies are important for a study of the mapping class groups of relative handlebodies. In this context we refer to the paper of McCullough-Miller-Zimmermann [MMZ] which is an extensive study of finite groups acting on handlebodies.

(4) Many questions remain open. For example, it is easy to see that every finite group acts freely on some handlebody but it is not known which finite groups occur as the mapping class group of relative handlebodies of genus two, three etc. with two, three etc. relator curves.

§9. Duality.

We continue our discussion of essential surfaces in relative handlebodies (M, k) as begun in section 7. But we are only interested in surfaces which do not meet k. We will make the basic observation (prop. 9.12) that these surfaces are generated by "piping" and "pruning". Using these operations, we will show how to interpret *non-separating* surfaces in relative handlebodies as elements of the kernel of a certain matrix, the "Heegaard matrix", associated to relative handlebodies (prop. 9.7). Based on this interpretation, we give a quick and elementary proof for the well-known existence of incompressible surfaces in 3-manifolds with boundary (see 9.8). In fact, we will show that Heegaard-matrices describe duality in 3-manifolds (thm. 9.22). In particular, we will derive a computational form of the Poincaré duality map for 3-manifolds.

9.1. *Piping, Pseudo-Piping and Pruning.*

Compressing surfaces along discs is a common operation in 3-manifold theory while its inverse is hardly ever being used. It turns out, however, that in a study of surfaces in relative handlebodies certain versions of both these operations are very useful. We call them "piping" and "pruning".

In order to introduce these operations in their proper generality, let S be any 2-manifold (possibly disconnected or non-orientable) in a 3-manifold M with $S \cap \partial M = \partial S$. Let t be an arc in ∂M with $t \cap S = \partial t$ and let $U(t)$ denote its regular neighborhood in M. Set $S' := (S - U(t))^- \cup A$, where A is the component of $\partial U(t) - (S \cup \partial M)$ whose closure intersects S in two arcs. Then S' is a 2-manifold in M which may or may not be orientable. We say S' is obtained from S by *piping* along t. Moreover, A is the *pipe* and t is the *piping-arc* of the process. We say a 2-manifold S' is obtained from S by *pruning* if there is a disc $D \subset \partial M$ such that $S' = (S - U(D))^- \cup D'$, where D' is the copy of D in $(\partial U(D) - \partial M)^-$.

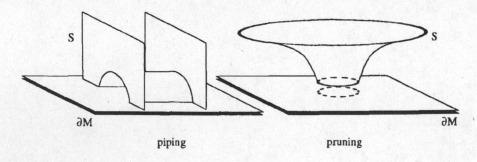

piping   pruning

(Figure 9.1.)

Given a system k of simple closed curves in ∂M we say that S' is obtained from S by *piping along* k if the piping-arc lies in k. We say S' is obtained from S by *pseudo-piping along* k if $U(t) \cap k$ is a non-empty system of arcs joining the two components of $U(t) \cap S$ (thus pipings are those pseudo-pipings for which $U(t) \cap k$ is connected). More generally, we say a 2-manifold S' is obtained from S by piping along k (or by pseudo-piping or pruning), if there is a finite sequence $S = S_0, S_1, ..., S_n = S'$ such that S_{i+1} is obtained from S_i by piping along k (resp. by pseudo-piping or pruning). Note that S' is obtained from S by piping and pruning if it is obtained by pseudo-piping.

Let S be an orientable (but possibly disconnected) 2-manifold in a 3-manifold M and let k be a system of simple closed curves in M (transverse to S). Then the following criterion holds.

9.2. Lemma. *Suppose the algebraic intersection number of S with every component of k vanishes. Then there is an orientable 2-manifold in M which is disjoint to k and which is obtained from S by piping along k.*

Remark. Observe that the converse holds trivially.

Proof. Consider an arbitrary (oriented) component of k, say k_i. Then $k_i \cap S$ is a finite set of points each of which is positively or negatively labeled according whether the intersection number of this point is positive or negative. Such a set of labeled points on k_i must necessarily have a pair of adjacent points with opposite labels since, by hypothesis, the algebraic intersection number of S with k_i equals zero. This pair of points separates an arc t from k_i with $t \cap S = \partial t$. Piping along this arc results in an oriented 2-manifold which intersects k in strictly less points. Thus, iterating the previous procedure, we eventually obtain the required 2-manifold. $\Diamond$

For later use observe the following immediate consequence of the piping-process:

9.3. Lemma. *Let S' be a 2-manifold obtained from another 2-manifold S by piping. Then S is the intersection of two codim zero submanifolds in M iff this holds for S'.*

Proof. Let M_1, M_2 denote two codim-zero sub-manifolds of M with $M_1 \cap M_2 = S$. Observe that a piping-arc t for S is contained either in M_1 or in M_2. W.l.o.g. it lies in M_1. Let U be a regular neighborhood of this arc in M_1. Define $M_1' := (M_1 - U)^-$ and $M_2' := M_2 \cup U$. Then the 2-manifold obtained from S by piping along t equals the intersection $M_1' \cap M_2'$; and the result follows by induction. $\Diamond$

9.4. *Piping in Relative Handlebodies.*

We wish to use pseudo-piping, or piping and pruning, for a study of embedding types of surfaces in relative handlebodies (M, k) which are *strongly essential* in the sense that they are essential and disjoint to k.

Let (M, k) be a relative handlebody of genus n and let $\mathcal{D}$ denote a system of oriented, essential and pairwise not parallel discs in (M, k). Then a *basis carried by* $\mathcal{D}$ is a system $\mathcal{B}$ of oriented discs in M which are all (admissibly) parallel to discs from $\mathcal{D}$.   The orientations of the discs from $\mathcal{B}$ are not necessarily coherent across the parallelity regions but if they all are then $\mathcal{B}$ is called a *coherent basis* (carried by $\mathcal{D}$).

If $\mathcal{D}$ is a meridian-system in (M, k), then any coherent basis $\mathcal{B}$ carried by $\mathcal{D}$ in (M, k) may be identified with an n-tuple $m(\mathcal{B}) = (m_1, m_2, ..., m_n) \in \mathbf{Z}^n$, $n = genus(M)$.   Here the positive integer $|m_i|$, $1 \le i \le n$, denotes the total number of discs from $\mathcal{B}$ which are parallel to the component D_i of $\mathcal{D}$. The sign of m_i is positive or negative according whether or not the orientation of these discs coincides with that of D_i.   In this way, the set of all coherent basis of (M, k), carried by $\mathcal{D}$, is parametrized by the lattice $\mathbf{Z}^n$.

9.5. Example. The coherent basis $(+2, -3, -1)$.

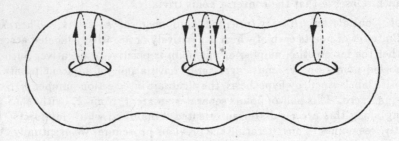

(Figure 9.2.)

A coherent basis may or may not be *realizable* in the sense that there is an admissible surfaces in (M, k) which is obtained from it by piping.   Our next goal is to find realizable, coherent basis (carried by meridian-systems), and to decide whether a coherent basis is realizable.   This can be achieved with the concept of *Heegaard-matrices*.

9.6. Definition. *Let $\mathcal{D}$ be an essential meridian-system for a relative handle-body (M, k). Let $k_1, k_2, ..., k_m$ (possibly $k_i = \emptyset$, for some $1 \le i \le m$) and let $D_1, D_2, ..., D_n$ denote the components of k and $\mathcal{D}$, respectively. Define*

$$a_{ij} := \delta_{ij} = \begin{cases} a(k_i, D_j), & \text{if } k_i \ne \emptyset \\ 0, & \text{otherwise,} \end{cases}$$

where $a(k_i, D_j)$ denotes the algebraic intersection number.   Then the matrix $A_{\mathcal{D}}(M, k) := (a_{ij})$ is called the <u>Heegaard-matrix</u> of (M, k).   If, in addition, $A_{\mathcal{D}}(M, k)$ is a square-matrix, then $\det A_{\mathcal{D}}(M, k)$ is the <u>Heegaard-determinant</u> of (M, k).

Remark. Note that the rank, and so the kernel, of the Heegaard-matrix $A(M, k)$ is independent of the actual choice of an oriented meridian-system.   To see this, recall from cor. 1.6 that any two meridian-systems differ by a finite sequence of disc-slides and note that under a disc-slide the Heegaard-matrix changes by some column operation alone. Similarly, the rank is invariant under relator-slides (an operation which will be considered more closely in chapter IV). Therefore we often write $A(M,k)$ instead of $A_D(M,k)$.

9.7. Proposition. *Let (M,k) be a relative handlebody and let D be an essential meridian-system in (M,k). Let $A(M,k)$ be the Heegaard-matrix of (M,k). Then the following holds:*

(1) If $A(M,k)$ has non-trivial kernel, then there is a coherent basis B carried by D and an orientable, connected and non-separating 2-manifold in (M,k) which is disjoint to k and which is obtained from B by piping along k.

(2) If $B = (x_1, x_2, ..., x_n)$ is a coherent basis carried by D which solves the Heegaard-equation $A(M,k)x = 0$, then B is realizable.

Proof. Suppose the kernel of $A(M,k)$ is non-trivial. Then there is a tupel $x = (x_1, x_2, ..., x_n)$ of rational integers which solves $A(M,k)x = 0$. Multiplying by the largest denominator if necessary, we may suppose $x \in \mathbf{Z}^n$. Dividing by the greatest common divisor of the entries, we may further suppose that at least one entry of x is odd. Now, every integer solution-tuple is realizable. Indeed, such a tuple gives rise to a coherent basis B carried by D whose algebraic intersection with any curve from k is zero. Thus, by lemma 9.2, we obtain, by piping along k, from this basis an orientable 2-manifold in (M,k) which is disjoint to k. This 2-manifold need not be connected. But there is a closed curve in M which intersects the 2-manifold in an odd number of points since one entry of x is odd. Thus at least one component of this 2-manifold has the same property. It follows that this component is a non-separating surface. Of course, this surface can be obtained from some sub-system of the above basis. This sub-system in turn corresponds to a realizable basis which is a solution of $A(M,k)x = 0$; and so (1) follows. (2) is an immediate consequence of lemma 9.2. ◇

Any orientable and non-separating surface in (M,k) disjoint to k can be turned into an orientable, non-separating and essential surface disjoint to k, by successively compressing along $(\partial\text{-})$ compression-discs. Thus the previous proposition tells us, in particular, that any relative handlebody (M,k) with $\#k < \text{genus } \partial M$ contains at least one orientable, non-separating and essential surface disjoint to k. Thus the following corollary is immediate consequence of this consideration (attach 2-handles along k).

9.8. Corollary. *If N is a 3-manifold whose boundary is non-empty and different from a union of 2-spheres, then N contains at least one orientable, non-separating and essential surface. ◇*

9.9. Remarks. (1) Corollary 9.8 is a well-known and basic property of 3-manifolds. It is usually deduced from Poincaré duality (see e.g. [Hem]). Notice that the above approach not only shows the existence of a non-separating surface, but actually constructs one. We will later explore the algorithmic consequences of this observation.

(2) It is usually much easier to construct *non-orientable* 2-manifolds in relative handlebodies. In fact, non-orientable 2-manifolds are much more frequent than orientable ones. Take e.g. a simple closed curve in the boundary of a solid torus which intersects the meridian-disc in an even number of points. Then this meridian-disc can be piped to a non-orientable 2-manifold avoiding the curve.

Pseudo-piping, and piping and pruning, are useful tools for establishing topological information concerning surfaces. Here is a typical example (for other examples see later). Let $\mathcal{D} = D_1 \cup ... \cup D_n$ be a meridian-system in the relative handlebody (M, k). Let $\mathcal{B}$ be a basis carried by $\mathcal{D}$. Denote by $\mathcal{B}_i$ the union of all discs from $\mathcal{B}$ parallel to D_i.

9.10. Proposition. *Let S be any 2-manifold in the relative handlebody (M, k), obtained from $\mathcal{B}$ by pseudo-piping along k. Then S is orientable if $\#\mathcal{B}_i$ is even, for all indices i, $1 \leq i \leq n$.*

Proof. Assume the converse. Then there is basis $\mathcal{B}$ carried by $\mathcal{D}$ with $\#\mathcal{B}_i$ even, for all $1 \leq i \leq n$, and a non-orientable 2-manifold S in (M, k) obtained from $\mathcal{B}$ by pseudo-piping. Since S is non-orientable and so one-sided in M, there is a closed curve l in M which intersects S in just one point. Now, for every index i, $1 \leq i \leq n$, we may choose a closed curve l_i in M which intersects every disc from $\mathcal{B}_i$ in exactly one point, but which is disjoint to $\mathcal{B} - \mathcal{B}_i$. Since S is obtained from $\mathcal{B}$ by pseudo-piping, i.e., by piping and pruning, we have that $\#(S \cap l_i)$ must equal $\#(\mathcal{B}_i \cap l_i)$, $1 \leq i \leq n$. So, by hypothesis, it must be even, for all $1 \leq i \leq n$. In particular, the intersection-points of $S \cap l_i$ come in pairs. Joining pairs of these intersection-points by tunnels, we obtain from S a surface S' which is homologous to S (mod 2) and which does not meet $\bigcup_{1 \leq i \leq n} l_i$ at all. But l is homologous (mod 2) to some linear combination of l_i's. Since the intersection number is a homology invariant, it therefore follows that the intersection number (mod 2) of S with l is zero. This in turn is impossible since l intersects the 2-manifold S in exactly one point. $\Diamond$

9.11. *Pseudo-Piping and Essential Surfaces.*

According to prop. 9.7 there is a non-separating and essential surface in a relative handlebody (M, k), avoiding k, provided the kernel of its Heegaard-matrix is non-trivial. Indeed, we get such a surface by piping an appropriate coherent basis. Compressing this surface along discs, we obtain an essential and non-separating surface disjoint to k. We next show that *every* essential 2-manifold, disjoint to k, is *supported* by some basis, in the sense that it is obtained from that basis by piping (along k) and pruning.

Recall from section 6 that to any Heegaard-diagram $\underline{D}$ there is associated a relative handlebody $M(\underline{D})$ and a canonical disc-system $\mathcal{D} \subset M(\underline{D})$ coming from the meridian-discs of $\underline{D}$.

9.12. Proposition. *Let $\underline{D} = (D, h, \underline{K})$ be an irreducible pant-diagram. Let $\mathcal{D}$ be the disc-system in the relative handlebody $(M, k) = M(\underline{D})$ associated to $\underline{D}$. Then, for every essential 2-manifold S in (M, k), disjoint to k, there is a basis carried by $\mathcal{D}$ such that S is obtained from $\mathcal{B}$ by pseudo-piping along k.*

Remarks. (1) Recall from prop. 6.7, that this proposition applies to all ∂-irreducible relative handlebodies. It also applies to non-orientable 2-manifolds S.

(2) If S is orientable, then $a(\mathcal{B}, k_j) = 0$, for all components k_j of k, since pseudo-piping does not change the relative homology-class.

(3) We do *not* claim that S can be pseudo-piped along a system of *disjoint* pseudo-pipes. But note that it follows that S can obtained by first piping along a system of disjoint pipes and then pruning.

Proof. Let S be any essential surface in $(M, k) = M(\underline{D})$ disjoint to k. Let S be (admissibly) isotoped so that its intersection with $\mathcal{D}$ is as small as possible. Then $(M - U(\mathcal{D}))^-$ is a system of 3-balls and $(S - U(\mathcal{D}))^-$ is a system of discs. More precisely, $\partial(S - U(\mathcal{D}))^-$ is a circle-pattern in $\underline{D}$.

It is easy to verify that this circle-pattern is a special case of a "very good circle-collection" in $\underline{D}$. Here we call a system C of simple closed curves in the underlying 2-manifold $F = F(\underline{D})$ of $\underline{D}$ (in general position to $\underline{K} \cup \partial D$) a *good circle-collection* if (1) $h(\overline{C} \cap D) = C \cap D$, (2) every component from C intersects every component from $\partial D - \partial \mathcal{K}$ in at most one point, and (3) every component from $(\partial C - D)^-$ intersects every component from $\underline{K}$ in at most one point. We call it *very good* if, in addition, every circle from C intersects every arc from $\underline{K}$ in at most one point.

Let t be an arc in $F - D$ such that $t \cap C = \partial t$ and ∂t is contained in one component of C. Set

$$C' := (C - U(t))^- \cup t_1 \cup t_2,$$

where $U(t)$ is the regular neighborhood in F and where t_1, t_2 denote the two copies of t in $\partial U(t)$. We say that C' is obtained from C by a *compression* *(along t)*.

As for circle-patterns, there is associated to any good circle-collection C in $\underline{D}$ a surface $S(C) \subset M(\underline{D})$ (obtained by attaching the discs-system bound by $\overline{C}$). If C' is a good circle-collection obtained by compressing a good circle-collection C, then $S(C)$ is obtained from $S(C)$ by pseudo-piping (along k). Moreover, every good circle-collection of ∂-parallel circles in $(F - D)^-$ bounds a basis carried by $\mathcal{D}$. Thus, given a circle-pattern C in $\underline{D}$ with $S(C)$ essential, it remains to show that C can be compressed into a good circle-collection of circles parallel to ∂D. We do this in the following steps.

Step 1. Let l be a component from C. Let A be a disc (or an annulus) in $(F - D)^-$ with $A \cap l = (\partial A - D)^-$ (and exactly one boundary curve in ∂D). Then $(\partial A - D)^-$ has at most three components since $(F - D)^-$ is a 2-sphere with three holes and since $(C - D)^-$ is a system of essential arcs in $(F - D)^-$. If A is an annulus, then $(\partial A - \partial D)^-$ is connected. If A is a disc, we say A is a *square* or *triangle* according whether $(\partial A - D)^-$ has two or three components, respectively. Thus l splits $(F - D)^-$ into squares, triangles and annuli. But note that at most two of these components are not squares. Therefore l bounds a disc in F which is *good* in the sense that at most one component of $(B - D)^-$ is not a square. Note further that every component of C, contained in a good disc B, bounds a good disc in B.

In particular, there is an innermost good disc for C, i.e., a good disc B with $B \cap C = \partial B$. For every square-component A from $(B - D)^-$ fix an arc, joining the two components of $(\partial A - D)^-$ and intersecting $\underline{K}$ in a minimal number of points. Compress C along the collection of all of them (Figure 9.3 illustrates this procedure). Do the same for $C - B$ and so on. The process has to stop after finitely many steps. Denote by C_1 the resulting circle-collection. By construction, every circle from C_1 bounds a disc B such that either $B \cap D$ is connected, or one component of $(B - D)^-$ is a triangle or annulus. Moreover, it is not hard to verify that C_1 is again a very good circle-collection (all compression arcs have to meet $\underline{K}$ since $S(C)$ is essential).

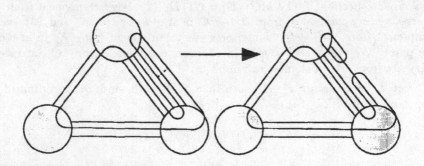

(Figure 9.3.)

Step 2. Let us call a circle in F *good* if it bounds a disc in F whose intersection with D is connected. Our goal is to compress C_1 into a very good circle-collection, C_2, of good circles. To this end let l be a circle from C_1 which is not good. Then l bounds a disc B such that one component of $(B - D)^-$ is a triangle or annulus. Suppose l is innermost in the sense that B contains only good circles. If a component A of $(B - D)^-$ is an annulus, then compress C_1 along the arc in A next to the arc-component of $A \cap D$ (i.e., the arc t in the following picture).

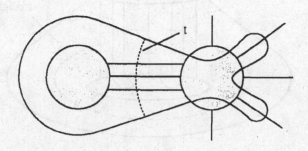

(Figure 9.4.)

If a component A of $(B - D)^-$ is a triangle, we have to consider two cases according whether A meets two or three discs of D. If A meets three discs from D, then fix the mid-point of the triangle A and join this mid-point by three (straight) arcs in A with the three components of $(\partial A - D)^-$. Let these three arcs denoted by t_1, t_2, t_3. Since C_1 is a good circle-collection, it follows that at least two arcs of them, say t_1, t_2, have to intersect $\underline{K}$. In this situation, compress first along the compression-arc $t_2 \cup t_3$ and, afterwards, the resulting curve-system along the compression-arc t_1.

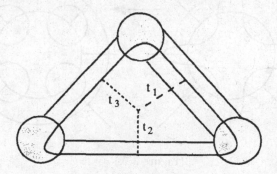

(Figure 9.5.)

If A meets only two discs from D, then compress along the dotted arcs indicated in the following picture.

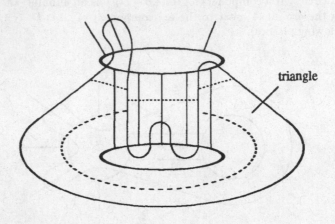

(Figure 9.6.)

It is easily checked that in this way C_1 can be compressed into a very good circle-collection C_2 of good circle.

Step 3. Every good circle can be compressed into one ∂-parallel circle and a collection of circles which are *very good* in the sense that their intersection with D and $(F-D)^-$ is connected (Figure 9.7 illustrates the relevant compressions). Thus C_2 can be compressed into a very good circle-collection C_3 of ∂-parallel circles and very good circles.

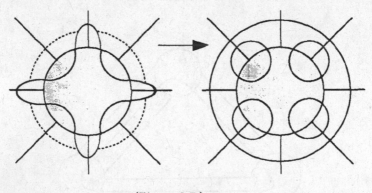

(Figure 9.7.)

Step 4. Finally, let C_4 denote the circle-collection obtained from C_3 by removing all very good circles. Note that only very good circles from C_3 meet D. Thus, in particular, the very good circles from C_3 are paired under the attaching homeomorphism $h : D \to D$.

It follows that $S(C)$ is obtained from $S(C_4)$ by (successive) pseudo-piping since it is obtained from $S(C_3)$ by pseudo-piping.

This finishes the proof of the proposition since $S(C_4)$ is a basis carried by $\mathcal{D}$. ◊

For numerical as well as theoretical purposes, it is worth mentioning that, under a slightly stronger hypothesis, prop. 9.12 allows a further refinement. Indeed, the following corollary tells us that, in relative handlebodies given by full rather than irreducible pant-diagrams, essential surfaces come from meridian-systems.

9.13. Corollary. *Let $\underline{D}$ be a full pant-diagram and let $\mathcal{D}^+ \subset M(\underline{D})$ be the disc-system associated to $\underline{D}$. Let $\mathcal{D} \subset \mathcal{D}^+$ be a meridian-system for the relative handlebody $M(\underline{D})$. Then, for any essential surface S in $(M,k) = M(\underline{D})$ disjoint to k, there is a basis B carried by $\mathcal{D}$ such that S is obtained from B by pseudo-piping.*

Proof. According to the previous proposition, there is a basis carried by $\mathcal{D}^+$, say B, such that the surface S is obtained from B by pseudo-piping.

Since the pant-diagram $\underline{D} = (D, h, \underline{K})$ is supposed to be full, its underlying 2-manifold $F = F(\underline{D})$ consists of 2-spheres which contain exactly three discs and any two of them are joined by at least one arc from $\underline{K}$. Given any component F_0 of $(F - D)^-$, we find that, in the terminology of the previous proof, compressing a circle in F_0 admissibly isotopic to *one* component of ∂F_0 along an appropriate admissible compression-arc, yields *two* curves admissibly isotopic to the other two components of F_0.

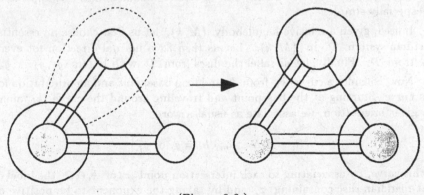

(Figure 9.8.)

To finish the proof, we need to do some bookkeeping. To this end recall from section 5 that $\mathcal{D}^+$ determines a branching complex, say Δ^+ (which incidentally also carries k). The discs from $\mathcal{D}^+$, as well as the discs from B, correspond to the vertices from Δ^+. Splitting at all vertices, corresponding to the given meridian-system $\mathcal{D}$, we obtain from Δ^+ a simply connected 2-complex Δ. The collection of outer vertices of Δ will be denoted by $\partial\Delta$. Now, let Δ_0 be any *free region for* $\mathcal{B}$, i.e., a connected subcomplex of triangles from Δ with the property that no disc from $\mathcal{B}$ corresponds to an inner vertex of Δ_0. Let Δ_0^+ be the largest free region containing Δ_0. If $\partial\Delta_0^+ = \partial\Delta$, then $\Delta_0^+ = \Delta$ and we are done. If not, let z be a vertex from $\partial\Delta_0^+ - \partial\Delta$. Then there is a non-empty system, $\mathcal{B}_z$, of discs from $\mathcal{B}$ corresponding to z. But z is also incident to some triangle E from Δ not contained in Δ_0^+. By the above observation, the system $\mathcal{B}_z$ may be replaced by a system, say $\mathcal{B}_z'$, of discs corresponding to the two vertices of E different from z. Set $\mathcal{B}_1 := (\mathcal{B} - \mathcal{B}_z) \cup \mathcal{B}_z'$ and $\Delta_1 := E \cup \Delta_0$. Then Δ_1 is a free region for $\mathcal{B}_1$. Thus, replacing $\mathcal{B}, \Delta_0$ by $\mathcal{B}_1, \Delta_1$, we may repeat the above process and so on. But this procedure must stop after finitely many steps since every step increases the free region. $\Diamond$

According to prop. 9.12, the obstruction for the existence of non-separating surfaces in (M, k) is understood: it is given by the Heegaard-matrix.

We next wish to show that the (homology classes) of surfaces in (M, k), disjoint to k, are in one-to-one correspondence with the (homology classes of) curves in M. This remarkable fact is known as Poincare-duality. In the following we will base this duality on our piping-construction. We will also point to certain numerical merits of the approach taken here.

9.14. *The First Homology Group.*

As indicated before, we wish to derive a computational form of the Poincaré map. It therefore suggests to present homology groups in the classical way, i.e. by their *incidence-matrices*. In the case of relative handlebodies this appears to be especially simple.

Indeed, given a relative handlebody (M, k), let us first choose an essential meridian-system $\mathcal{D}$ in (M, k). Let us then fix a normal direction for every disc from $\mathcal{D}$. Finally, let us label the discs from $\mathcal{D}$ with letters $g_1, g_2, ..., g_n$.

Now, consider a curve k_i from k and fix a basepoint and an orientation for this curve. Starting at the basepoint and traveling around the curve k_i along the given orientation, we associate as usual a word

$$R_i = h_1 h_2 ... h_m, \quad h_i = g_j \text{ or } g_j^{-1}$$

to this curve, by associating to each intersection point z of $k_i \cap \mathcal{D}$ the label of that meridian-disc containing z, and by taking the exponent to be positive or negative according whether or not k_i intersects $\mathcal{D}$ at z in the given normal direction.

In this way, we associate to (M, k) a set of generators g_i and relations R_j which together form a presentation

$$< g_1, ..., g_n \mid R_1, ..., R_m > .$$

This presentation defines a group $\pi(M, k)$ which may be called the *fundamental group of* (M, k). It is in fact isomorphic to the fundamental group of $M^+(k)$ (but different from the relative first homotopy group of (M, k)). In particular, it is an invariant of the admissible homeomorphisms-type of the relative handlebody (M, k). However, the fundamental group of (M, k) does *not* determine the admissible homeomorphism-type of (M, k) since the latter type is in general not determined by the homeomorphism-type of $M^+(k)$ (see section 26 for some concrete examples). In this context, it is interesting to note that the admissible homeomorphism type of (M, k) is determined by its admissible homotopy type - at least when (M, k) is simple (see [Joh 1, thm. 24.2]). Thus the fundamental group does not determine the admissible homotopy type of (M, k); a fact which is in strong contrast to aspherical complexes without boundary-patterns, such as irreducible 3-manifolds.

The fundamental groups of relative handlebodies are in general not abelian and therefore difficult to handle. Historically, this was one of the reasons for considering the abelianized groups instead. The *first homology group* $H(M, k)$ of (M, k) may thus be defined to be the quotient of the fundamental group $\pi(M, k)$ under its commutator subgroup. From a computational point of view, it is worth mentioning that, given a presentation of $\pi(M, k)$ as above, the abelianization is not only very easy to carry out, but also leads to a canonical matrix, describing the homology group, namely its "incidence-matrix". Here the *incidence-matrix* (a_{ij}) of the presentation of an abelian group is defined by setting a_{ij} equal to the coefficient of the j-th generator in the i-th relator (see [ST, p.310]). Thus, given a labeled meridian-system for (M, k), or equivalently a presentation of its fundamental group, the previous process associates to (M, k) a matrix, the *incidence-matrix* of $H(M, k)$. Observe that this matrix is the same as the Heegaard-matrix of (M, k) with respect to the given meridian-system, and, in light of prop. 9.12, this gives a first hint toward our envisioned special connection of surfaces and homology via piping.

9.15. *The Second Homology Group.*

We next introduce two abelian groups which both may be considered as the torsion-free part of the second homology of relative handlebodies. They both will be important in our computational form of Poincaré duality for relative handlebodies.

To begin with, consider first the solution-set

$$S(M, k) := \{x \in \mathbf{Z}^n \mid A(M, k)x = 0\},$$

of the Heegaard-matrix $A(M, k)$ in the integer-lattice. This set is certainly closed under addition. So it forms an abelian group. We call it the *group of integer-solutions* (of $A(M, k)$).

Let us now turn to the second group, the "cobordism group" $F(M, k)$. To this end consider *admissible, oriented 2-manifolds* in (M, k), i.e., oriented 2-manifolds S in M with $S \cap \partial M = \partial S \subset \partial M - k$. In the remainder of this section all admissible 2-manifolds are supposed to be oriented. Two disjoint admissible 2-manifolds S and S' in (M, k) are called *cobordant*, provided there is an oriented submanifold N in M with $S \cup S' = (\partial N - \partial M)^-$ and provided the orientations of S and $-S'$ are the orientations induced by N. (Notice that the piping-procedure provides us with a special sort of cobordism.) Two admissible 2-manifolds S and S' are *strongly homologous* (or *weakly cobordant*), if there is a finite sequence

$$S = S_1, S_2, ..., S_n = S'$$

of admissible 2-manifolds so that S_i and S_{i+1} is disjoint and cobordant, for every $1 \leq i \leq n-1$. (This is the classical definition of homology, which may be traced back to Riemann; another interesting equivalence relation for admissible 2-manifolds may be generated by setting equal *any* two disjoint admissible 2-manifolds).

Given an admissible 2-manifold S in (M, k), denote by $[S]$ its strong homology class, and let $F(M, k)$ denote the set of all strong homology classes. Given two admissible 2-manifolds S and S', we define the *sum* $S + S'$ as follows. First, after a small general position isotopy (which does not change the homology class), we may suppose that $S \cap S'$ is a system of curves. Every curve gives rise to two different cut-and-paste operations. But there is exactly one of them which transforms the two *oriented* 2-manifolds S and S' into a new *oriented*, possibly singular 2-manifold (without changing the orientation of the 2-simplices involved). Moreover, any such operation reduces the number of (self-) intersection curves, and so, after finitely many of the above procedures, we eventually end up with a new 2-manifold. This 2-manifold is well-defined (modulo homology), and is denoted by $S + S'$.

Now we claim the assignment $(S, S') \mapsto S + S'$ turns the set $F(M, k)$ into an abelian group and that this abelian group may be identified with $S(M, k)$. To verify this claim, we first need to collect some facts about algebraic intersection numbers, $a(S, l)$, of oriented 2-manifolds S with oriented curves l in M.

Algebraic intersection numbers are easily defined in the present context. Indeed, recall that to any intersection-point p from the intersection of an oriented surface, S, and an oriented, transversal curve, l, in the oriented 3-manifold, M, there is associated a *label*, $a(p, l, S)$, which is $+1$ or -1 (according whether or not the orientation of l and S define an orientation of the neighborhood $U(p)$ which coincides with that of M). The *algebraic intersection number* of S and l is given by $a(S, l) := \sum a(p, l, S)$, where the sum is taken over all $p \in S \cap l$.

Given two admissible 2-manifolds S, S' in (M, k), together with a simple closed curve l, observe that (1) the intersection points from $(S \cap l) \cup (S' \cap l)$ are in one-to-one correspondence with those from $(S + S') \cap l$ and that (2) corresponding intersection-points have the same labels. Thus:

9.16. Lemma. *Let* S *and* S' *be two admissible 2-manifold in the relative handlebody* (M, k). *Then* $a(S, l) + a(S', l) = a(S + S', l)$, *for any simple closed curve* l *in* M. $\diamond$

We also need that the algebraic intersection number is an invariant of strong homology. Of course, this is known to be true for much more general homology relations. But for our special definition of homology, this property has a very elementary proof.

9.17. Lemma. *Let* S *and* S' *be two admissible 2-manifolds in the relative handlebody* (M, k). *Suppose that* S *and* S' *are strongly homologous. Then* $a(S, l) = a(S', l)$, *for every simple closed curve* l *in* M.

Proof. According to the definition of strong homology, there is a finite sequence of cobordant admissible 2-manifolds $S = S_1, S_2, ..., S_n = S'$. Thus, in order to prove the lemma, we may suppose that S and S' themselves are cobordant. Then there is a 3-manifold N in M such that $S \cup S' = (\partial N - \partial M)^-$. Consider a component l_0 from $l \cap N$, and let p_1, p_2 be its two end-points (possibly empty, if l_0 is closed). Then p_1, p_2 are intersection-points of $l \cap \partial N$, and we have $a(p_1, l, \partial N) = -a(p_2, l, \partial N)$. If both end-points of l_0 lie in S, then $a(p_1, l, S) = -a(p_2, l, S)$, and the same with S'. If p_1 lies in S and p_2 in S', then $a(p_1, l, S) = a(p_2, l, S')$. Thus lemma 9.17 follows immediately from a counting-argument. $\diamond$

For the next lemma, fix a meridian-system $\mathcal{D}$ in (M, k). For every disc D_i, $1 \le i \le n$, from $\mathcal{D}$, fix a dual curve, i.e. a simple closed curve l_i with

$$\#(l_i \cap D_j) := \delta_{ij} = \begin{cases} 1, & \text{if } i = j; \\ 0, & \text{otherwise.} \end{cases}$$

Then the following characterization for strongly homologous 2-manifolds holds:

9.18. Lemma. *Let* S *and* S' *be two admissible 2-manifolds in* (M, k). *Then* S *is strongly homologous to* S' *if and only if* $a(l_i, S) = a(l_i, S')$, *for all* $1 \le i \le n$.

Proof. One direction is an immediate consequence of the previous lemma. For the other direction, suppose $a(l_i, S) = a(l_i, S')$, for all $1 \le i \le n$. Let $-S$ denotes the 2-manifold obtained from S by reversing its orientation. Then it remains to show that the 2-manifold $S' + (-S)$ is strongly null-homologous. But $a(l_i, -S) = -a(l_i, S)$. So, by lemma 9.16, $S' + (-S)$ is a 2-manifold with $a(l_i, S' + (-S)) = 0$, for all $1 \le i \le n$. In particular, by piping along $\bigcup l_i$, we obtain from $S' + (-S)$ a 2-manifold S^* with $S^* \cap l_i = \emptyset$, for all $1 \le i \le n$. Clearly, S^* is strongly homologous to $S' + (-S)$. So it suffices to show that S^* is strongly null-homologous. For this observe, that for every index i, $1 \le i \le n - 1$, there is an arc t_i which joins l_i with l_{i+1} such that $(M - U)^-$ is a product I-bundle, where $U = U(\bigcup l_i \cup \bigcup t_j)$. Consider $S^* \cap U$. Observe that $S^* \cap U$ consists of discs and that every such disc separates U.

Thus, replacing every disc A from $S^* \cap U$ by one of the two 2-manifolds in which ∂U is split by ∂A, we obtain a 2-manifold $\hat{S}$ with

(1) $\hat{S}$ is strongly homologous to S^*,

(2) $\hat{S} \subset (M - U)^-$, and

(3) $\partial \hat{S} \subset \partial M$.

Now, cutting $\hat{S}$ along compression discs does not change the homology class, and in this way $\hat{S}$ is strongly homologous to some incompressible 2-manifold in $(M - U)^-$. But an incompressible surface in the product I-bundle $(M - U)^-$ whose boundary is contained in ∂M is ∂-parallel in M (see e.g. [Joh 1, prop. 5.6]), and so strongly null-homologous. This finishes the proof. $\Diamond$

We are now ready to prove

9.19. Proposition. *The assignment* $([S_1], [S_2]) \mapsto [S_1 + S_2]$ *turns the set* $F(M, k)$ *into an abelian group whose zero-element is given by the empty set and whose inverses are given by changing orientations.*

Remark. The abelian group $F(M, k)$ may be called the *cobordism group* of (M, k). Using thm. 9.22 below as well as the Poincaré duality theorem, it may further be shown that $F(M, k)$ is isomorphic to the torsion-free part of $H_2(M^+(k), \partial M^+(k); \mathbf{Z})$.

Proof. We have to show that the composition $([S_1], [S_2]) \to [S_1 + S_2]$ is well-defined, commutative and associative. For this fix, as for lemma 9.18, a meridian-system $\mathcal{D}$ for (M, k) and a curve-system l_i, $1 \le i \le n$, in M dual to $\mathcal{D}$.

To show that the composition is well-defined, let S_i' be a 2-manifold with $[S_i] = [S_i']$, for $i = 1, 2$. Then, for every index i, $1 \le i \le n$, we have

$$
\begin{aligned}
a(l_i, S_1 + S_2) &= a(l_i, S_1) + a(l_i, S_2), \quad \text{by lemma 9.16,} \\
&= a(l_i, S_1') + a(l_i, S_i'), \quad \text{by lemma 9.17,} \\
&= a(l_i, S_1' + S_2'), \quad \text{by lemma 9.16,}
\end{aligned}
$$

and so $[S_1 + S_2] = [S_1' + S_2']$, by lemma 9.18.

A similar argument shows that the composition is commutative and associative. This finishes the proof. $\Diamond$

The two abelian groups $S(M, k)$ and $F(M, k)$ introduced above are actually closely related. The relation may be established after fixing a minimal meridian-system $\mathcal{D}$ for (M, k). Indeed, recall from prop. 9.7, that to any element m of the solutions-group $S(M, k)$ there is associated a realizable coherent basis $\mathcal{B}$ (with respect to $\mathcal{D}$). A realization of this basis $\mathcal{B}$ is a 2-manifold S_m obtained from $\mathcal{B}$ by piping along k. This realization is not unique since the choice of the pipes is not uniquely determined. But no such realization meets k and any two realizations of B are easily seen to be strongly homologous.

9.20. Proposition. *The map* $\varphi : S(M,k) \to F(M,k)$, *defined by the assignment* $m \mapsto [S_m]$, *is an isomorphism.*

Proof. As previously, fix a system of curves l_i, $1 \le i \le n$, in M which is dual to the meridian-system $\mathcal{D}$.

To verify that φ is a homomorphism, let $a = (a_1, ..., a_n)$ and $b = (b_1, ..., b_n)$ be two arbitrary elements from $S(M,k)$, and consider the algebraic intersection numbers of the 2-manifolds S_a, S_b and S_{a+b} with the curves l_j, $1 \le j \le n$. By definition of the previous 2-manifolds, we have

$$a(S_a, l_j) = a_j, \ a(S_b, l_i) = b_j \ \text{and} \ a(S_{a+b}, l_j) = a_j + b_j,$$

for all indices i, $1 \le i \le n$. Moreover, by lemma 9.16, we have

$$a(S_a + S_b, l_j) = a(S_a, l_j) + a(S_b, l_j),$$

for all $1 \le j \le n$. Thus, altogether,

$$a(S_a + S_b, l_j) = a(S_{a+b}, l_j),$$

for all $1 \le j \le n$, and so, by lemma 9.18,

$$\varphi(a) + \varphi(b) = [S_a] + [S_b] = [S_{a+b}] = \varphi(a + b).$$

In order to show that φ is injective, let $a = (a_1, ..., a_n)$ be some element from $S(M,k)$ with $\varphi(a) = 0$. Then, by definition of φ, the 2-manifold S_a is strongly null-homologous. So, by lemma 9.17, we have

$$a_j = a(S_a, l_j) = 0,$$

for all indices j, $1 \le j \le n$. Thus $a = 0$.

It remains to verify that φ is surjective. For this let S be any admissible 2-manifold in (M, k). W.l.o.g., we may suppose that S is strongly essential in (M, k), for otherwise we may reduce the complexity of S by disc-compressions which neither meet k nor change the homology class of S. Then, by prop. 9.12, there is some basis B such that S is obtained from B by pseudo-piping along k. Every disc from B is contained in the 3-ball $(M - U(\mathcal{D}))^-$. So it is cobordant in M to a collection of discs from $(\partial U(\mathcal{D}) - \partial M)^-$. Thus B is cobordant to a system B_0 of discs which are parallel to discs from $\mathcal{D}$. By construction, S is strongly homologous to B_0 in M. So $a(B_0, k_j) = 0$, for every component k_j of k, since $S \cap k = \emptyset$ and since the algebraic intersection number is a strongly homology invariant (see proof of 9.17). Forgetting parallel discs from $\mathcal{B}_0$ with opposite orientation, we obtain from $\mathcal{B}_0$ a coherent basis $\mathcal{B}_0^*$ with the same properties. Set

$$x_j := a(\mathcal{B}_0, l_j) = a(\mathcal{B}_0^*, l_j),\ 1 \le j \le n.$$

Then, by definition of the Heegaard-matrix $A(M, k)$ and since $a(\mathcal{B}_0^*, k_j) = 0$, for all components k_j of k, we find that $\mathcal{B}_0^*$ is a coherent basis which solves the Heegaard-equation $A(M, k)x = 0$. Thus $x = (x_1, ..., x_n) \in S(M, k)$. Moreover, by lemma 9.18, it follows that S and S_x are strongly homologous since S is obtained from $\mathcal{B}$ by pseudo-piping and the algebraic intersection numbers of $\mathcal{B}$ and $\mathcal{B}_0^*$ with any curve l_j are the same. Thus, finally, $\varphi(x) = [S_x] = [S]$. So φ is surjective. $\Diamond$

9.21. Duality in Relative Handlebodies.

Recall that any finitely generated abelian group G is a direct sum of a torsion-free part $F(G)$ and a torsion-part $T(G)$. In fact, given a pesentation of G, it turns out that $F(G)$ is the kernel of the incidence-matrix of this presentation (see e.g. [ST,pp. 304]).

In particular, the first homology group $H(M, k)$ of any relative handlebody (M, k) has a decomposition

$$H(M, k) \cong F(H(M, k)) \oplus T(H(M, k)).$$

since it is a finitely generated abelian group. A presentation

$$< g_1, ..., g_m \mid R_1, ..., R_n >$$

for $H(M, k)$ is easily obtained, after having fixed a minimal and oriented meridian-system for (M, k) (see 9.14). So we may associate the incidence-matrix $I(H(M, k))$ to $H(M, k)$ (incidentally, the similarity class of this matrix is independent of the actual choice of the presentation). But $F(H(M, k))$ is isomorphic to the kernel of $I(H(M, k))$. So

$$g^* = \sum_{1 \le i \le n} x_i g_i \in F(H(M, k))$$

if and only if

$$x = (x_1, ..., x_n)\ \text{is a solution of}\ I(H(M, k))x = 0.$$

if and only if

$$x \in S(M, k) = \{x \in \mathbf{Z}^n \mid A(M, k)x = 0\}$$

since the incidence-matrix $I(H(M, k))$ equals the Heegaard-matrix $A(M, k)$.

Thus the assignment

$$g^* = \sum_{1 \le i \le n} x_i g_i \mapsto D(g^*) := (x_1, ..., x_n)$$

defines a map $D : F(H(M, k)) \to S(M, k)$, which clearly is an isomorphism of free abelian groups. This, together with prop. 9.20, shows

9.22. Theorem. *There is a computable isomorphism* $F(H(M, k)) \to F(M, k)$. ◇

9.23. Example. Let us illustrate the previous method for finding surfaces (and some of its intricacies) by means of an example. For this consider the trefoil knot.

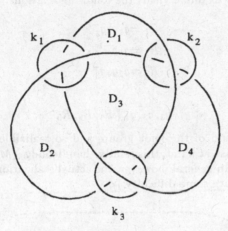

(Figure 9.9.)

It is well-known how to read off a presentation for the knot group, from a given knot-diagram as above. The method chosen below ensures that the resulting presentation is simple, in the sense that it is realized by some relative handlebody.

To describe the method, consider a knot-diagram as well as the graph G_K of the knot K (recall that the *graph* is the image of the given projection of the knot to the 2-sphere and that the *diagram* is obtained from the graph by indicating over- and undercrossings). Observe that a knot-graph with n vertices separates the underlying 2-sphere into a set of $n + 2$ discs; $D_1, ..., D_5$ in the above figure. Note further that

$$M := (S^3 - U(G_K))^-$$

is a handlebody, and that any collection of $n + 1$ of the above discs forms a meridian-system of this handlebody. Pick one of these collections (preferably that one which does not contain the point at infinity); $D_1, ..., D_4$ in our example. Around every vertex of G_K choose a little circle k_i as shown in the above figure. This circle lies in the boundary ∂M, and bounds a disc (2-handle) in $U(G_K) - K$. It then is easy to deduce that the knot-space $(S^3 - U(K))^-$ equals the n-relator 3-manifold $M^+(k)$, where $k := \bigcup_{1 \le i \le n} k_i$.

Given the above situation, we now may read off a presentation of the fundamental group (M,k), i.e. the knot group of K. For this, we first fix a normal direction for the underlying 2-sphere of the graph G_K. Next, for every disc D_j, we fix the induced normal direction and we associate to D_j a letter g_j. Finally, note that every curve k_i gives rise to a word R_i in the generators g_j subject to the requirement that we associate to any intersection of k_i with the disc D_j a symbol $+g_j$ or $-g_j$ according whether or not the curve k_i intersects the D_j in the normal direction. It follows that $< g_1, ..., g_{n+1} \mid R_1, ..., R_n >$ is the required presentation of the knot-group of K.

Observe that our example yields the following relations:

$$R_1 = g_1 g_2 g_3^{-1}$$
$$R_2 = g_1 g_3^{-1} g_4$$
$$R_3 = g_2 g_4 g_3^{-1}$$

By construction,

$$< g_1, g_2, g_3, g_4 \mid R_1, R_2, R_3 >$$

is a simple presentation of the knot group, and so realizible by some relative handlebody. In the case at hand, this relative handlebody (M,k) may be given by the figure below. (In general, however, the actual realization of a given simple presentation may create some difficulty.)

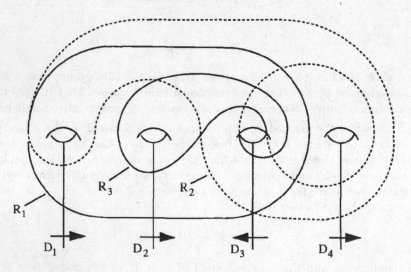

(Figure 9.10.)

Now, note that the admissible surfaces in (M,k) are in fact surfaces in the knot space of K, too (but not vice versa). In order to construct the homology classes of admissible surfaces in (M,k), we now have to consider the incidence-matrix of $H(M,k)$ as given by the above presentation of the knot group $\pi(M,k)$ (see 9.15).

$$\begin{pmatrix} & g_1 & g_2 & g_3 & g_4 \\ R_1 & 1 & 1 & -1 & 0 \\ R_2 & 1 & 0 & -1 & 1 \\ R_3 & 0 & 1 & -1 & 1 \end{pmatrix}$$

The normal form of this matrix is given by

$$\begin{pmatrix} & g_1^* & g_2^* & g_3^* & g_4^* \\ R_1^* & 1 & 0 & 0 & 0 \\ R_2^* & 0 & 1 & 0 & 0 \\ R_3^* & 0 & 0 & 1 & 0 \end{pmatrix}$$

where $g_1^* = g_1 - g_4, g_2^* = g_2 - g_4, g_3^* = g_3 - 2g_4$ and $g_4^* = g_4$. It follows that the kernel of the incidence matrix, and so $F(H(M,k))$ (see 9.22), is isomorphic to $\mathbb{Z}$, and generated by the solution of the following system of linear equations:

$$g_1^* = g_1 - g_4 = 0$$
$$g_2^* = g_2 - g_4 = 0$$
$$g_3^* = g_3 - 2g_4 = 0$$
$$g_4^* = g_4 = 1$$

This solution is given by $g = (g_1, g_2, g_3, g_4) = (1, 1, 2, 1)$. Now, let $\mathcal{B}$ be the system obtained by taking g_i copies of the meridian-disc D_i. Then we obtain an admissible surfaces from $\mathcal{B}$ by piping. These surfaces are not uniquely given, but all of them are punctured tori (every one of them is isotopic to the Seifert surface of the trefoil knot K). To check the latter, we need to know how the various pipes connect the components of $\mathcal{B}$. This is very tedious, if done by hand. Fortunately, this mechanical task can be carried out by a computer. In the appendix, we illustrate this with a more interesting example.

§10. Presentations for Surfaces.

According to prop. 9.12, every essential surface in a relative handlebody can be obtained from some basis by piping and pruning. In the last section we have seen that the *homology class* of such a surface is already completely described by the distribution of its basis-discs. For the *admissible isotopy-type* of a surface, however, one also needs to take into account the distribution of its pipes. It turns out that the concept of "marked basis" provides a convenient way for controlling pipes. Here we introduce this concept and establish its close relation to surfaces. The main result (cor. 10.17) is that basically every essential surface in a simple relative handlebodies is completely given by a marked basis.

10.1. *Marked Basis.*

Marked basis are designed to represent 2-manifolds in relative handlebodies. It is somewhat easier to visualize them in Heegaard diagrams. This is the reason why we introduce the concept for diagrams. But it is no problem to translate the upcoming discussion into the context of relative handlebodies.

Let $\underline{D} = (D, h, \underline{K})$ be a Heegaard-diagram with underlying 2-manifold F and recall from section 6 that $\mathcal{C}(\underline{D})$ denotes the set of all pattern-isotopy classes of circle-patterns in $\underline{D}$. Let D' be a collection of discs from D with $D' \cap h(D') = \emptyset$ and $D' \cup h(D') = D$ (this always exists since h is fix-point free). A *basis* for $\underline{D}$ is a system $\mathcal{B}$ of simple closed curves in $U(D') - D'$ which are admissibly boundary-parallel in $(F - D)^-$. Let Q be the collection of midpoints of all arcs from $\underline{K}$. A *marking* $\underline{b}$ *(for $\mathcal{B}$)* is a collection of components from $\underline{K} - (Q \cup \mathcal{B})$ such that, for every component l from $\mathcal{K} - Q$, the following holds:

(1) $l \cap \underline{b}$ is connected, and

(2) $l \cap \underline{b} \neq \emptyset$ if and only if $l \cap \mathcal{B} \neq \emptyset$.

10.2. Example. A marked basis.

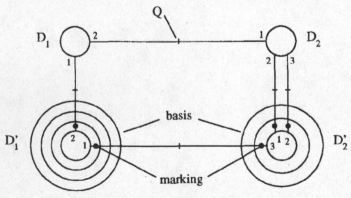

(Figure 10.1.)

A marking for a basis $\mathcal{B}$ may or may not be "realizable". To define this notion, observe that any marking $\underline{b}$ of $\mathcal{B}$ gives rise to a partition of $\underline{K} \cap \mathcal{B}$ into "positive" and "negative" points. Indeed, fix an orientation for $\underline{K}/h$ and suppose $\underline{K}$ carries the induced orientation. Then call a point from $\mathcal{B} \cap \underline{K}$ *positive* or *negative* according whether it lies to the right resp. left of $\underline{b} \cap \underline{k}$, in the component k of $\underline{K}$ containing it. Given this notation, define $\underline{b}$ to be *realizable*, if, for every relator-cycle r_j of $\underline{D}$, the number of positive and negative points from $r_j \cap \mathcal{B}$ are the same. The marking from 10.2 is an example of a realizable marking.

10.3. Definition. *A _marked basis_ for the diagram $\underline{D}$ is a pair $(\mathcal{B}, \underline{b})$, where $\mathcal{B}$ is a basis for $\underline{D}$ and where $\underline{b}$ is a marking for $\mathcal{B}$. A marked basis is _realizable_ if its marking is realizable. $\mathcal{B}(\underline{D})$ denotes the set of all realizable, marked basis for $\underline{D}$.*

10.4. *Turning Realizable Marked Basis into Surfaces.*

We next show how to turn a realizable, marked basis for $\underline{D} = (D, h, \underline{K})$ into a circle-pattern. We do this by invoking two reduction-processes. The first process turns the marked basis into a circle-system by systematically trading intersections with $\underline{K}$ for intersections with ∂D. The second process turns a circle-system into a circle-pattern by removing "inessential" intersections with ∂D (see 6.5 for the difference between circle-systems and circle-patterns).

10.5. *Reduction-process for marked basis.* Let $(\mathcal{B}, \underline{b}) \in \mathcal{B}(\underline{D})$ be a realizable, marked basis. Define $C_0 := \mathcal{B}$.

Step 1. Suppose there is an arc t in $\underline{K} - \underline{b}$ which joins a positive with a negative point from $k \cap C_0$. Then replace C_0 by $C_0' := (C_0 - U(t)) \cup t' \cup t''$, where t', t'' are the two copies of t in $\partial U(t)$. Repeat the process for C_0' and so on. Denote by C_1 the resulting system of circles.

Step 2. For every arc k from $\underline{K}$ there is now a point $z \in k - C_1$ such that every component from $k - z$ intersects C_1 either in positive, or in negative points alone. Let α_t, $t \in [0, 1]$, be the linear extension of an isotopy which expands $U(z)$ to $U(k)$. Set C_1 to be $\alpha_1(C)$. Apply this process for every arc k from $\underline{K}$ and denote the resulting system of circles by C_2.

Step 3. Set $C(\mathcal{B}, \underline{b})$ to be the union of C_2 with that system of circles in $F - \underline{K}$, around the arcs from $\underline{K}$, for which $D \cap C(\mathcal{B}, \underline{b}) = h(D \cap C(\mathcal{B}, \underline{b}))$ (this system always exists; see 10.8).

Let $\beta(B, \underline{b})$ denote the resulting circle-system.

10.6. Example. The following picture shows the effect of step 1 and 2 applied to the realizable, marked basis of 10.2. Step 3 turns this collection of circles into a circle-system by adding two circles around k_1 and three circles around k_2.

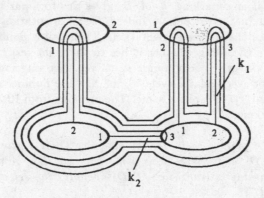

(Figure 10.2.)

10.7. *Reduction-process for circle-systems.* Let C be a circle-system in $\underline{D}$. Suppose C is not a circle-pattern. Then C separates a disc, say D_0, from D or $(F - D)^-$ with $D_0 \cap \underline{K} = \emptyset$. $l := D_0 \cap \partial D$ is an arc in ∂D which does not meet $\partial \underline{K}$. Define

$$C' := (C - U(l \cup hl))^- \cup (l' \cup l'') \cup h(l' \cup l''),$$

where l', l'' are the copies of l in $\partial U(l)$. Forget all circles form C' contained in $F - (\partial D \cup \underline{K})$. Of course, the result is a circle-system again. Apply this procedure to the new circle-system and so on. The process must stop after finitely many steps since each step reduces the intersection with ∂D. The result is a circle-pattern.

Let σC denote the resulting circle-pattern.

10.8. Proposition. *The assignment* $(\mathcal{B}, \underline{b}) \mapsto \sigma\beta(\mathcal{B}, \underline{b})$ *defines a map* $\mathcal{B}(\underline{D}) \to \mathcal{C}(\underline{D})$.

Proof. We have to show that $\sigma\beta(\mathcal{B}, \underline{b})$ is a circle-pattern in $\underline{D}$. Now, $(\mathcal{B}, \underline{b}) \in \mathcal{B}(\mathcal{B})$. So $(\mathcal{B}, \underline{b})$ is a realizable, marked basis. Thus its marking is realizable. Therefore, by definition, the number of positive and negative points in $\mathcal{B} \cap r_j$ are the same, for every relator-cycle r_j. But the first reduction-process cancels successively pairs of adjacent positive and negative points of r_j until no points are left over. So $\beta(\mathcal{B}, \underline{b})$ must be a circle-system. The second reduction-process turns the circle-system $\beta(\mathcal{B}, \underline{b})$ into a circle-system which does not separate a disc from D or $(F - D)^-$ disjoint to $\underline{K}$. By definition, such a circle-system is a circle-pattern. $\Diamond$

Recall from section 6 that to any circle-system C in $\underline{D}$ there is associated an admissible surface $S(C)$ in the relative handlebody $M(\underline{D})$. The first reduction-process emulates the piping-process while the second reduction-process takes care of pruning. Indeed, notice that any basis $\mathcal{B}$ for $\underline{D}$ bounds a system of discs in $M(\underline{D})$ and that $S(\beta(\mathcal{B},\underline{b}))$ is obtained from that disc-system by piping. The next proposition tells us that the surface $S(\sigma\beta(\mathcal{B},\underline{b}))$ is obtained from $S(\beta(\mathcal{B},\underline{b}))$ by pruning.

10.9. Proposition. *Let* $\underline{D}$ *be a Heegaard diagram and let* $C \in \mathcal{C}(\underline{D})$ *be a circle-pattern. Then no component of* $\partial S(C)$ *bounds a disc in* $\partial M(\underline{D})$.

Proof. Assume there is a disc $A \subset \partial M(\underline{D})$ with $\partial A \subset \partial S(C)$. Then there is also such a disc with $A \cap \partial S(C) = \partial A$. Let $\mathcal{D}$ be the disc-system in $M(\underline{D})$ associated to $\underline{D}$. Then there is a disc $A_0 \subset A$ such that $A_0 \cap \mathcal{D}$ is connected and equal to $(\partial A_0 - \partial A)^-$. A_0 lies also in $(F - D)^-$, where F denotes the underlying 2-manifold of $\underline{D}$. Moreover, $A_0 \cap \partial A$ is a component of $(C - D)^-$. Thus C cannot be a circle-pattern in $\underline{D}$. Contradiction. $\Diamond$

10.10. Corollary. *Let* $\underline{D}$ *be a Heegaard diagram such that* $M(\underline{D})$ *is a ∂-irreducible relative handlebody. Let* $C \in \mathcal{C}^+(\underline{D})$. *Then* $S(C)$ *is not totally compressible.*

Proof. Assume $S = S(C)$ is totally compressible. Then S can be disc-compressed into a system S' of discs and 2-spheres. Set $(M,k) = M(\underline{D})$. Then $\partial S' = \partial S \subset (\partial M - k)^-$. So every component of ∂S bounds a disc in $\partial M - k$ since $M(\underline{D})$ is supposed to be ∂-irreducible. This contradicts the previous proposition. $\Diamond$

In the present context we further mention the following feature which is well-known for incompressible surfaces in handlebodies.

10.11. Proposition. *Let* $\underline{D} = (D, h, \underline{K})$ *be a Heegaard diagram and let* $C \in \mathcal{C}(\underline{D})$. *Then* $\tilde{M} := (M(\underline{D}) - U(S(C)))^-$ *consists of handlebodies.*

Proof. Let $\mathcal{D}$ be the system of meridian-discs in the handlebody $M := M(\underline{D})$ associated to $\underline{D}$. C bounds discs in $(M - U(\mathcal{D}))^-$ and intersects D in arcs. Thus, in particular, the surface $S(C)$ splits the system $(M - U(\mathcal{D}))^-$ into 3-balls. We therefore obtain $\tilde{M}$ by attaching the latter 3-balls along discs. Thus it must be a handlebody. $\Diamond$

The up-shot is that even compressible surfaces in relative handlebodies share a few properties with incompressible surfaces as long as they come from circle-patterns.

10.12. *Presenting Essential Surfaces.*

Every marked basis $(\mathcal{B}, \underline{b}) \in \mathcal{B}(\underline{D})$ gives rise to a (unique) admissible surface $S(\sigma\beta(\mathcal{B},\underline{b})) \subset M(\underline{D})$ and every admissible surface in $M(\underline{D})$ extends to a (unique) surface in $M^+(\underline{D})$. In this way marked basis give rise to surfaces in 3-manifolds. For instance, the marked basis from 10.2 yields the Seifert surface of the trefoil knot. What about the converse? It turns out that vice versa almost all essential surfaces come from marked basis. The remainder of this section will make this statement precise for essential surfaces in relative handlebodies. The result will be a useful tool for studying surfaces in 3-manifolds in general.

We are interested in essential surfaces in relative handlebodies (not meeting the relator curve), especially in those which are not boundary parallel. We will approach these surfaces by studying a slightly more restricted, but technically more accessible class of surfaces given as follows.

10.13. Definition. *An essential surface* S *in a relative handlebody* (M, k) *with* $S \cap k = \emptyset$ *is ∂-<u>reduced</u> if*

(1) S admits no compression annulus and if

(2) every surface $S' \subset M$, not meeting k and isotopic to S, is essential.

An essential annulus A in the complement of S is a <u>compression-annulus</u> for S if it joins S with $\partial M - k$ and if the curve $A \cap S$ is not ∂-parallel in S.

Remarks. (1) To make the notion "essential annulus" more precise, consider (M, k) as a manifold with boundary-pattern, where the faces are given by the components of $U(k)$ and $(\partial M - U(k))^-$. Then "essential" refers to the 3-manifold with boundary-patterns obtained from (M, k) by splitting along S (see [Joh 1, §1 and §4]).

(2) A compression-annulus for S is sometimes called an *accidental parabolic* for S.

Condition (1) and (2) of 10.13 are not unrelated:

10.14. Lemma. *Let S be an essential surface in (M, k) with $S \cap k = \emptyset$ but not the disc with one or two holes. Then S has a compression annulus if S is isotopic (in M) to some inessential surface S' in (M, k) with $S' \cap k = \emptyset$.*

Proof. Let D be a ∂-compression disc for S'. Then the closure of at least one component of $\partial U(D \cup \partial S') - S' \cup \partial M$ is a compression annulus for S' since S' is not a disc with one or two holes. The proof is finished by the observation that S has a compression annulus if and only if S' does. ◇

The following proposition tells us that in a search for strongly essential surfaces which are not boundary-parallel it suffices to search for ∂-reduced surfaces.

10.15. Proposition. *Let (M, k) be a ∂-irreducible relative handlebody. If there is an orientable and essential surface in (M, k) which is disjoint to k and which is not ∂-parallel in M, then there is also such a ∂-reduced surface in (M, k).*

Proof. Let S be an (orientable) essential surface in (M, k), disjoint to k and not ∂-parallel. Suppose S is chosen so that the above holds and so that, in addition, χS is as large as possible. It follows that every surface, isotopic to S, is essential in (M, k) since S is not ∂-parallel. If S is

∂-reduced we are done. If not, there is a compression-annulus A for S. Define $S' := (S - U(A))^- \cup (A' \cup A'')$, where A', A'' are the copies of A in $\partial U(A)$. S' is incompressible, not ∂-parallel and disjoint to k since S is. The proposition follows by induction when we have show that S' is also essential. To see the latter assume the converse. Then there is a system B of ∂-compression discs such that $S'' := (S' - U(B))^- \cup (B' \cup B'')$ is essential (and disjoint to k). Here B', B'' denote the copies of B in $\partial U(B)$. Note that every disc from B meets $A' \cup A''$ but not $int(U(A))$, for S is essential. Now, by our maximality condition on χS, it follows that S'' is ∂-parallel in M. Let N be the parallelity region for S''. If $U(A) \not\subset N$, then $N \cup U(A) \cup U(B)$ is a parallelity region for S. If $U(A) \subset N$, then $U(B) \subset (M - N)^-$. So S must be ∂-parallel in M since it is contained in the parallelity region of some other surface. In any case we get a contradiction and the proof is finished. $\lozenge$

10.16. Proposition. *Let (M, k) be a ∂-irreducible, relative handlebody and let $\mathcal{D} \subset M$ be a system of essential discs in (M, k). Let S be a ∂-reduced surface in (M, k) which is obtained from a basis $\mathcal{B} \subset U(\mathcal{D})$ by piping and pruning. Suppose at least one of the pipes lies in $U(\mathcal{D})$.*

Then there is a ∂-reduced 2-manifold S^* in (M, k) which is is obtained, by piping and pruning, from a basis $\mathcal{B}^* \subset \mathcal{B}$ with $\#\mathcal{B}^* < \#\mathcal{B}$ and for which the following holds:

(1) S^* is homeomorphic to S if S is not the disc with one or two holes.

(2) S^* is an annulus if S is and it is a disc with one or two holes if S is.

(3) S^* is homologous to S and $\chi S^* \geq \chi S$, if S is non-separating.

(4) S^* is isotopic to S, if (M, k) is a simple relative handlebody.

Remark. In (2) we do *not* say that S^* is homeomorphic to S, and in (4) we do *not* say that S^* is admissibly isotopic to S, i.e., we allow the isotopy to cross the relator curves.

Proof of Proposition. By hypothesis, S is obtained from the basis $\mathcal{B}$ by first piping along a system $\mathcal{P}$ of pipes and then pruning the resulting surface along a system $\mathcal{A}$ of discs in $U(\partial M)$. Let $\mathcal{B}' \subset \mathcal{B}$ denote the collection of sub-discs characterized by the properties that $\mathcal{B}' \cap \mathcal{P} = (\partial \mathcal{B}' - \partial \mathcal{B})^-$ and every component of $\mathcal{B} - \mathcal{B}'$ contains exactly one point from $\mathcal{B} \cap k$ (and set $B' := B \cap \mathcal{B}'$, for every B from $\mathcal{B}$). Define

$$S' := \mathcal{B}' \cup \mathcal{P}.$$

Then S' is the intermediate surface in M with $S' \cap \partial M = \partial S' \subset \partial M - k$ and $\partial \mathcal{A} \subset S'$ and the property that S is obtained from S' by pruning along $\mathcal{A}$. To be even more precise, fix a product structure $U(\partial M) = \partial M \times I$ for $U(\partial M)$ in such a way that $S' \cap U(\partial M) = \partial S' \times I$, i.e., $S' \cap U(\partial M)$ is vertical in $U(\partial M)$. Let $p : \partial M \times I \to \partial M$ be the canonical projection onto one factor. Then we may suppose that $\mathcal{A}$ is chosen so that the above holds and that, in addition, $\mathcal{A}$ is a collection of horizontal discs which lie on different levels of

$\partial M \times I$. Note that $S \subset S' \cup \mathcal{A}$ and that $(S' - S)^-$ is a collection of annuli. For later use it is also important to remember that the restriction $p|\partial\mathcal{A}$ (but not necessarily $p|\mathcal{A}$) is an embedding into $\partial S'$.

By assumption, there is at least one pipe from $\mathcal{P}$ which is entirely contained in $U(\mathcal{D})$. Thus there are two discs B_1, B_2 from $\mathcal{B}$ which lie in one component of $U(\mathcal{D})$ and which are joined by a pencil from $\mathcal{P}$ contained in $U(\mathcal{D})$. Let $E \subset U(\mathcal{D})$ denote the parallelity region between B_1 and B_2 and let P denote the union of all pencils from $\mathcal{P}$ contained in E. Then $(\partial E - \partial M)^- = B_1 \cup B_2$ and $(\partial P - \partial E)^- \subset (\partial B'_1 \cup \partial B'_2 - \partial M)^-$. We next use P to construct another 3-ball $E' \subset E$. To do this note that P splits E into 3-balls and that one of them, say E'_1, contains $B'_1 \cup B'_2$. Note also that $(\partial E'_1 - \partial E)^- = P \subset S$. If P is connected, then $E' := E'_1$ is the desired 3-ball. If not, then let $C' \subset E'$ be a square with two opposite faces in $B_1 \cup B_2$ and the other faces in two different pipes from P. Note that $\partial C'$ lies in S and so it bounds a disc C in S (recall S is essential). Now, C' splits E' into two 3-balls and one of them lies in the 3-ball bound by $C \cup C'$ (recall M is irreducible). Let E'_2 be the other one. Isotope S (rel $(S - C)^-$) so that afterwards $C = C'$ and then apply the above procedure to E'_2 instead of E'_1, and so on. This process stops after finitely many steps and the result is a 3-ball $E' \subset E$ with $P \cap int(E') = \emptyset$, $(\partial E' - \partial E)^- \subset S$ and $E' \cap \partial M$ connected.

We will use E' to change the collections $\mathcal{B}, \mathcal{P}, \mathcal{A}$ into new collections $\mathcal{B}^*, \mathcal{P}^*, \mathcal{A}^*$ with better properties. First define

$$\mathcal{B}^* := \mathcal{B}' - \{\, B'_1, B'_2 \,\}.$$

To construct $\mathcal{P}^*$ note that the intersection $E' \cap \partial M$ is a square with two opposite faces in $B_1 \cup B_2$. Fix a system of arcs in $E' \cap \partial M$ joining B_1 and B_2, by selecting exactly one such arc in any one of those components from $(E' \cap \partial M) - k$ which do not meet a pencil from $\mathcal{P}$ contained in E. Next let $\mathcal{C}$ be a system of squares in E' which have one face in $(\partial E' - \partial E)^-$ and two opposite faces in B'_1, B'_2, and such that the rest of all faces forms the arc-system we have just specified. Set $B''_i := B' \cap E'$, for $i = 1, 2$, and $Q := B''_1 \cup (\partial E' - \partial E)^- \cup B''_2$. Define

$$\mathcal{P}^* := \mathcal{P} \cup (Q - U(\mathcal{C}))^- \cup (\partial U(\mathcal{C}) - \partial E')^-,$$

where the regular neighborhood $U(\mathcal{C})$ is taken in E'. Given $\mathcal{P}^*$, define a new surface

$$S^+ := \mathcal{B}^* \cup \mathcal{P}^*.$$

Note that $p(\partial \mathcal{A}) \subset \partial S^+ \cup (B''_1 \cap \partial B_1) \cup (B''_2 \cap \partial B_2)$.

Finally, we construct $\mathcal{A}^*$. To this end consider $\mathcal{F} := U(\mathcal{C}) \cap (\partial E' - \partial M)^-$. Note that $\mathcal{F}$ is a system of bands joining ∂B_1 and ∂B_2. Notice further that $\mathcal{A} \cap B''_1 = (\partial U(B''_1 \cap \partial B_1) - \partial B'_1)^-$, where the regular neighborhood is taken in B''_1 (to see this recall that $p|\partial\mathcal{A}$ is an embedding and that $p(\partial\mathcal{A}) \subset \partial S^+ \cup (B'_1 \cap \partial B_1) \cup (B'_2 \cap \partial B_2)$). The same with $\mathcal{A} \cap B''_2$. Hence, in particular, $\mathcal{A}$

intersects every component of $\mathcal{F}$ in at most two arcs. Let $\mathcal{F}^*$ be the collection of all those squares into which $\mathcal{F}$ is split by $\mathcal{A}$ and which meet $(\partial E' - \partial E)^-$. Note that $\mathcal{A} \cap \mathcal{F} = (\partial \mathcal{F}^* - \partial \mathcal{F})^-$. But note also that the components from $\mathcal{F}^*$ may or may not meet ∂M. For later notational convenience, we suppose w.l.o.g. that $\mathcal{F}^*$ is pushed into a system of horizontal squares in $E \cap U(\partial M)$ (using an isotopy which fixes $\mathcal{F}^* \cap (B_1 \cup B_2)$ and which keeps arcs in the frontier $(\partial U(\mathcal{C}) - \partial E')^-$). Define

$$\mathcal{A}^* := \mathcal{A} \cup \mathcal{F}^*.$$

As an immediate consequence of the construction we have that $\mathcal{A}^*$ is a collection of pairwise disjoint surfaces (recall $\mathcal{A} \cap \mathcal{F}^* = \mathcal{A} \cap \mathcal{F}$) and that, after splitting S^+ along $\mathcal{A}^*$, we get S (plus an additional collection of trivial discs). Indeed, we are just switching the decomposition of S; roughly speaking, we switch from $S = S' \cup \mathcal{A}$ to $S = (S' - \mathcal{F}^*)^- \cup (\mathcal{A} \cup \mathcal{F}^*) = S^+ \cup \mathcal{A}^*$. Note also that $p|\partial \mathcal{A}^*$ is still an embedding, for $p|\partial \mathcal{A}$ and $p|\mathcal{B}^*$ are embeddings and $p(\mathcal{F}^* - B_1 \cup B_2) \cap p(\partial \mathcal{A}) = \emptyset$. Moreover, $p(\mathcal{A}^*) \subset \partial M - k$ since $p(\mathcal{A})$ as well as $p(\mathcal{F}^*)$ lies in $\partial M - k$. We next establish two more properties of $\mathcal{A}^*$.

Let A^* be a surface from $\mathcal{A}^*$ and keep in mind that $p|\partial A^*$ is an embedding. If A^* happens to be compressible in $U(\partial M)$, then let C' be a compression disc for A^* in $U(\partial M)$. Since $A^* \subset S$ and since S is supposed to be essential in (M, k), it follows that $\partial C'$ must bound a disc C in S. Now C' and C are isotopic (rel boundary) since M is irreducible. So extending the latter isotopy to S and replacing A^* by $A^* \cup C'$ if necessary, we may suppose w.l.o.g that A^* is incompressible in $U(\partial M)$. Then, using the argument from [Wa 2, prop. 3.1], it is not hard to see that A^* can be isotoped (rel boundary) so that afterwards first $A^* \cap (p(\partial A^*) \times I) = \partial A^*$ and then A^* is horizontal. Thus $p(\partial A^*)$ bounds a surface $A' \subset \partial M$ such that A^* can be isotoped (rel boundary) next to A'. In fact, $A' \subset \partial M - k$. To see the latter recall from above that $p(\mathcal{A}^*) \subset \partial M - k$. So $\mathcal{A}^* \cap (k \times I) = \emptyset$ and the parallelity region between A^* and A' cannot meet $k \times I$. Thus we may suppose the components of $\mathcal{A}^*$ are horizontal surfaces next to surfaces in $\partial M - k$.

Suppose A^* is not a disc. Then there is an essential annulus joining A^* with A' since A^* is parallel to A'. Therefore there is such an annulus joining S with $\partial M - k$. Indeed, every essential curve in A^* is the boundary curve of an annulus joining S with $\partial M - k$ and every curve which is essential in A^* is essential in S (see above). Hence and since S is ∂-reduced, it follows that A^* must be a disc with one or two holes and that S is a disc with one or two holes if A^* is.

We now finish the proof as follows. First note that all discs from $\mathcal{P}^*$ may be viewed as bands and that all these bands can be turned into pipes by pulling them tight. It follows that, if $\mathcal{P}^*$ consists entirely of bands, then S^+ is admissibly isotopic to a surface which is obtained from $\mathcal{B}^*$ by piping (along the new pipes coming from $\mathcal{P}^*$).

Now suppose that both $\mathcal{P}^*$ and $\mathcal{A}^*$ consists of discs. Then, more precisely, we may suppose that all these discs are horizontal (see above). We obtain S

from S^+ by pruning along $\mathcal{A}^*$, provided $p(\partial A^*) \subset \partial S^+$. In this case we are done since $\#\mathcal{B}^* < \#\mathcal{B}$ and since S^+ is obtained from $\mathcal{B}^*$ by piping. So let A^* be a disc from $\mathcal{A}^*$ with $p(\partial A^*) \not\subset \partial S^+$. Then $A^* \cap \partial M \neq \emptyset$. But note that, by construction, $A^* \cap \partial M \subset \partial S$ (since $A^* \cap \partial M \subset U(\mathcal{C}) \cap (\partial B_1 \cup \partial B_2)$). If $A^* \cap \partial M$ is disconnected, then there is an arc c in A^* which joins two different components of $A^* \cap \partial M$. This arc is the face of a 2-faced disc, C', whose other face lies in $\partial M - k$ (A^* is parallel to a disc in $\partial M - k$). Since S is essential and (M, k) is ∂-irreducible, it follows that c separates a disc from S which is admissibly isotopic to C' (rel c). Now c can be chosen in A^* close to ∂S^+ and so the above isotopy changes S^* into another surface with similar properties but reduced basis $\mathcal{B}^*$. Hence w.l.o.g. $A^* \cap \partial M$ is connected. Then let $\mathcal{A}^{**}$ be the collection of all discs from $\mathcal{A}^*$ which do not meet ∂M at all. Using the fact that $\mathcal{A}^*$ consists of discs, we see that we get a surface admissibly isotopic to S, by first piping $\mathcal{B}^*$ along $\mathcal{P}^*$ and then pruning the resulting surface along $\mathcal{A}^{**}$. So the proposition follows as before.

Suppose $\mathcal{P}^*$ consists of discs but not $\mathcal{A}^*$. Then let $\mathcal{A}^{**} \subset \mathcal{A}^*$ be the set of all discs from $\mathcal{A}^*$. By what we have seen above, it follows that, after piping and pruning along $\mathcal{P}^*$ and $\mathcal{A}^{**}$, we obtain from $\mathcal{B}^*$ the disjoint union of a surface S^* with a collection of annuli (see the properties of $\mathcal{A}^*$ above). This 2-manifold is homologous to S and (3) of the proposition follows if S is non-separating. If (M, k) is simple, then S^* is isotopic to S (not necessarily admissibly isotopic though), and (4) of the proposition follows. In general, however, S^* is only homeomorphic to S. But if S is not a disc with one or two holes, then, using an argument from 10.14, we see that the surface S^* must be ∂-reduced. Moreover, S^* is obtained from some sub-basis of $\mathcal{B}^*$ by piping and pruning. Hence and since $\#\mathcal{B}^* < \#\mathcal{B}$, the conclusion of the proposition follows if S is not a disc with one or two holes. In the other case, either S^* or one of the annuli above is ∂-reduced and obtained from some sub-basis of $\mathcal{B}^*$ by piping and pruning (and non-separating if S is). Choosing this surface, the proposition follows again.

Suppose finally that there is a component P_0^* of $\mathcal{P}^*$ which is not a band. Then P_0^* must be a ∂-parallel annulus. Moreover, it is easily verified that every component of ∂P_0^* is either disjoint to or entirely contained in $\mathcal{A}^*$ (for notational convenience we here assume that $\partial \mathcal{A}^* \subset \partial M$). Let A_1^*, A_2^* be all the components from $\mathcal{A}^*$ (possibly empty) which contain a component from ∂P_0^*. None of them can be a disc since $\partial A_1^* \cup \partial A_2^* \subset \partial M - k$ and since $\mathcal{B}$ is supposed to be essential in (M, k). So S must be a disc with two holes if either A_1^* or A_2^* is not an annulus (see above). We now forget P_0^*, A_1^*, A_2^* and apply the previous procedure to the 2-manifold of $(S^* - (A_1^* \cup P_0^* \cup A_2^*))^-$. Since every component from $\mathcal{A}^*$ is ∂-parallel but not S itself, we end-up with a 2-manifold which satisfies (1)-(4) from the proposition and which is obtained from some sub-basis of $\mathcal{B}^*$ by piping and pruning. So the proposition follows again.

This finishes the proof. $\Diamond$

As a corollary we obtain the main result of this section which says that basically all ∂-reduced surface in a simple relative handlebody (not meeting the relator curves) are supported by a marked basis. Here we say that a surface S in $M(\underline{D})$ is *supported by a marked basis* $(\mathcal{B}, \underline{b}) \in \mathcal{B}(\underline{D})$ if $S(\sigma\beta(\mathcal{B}, \underline{b}))$ is admissibly isotopic to S (where β and σ are from 10.5 and 10.7).

10.17. Corollary. *Let* $\underline{D} = (D, h, \underline{K})$ *be a useful pant diagram. Let* S *be a ∂-reduced surface in* $M(\underline{D})$. *Then there is a marked basis* $(\mathcal{B}, \underline{b}) \in \mathcal{B}(\underline{D})$ *such that* $S(\sigma\beta(\mathcal{B}, \underline{b}))$ *is a ∂-reduced 2-manifold and that the following holds:*

(1) $S(\sigma\beta(\mathcal{B}, \underline{b}))$ *is homeomorphic to* S *if* S *is not the disc with one or two holes.*

(2) $S(\sigma\beta(\mathcal{B}, \underline{b}))$ *is an annulus if* S *is and it is a disc with one or two holes if* S *is.*

(3) $S(\sigma\beta(\mathcal{B}, \underline{b}))$ *is homologous to* S *and* $\chi S(\sigma\beta(\mathcal{B}, \underline{b})) \geq \chi S$, *if* S *is non-separating.*

(4) $S(\sigma\beta(\mathcal{B}, \underline{b}))$ *is isotopic to* S, *if* $\underline{D}$ *is simple (and so* $M(\underline{D})$ *is simple).*

Remarks. (1) In (2) we do *not* say that $S(\sigma\beta(\mathcal{B}, \underline{b}))$ is homeomorphic to S, and in (4) we do *not* say that $S(\sigma\beta(\mathcal{B}, \underline{b}))$ is admissibly isotopic to S, i.e., we allow the isotopy to cross the relator curves.
(2) If $\underline{D}$ is a full pant-diagram, then the basis $\mathcal{B}$ can be chosen in $U(\mathcal{D})$ for any meridian-system $\mathcal{D} \subset M(\underline{D})$ contained in the disc-system associated to $\underline{D}$.

Proof of Corollary. Let $\mathcal{D} \subset M(\underline{D}) = (M, k)$ be the disc-system associated to $\underline{D}$. Then $(M, k) := M(\mathcal{D})$ is a ∂-irreducible, relative handlebody and $\mathcal{D}$ is an essential disc-system in (M, k) since $\underline{D}$ is a useful pant diagram. By prop. 9.12, there is a basis, say $\mathcal{B} \subset U(\mathcal{D})$, such that S is obtained from $\mathcal{B}$ by piping and pruning (see 9.12 remark (3)). By prop. 10.16, we may suppose that none of the pipes lie in $U(\mathcal{D})$ (for otherwise we replace $\mathcal{B}$ by a smaller basis). Then, for every component D_i, $i \geq 1$, from $\mathcal{D}$, denote by $\mathcal{B}_i$ the collection of all those discs from $\mathcal{B}$ which are admissibly parallel to D_i. W.l.o.g. we may suppose that $\mathcal{B}_i$ is entirely contained in one of the two components of $U(D_i) - D_i$, for all $i \geq 1$. Let ℓ be a component of $k \cap U(D_i)$. Since $\mathcal{P}$ has no pipes in $U(\mathcal{D})$, the intersection-points from $\ell \cap \partial\mathcal{B}_i$ fall into two classes according to the directions in which the pipes, starting at the intersection points, leave $U(D_i)$. Thus there is a unique component of $k - \partial\mathcal{B}$ meeting $U(D_i)$ and separating these two classes of intersection points. The collection of all these components forms a marking, say $\underline{b}$, for the basis $\partial\mathcal{B}$. As explained in 10.5, this marked basis defines a reduction-process which results in a circle-system $C = \beta(\partial\mathcal{B}, \underline{b})$, and so in an admissible surface $S' := S(C)$. Furthermore, we obtain S from S' by pruning, i.e., compressing along compression discs near $\partial M - k$. The map σ simply emulates this compression-process on the level of Heegaard-diagrams (see 10.7). Therefore $S = S(\sigma\beta(\partial\mathcal{B}, \underline{b}))$. This finishes the proof of the corollary. $\Diamond$

§11. The Seifert Characteristic.

We define the Seifert characteristic of a relative handlebody (M, k) to be the minimal absolute Euler characteristic taken over all non-separating surfaces in M not meeting k. In this section we relate the Seifert characteristic to other, more accessible invariants. Specifically, we derive an estimate for the Seifert characteristic in terms of the minimal absolute norm taken over all integer solution vectors of the Heegaard equation associated to the relative handlebody (see thm. 11.12). Given this estimate, it is then an easy matter to construct, e.g., examples of sequences of relative handlebodies whose Seifert characteristics tend to infinity (see ex. 11.13)

11.1. *Complexities.*

We discuss three different invariants for relative handlebodies. The first one is familiar. The other two use, in an essential way, piping decompositions of surfaces.

Throughout this section $\underline{D}$ denotes a Heegaard diagram and $\mathcal{D}^+$ denotes the disc-system in the relative handlebody $(M, k) = M(\underline{D})$ associated to $\underline{D}$. By a 2-manifold in (M, k) we always mean a proper, orientable 2-manifold in M which is disjoint to k.

11.2. *Thurston norm and Seifert characteristic.* The *Thurston-norm* of a 2-manifold S in (M, k) is defined to be

$$\tau[S] := \min \{ -\chi(S') \mid S' \subset M \text{ is a 2-manifold in } (M, k) \text{ homologous to } S \}.$$

The *Seifert characteristic* of (M, k) is given by

$$\sigma M(\underline{D}) = \sigma(M, k) := \min \{ -\chi(S) \},$$

where the minimum is taken over all non-separating surfaces S in (M, k).

11.3. *Algebraic length.* Suppose $\underline{D}$ is a full pant-diagram and let $\mathcal{D} \subset \mathcal{D}^+$ be a meridian-system for $(M, k) = \overline{M}(\underline{D})$. Denote by $A_{\mathcal{D}}(M, k)$ the Heegaard-matrix of (M, k) (see def. 9.6). To every $x \in \mathbf{Z}^n \cap \operatorname{kern} A_{\mathcal{D}}(M, k)$, $n = $ genus ∂M, there is associated a non-separating surface $S_x \subset M$ with $S_x \cap k = \emptyset$ (see prop. 9.7). Vice versa, for any non-separating surface S in (M, k), there is a unique $x = (x_1, x_2, ..., x_n) \in \mathbf{Z}^n \cap \operatorname{kern} A_{\mathcal{D}}(M, k)$, such that S_x is (strongly) homologous to S (see prop. 9.20). The *algebraic length* of the surface S is given by

$$\| S \|_{\mathcal{D}} := \| S_x \|_{\mathcal{D}} := \| x \|_{\mathcal{D}} := \sum_{1 \leq i \leq n} |x_i|.$$

Using this notation, let the *algebraic length* of (M, k) (w.r.t. $\mathcal{D}$) defined to be

$$\| M(\underline{D}) \|_{\mathcal{D}} = \| M, k \| := \min \{ \| x \|_{\mathcal{D}} \mid x \in \mathbf{Z}^n \cap \operatorname{kern} A_{\mathcal{D}}(M, k) \}.$$

This complexity is relatively easy to calculate, especially if k is connected. Of course, it depends on the choice of $\mathcal{D}^+$ but for the applications we have in mind this dependence causes no problem. To turn the algebraic length into an invariant of the admissible homeomorphism type, define

$$\| M(\underline{D}) \| = \| M, k \| := \min \| M, k \|_{\mathcal{D}},$$

where the minimum is taken over all meridian-systems $\mathcal{D}$ in (M, k) which extend to a full disc-system (a disc-system is *full* if its associated Heegaard-diagram is a full pant-diagram).

11.4. *Geometric length.* Again suppose $\underline{D}$ is a full pant-diagram and let $\mathcal{D} \subset \mathcal{D}^+$ be a meridian-system for (M, k). Recall from cor. 9.13 that every essential surface S in (M, k) is supported by some basis (in the sense that it is obtained from this basis by piping along k and pruning). More precisely, S is supported by a basis which in turn is carried by $\mathcal{D}$. In the following we only consider basis carried by $\mathcal{D}$. Under this convention define the *geometric length* of the surface S by

$$l_{\mathcal{D}}(S) := \min \{ \, \#\mathcal{B} \mid \mathcal{B} \text{ supports } S \, \}$$

(recall a basis may or may not be coherent). Two essential 2-manifolds are said to be *Dehn-equivalent* if they are equivalent under the equivalence relation generated by admissible isotopy and Dehn twists along essential annuli in (M, k). Define the *geometric length* of the Dehn-equivalence class $[S]$ of S to be

$$l_{\mathcal{D}}[S] := \min \{ \, l_{\mathcal{D}}(S') \, \}$$

where the minimum is taken over all essential 2-manifolds S' which are Dehn-equivalent to S. Finally, set

$$l_{\mathcal{D}}(M(\underline{D})) = l_{\mathcal{D}}(M, k) := \min \{ \, l_{\mathcal{D}}[S] \, \},$$

where the minimum is taken over all essential surfaces S in (M, k). Define the *geometric length* of (M, k) by

$$l(M(\underline{D})) = l(M, k) := \min \{ \, l_{\mathcal{D}}(M(\underline{D})) \, \},$$

where the minimum is taken over all meridian-systems $\mathcal{D}$ in $(M, k) = M(\underline{D})$ which extend to a full disc-system.

The complexities $\sigma M(\underline{D})$, $\| M(\underline{D}) \|$ and $l(M(\underline{D}))$ are all invariants of the admissible homeomorphism type of the relative handlebodies, $M(\underline{D})$, but they are not necessarily invariants of the completed 3-manifolds, $M^*(\underline{D})$. Thus, potentially, they may be of use e.g. in distinguishing one-relator structures for 3-manifolds.

11.5. *Estimates.*

We are interested in estimating the Seifert characteristic. For this we utilize a relationship between Seifert characteristics on the one hand and algebraic lengths on the other. To make this relationship explicit we first establish an intermediate estimate involving geometric lengths (see props. 11.9 and 11.10). From this we then derive the estimate for the Seifert characteristic in terms of the algebraic length. This often yields reasonable estimates for the Seifert characteristic since algebraic lengths are comparatively easy to calculate.

We begin with an observation concerning piping decompositions of essential surfaces.

Let $\mathcal{B} \subset U(\mathcal{D}^+)$ be a basis supporting an essential 2-manifold S in $(M, k) = M(\underline{D})$. Then there is a 2-manifold S' in (M, k) which is obtained from $\mathcal{B}$ by piping along a system $\mathcal{P}$ of pipes. To be more precise, let $\mathcal{B}' \subset \mathcal{B}$ denote the disc system characterized by the properties that $\mathcal{B}' \cap \mathcal{P} = (\partial\mathcal{B}' - \partial\mathcal{B})^-$ and every component from $\mathcal{B} - \mathcal{B}'$ contains exactly one point from $k \cap \mathcal{B}$. Then $S' = \mathcal{B}' \cup \mathcal{P}$. The surface S is obtained from S' by pruning along a collection $\mathcal{A}$ of discs. We say

$$(\mathcal{B}, \mathcal{P}, \mathcal{A})$$

is a *piping decomposition* of S. Given $\mathcal{B}$ and $\mathcal{P}$, the collection $\mathcal{A}$ is uniquely determined. However, the piping decomposition itself is not unique, i.e., there are usually various surfaces admissibly isotopic to S which have a piping decomposition different from $(\mathcal{B}, \mathcal{P}, \mathcal{A})$. It is a challenge to find an optimal piping decomposition for (the admissible isotopy class of) a given surface. We next establish some properties for piping decompositions to help in this direction.

Let B be a disc from $\mathcal{B}$. Set $B' = B \cap \mathcal{B}'$. Denote by $\mathcal{P}_B$ the system of all pipes starting or ending at B. Let b be a component of $B' \cap \partial M$. Then b is an arc (contained in $\partial S'$) joining two (possibly equal) pipes from $\mathcal{P}_B$, say P_1, P_2. Notice that P_1, P_2 may or may not leave B in the same direction. b is called a *switch* for $(\mathcal{B}, \mathcal{P}, \mathcal{A})$ if they leave in opposite directions. It is called a *bounded switch* if it also lies in a boundary component of $\partial S'$ which bounds a disc in $\partial M - k$ (see fig. 11.2).

11.6. Lemma. *Let (M, k) be a ∂-irreducible, relative handlebody. Let S be an essential 2-manifold in (M, k) with piping decomposition $(\mathcal{B}, \mathcal{P}, \mathcal{A})$. Suppose $(\mathcal{B}, \mathcal{P}, \mathcal{A})$ has a bounded switch. Then S is admissibly isotopic to a 2-manifold which has a piping decomposition $(\mathcal{B}^*, \mathcal{P}^*, \mathcal{A}^*)$ with $(\#\mathcal{B}^*, \#\mathcal{P}^*, \#\mathcal{A}^*) < (\#\mathcal{B}, \#\mathcal{P}, \#\mathcal{A})$ (w.r.t. the lexicographical ordering).*

Proof. We actually prove a somewhat stronger result. To formulate it we need a little preparation. Consider the piping decomposition $(\mathcal{B}, \mathcal{P}, \mathcal{A})$ of S. Let B be a disc from the basis $\mathcal{B}$. Then $\mathcal{P} \cap B$ is a collection of arcs. W.l.o.g. we suppose $\mathcal{P} \cap B \subset U(k \cap \partial B)$, where the regular neighborhood is taken in B). To fix ideas let x_0 be a point from $k \cap \partial B$ and let $\gamma_1, ..., \gamma_n$ be all arcs from $\mathcal{P} \cap B$ contained in the component $U(x_0)$ of $U(k \cap \partial B)$. Let $C_i \subset U(x_0)$

be the unique disc given by $(\partial C_i - \partial B)^- = \gamma_i$ and $x_0 \in C_i$. Let the indices be chosen so that $C_{i+1} \subset C_i$, for all $1 \le i \le n - 1$. Now consider a disc A_0 from $\mathcal{A}$. W.l.o.g. $A_0 \subset \partial M - k$. So $A_0 \cap \partial B$ is a collection of pairwise disjoint arcs in ∂B not meeting $k \cap \partial B$.

We claim the conclusion of the lemma holds whenever there is component from $A_0 \cap \partial B$ which is entirely contained in $U(x_0)$. To see the claim assume there is a component a_0 of $A_0 \cap \partial B$ with $a_0 \subset U(x_0) \cap \partial B$. Then w.l.o.g. we may suppose A_0 and a_0 are chosen so that the above holds and that, in addition, $a_0 \cap (\mathcal{P} \cap B) = \partial a_0$. In particular, a_0 joins two arcs, say γ_i, γ_{i+1}, from $(\mathcal{P} \cap B) \cap U(x_0)$. Let a_0 be chosen so that the above holds and that, in addition, the index i is as large as possible (i.e., there is no component such as a_0 which is closer to x_0). Let z_i, z_{i+1} denote those end-points of γ_i, γ_{i+1}, respectively, which do not lie in ∂a_0. Let r_i, r_{i+1} be the (possibly equal) boundary curves from $\partial S'$ containing z_i, z_{i+1}, respectively. Here S' denotes the surface obtained from $\mathcal{B}$ by piping along $\mathcal{P}$, i.e., $S' = \mathcal{B}' \cup \mathcal{P}$. Now we have several possibilities according whether or not either r_i or r_{i+1} or both of them are contractible boundary curves.

Suppose neither r_i nor r_{i+1} is contractible. Then r_i and r_{i+1} are boundary curves of S (and not just S'). The disc $Q_i := (C_i - C_{i+1})^-$ has the property that $Q_i \cap S = (\partial Q_i - \partial M)^-$ and $Q_i \cap k = \emptyset$. But S is essential in (M, k) and so Q_i cannot be a ∂-compression disc for S. Thus the arc $Q_i \cap S$ separates a disc Q_i' from S with $Q_i' \cap k = \emptyset$. The union $Q_i \cup Q_i'$ is a disc in M whose boundary lies in $\partial M - k$, and this disc separates a 3-ball from M which does not meet k (recall (M, k) is ∂-irreducible). Thus there is an admissible isotopy across this 3-ball which deforms S into the surface $(S - Q_i')^- \cup Q_i$. Clearly, the latter surface has a piping decomposition $(\mathcal{B}^*, \mathcal{P}^*, \mathcal{A}^*)$ with $(\#\mathcal{B}^*, \#\mathcal{P}^*, \#\mathcal{A}^*) < (\#\mathcal{B}, \#\mathcal{P}, \#\mathcal{A})$, and we are done.

Suppose r_i or r_{i+1} is contractible. Then it follows from our maximality condition on the index i that r_i must be contractible.

Suppose r_i but not r_{i+1} is contractible. Then consider the surface S^+ obtained from S' by pruning along A_0. Note that r_i and r_{i+1} are still boundary curves of S^+ (otherwise S would be one sided and so non-orientable). So Q_i is a disc with $Q_i \cap S^+ = (\partial Q_i - \partial M)^-$ and $Q_i \cap k = \emptyset$. But Q_i may well be (and in fact is) a ∂-compression disc for S^+ and so we cannot use the argument above. The trick in this situation is to observe that, modulo admissible isotopy, pruning the curve r_i is the same as ∂-compressing along Q_i (the picture on the left of figure 11.1 illustrates this curious fact). But as above we see that splitting along Q_i reduces $(\#\mathcal{B}, \#\mathcal{P}, \#\mathcal{A})$, and so we are done.

Suppose both r_i and r_{i+1} are contractible. Let A_i, A_{i+1} denote the discs from $\mathcal{A}$ with $r_i = \partial A_i$ and $r_{i+1} = \partial A_{i+1}$. We know that $A_i \ne A_0$ (see above). If also $A_{i+1} \ne A_0$, then we can argue as before. So suppose $A_{i+1} = A_0$. Let a_0' be the component of $A_0 \cap \partial B$ joining z_i with z_{i+1}. Note a_0 can be pushed (rel ∂a_0) across A_0 and into ∂A_0. Thus the disc Q_i may be viewed as a ∂-compression disc for S'. Moreover, r_i and r_{i+1}

are two concentric circles in the disc A_i joined by the arc a_0'. The picture on the right of figure 11.1 gives a schematic illustration of this situation. Note that pruning S' along A_i and A_{i+1} gives the same surface (mod admissible isotopy) than first splitting S' along Q_i and then pruning the result along the disc $(A_{i+1} - (A_i \cup U(a_0')))^-$. Thus we obtain S by first splitting S' along Q_i and then pruning the resulting surface. Hence, as above, S is again admissibly isotopic to a surface which has a piping decomposition which is smaller than $(\mathcal{B}, \mathcal{P}, \mathcal{A})$. Altogether, the claim is therefore established.

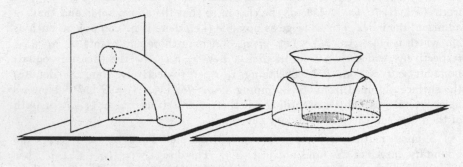

(Figure 11.1.)

The proof of lemma 11.6 is now easy to finish. We only have to verify that the existence of a bounded switch produces the situation described above. So let b be a bounded switch for $(\mathcal{B}, \mathcal{P}, \mathcal{A})$. Let B be the disc from $\mathcal{B}$ containing the arc b and let P_1, P_2, be the two adjacent pipes from $\mathcal{P}_B$ joined by b. Let r_0 be the component of ∂S_0 containing b. Then r_0 bounds a disc, say A_0, in $\partial M - k$ (i.e., A_0 is from $\mathcal{A}$). Since A_0 does not meet k, there must be a pipe P_3 meeting A_0 and passing through P_2, say. Hence we have indeed the above situation. The following picture illustrates this situation.

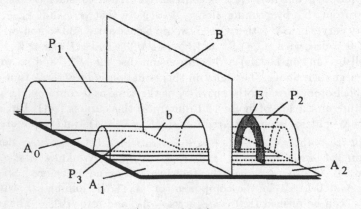

(Figure 11.2.)

By the argument above the conclusion of the lemma follows. ◇

11.7. Before we proceed we need to refine the complexity $l_{\mathcal{D}}[S]$ somewhat. The refined complexity will be a tuple of integers rather than a single integer. To define it let us be given an essential 2-manifold S in (M, k) with piping decomposition $(\mathcal{B}, \mathcal{P}, \mathcal{A})$ say. Let $\mathcal{B}' \subset \mathcal{B}$ be the disc-system defined in 11.5. Recall that $\mathcal{P} \cap \mathcal{B}' = (\partial \mathcal{B}' - \partial B)^-$ and that therefore $S' := \mathcal{B}' \cup \mathcal{P}$ is a 2-manifold in (M, k). We get S from S' by pruning (along $\mathcal{A}$). Set $B' := B \cap \mathcal{B}'$, for every disc B from $\mathcal{B}$. Notice that every component of $B' \cap \partial B$ lies in some boundary component of S' and this boundary component may or may not be compressible. Let b' be the union of all those components of $B' \cap \partial B$ which lie in *compressible* components of $\partial S'$. Denote

$$B'' := (B' - U(b'))^-,$$

where the regular neighborhood is taken in B. Given this notation, set

$$\mathcal{B}_i := \{\, B \mid \#(B'' \cap \partial B) = i \,\}$$

and define

$$c(\mathcal{B}) := (\ \#\mathcal{B}_0, \ \#\mathcal{B}_1, \ \dots \ ).$$

Of course, $c(\mathcal{B}^*) < c(\mathcal{B})$ (w.r.t. the lexicographical order) if $\mathcal{B}^* \subset \mathcal{B}$ is a basis, supporting S, with $\#\mathcal{B}^* < \#\mathcal{B}$ - but not vice versa. Set

$$c(S) := \min c(\mathcal{B}),$$

where the minimum is taken over all basis $\mathcal{B} \subset U(\mathcal{D}^+)$ supporting S. Define

$$c[S] := \min c(S'),$$

where the minimum is taken over all 2-manifolds S' in $M(\underline{D})$ Dehn-equivalent to S.

11.8. Lemma. *Let $\underline{D}$ be a full pant-diagram. Let S be an essential 2-manifold in $M(\underline{D})$. Suppose $\mathcal{B} \subset U(\mathcal{D}^+)$ is a basis, supporting S, with $c(\mathcal{B}) = c[S]$. Then*

$$\#\mathcal{B}_0 = 0, \quad \#\mathcal{B}_1 = 0, \quad \#\mathcal{B}_2 \le \varphi(-\chi S, -\chi \partial M), \quad \text{and} \quad \#(\mathcal{B} - \mathcal{B}_2) \le -2\chi S,$$

where $\varphi : \mathbf{N} \times \mathbf{N} \to \mathbf{N}$ is a map which is strictly increasing in both variables. For instance, we may take $\varphi(x, y) := (1 + x)y \cdot 3^{2+y}$.

Proof. Since S is supported by $\mathcal{B}$, there is piping decomposition $(\mathcal{B}, \mathcal{P}, \mathcal{A})$ for S (see above).

Assume $\mathcal{B}_0 \cup \mathcal{B}_1 \ne \emptyset$. Then we claim there is a basis $\mathcal{B}^* \subset U(\mathcal{D}^+)$ with $c(\mathcal{B}^*) < c(\mathcal{B})$ which supports S. Here is the construction (for the upcoming discussion consult fig. 11.3). Let B be a disc from $\mathcal{B}_0 \cup \mathcal{B}_1$. Denote by $\mathcal{P}_B$ the system of all pipes from $\mathcal{P}$ starting (or ending) in B. Since $c(\mathcal{B}) = c[\mathcal{S}]$ and since B is from $\mathcal{B}_0 \cup \mathcal{B}_1$, it follows from lemma 11.6 that all pipes from $\mathcal{P}_B$ must leave B in the same direction. Since $\underline{D}$ is a pant-diagram, $\mathcal{D}^+$ splits ∂M into pairs of pants (i.e., into thrice punctured 2-spheres). Let Q be that one of them which meets B from the same side as $\mathcal{P}_B$. Let B_1, B_2 be the two discs from $\mathcal{D}^+$ bound by the two curves from $\partial Q - \partial B$. Denote by $E(Q)$ the 3-ball in M bound by the 2-sphere $Q \cup B \cup B_1 \cup B_2$. W.l.o.g. $B \cap E(Q) = B$. The intersection of the surface S with the disc B is the union of one disc, namely $B' \in \mathcal{B}'$, and a collection of arcs. Let S_B be that component of $S \cap E(Q)$ which contains B'. Note that $(\partial U(S_B) - \partial E(Q))^-$ consists of two discs in $E(Q)$, where $U(S_B)$ is the regular neighborhood in $E(Q)$. Let F_B be that one of them which does not meet B. Let E', E'' be the two 3-balls into which F_B splits $E(Q)$. W.l.o.g. $S_B \subset E'$. Now choose a disc $D \subset E(Q)$ with $D \cap \partial E(Q) = \partial D$ and $D \cap \mathcal{P}_B = \emptyset$ which separates B_1 and B_2 and whose intersection with B and every component of $k \cap Q$ is connected. Let D_1, D_2 denote the two components of $(\partial U(D) - \partial E(Q))^-$, where $U(D)$ is the regular neighborhood in $E(Q)$. Let $B_i^* := D_i \cup (B - U(D))^-$, $i = 1, 2$. Then B_1^*, B_2^* are discs admissibly parallel to B_1, B_2, respectively.

Suppose E'' does not contain a component of $k \cap Q$. Then $D \cap S = D \cap B$ is an arc in S which separates a disc from S (S is essential in $M(\underline{D})$). Thus one component of $S^* := (S - U(D))^-$ is a disc, say S_1^*. W.l.o.g. $B_1^* \subset S_1^*$. But S_1^* must be ∂-parallel in $M(\underline{D})$ since $M(\underline{D})$ is ∂-irreducible ($\underline{D}$ is a full pant-diagram). Therefore $S^* - S_1^*$ is admissibly isotopic to S. Moreover, $S^* - S_1^*$ is supported by some basis $\mathcal{B}^* \subset (\mathcal{B} - B) \cup B_2^*$ with $\#\mathcal{B}^* < \#\mathcal{B}$. So $c(\mathcal{B}^*) < c(\mathcal{B})$.

Suppose E'' does contain at least one component of $k \cap Q$. Then $F_B \cap U(D)$ may be interpreted as a pseudo-pipe, say P', joining B_1^* with B_2^* (to see a typical situation pull along the little arrows in fig. 11.3). It follows that S is supported by $\mathcal{B}^* := (\mathcal{B} - B) \cup B_1^* \cup B_2^*$. But $B \in \mathcal{B}_0 \cup \mathcal{B}_1$ and $B_1^*, B_2^* \notin \mathcal{B}_0^* \cup \mathcal{B}_1^*$. Hence $c(\mathcal{B}^*) < c(\mathcal{B})$.

Thus, in any case, we get a contradiction to the hypothesis that $c(\mathcal{B}) = c[\mathcal{S}]$. Therefore our assumption is false and so $\mathcal{B}_0 \cup \mathcal{B}_1 = \emptyset$.

We next estimate $\#\mathcal{B}_2$. To do this, recall $\mathcal{B} \subset U(\mathcal{D}^+)$ and note that $\mathcal{D}^+$ consists of $-\frac{3}{2}\chi\partial M = 3 \cdot \text{genus}(\partial M) - 3$ discs. Thus it suffices to show that the maximal number of pairwise admissibly parallel discs from $\mathcal{B}_2$ cannot exceed $2 \cdot (1 - \chi S) \cdot 3^{1 - \chi \partial M}$. To show this, let $B_1, B_2, ..., B_m$ be any collection of pairwise admissibly parallel discs from $\mathcal{B}_2$. For every $1 \le i \le m$, choose a proper arc b_i in B_i'' (see 11.7 for the definition of B'') which joins the two components from $B_i'' \cap \partial B_i$ (recall B_i is from $\mathcal{B}_2$). All these arcs are certainly parallel in (M, k). More precisely, there is a square A in M for which the following holds w.l.o.g.:

(1) $A \cap \partial M$ consists of two arcs and both of them are contained in $\partial M - k$,

(2) $(\partial A - \partial M)^- = b_1 \cup b_m$, and

(3) $A \cap (B_1 \cup ... \cup B_m) = b_1 \cup ... \cup b_m$.

The arcs $b_1, ..., b_m$ are proper arcs in S. In fact, they are all essential in S since $\underline{D}$ is supposed to be a full pant-diagram. The intersection cannot be reduced by some isotopy of S which is constant on $(\partial A - \partial M)^-$. (Otherwise we would find, by studying the pre-image of A under the isotopy $S \times I \to M$, that $S \cap A$ separates a square from S which, together with some square in A, forms an inessential annulus in (M, k). This in turn is impossible since $\underline{D}$ is full). Now if $m > 2 \cdot (1 - \chi S) \cdot 3^{1 - \chi \partial M}$, then, by lemma 7.17, there is a Dehn twist (along some appropriate strongly essential annulus in $U(A \cup S)$) which turns S into a 2-manifold S^* with $\#(S^* \cap \mathcal{D}^+) < \#(S \cap \mathcal{D}^+)$. By a similar argument as used before, S^* is supported by some basis $\mathcal{B}^* \subset U(\mathcal{D}^+)$ with $\#\mathcal{B}^* < \#\mathcal{B}$. Thus $c(\mathcal{B}^*) < c(\mathcal{B})$. But this is impossible since $c(\mathcal{B}) = c[S]$. Therefore our assumption is wrong and so $\#\mathcal{B}_2 \leq (\chi S - 1)\chi \partial M \cdot 3^{2 - \chi \partial M}$.

Finally we prove $\#(\mathcal{B} - \mathcal{B}_2) \leq -2\chi S$. To show this claim we may suppose S is connected. Let $\mathcal{B}'' = \bigcup_{B \in \mathcal{B}} B''$ (see 11.7 for the definition of $B'' \subset B$). By definition, $\mathcal{B}'' \subset S$. Moreover, $\mathcal{E} := (S - \mathcal{B}'')^-$ is also a system of discs (recall $\mathcal{B}_0 \cup \mathcal{B}_1 = \emptyset$) and $\mathcal{L} := \mathcal{B}'' \cap \mathcal{E}$ is a system of proper arcs in S (since $\mathcal{E}$ is the union of all pipes and compression discs from the piping decomposition of S). Now consider the decomposition $S = \mathcal{B}'' \cup \mathcal{E}$. Note that there is a sub-system $\mathcal{L}' \subset \mathcal{L}$ of arcs such that $\tilde{S} := (S - U(\mathcal{L}'))^-$ is a single disc. If $\mathcal{L} \cap \tilde{S} \neq \emptyset$, there is a disc $D \subset \tilde{S}$ such that $\ell := D \cap \mathcal{L} = (\partial D - \partial \tilde{S})^-$. Clearly, $\#(U(\mathcal{L}') \cap \partial \tilde{S}) \geq \#(U(\mathcal{L}' \cup \ell) \cap \partial(S - D)^-)$ with strict inequality whenever D contains a disc from $\mathcal{B} - \mathcal{B}_2$ (recall again that $\mathcal{B}_0 \cup \mathcal{B}_1 = \emptyset$). Thus, inductively, $\#(U(\mathcal{L}') \cap \partial \tilde{S}) \geq 2 + \#(\mathcal{B} - \mathcal{B}_2)$. So $-2\chi S = 2 \cdot \#\mathcal{L}' - 2 = \#(U(\mathcal{L}') \cap \partial \tilde{S}) - 2 \geq 2 + \#(\mathcal{B} - \mathcal{B}_2) - 2 = \#(\mathcal{B} - \mathcal{B}_2)$. The claim is therefore established.

This finishes the proof of the lemma. $\Diamond$

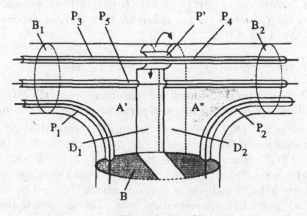

(Figure 11.3.)

After this preparation we are now ready to establish the following estimate concerning the geometric length of 2-manifolds.

11.9. Proposition. *There is a map* $\psi : \mathbf{N} \times \mathbf{N} \to \mathbf{N}$, *strictly increasing in both variables, such that*

$$l_D[S] \leq \psi(-\chi S, -\chi \partial M),$$

for every every full pant-diagram $\underline{D}$ *and every essential 2-manifold* S *in* $M(\underline{D})$. *More precisely, we may take* $\psi(x, y) := 6(1 + x)y \cdot 3^{3+y}$.

Remark. $\mathcal{D} \subset \mathcal{D}^+$ denotes some fixed meridian-system for $M(\underline{D})$.

Proof. Suppose S has been chosen within its Dehn-equivalence class and a basis $\mathcal{B} \subset U(\mathcal{D})$, supporting S, has been chosen so that $c(\mathcal{B}) = c[S]$. Then, by an argument from cor. 9.13, it follows that S is also supported by some basis $\mathcal{B}^* \subset U(\mathcal{D})$ with $\#\mathcal{B}^* \leq -\frac{3}{2}\chi \partial M \cdot \#\mathcal{B}$, where $\mathcal{D} \subset \mathcal{D}^+$ is a meridian-system in $\mathcal{D}^+$. This proves that $l_D[S] \leq -\frac{3}{2}\chi \partial M \cdot \#\mathcal{B}$. Using this and the estimates from the previous lemma, we have

$$l_D[S] \leq -\tfrac{3}{2}\chi \partial M \cdot \#\mathcal{B} \leq -\tfrac{3}{2}\chi \partial M \cdot (-2\chi S + \#\mathcal{B}_2)$$
$$\leq \tfrac{3}{2}(-\chi \partial M) \cdot (-2\chi S + \varphi(-\chi S, -\chi \partial M)),$$

where $\varphi : \mathbf{N} \times \mathbf{N} \to \mathbf{N}$ is the map from 11.8. This finishes the proof. $\Diamond$

Recall that a basis $\mathcal{B}$ for an essential 2-manifold S comes equipped with a normal orientation (induced by one for S). The normal vectors of parallel discs from $\mathcal{B}$ may or may not point in the same direction, but recall that if they all do, the basis $\mathcal{B}$ is called *coherent*. Any essential 2-manifold coming from a solution of $A(M, k)x = 0$ has, by definition, a coherent basis. The following observation concerning non-coherent basis is crucial for our purpose.

11.10. Proposition. *Let* $\underline{D}$ *be a full pant-diagram. Let* S *be a* ∂-*reduced, non-separating 2-manifold in* $M(\underline{D})$. *Let* $\mathcal{B} \subset U(\mathcal{D})$ *be a basis, supporting* S, *with* $\#\mathcal{B} = l_D(S)$. *If* $\mathcal{B}$ *is not coherent, then there is a 2-manifold* S^+, *homologous to* S, *such that*

$$l_D(S^+) \leq l_D(S) - 2, \quad and \quad \chi(S^+) \geq \chi(S) - 2.$$

Remark. (1) $\mathcal{D} \subset \mathcal{D}^+$ denotes some fixed meridian-system for $M(\underline{D})$.
(2) A 2-manifold in a 3-manifold is non-separating, if one of its components is.

Proof. To fix ideas let $(\mathcal{B}, \mathcal{P}, \mathcal{A})$ be the piping decomposition of S. Recall from 11.5 the definition of the disc-system $\mathcal{B}' \subset \mathcal{B}$ with $\mathcal{P} \cap \mathcal{B}' = (\partial \mathcal{B}' - \partial \mathcal{B})^-$. Define $B' := B \cap \mathcal{B}'$, for every $B \in \mathcal{B}$, and set $S' := \mathcal{B}' \cup \mathcal{P}$. Since $\mathcal{B}$ is supposed to be non-coherent, there must be at least two discs from $\mathcal{B}$, say B_1, B_2, which are parallel and whose normal vectors (induced by some normal orientation of S) point into opposite directions. Let E denote the parallelity region between B_1 and B_2. Then $E \cap \partial M$ is an annulus and $k \cap E$ is a system of essential arcs in this annulus. Let t be a component of $k \cap E$. Since

$M(\underline{D})$ is ∂-irreducible ($\underline{D}$ is a full pant-diagram) and since $\#\mathcal{B}$ is minimal, it follows from prop. 10.16 (3) that t cannot be a piping arc. Therefore there are two piping-arcs, $t_1, t_2 \subset k$ (possibly equal), with $t \cap t_i = t \cap \partial t_i \subset \partial t$, $i = 1, 2$. Let P_i, $i = 1, 2$, denote the pipe associated to t_i.

t may or may not be contained in $t_1 \cup t_2$ (see fig. 11.4). In any case, however, we can construct a pipe, say P'_t, along t which joins B'_1 with B'_2. Let $P'_E = \bigcup P'_t$, where the union is taken over all components t from $k \cap E$. Set

$$S^* := (S' - U(P'_E))^- \cup (\partial U(P'_E) - S')^-,$$

minus the component contained in E (see fig. 11.4). Denote by S^+ the 2-manifold obtained from S^* by pruning. Note that S^*, and so S^+ as well, is an orientable 2-manifold since, by our choice of B_1 and B_2, all pipes P_t meet the (orientable) 2-manifold S' from one side only. Of course, S^+ is (strongly) homologous to S. We claim S^+ is the required 2-manifold.

Type I Process

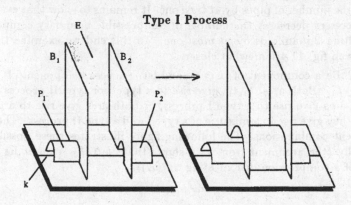

Type II Process

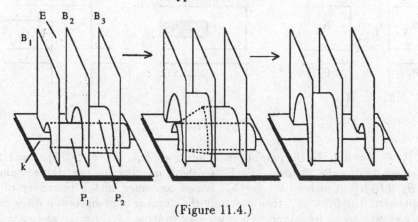

(Figure 11.4.)

By construction, S^+ is supported by the basis $\mathcal{B} - (B_1 \cup B_2)$ and, by hypothesis, $\#\mathcal{B} = l_{\mathcal{D}}(S)$. So $l_{\mathcal{D}}(S^+) \leq l_{\mathcal{D}}(S) - 2$.

It remains to estimate χS^+ in terms of χS. To this end note that the Euler characteristic of S can be calculated from the piping decomposition $(\mathcal{B}, \mathcal{P}, \mathcal{A})$ of S. Explicitly, we have

$$\chi(S) = \#\mathcal{B} - \#\mathcal{P} + \#\mathcal{A}$$

To calculate χS^+ we therefore have to describe the piping decomposition of S^+. But S^+ is obtained from S^* by pruning and S^* is obtained from S' by successively applying one of the two processes indicated in fig. 11.4. Each one of the latter two processes merges two (possibly equal) pipes from $\mathcal{P}$ into one band and then straightens this band into a pipe (or an annulus which we then forget). Thus a piping decomposition for S^+, say $(\mathcal{B}^+, \mathcal{P}^+, \mathcal{A}^+)$, is uniquely determined by the processes at hand. Every one of the above processes diminishes the number of pipes by *at least* one. It remains to show that every one of these processes decreases the number of compressible boundary components of the resulting 2-manifolds by *at most* one. To this end we examine the two processes from fig. 11.4 somewhat closer.

Let t_0 be a component of $k \cap E$ and let t_1 be a component of $k \cap E$ adjacent to t_0. Both arcs t_0, t_1 give rise to a type I or type II process. Thus either both arcs give rise to a type I process, or both arcs give rise to a type II process, or they give rise to a mixture of a type I and a type II process. These are three different combinations. The following figure illustrates these possibilities schematically (the argument from 11.6 shows that $b_1 \cup b_2$ cannot lie in the boundary of a compression disc meeting $int(E)$).

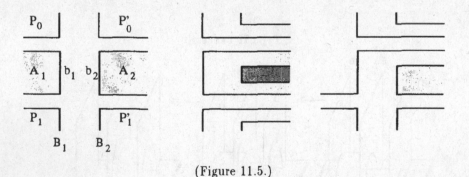

(Figure 11.5.)

Let P_i, P_i', $i = 0, 1$, be the pair of pipes ending at ∂t_i. Let $b_j \subset \partial B_j$, $j = 1, 2$, be the arc joining P_0, P_1 resp. P_0', P_1' (without meeting any other pipe ending in $B_1 \cup B_2$). If either b_1 or b_2 is not contained in the boundary of a compression disc from $\mathcal{A}$, then certainly the number of compression discs can only go down by at most one. Thus we may suppose b_j, $j = 1$ *and* $j = 2$, lies in the boundary of some compression-disc, say A_j, from $\mathcal{A}$. Then, in particular, t_0 and t_1 have to give rise to the same process (either I or II). Otherwise b_1, say, had to be a bounded switching arc. But, according to lemma 11.6, this is impossible since $\#\mathcal{B} = l_{\mathcal{D}}(S)$. Now suppose both t_0 and t_1 give rise to a type I process. Then either (1) $A = A'$ and A will be forgotten, or (2) $A \neq A'$ and then A and A' will be merged to one compression-disc for the new surface. In any of these two case the number of compression discs goes down by at most one. The same is true if both t_0, t_1 give rise to a type II process.

This shows that in any case the number of compression discs goes down by at most one (it might even go up). Applying the same argument to the process associated to the arc next to t_1 and so on, it follows inductively that $\mathcal{P}^+ - \mathcal{A}^+ \leq \mathcal{P} - \mathcal{A}$. Therefore, we have

$$\chi(S^+) = \#(\mathcal{B} - (B_1 \cup B_2)) - \#\mathcal{P}^+ + \#\mathcal{A}^+ \geq (\#\mathcal{B} - \#\mathcal{P} + \#\mathcal{A}) - 2 = \chi(S) - 2.$$

This proves the proposition. $\Diamond$

11.11. Applications.

As an application of the calculations made so far, we can now give a useful estimate for the Seifert characteristic and the Thurston norm. The following theorem is our main result in this direction. It estimates the Seifert characteristic and the Thurston norm in terms of the algebraic length. This estimate allows us to identify certain relative handlebodies (and Heegaard diagrams) as "complicated" because they contain only complicated non-separating surfaces. However, the proposition says nothing about those relative handlebodies whose homology is trivial, i.e., those relative handlebodies which contain no non-separating surfaces at all.

11.12. Theorem. *There is a known function* $\varphi : \mathbf{N} \times \mathbf{N} \to \mathbf{N}$, *strictly increasing in both variables, such that*

$$\| M(\underline{D}) \|_{\mathcal{D}} \leq \varphi(\sigma M(\underline{D}), -\chi \partial M(\underline{D})),$$

for every full pant-diagram $\underline{D}$, *and that*

$$\| S \|_{\mathcal{D}} \leq \varphi(\tau(S), -\chi \partial M(\underline{D})),$$

for every full and simple pant-diagram $\underline{D}$ *and every 2-manifold* S *in* $M(\underline{D})$.
Proof. $\mathcal{D}$ denotes a meridian-system for $M(\underline{D})$ contained in the disc-system associated to $\underline{D}$.
Proof. We first prove the second statement. For this let S be chosen within its homology class so that $-\chi S = \tau(S)$. Then S is essential in $M(\underline{D})$ since χS is maximal. Since $\underline{D}$ is a simple diagram, there are no Dehn twists along essential annuli. Thus we may suppose that S is chosen so that the above holds and that, in addition, $l_{\mathcal{D}}(S) = l_{\mathcal{D}}[S]$. Applying prop. '11.10, we get

a 2-manifold S^*, homologous to S and supported by some *coherent* basis $\mathcal{B} \subset U(\mathcal{D})$ such that

$$l_{\mathcal{D}}[S^*] = \#\mathcal{B} \quad \text{and} \quad \chi S^* \geq \chi S - l_{\mathcal{D}}(S) = \chi S - l_{\mathcal{D}}[S].$$

Thus, by prop. 11.9, we have

$$\chi S^* \geq \chi S - \psi(-\chi S, -\chi \partial M),$$

for some known function $\psi : \mathbf{N} \times \mathbf{N} \to \mathbf{N}$, strictly increasing in both variable. Since $\mathcal{B}$ is a coherent basis, it represents a solution of $A(M, k)x = 0$. Thus, by definition of the algebraic length and by prop. 11.9 again, we obtain

$$\| S^* \| \leq l_{\mathcal{D}}[S^*] \leq \psi(-\chi S^*, -\chi \partial M) \leq \psi(-\chi S + \psi(-\chi S, -\chi \partial M), -\chi \partial M).$$

So $\varphi(-\chi S, -\chi \partial M) := \psi(-\chi S + \psi(-\chi S, -\chi \partial M), -\chi \partial M)$ satisfies the second statement.

To prove the first statement let a non-separating surface S in (M, k) be chosen so that $-\chi S = \sigma M(\underline{D})$. Since the Euler characteristic of 2-manifolds is invariant under Dehn twists along annuli, we may further suppose that S is chosen within its Dehn-equivalence class so that the above holds and that, in addition, $l_{\mathcal{D}}(S) = l_{\mathcal{D}}[S]$. Then it follows as above that $\| M(\underline{D}) \| \leq \| S \| \leq \varphi(-\chi S, -\chi \partial M) \leq \varphi(\sigma M(\underline{D}), -\chi \partial M)$. $\Diamond$

11.13. Example. In view of the previous proposition the Seifert characteristic of a relative handlebody (M, k) can be estimated by algebraic means. For this we simply have to determine those integer lattice-vectors in the kernel of the associated Heegaard-matrix $A_{\mathcal{D}}(M, k)$ which have minimal norm $\| \cdot \|_{\mathcal{D}}$. To illustrate this consider the following relative handlebodies.

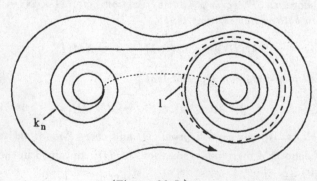

(Figure 11.6.)

More precisely, consider the sequence of curves k_n, on the above handlebody M of genus two, obtained by twisting around l. The Heegaard-matrix $A_{\mathcal{D}}(M, k_n)$ is obviously given by $[2, n]$. So minimal solution-vectors are given by $(n, -2)$.

Thus $\| (M, k_n) \|_{\mathcal{D}} = 2 + n$. So $\| (M, k_n) \|_{\mathcal{D}}$ tends to infinity, for $n \to \infty$. By thm. 11.12, this implies that also the Seifert characteristic $\sigma(M, k_n)$ tends to infinity.

Incidentally, in the special situation at hand, the latter property can be verified independently by appealing to known facts instead. Indeed, notice that we get Seifert fiber spaces by attaching a 2-handle along the curves k_n in the above relative handlebody. More precisely, we get Seifert fiber spaces over the disc with two exceptional fibers (of order 2 and n, respectively). So the above property follows also from the fact [Wa 1] that every incompressible and non-separating surface in such a Seifert fiber space is horizontal (mod isotopy).

11.14. Remarks. (1) See [Ko 5] for yet another criterion which forces the Seifert characteristics of relative handlebodies to go to infinity. In [Ko 5] it is also shown that the iterates of a curve on the boundary of a handlebody M under a pseudo-Anosov diffeomorphism $h : \partial M \to \partial M$ often yield sequences of relative handlebodies for which even the minimal genera (and not just the Euler characteristics) of admissible surfaces in $(M, h^n k)$ tend to infinity.

(2) See chapter III and IV for related estimates for Seifert characteristics of 3-manifolds.

11.15. *Surface-Systems.*

Given a 2-manifold, S, let $\#S$ be the number of components of S. Set $\chi_{min} S := \min \chi S$, where the minimum is taken over all components of S.

11.16. Proposition. *There is an arithmetic function $\varphi : \mathbf{N} \times \mathbf{N} \to \mathbf{N}$ such that the following holds. Let M be a handlebody and let S be an incompressible 2-manifold in M, $S \cap \partial M = \partial S$. Suppose that no component of S is ∂-parallel and that no two components of S are parallel. Then $\#S \leq \varphi(-\chi_{min} S, -\chi \partial M)$.*

Proof. Let S be as in the proposition. Throughout this proof a 2-manifold $S' \subset S$ consists of components of S. A 2-manifold $S' \subset M$ is *good* if there is a system $A \subset \partial M$ of annuli with the property that every component of A contains a boundary curve of every component of S. A is then called an *annulus system associated to* S.

There is a good 2-manifold $S' \subset S$ with $\#S \leq k(M) \cdot \#(S')$, $k(M) := 2^n$, $n = -\frac{3}{2}\chi \partial M$. To see this let A be any system of non-contractible annuli in ∂M with $\partial S \subset A$. Let A^+ be the union of A with all annulus-components of $(\partial M - A)^-$. Then $\#A^+ \leq -\frac{3}{2}\chi \partial M = n$. Consider S and A as *sets* of components. Define a map $f : S \to \mathcal{P}(A^+)$ (power set of A^+) by assigning to any component, F, of S the set of all components of A^+ containing components from ∂F. We have $S = \bigcup f^{-1}A$, $A \in \mathcal{P}(A)$. Let $A' \in \mathcal{P}(A)$ with $\#f^{-1}A'$ maximal. Then $\#S \leq \#\mathcal{P}(A) \cdot \#f^{-1}A' = 2^n \cdot \#f^{-1}A'$. Thus $S' := f^{-1}A'$ satisfies the claim.

Every incompressible surface in a handlebody is either a disc or has at least one ∂-compression disc. We use this fact to define a complexity for S as follows.

First, we call a disc $B \subset M$ an *adjusted ∂-compression disc for* $\mathcal{S}$, provided (1) there is a component $\mathcal{S}_0$ from $\mathcal{S}$ so that B is a ∂-compression disc for $\mathcal{S}$, and that (2) B is isotoped (rel $B \cap \mathcal{S}_0$) so that $\#(B \cap \mathcal{S})$ is as small as possible. Let $c(B, \mathcal{S})$ be the total number of components from $\mathcal{S}$ meeting B. Finally set $c(\mathcal{S}) := \max c(B, \mathcal{S})$, where the maximum is taken over all adjusted ∂-compression discs, B, for $\mathcal{S}$.

We claim $\#\mathcal{S} \le -3\chi\partial M \cdot c(\mathcal{S})$. To see this recall that $\partial\mathcal{S}$ is contained in an annulus-system $\mathcal{A}$ with $\#\mathcal{A} \le -\frac{3}{2}\chi\partial M$. At most $-3\chi\partial M$ surfaces from $\mathcal{S}$ can have a ∂-compression disc B with $c(B) = 1$. To see this notice that any such surface must have a boundary curve next to a boundary curve of $\mathcal{A}^+$ and that $\mathcal{A}^+$ has at most $-2 \cdot \frac{3}{2}\chi\partial M$ boundary curves. Thus, forgetting at most $-3\chi\partial M$ components from $\mathcal{S}$, we obtain a 2-manifold $\mathcal{S}'$ with $c(\mathcal{S}') = c(\mathcal{S}) - 1$. The estimate for $\#\mathcal{S}$ then follows by induction.

Suppose $\mathcal{S}$ is good and suppose B is an adjusted ∂-compression disc for $\mathcal{S}$. Then we claim $B \cap \mathcal{S}$ is a system of pairwise parallel arcs in B. Assume the converse. Consider an annulus-system $\mathcal{A}$ associated to $\mathcal{S}$. It follows from the choice of B and $\mathcal{A}$ that $B \cap \mathcal{A}$ consists of exactly two arcs. Since $\mathcal{S}$ is incompressible and since B is adjusted, it follows further that $B \cap \mathcal{S}$ consists of arcs (whose end-points lie in $B \cap \mathcal{A}$). By assumption, there is at least one of these arcs which separates a disc B' from B with $B' \cap \partial B \subset \mathcal{A}$. W.l.o.g. $B' \cap \mathcal{S} = (\partial B' - \partial B)^-$. There is an annulus $A' \subset \mathcal{A}$ with $A' \cap \mathcal{S} = \partial A'$ and such that $B' \cap \mathcal{A}$ is an essential arc in A'. $(A' - U(B'))^- \cup B_1' \cup B_2'$ is a disc, where B_1', B_2' are the two copies of B' in $\partial U(B')$. Since $\mathcal{S}$ is incompressible and M is irreducible, it follows from the existence of this disc that at least one component of $\mathcal{S}$ is a ∂-parallel annulus; contradicting the hypothesis that no component of a good 2-manifold is ∂-parallel.

Now select a good 2-manifold $\mathcal{S}_1 \subset \mathcal{S}$ whose number of components is as large as possible. Let B_1 be an adjusted ∂-compression disc for $\mathcal{S}_1$ with $c(B_1) = c(\mathcal{S}_1)$. Let $\mathcal{S}_1'$ be the collection of all surfaces from $\mathcal{S}_1$ meeting B_1. Set $\ell_1 := (\partial B_1 - \partial M)^-$ and $M_1 := (M - U(\ell_1))^-$. Then M_1 is a handlebody whose genus is $1 + \text{genus}(M)$. W.l.o.g. we may suppose that $\mathcal{S}_1' \cap B \subset U(\ell_1)$. Select a submanifold $\mathcal{S}_2$ from $\mathcal{S}_1'$ such that $\mathcal{S}_2 \cap M_1$ is good in M_1 and such that the number of components of $\mathcal{S}_2$ is as large as possible. Repeating the above process with $\mathcal{S}_2$ instead of $\mathcal{S}_1$ we get a handlebody M_2 and a sub-system $\mathcal{S}_2 \subset \mathcal{S}_1 \subset \mathcal{S}$. Thus, after at most $m = -\chi_{min}\mathcal{S}$ steps, we get a good 2-manifold $\mathcal{S}_m \subset \mathcal{S}$ and a handlebody M_m such that (1) $\text{genus}(M_m) = \text{genus}(M) + m$ and that (2) $\mathcal{S}_m \cap M_m$ consists of discs.

There is a system $\mathcal{D}_m$ of at most $-\frac{3}{2}\chi\partial M_m$ discs in M_m with $\mathcal{S}_m \cap M_m \subset X_m := U(\mathcal{D}_m)$. We consider X_m as an I-bundle over a (possibly disconnected) 2-manifold. It is easily verified that X_m fits over $U(\ell_m)$ to form an I-bundle X_{m-1} with $\mathcal{S}_m \cap M_{m-1} \subset X_{m-1}$. Inductively, it follows that $\mathcal{S}_m$ is contained (as a horizontal 2-manifold) in an I-bundle X_1 with at most $-\chi_{min}\mathcal{S}$ components. This proves that $\#\mathcal{S}_m \le -\chi_{min}\mathcal{S}$. A function φ as required for the proposition is now easy to manufacture from the above estimates. $\diamond$

Taking the above proposition together with the Rigidity Theorem 31.22, we get

11.17. Corollary. *There is an arithmetic function* $g : \mathbf{N} \times \mathbf{N} \to \mathbf{N}$ *with the following property. If* S *is an incompressible 2-manifold in an irreducible 3-manifold* M, *then* $\#S \leq g(\mathrm{genus}(M), \chi_{min}S)$. ◊

11.18. Remarks. (1) This corollary is related to the finiteness result of [Ha 3].

(2) For systems S of tori the function g can be taken to be $3 \cdot \mathrm{genus}(M) - 3$ [Ko 4].

(3) According to [Ja 2,III.24], g can always be taken to be a function in the Heegaard genera, $\mathrm{genus}(M)$, alone.

11.19. Example.

We conclude this section with an example which shows that there may well be homologous surfaces S, S' in a relative handlebody with

$$l_{\mathcal{D}}(S) > l_{\mathcal{D}}(S') \quad \text{and} \quad |\chi S| < |\chi S'|,$$

i.e., we may decrease the absolute Euler characteristic of a surface by increasing the basis supporting it; provided we allow non-coherent basis.

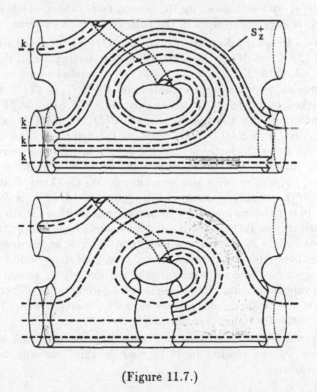

(Figure 11.7.)

§12. Algorithms for Surfaces.

This section is concerned with algorithmic problems. Specifically, it addresses the problem whether a given surface in a relative handlebody is essential, boundary-parallel or isotopic to another given (essential) surface. It turns out that all these questions can be settled relatively easy and relatively fast, provided the surfaces are given in terms of circle-patterns. In fact, this section offers algorithms which decides these question in at most quadratic time (12.21, 12.30 and 12.24).

12.1. *Preliminaries.*

In this section we begin a careful analysis of the compression problem in its most basic form. The problem will resurface in various other forms throughout this book. In the end we will deal with the extreme case of Heegaard surfaces, i.e. with surfaces which are totally compressible on both sides. In the present setting, we will distinguish between disc-compressions and ∂-compressions for surfaces. More precisely, we will first study disc-compressions for ∂-incompressible surfaces and then ∂-compressions. The reason for this being technical and due to the fact that disc-compressions and ∂-compressions behave somewhat differently. In dealing with potentially compressible surfaces, as we do here, one has to be especially careful with isotopies. By this reason (and others which will become obvious later), it suggests to work in the context of circle-patterns.

Let $\underline{D}$ be a Heegaard-diagram. Denote by $SM(\underline{D})$ the set of all *admissible surfaces* in $(M,k) = M(\underline{D})$ modulo *admissible isotopy*, i.e., the set of all surfaces $S \subset M$ with $S \cap \partial M = \partial S \subset \partial M - k$ modulo isotopy preserving k. The piping-process defines a constructable map $P : \mathcal{B}(\underline{D}) \to \mathcal{C}(\underline{D})$ which maps realizable marked basis $(\mathcal{B}, \underline{b})$ to circle-patterns $P(\overline{\mathcal{B}}, \underline{b}) := \sigma\beta(\mathcal{B}, \underline{b})$ (prop. 10.8). Composition with the canonical map $\mathcal{C}(\underline{D}) \to SM(\underline{D})$, $C \mapsto S(C)$, defines a map $\mathcal{B}(\underline{D}) \to SM(\underline{D})$. Of course, a marked basis from $\mathcal{B}(\underline{D})$ may well be mapped under the latter map to the empty set. But we know from cor. 10.17 that the image of that map contains the set of all ∂-reduced surfaces in $M(\underline{D})$ (except for once or twice punctured discs). We also know that no surface $S(C)$, $C \in \mathcal{C}(\underline{D})$, is totally compressible, provided $M(\underline{D})$ is ∂-irreducible (cor. 10.10). It remains to look for a compression-algorithm for circle-patterns. Such an algorithm, would transform admissible surfaces into essential ones. It turns out that, while it may take a long time to turn an admissible surface into a ∂-incompressible one, it takes only a relatively short time to turn a ∂-incompressible surface into an essential one (the underlying reason being that Dehn twists along annuli may complicate the embedding types of ∂-compressible surfaces). This observation yields the desired polynomial procedure for deciding for any given surface whether it is essential.

As indicated before, we first deal with disc-compressions of surfaces. ∂-compressions will be treated in 12.17; and in 12.27 we will deal with ∂-parallelity of surfaces.

12.2. *The Compression-Process.*

We first describe the compression-process and worry about its efficiency later. The idea is to introduce a "compression complexity" for circle-patterns and a compression-process for (rapidly) reducing it. For this we need the notion of "compression regions".

12.3. *Compression-regions.* We introduce the concept of compression-regions for the slightly bigger class of circle-systems, rather than circle-patterns (see 6.5 for definitions).

To begin with let $\underline{D} = (D, h, \underline{K})$ be a Heegaard diagram and let F denote its underlying 2-manifold. Let C be a circle-system in $\underline{D}$. C may or may not be *orientable*, i.e., it may or may not have an orientation. Here we define an *orientation* of C to be a choice of normal vector for each of the components of C which is coherent under the attaching homeomorphism h. Then C is orientable if and only if the surface $S(C) \subset M(\underline{D})$ has trivial normal bundle if and only if $S(C)$ is orientable. Set

$$\mathcal{C}^+(\underline{D}) := \{ \; C \in \mathcal{C}(\underline{D}) \mid C \in \mathcal{C}(\underline{D}) \;\; \text{is orientable} \; \},$$

i.e., $\mathcal{C}^+(\underline{D})$ is the set of admissible isotopy classes of all orientable circle-patterns in $\underline{D}$.

Let C be an oriented circle-system in $\underline{D}$ and let C_0 be a circle from C. Note that the closure $\bar{A}$ of every component, A, of $D - C$ is not only a disc, but a disc with (complete) boundary-pattern. Indeed, the faces of $\bar{A}$ are given by the closure of the components of $\bar{A} \cap C$ and $\bar{A} \cap (\partial D - \partial \underline{K})$. In particular, we may distinguish between square regions and non-square regions (where a square region has exactly four faces). Notice that, in contrast to square regions, the total number of non-square regions and their faces is limited by some function depending on $\underline{D}$ alone, namely by $4 \cdot c(\underline{D}) := 2 \cdot \#\partial\mathcal{K}$.

An *essential region* in $\bar{A}$ is a disc $R \subset \bar{A}$ such that

(1) $R \cap \partial\bar{A} \subset C$, and

(2) $(\partial R - \partial\bar{A})^-$ consists of essential arcs in $\bar{A}$.

An essential region $R := R(C_0, A)$ in a *non-square* region A from $D - C$ is called a *compression-region* (*for* C_0 *in* $\bar{A}$), if, in addition,

(3) $R \cap C \subset C_0$, and

(4) R contains every region with (1)-(3), modulo admissible isotopy in $\bar{A}$.

More precisely, we write $R^+(C_0, A)$, $R^-(C_0, A)$, respectively, when we want to indicate whether the normal vector of C_0 points into or out of $R(C_0, A)$. Note that, for every non-square region A from $D - C$, the compression-regions $R^+(C_0, A)$ and $R^-(C_0, A)$ are both connected, and that $R^+(C_0, A) \cap R^-(C_0, A)$ may well be non-empty.

12.4. Lemma. *Let C_1, C_2 be circles from C. Then, in the notation above, $R^+(C_1, A) \cap R^+(C_2, A) = \emptyset$ (mod admissible isotopy) if and only if $C_1 \cap C_2 = \emptyset$.*

Proof. One direction is obvious. For the other direction assume $C_1 \cap C_2 = \emptyset$ and $R^+(C_1, A) \cap R^+(C_2, A) \neq \emptyset$. It follows the existence of arcs $c_i \subset R^+(C_i, A) \subset \bar{A}$, $i = 1, 2$, with $c_i \cap C_i = \partial c_i$ and such that c_1 and c_2 intersect transversally in a single point. But ∂c_i can be joined in C_i by some arc. Taking the union of c_i, $i = 1, 2$, with that arc, we obtain two *closed* curves in the underlying 2-manifold, F, of $\underline{D}$ which intersect transversally in a single point. This however is impossible since F is a system of 2-spheres. ◊

Set $R^+(C, A) := \bigcup R^+(C_i, A)$, where the union is taken over all circles C_i from C (it follows from the previous lemma that this union is a disjoint union). Given a collection D' of discs from D, we define the *positive compression-region* $R^+(C, D')$ of C to be the union $\bigcup R^+(C, A)$, taken over all regions A from $D' - C$ (this union is also a disjoint union). In the same way, define the *negative* compression-region $R^-(C, D')$. Notice that both $R^+(C, D')$ and $R^-(C, D')$ may be disconnected.

12.5. *Compression-process.* The compression-process consists of two steps. To describe them, let C be an oriented circle-system in $\underline{D}$.

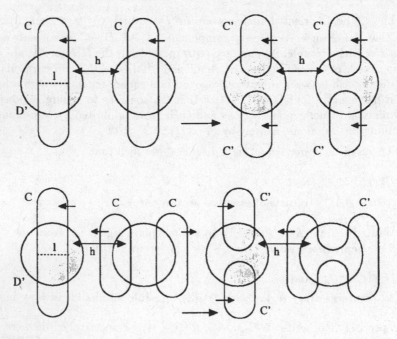

(Figure 12.1.)

Step 1. Fix a collection D' of discs from D with $h(D') \cap D' = \emptyset$ and $h(D') \cup D' = D$, and construct the positive compression-region $R^+(C, D')$.

Step 2. Let $R^+ := R^+(C, D') \cup hR^+(C, D')$ and define $C' := (C - R^+)^- \cup (\partial R^+ - C)^-$.

We say C' is obtained from C by the compression-process along R^+. Figure 12.1 illustrates schematically the effect of such a compression-process. In those pictures the positive compression-region $R^+(C, D')$ is the neighborhood of the arc l, i.e., it is a square. Notice that this is no restriction since the result of a compression-process along any compression-region is the product of compression-processes along squares. Notice further that the result of the two compression-processes illustrated in figure 12.1 are drastically different on the surface level. Indeed, in the first process we have $\chi S(C') > \chi S(C)$, while in the second process $S(C')$ is isotopic to $S(C)$ in the relative handlebody $M(\underline{D})$.

12.6. The previously described compression-process turns a given oriented circle-system C in $\underline{D}$ into another circle-system $\Gamma_r^+ C := C'$. In exactly the same way, we get a circle-system, notation: $\Gamma_l^+ C$, if we use the compression-region $R^+(C, D - D')$ instead of $R^+(C, D')$. Define $\Gamma^+ C := \Gamma_l^+ \Gamma_r^+ C$. Similarly, using the negative compression-regions $R^-(C, D')$ and $R^-(C, D - D')$, we define $\Gamma^- C$. Note that, strictly speaking, $\Gamma^+ C$ (and $\Gamma^- C$) depends on the choice of D' since $\Gamma_l^+ (\Gamma_r^+ C)$ need not be the same as $\Gamma_l^+ (\Gamma_r^+ C)$. But this dependency will not cause trouble.

The following proposition describes the result of the compression-process for ∂-incompressible surfaces. Here we say that an admissible surface in a relative handlebody $(M, k) = M(\underline{D})$ is ∂-*compressible* (in $M(\underline{D})$ or in (M, k)), if it admits a ∂-compression disc which avoids k (there is of course always a ∂-compression disc meeting k unless the surface itself is a disc). We set

$$\mathcal{C}_\partial^+(\underline{D}) := \{\, C \in \mathcal{C}^+(\underline{D}) \mid S(C) \text{ is } \partial\text{-incompressible in } M(\underline{D}) \,\}$$

12.7. Proposition. *Let* $\underline{D} = (D, h, \underline{K})$ *be a Heegaard diagram. Suppose* $M(\underline{D})$ *is* ∂-*irreducible. Then the following holds, for every* $C \in \mathcal{C}_\partial^+(\underline{D})$:

(1) $\Gamma^+ C \in \mathcal{C}_\partial^+(\underline{D})$ *and* $\chi(\Gamma^+)^d C \geq \chi C$, *for all* $d \geq 0$.

(2) $S((\Gamma^+)^d C)$, $d \geq 0$, *is isotopic (fixing boundary) to* $S(C)$ *iff* $\chi(\Gamma^+)^d C = \chi C$, *and*

(3) $S((\Gamma^+)^d C)$, $d \geq 0$, *is obtained from* $S(C)$ *by a disc-compression if* $\chi(\Gamma^+)^d C > \chi C$.

Remarks. (1) Similar statements hold true for $\Gamma^- C$.
(2) Recall $M(\underline{D})$ is ∂-irreducible if $\underline{D}$ is a useful diagram.

Proof. We only show $\Gamma_\partial^+ C \in C^+(\underline{D})$. All other statements are immediate consequences of the way the compression-process is defined. In fact, (2) and (3) holds for all $C \in C^+(\underline{D})$.

Assume $\Gamma^+ C \notin C_\partial^+(\underline{D})$. Since $C \in C_\partial^+(\underline{D})$ it then follows from (2) and (3) that $\Gamma^+ C \notin C^+(\underline{D})$, i.e., $\Gamma^+ C$ is not an orientable circle-pattern. Let F denote the underlying 2-manifold of $\underline{D}$ and let $\mathcal{D} \subset M(\underline{D})$ denote the disc-system associated to $\underline{D}$. $\Gamma^+ C$ is obtained from C by successive application of the compression process. The compression process turns orientable circle-systems into orientable circle-systems. So, by assumption, $\Gamma^+ C$ cannot be a circle-pattern. Thus it violates either (2) or (3) from def. 6.5. In fact, it must violate (2) since the compression-process leaves $(C - D)^-$ untouched. Hence there is a component, say l, from $D \cap \Gamma^+ C$ which separates a disc $E \subset D$ from D disjoint from $\partial \mathcal{K}$. W.l.o.g. $E \cap \Gamma^+ C = (\partial E - \partial D)^-$. Now l is an arc in $S(\Gamma^+ C)$. It has to separate a disc from $S(\Gamma^+ C)$, for $S(\Gamma^+ C)$ is ∂-incompressible. The union of E with this disc is a proper disc in $M(\underline{D})$ disjoint to the relator curves. Thus its boundary bounds a disc, say G, in $\partial M(\underline{D})$ since $M(\underline{D})$ is supposed to be ∂-irreducible. $G \cap \mathcal{D}$ is a system of arcs. So there is at least one disc $G' \subset G$ with $G' \cap \mathcal{D} = (\partial G' - \partial G)^-$ connected and $G' \cap \partial G \subset \partial S(C)$. Set $b := G' \cap \partial G$. Then b is a component of $(C - D)^-$ which separates a disc from $(F - D)^-$, namely G'. It follows that C itself is not a circle-pattern, i.e., $C \notin C_\partial^+(\underline{D})$. This contradiction finishes the proof. $\Diamond$

12.8. *Compression-Complexity.*

The compression-process is designed to reduce a "compression-complexity" and to do this effectively.

We begin with an observation.

12.9. Lemma. *Let $\underline{D} = (D, h, \underline{K})$ be a Heegaard-diagram such that $M(\underline{D})$ is ∂-irreducible. Let $\mathcal{D} \subset M(\underline{D})$ be the disc-system associated to $\underline{D}$. Let B be a compression-disc for a surface $S = S(C)$, $C \in C_\partial^+(\underline{D})$. Suppose B is chosen so that, in addition, $B \cap \mathcal{D}$ is a system of arcs whose number is as small as possible. Then no arc from $B \cap \mathcal{D}$ is contained in a square region from $\mathcal{D} - S$.*

Proof. Assume the converse, i.e., assume there is a square-region, A, from $\mathcal{D} - S$ whose closure, $\bar{A}$, contains an arc l from $B \cap \mathcal{D}$. Then l must join the two components from $(\partial \bar{A} - \partial \mathcal{D})^-$ since $B \cap \partial M(\underline{D}) = \emptyset$ and since $\#(B \cap \mathcal{D})$ is minimal. Denote by A', A'' the two squares into which $\bar{A}$ is split by l, and by B', B'' the two discs into which B is split by l. W.l.o.g. the choices are made so that the above holds and that, in addition, $A' \cap B' = l$. Set $E := A' \cup B'$. Then E is a disc and $E \cap S$ is an arc in S. This arc must separate a disc from S since S is supposed to be ∂-incompressible. The union of this disc with E is a proper disc in $M(\underline{D})$ disjoint to the relator curves. Thus its boundary bounds a disc in $\partial M(\underline{D})$ since $M(\underline{D})$ is ∂-irreducible. It therefore follows, as in 12.7, that C cannot be a circle-pattern. $\Diamond$

As a first application of this lemma we show that the minimal "length" of compression discs for 2-manifolds in relative handlebodies does not depend on the embedding or the topological type of the 2-manifold, but only on the topological type of the relative handlebody.

Recall every 2-manifold $S = S(C)$, $C \in C^+(\underline{D})$, has a trivial normal bundle. Thus, after fixing one of the two normal directions, we may distinguish between a "positive" and a "negative" side of S. We call a compression-disc B for S *positive* or *negative* according whether it meets S from the positive or the negative side. Define

$$L^+(C) := \min\{ \#(B \cap \mathcal{D}) \},$$

where the minimum is taken over all *positive* compression-discs B for $S(C)$ which are in general position with respect to the disc-system $\mathcal{D} \subset M(\underline{D})$ associated to $\underline{D}$. Set $L^+(C) := 0$, if there is no positive compression-disc for $S(C)$ (note every compression-disc for $S(C)$ has to meet $\mathcal{D}$). In a similar way, define $L^-(C)$ by using *negative* compression discs instead.

12.10. Proposition. *Let $\underline{D} = (D, h, \underline{K})$ be a Heegaard diagram such that $M(\mathcal{D})$ is ∂-irreducible. Then there is a constant $d(\underline{D})$ with*

$$L^+(C) \leq d(\underline{D}) \quad and \quad L^-(C) \leq d(\underline{D}), \quad for \; all \; C \in C_\partial^+(\underline{D}).$$

Proof. Given a handlebody N, denote by $\mathcal{P} = \mathcal{P}(N)$ the set of all systems of pairwise non-parallel discs in N which split N into 3-balls, and denote by $\mathcal{F} = \mathcal{F}(N)$ the set of all (possibly compressible) 2-manifolds $F \subset \partial N$ for which no component of F is a disc. For every $(F, \mathcal{B}) \in \mathcal{F}(N) \times \mathcal{P}(N)$, set $L(F, \mathcal{B}) := \min \#(F \cap \mathcal{B})$, where the minimum is taken over all essential discs $B \subset N$, $\partial B \subset F$ and transversal to $\mathcal{B}$. Define

$$\mathcal{S}(N, \alpha) := \{ (F, \mathcal{B}) \in \mathcal{F} \times \mathcal{P} \mid \#(F \cap \mathcal{B}) \leq \alpha \; and \; L(F, \mathcal{B}) > 0 \}.$$

The mapping class group $\pi_0 \text{Diff } N$ acts on $\mathcal{S}(N, \alpha)$ by $h(F, \mathcal{B}) := (hF, h\mathcal{B})$. The quotient $\mathcal{S}(N, \alpha)/\pi_0\text{Diff } M$ is finite. To see this recall $\mathcal{P}(N)$ is finite modulo handlebody diffeomorphisms. Thus it remains to show that, for every $\mathcal{B} \in \mathcal{P}(N)$, the set $\mathcal{S}(N, \mathcal{B}, \alpha) := \{ F \in \mathcal{F}(N) \mid \#(F \cap \mathcal{B}) \leq \alpha \; and \; L(F, \mathcal{B}) > 0 \}$ is finite modulo isotopy in ∂N preserving $\partial \mathcal{B}$. But, for every $F \in \mathcal{S}(N, \mathcal{B}, \alpha)$, $\partial F \cap (\partial N - U(\mathcal{B}))^-$ is a system of arcs in $(\partial N - U(\mathcal{B}))^-$ since $L(F, \mathcal{B}) > 0$. Moreover, the set of all systems of essential and pairwise non-parallel arcs in $(\partial N - U(\mathcal{B}))^-$ is finite modulo diffeomorphisms of $(\partial N - U(\mathcal{B}))^-$. So the finiteness of $\mathcal{S}(N, \alpha)/\pi_0\text{Diff } M$ follows from the fact that every diffeomorphism of $(\partial N - U(\mathcal{B}))^-$, preserving each boundary component, extends to N. Define

$$L(N, \alpha) := \max L(F, \mathcal{B}).$$

where the maximum is taken over all $(F, \mathcal{B}) \in \mathcal{S}(N, \alpha)$. Note that $L(N, \alpha) < \infty$ since $\mathcal{S}(N, \alpha)/\pi_0\text{Diff } (N)$ is finite and since $L(hF, h\mathcal{B}) = L(F, \mathcal{B})$ for all diffeomorphisms $h : N \to N$. Since the diffeomorphism type of N is determined by genus ∂N, there is a function $d : \mathbf{N} \times \mathbf{N} \to \mathbf{N}$, strictly increasing in both variables, with $L(N, \alpha) \leq d(\text{genus } \partial N, \alpha)$. In particular, $L(F, \mathcal{B}) \leq d(\text{genus } \partial N, \alpha)$, for every $(F, \mathcal{B}) \in \mathcal{S}(N, \alpha)$.

Now let $C \in \mathcal{C}^+(\underline{D})$. Recall $L^+(C) > 0$. Let $M := M(\underline{D})$ and define

$$\hat{M} := (M - U(S))^-, \quad \hat{\mathcal{D}} := \mathcal{D} \cap \hat{M} \quad \text{and} \quad \hat{S} := U(S) \cap \partial \hat{M}.$$

Let $\hat{\mathcal{D}}_0$ be the union of all square regions from $\hat{\mathcal{D}} = (\mathcal{D} - U(S))^-$. Define

$$\tilde{M} := (\hat{M} - U(\hat{\mathcal{D}}_0))^-, \quad \tilde{\mathcal{D}} := \hat{\mathcal{D}} \cap \tilde{M} \quad \text{and} \quad \tilde{S} := \hat{S} \cap \partial \hat{M} = (\hat{S} - U(\mathcal{D}_0))^-.$$

Then $\tilde{S} \subset \partial \tilde{M}$ and $\tilde{M}$ is a handlebody. In fact, $\tilde{\mathcal{D}}$ splits $\tilde{M}$ into a system of 3-balls. Define S^* by forgetting all disc components from $\tilde{S}$. It follows from lemma 12.9 that $L(C) \leq L(S^*, \tilde{\mathcal{D}})$. So it remains to estimate $L(S^*, \tilde{\mathcal{D}})$. For this we utilize the estimates

$$\#(S^* \cap \tilde{\mathcal{D}}) \leq \#(\tilde{S} \cap \hat{\mathcal{D}}) \leq 4 \cdot c(\underline{D}) \quad \text{and} \quad \text{genus } \partial \tilde{M} \leq \# \tilde{\mathcal{D}} \leq 4 \cdot c(\underline{D}).$$

We have $(S^*, \tilde{\mathcal{D}}) \in \mathcal{S}(\tilde{M}, 4 \cdot c(\underline{D}))$. Thus $L^+(C) \leq L(S^*, \tilde{\mathcal{D}}) \leq d(\text{genus } \partial \tilde{M}, 4 \cdot c(\underline{D}))$. The same with $L^-(C)$. Since $C \in \mathcal{C}^+(\underline{D})$ was arbitrary, this show that $d(\underline{D}) := d(4 \cdot c(\underline{D}), 4 \cdot c(\underline{D}))$ satisfies the conclusion of the proposition. $\Diamond$

12.11. *The Compression-Map.*

We wish to estimate the time it takes to turn a given ∂-incompressible surface into an essential one. To formalize the process, recall from prop. 12.7 (1) that Γ^+ defines a map $\mathcal{C}_\partial^+(\underline{D}) \to \mathcal{C}_\partial^+(\underline{D})$. Thus composition is defined. So we may set

$$\Gamma_d := (\Gamma^-)^d \circ (\Gamma^+)^d, \ d \geq 0.$$

12.12. Proposition. *Let $\underline{D} = (D, h, \underline{K})$ be a Heegaard diagram such that $M(\underline{D})$ is ∂-irreducible. Then $S(\Gamma_d^{|\times C|} C)$ is an essential 2-manifold, provided $C \in \mathcal{C}_\partial^+(\underline{D})$ and $d = 4 \cdot c(\underline{D})$.*

Remark. Recall from section 6 that $c(\underline{D}) = \frac{1}{2} \#(\partial \underline{K})$.

Proof. Let $d := 4 \cdot c(\underline{D})$ and $C \in \mathcal{C}_\partial^+(\underline{D})$. Suppose $S = S(C)$ is compressible in $M(\underline{D})$. Then it suffices to show that $\chi(\Gamma^-)^d (\Gamma^+)^d C > \chi C$.

To prove this we need a complexity for circle-patterns which is being reduced by the compression-process. It turns out that the complexities $L^+ C$ and $L^- C$ introduced above, do not quite serve our purpose, for the compression-process sometimes enlarges $L^+ C$ or $L^- C$ rather than reducing it. We obtain a more adequate complexity if we first slightly generalize the notion of positive (and negative) compression-regions as follows.

Fix an orientation for C. This orientation induces a selection of normal vectors for the arcs from $C \cap D$. For every component A from $D - C$, denote by r^+A, resp. r^-A, the union of all arcs from $(\partial \bar{A} - \partial D)^-$ whose normal vector points into, resp. out of, the region $\bar{A}$. Denote $\Omega^+ A := (\bar{A} - U(r^- A \cup \partial D))^-$. Then $\Omega^+ A \cap \partial \bar{A} \subset r^+ A$. Define

$$\Omega^+ C := \bigcup_A \Omega^+ A, \quad \text{and} \quad c^+(C) := \#(\partial \Omega^+ C - C)^- - \#\Omega^+ C,$$

where the union is taken over all non-square regions, A, from $D - C$. In the same way, define $\Omega^- C$ and $c^-(C)$. Note that $h(\Omega^+ C) = \Omega^+ C$ and that $0 \leq c^+(C) \leq 4 \cdot c(\underline{D})$. The same with $\Omega^- C$ and $c^-(C)$.

Consider $R^+ = R^+(C, D') \cup hR^+(C, D')$ (see 12.5). W.l.o.g. $R^+ \subset \Omega^+ C$ and $R^+ \cap (\partial \Omega^+ C - C)^- = \emptyset$. By definition, $\Gamma_r^+ C = (C - R^+)^- \cup (\partial R^+ - C)^-$. Set

$$\Omega^+ \Gamma_r^+ C := (\Omega^+ C - R^+)^- \quad \text{and} \quad c^+(\Gamma_r^+ C) := \#(\partial \Omega^+ \Gamma_r^+ C - C)^- - \Omega^+ \Gamma_r^+ C.$$

In the same way, define $\Omega^+ \Gamma_l^+ \Gamma_r^+ C$ and $c^+(\Gamma^+) = c^+(\Gamma_l^+ \Gamma_r^+ C)$ etc. Of course, $c^+(C) \geq c^+(\Gamma^+ C) \geq c^+((\Gamma^+)^2 C) \geq \dots$.

Now let B be any positive compression disc for $S(C)$ (recall $S(C)$ is compressible). Suppose B is chosen so that, in addition, B intersects $\mathcal{D}$ in arcs whose number is as small as possible (here $\mathcal{D} \subset M(\underline{D})$ denotes the disc-system associated to $\underline{D}$). $B \cap \mathcal{D}$ induces, in a canonical way, an arc-system in D; denoted by $B \cap D$. It follows from lemma 12.9 and the construction of $\Omega^+ C$ that $B \cap D \subset \Omega^+ C$. Let $\mathcal{B}_0$ be the (disjoint) union of all those discs, B_0, in which B is split by $\mathcal{D}$ and for which $B_0 \cap \partial B$ is connected. $(\mathcal{B}_0 - \partial \mathcal{D})^-$ is a *non-empty* system of arcs since every compression disc of $S(C)$ has to meet $\mathcal{D}$. Moreover, $\mathcal{B}_0 \cap D \subset \Omega^+ C$ (see above) and $\mathcal{B}_0 \cap D \subset R^+(C, D') \cup R^+(C, D - D')$.

We first suppose $\mathcal{B}_0 \cap D \subset R^+(C, D')$. Then $\mathcal{B}_0 \cap D \subset R^+ \subset \Omega^+ C$. In particular, $R^+ \subset \Omega^+ C$ is non-empty. As a consequence we have $c^+(\Gamma_r^+ C) < c^+(C)$. Moreover, either $\chi \Gamma_r^+ C > \chi C$ or $S(\Gamma_r^+ C)$ is isotopic to $S(C)$. In the first case, we are done. In second case, the isotopy deforms the compression-disc B into a compression-disc B' for $S(\Gamma_r^+ C)$. It may or may not be that $\#(B' \cap \mathcal{D}) \leq \#(B \cap \mathcal{D})$. But, in any case, $B' \cap D \subset \Omega^+ \Gamma_r^+ C$. Thus we can repeat the same argument for $\Gamma_l^+(\Gamma_r^+ C)$, using $\Omega^+ \Gamma_r^+ C$ instead of $\Omega^+ C$. Then $c^+(\Gamma_l^+(\Gamma_r^+ C)) \leq c^+(\Gamma_r^+ C)$. Altogether, $c^+(\Gamma^+ C) < c^+(C)$.

If $\mathcal{B}_0 \cap D \not\subset R^+(C, D')$, then $\mathcal{B}_0 \cap D \subset R^+(C, D - D')$. By the argument used before, we get $c^+(\Gamma_r^+ C) \leq c^+(C)$ and $c^+(\Gamma_l^+(\Gamma_r^+ C)) < c^+(C)$. Thus, in any case, $c^+(\Gamma^+ C) < c^+(C)$.

So, inductively, it follows that $c^+((\Gamma^+)^d C) = 0$ if $d := 4 \cdot c(\underline{D})$, for $4 \cdot c(\underline{D}) \geq c^+(C)$. Thus $\Omega^+(\Gamma^+)^d C$ consists of a system of inessential discs, i.e., discs which meet C in a connected arc. In particular, the compression disc B disappeared in the process. This implies $\chi(\Gamma^+)^d C > \chi C$. The proof is therefore finished. $\Diamond$

The next corollary is an immediate consequence of the previous proposition.

12.13. Corollary. *Let $\underline{D}$ be a Heegaard diagram such that $M(\underline{D})$ is ∂-irreducible. Let $C \in \mathcal{C}^+(\underline{D})$ such that $S(C)$ is ∂-incompressible in $M(\underline{D})$. Then $S(C)$ is incompressible in $M(\underline{D})$ if and only if $\chi \Gamma_d^{|\chi C|} C = \chi C$, for $d = 4 \cdot c(\underline{D})$.* ◊

To determine the computation-complexity of Γ_d, we need the following estimate.

12.14. Lemma. *Let $\underline{D} = (D, h, \underline{K})$ be a Heegaard diagram and let F be its underlying 2-manifold. Suppose $M(\underline{D})$ is ∂-irreducible. Then $\#(A \cap D) \leq 2 \cdot c(\underline{D})$, for every $C \in \mathcal{C}(\underline{D})$ with $S(C)$ ∂-incompressible and all disc-systems $A \subset F$ with $\partial A \subset C$.*

Proof. Let C and A be given as in the lemma. Let E be the system of 3-balls bounded by F. Every arc from $D \cap \partial A$ is an arc from $D \cap C$ and so joins two different components from $\partial D - \underline{K}$. Assume $\#(A \cap D) > 2 \cdot c(\underline{D})$. Then there is at least one component, say A_0, from $D \cap A$ which is a square in the sense that (1) $A_0 \cap \partial D \subset \partial D - \underline{K}$ and that (2) $\#(\partial A_0 - \partial D)^- = 2$. $(\partial A_0 - \partial D)^-$ is contained in one component of C. Therefore there is an arc l in $S(C)$ joining the two components of $(\partial A_0 - \partial D)^-$. In fact, this arc can be chosen so that ∂l bounds a component of $A_0 \cap \partial D$. Thus w.l.o.g. there is a disc $B \subset M(\underline{D})$ with $B \cap S(C) = l = (\partial B - \partial M(\underline{D}))^-$. It follows that l separates a disc, say B', from $S(C)$ since $S(C)$ is supposed to be ∂-incompressible. $B \cup B'$ forms a disc whose boundary lies in $\partial M(\underline{D})$ and avoids all relator curves. The boundary of the latter disc bounds a disc in $\partial M(\underline{D})$ since $M(\underline{D})$ is supposed to be ∂-irreducible. By an argument used in 12.7, we then get a contradiction to the fact that C is a circle-pattern. ◊

12.15. Proposition. *Let $\underline{D} = (D, h, \underline{K})$ be a Heegaard diagram such that $M(\underline{D})$ is ∂-irreducible. Then, there is a constant $n(\underline{D})$ such that, for every $C \in \mathcal{C}^+_{\partial}(\underline{D})$, the circle-pattern $\Gamma_d C$ can be constructed in at most $n(\underline{D})$ steps.*

In particular, there is a linear polynomial $l(x)$ such that $\Gamma_d^{|\chi C|}$ can be constructed in at most $l(\chi C)$ steps.

Remark. We say Γ_d can be constructed in constant time (i.e. in a time which does not depend on the choice of C) and that $\Gamma_d^{|\chi C|} C$ can be constructed in linear time.

Proof. To get a mechanical process (i.e. a Turing machine) started, we need the circle-patterns $C \in \mathcal{C}^+_{\partial}(\underline{D})$ to be represented by some string of symbols, say. Here is a way of doing this. Let us call two arcs from $C \cap D$ or $C \cap (F - D)^-$ *admissibly parallel* if they are isotopic in $D - \mathcal{K}$ resp. $(F - D)^- - \mathcal{K}$. Every parallelity class is represented by a single arc, and the union of all the latter arcs forms a 1-dimensional complex $\mathcal{G}$; the *branching complex*. We say that C is carried by this branching complex. Any circle-pattern carried by a given branching complex is given by a vector of labels for the edges of this branching complex. These vectors are, as usual, integer solutions of the system of linear equations associated to the branching complexes (as given by the switching conditions). Moreover, there are only finitely many branching complexes. Thus we may suppose that circle-patterns are "given" to us as solutions of a finite set of systems of linear equations. This yields the desired symbolic representation of circle-patterns.

The regions of $D - \mathcal{G}$ correspond to the non-square regions of the circle-patterns fully carried by $\mathcal{G}$. Thus the non-square regions of circle-patterns C can be found in constant time (i.e. in a time depending on $\underline{D}$, but not on C). In order to construct compression regions in these non-square regions, we need to check when two given arcs from $C \cap D$ belong to the same circle from C. To do this we start at one of the two arcs, follow C along a fixed direction and check whether we ever reach the other arc. The time needed to find out the latter does not depend on C. Indeed, we know from lemma 12.14 that $\#(C_0 \cap D) \leq 2 \cdot c(\underline{D})$, for every circle C_0 from every $C \in \mathcal{C}^+(\underline{D})$. Thus the positive (negative) region for C can be constructed in constant time. By definition of Γ_d, it therefore follows that, for every $C \in \mathcal{C}_\partial^+(\underline{D})$, the circle-pattern $\Gamma_d C$ can be constructed in constant time. If $\Gamma_d C$ can be constructed in at most $n(\underline{D})$ steps say, then $\Gamma_d^{|\chi C|}$ can be constructed in at most $n(\underline{D}) \cdot |\chi C|$ steps. Thus $\Gamma_d^{|\chi C|}$ can be constructed in linear time. $\Diamond$

The following corollary is an immediate consequence of prop. 12.12 and 15.

12.16. Corollary. *Let $\underline{D}$ be a Heegaard-diagram such that $M(\underline{D})$ is ∂-irreducible. Then, for every $C \in \mathcal{C}_\partial^+(\underline{D})$, the surface $S(C)$ can be turned in linear time into an essential surface.* $\Diamond$

12.17. *Compression-Lines.*

Having dealt so far with disc-compressions for ∂-incompressible surfaces, we now turn our attention to ∂-compressions. Recall that an admissible surface in a relative handlebody (M, k) is ∂-compressible if it admits a ∂-compression disc avoiding k. It turns out that ∂-compressible surfaces too have a compression-complexity whose size can be estimated. Indeed, this complexity is given by the length of the shortest compression-line as introduced next.

12.18. Definition. *Let $\underline{D} = (D, h, \underline{K})$ be a Heegaard diagram with underlying 2-manifold F. Let C be a circle-system in $\underline{D}$. A compression-line (of length n) for C is a system $\alpha = \alpha_1 \cup ... \cup \alpha_n$, $n \geq 1$, of pairwise disjoint arcs in F with the following properties:*

(1) $\alpha_i \cap C = \partial \alpha_i$, and $\partial \alpha_i$ is contained in one circle of C, for every $1 \leq i \leq n$.

(2) every component of $D \cap \alpha$ (resp. $(\alpha - D)^-$) is essential in the region from $D - C$ (resp. $F - D \cup \underline{K} \cup C$) containing it.

(3) $\#(\alpha_i \cap D) \leq 1$, *for* $i = 1, n$; *and* $\#(\alpha_i \cap D) = 2$, *otherwise.*

(4) $h(\alpha \cap D) = \alpha \cap D$, *and, for every* $1 \leq i \leq n - 1$, *h maps one component of* $\alpha_i \cap D$ *onto one component of* $\alpha_{i+1} \cap D$.

If C has a fixed orientation, we say $\alpha = \alpha_1 \cup ... \cup \alpha_n$ is a *positive* (resp. *negative*) compression-line if α_1 meets C from the positive (resp. negative) side. Notice that in this case *all* α_i, $1 \leq i \leq n$, meet C from the positive (resp. negative) side.

12.19. Proposition. *Let $\underline{D}$ be a Heegaard diagram such that $M(\underline{D})$ is ∂-irreducible. A surface $S(\overline{C})$, $C \in \mathcal{C}^+(\underline{D})$, is ∂-compressible in the relative handlebody $M(\underline{D})$ if C admits a compression-line.*

Proof. Suppose $\alpha = \alpha_1 \cup ... \cup \alpha_n$ is a compression-line for C. It is our goal to associate a ∂-compression disc B to α. We do this as follows. First recall that, by property (1), $\partial \alpha_i$ is contained in a circle C_i from C. Thus $\partial \alpha_i$ can be joined in $S := S(C)$ by some arc. This arc, together with α_i bounds a disc B_i in $S \cap E$, where E is the system of 3-balls bound by the 2-manifold, F, underlying $\underline{D}$. It follows from property (4) that $B := (\bigcup B_i)/h$ is a 2-manifold, and it follows from (3) and (4) that it is actually a disc.

We claim B is a ∂-compression disc for S.

Assume the converse. Let $(M, k) = M(\underline{D})$ and let $\mathcal{D} \subset M(\underline{D})$ denote the disc-system associated to $\underline{D}$. It follows from the construction of B that $\beta := B \cap S$ is a proper arc in $S = S(C)$. By assumption, this arc must separate a disc, say B', from S. W.l.o.g. $B \cap B' = (\partial B - \partial M(\underline{D}))^- = (\partial B' - \partial S)^-$. Thus $B \cup B'$ is a disc whose boundary lies in $\partial M - k$. Therefore $\partial(B \cup B')$ bounds a disc, say A, in $\partial M - k$ since $M(\underline{D})$ is supposed to be ∂-irreducible. By the innermost-disc-argument, there is a disc A_0, $A_0 \subset A$ such that $A_0 \cap \mathcal{D}$ is connected and equal do $(\partial A_0 - \partial A)^-$. Now, $\partial A \subset B \cup B'$. But $A_0 \cap B' = \emptyset$, for $A \cap B' \subset C$ and C is a circle-pattern. Moreover, $A_0 \cap B = \emptyset$, for $A_0 \cap \partial A$ is a component of $(\alpha - D)^-$ and every component of $(\alpha - D)^-$ is essential in $F - D \cup \underline{K} \cup C$. Thus $A_0 \cap \partial A = \emptyset$. But this is impossible since $A \cap \mathcal{D}$ is a non-empty system of arcs (no disc from $\mathcal{D}$ is ∂-parallel in $M(\underline{D})$). Thus our assumption is false, and B is indeed a ∂-compression disc. Therefore $S(C)$ is ∂-compressible. $\Diamond$

12.20. Proposition. *Let $\underline{D}$ be a Heegaard diagram such that $M(\underline{D})$ is ∂-irreducible. Then there are constants $d = d(\underline{D})$, $e = e(\underline{D})$ such that the surface $S(C)$, $C \in \mathcal{C}^+(\underline{D})$, is essential if and only if, for both orientations of C, the following holds:*

(1) $\chi(\Gamma^+)^d C = \chi C$ *and*

(2) $(\Gamma^+)^d C$ *has no compression-line of length* $\leq e$.

Remark. We may take $d := 4 \cdot c(\underline{D})$ and $e := 48 \cdot c(\underline{D})$.

Proof. Suppose either (1) or (2) is violated. If $\chi(\Gamma^+)^d C \neq \chi C$, for any $d \geq 1$, then it easily follows from the way the compression process is defined that $S := S(C)$ is compressible. If C has a compression-line, of any length, then, by prop. 12.19, S is inessential.

For the other direction suppose S is inessential. Set $d = 4 \cdot c(\underline{D})$. W.l.o.g. we may also suppose $\chi(\Gamma^+)^d C = \chi C$ (for both of the two orientations of S), for otherwise we are done.

If $S(C)$ is ∂-incompressible but compressible, then it follows from prop. 12.7 that $\chi(\Gamma^+)^d C \neq \chi C$ for one of the two orientations of C.

Thus we may suppose S is ∂-compressible. Let B be a ∂-compression disc for S. Then, more precisely, we may suppose the orientation of C is chosen so that B is a positive ∂-compression disc. Moreover, we may suppose B is chosen so that the above holds and that, in addition, $B \cap \mathcal{D}$ consists of arcs whose number is as small as possible ($\mathcal{D}$ denotes the disc-system associated to $\underline{D}$). Of course, every arc from $B \cap \mathcal{D}$ lies in (the closure of) some component from $\mathcal{D} - S$.

We claim no component from $B \cap \mathcal{D}$ lies in a component of $\mathcal{D} - S$ which is a square region. Assume the converse, and let l be an arc from $B \cap \mathcal{D}$ which is contained in some square region, say A, from $\mathcal{D} - S$. Suppose l joins the two arcs from $(\partial A - \partial \mathcal{D})^-$ (in the other cases we argue similarly). Then l separates A into two squares, say A_1, A_2. W.l.o.g. l is chosen so that $A_1 \cap B = l$. But l separates B into two discs too. One of them, say B_1, contains the arc $B \cap S$. Thus the union $B_1 \cup A_1$ is a ∂-compression disc for S which (after a small general position isotopy if necessary) intersects $\mathcal{D}$ in strictly fewer arcs than B. This in turn contradicts our minimal choice of B. So the claim is established.

Since $\chi(\Gamma^+)^d C = \chi C$, we know from prop. 12.7 that $S(C)$ is isotopic to $S((\Gamma^+)^d C)$, using an ambient isotopy of $M(\underline{D})$ fixing the boundary. The same isotopy deforms B into a ∂-compression disc for $S((\Gamma^+)^d C)$, say B'. Moreover, by an argument used in 12.12, it follows that $B' \cap \mathcal{D}$ consists of arcs which join $B' \cap S'$ with $B' \cap \partial M(\underline{D}) = (\partial B' - S)^-$. Let B' be isotoped so that the above holds and that, in addition, $\#(B' \cap \mathcal{D})$ is as small as possible. Let $B'_1, ..., B'_n$ be the discs in which B' is split by $\mathcal{D}$. Let the indices be chosen so that $B'_i \cap B'_{i+1} \neq \emptyset$. Then $B'_2, ..., B'_{n-1}$ are squares (and B'_1, B'_n are triangles). The arcs from $B' \cap \mathcal{D}$ are contained in positive *non-square* regions of $\mathcal{D} - S((\Gamma^+)^d C)$, for the arcs from $B \cap \mathcal{D}$ are contained in positive non-square regions of $\mathcal{D} - S(C)$. The total number of non-square regions from $\mathcal{D} - S((\Gamma^+)^d C)$ as well as the total number of their faces is smaller than $2 \cdot c(\underline{D})$. Thus if $n > 4 \cdot c(\underline{D})$, then there is a region A from $\mathcal{D} - C$ and at least three arcs from $A \cap B'$ which are admissibly parallel in A. Let A_0 be a square in A which contains at least three arcs from $A \cap B'$ and which contains all arcs from $A \cap B'$ parallel to them. The regular neighborhood $X := U(A_0 \cup B'_1 \cup ... \cup B'_{n-1})$ (taken in $M(\mathcal{D})$) is an I-bundle over some surface G and $X \cap \mathcal{D}$ is a vertical square in X. X is contained in an essential I-bundle X' in $M(\underline{D})$ over some surface G' say. $\chi G' \geq \chi \partial M$. It then follows from lemma 7.16 the existence of a constant $a = a(\underline{D})$ and the existence of a product, g, of Dehn twists in X' so that $\#(gB' \cap \mathcal{D}) \leq a$. But gB' is a ∂-compression disc for S again. So $B' = gB'$ (see our minimal choice of B'). Thus $n \leq e := a \cdot 4c(\underline{D})$. Now $\alpha := B' \cap F$ is easily seen to be a compression-line (recall our minimal choice of B'). Moreover, the length of α is smaller than the constant e, for $n \leq e$. Thus we have shown the existence of some compression-line for $(\Gamma^+)^d C$ with length $\leq e$, provided $S(C)$ is ∂-compressible. In fact, e may be taken to be $e = 48 \cdot c(\underline{D})$ since, by 7.16, we may take $a = 12$.

This finishes the proof of the proposition. $\Diamond$

12.21. Corollary. *Let $\underline{D}$ be a Heegaard-diagram such that $M(\underline{D})$ is ∂-irreducible. Then it can be decided in constant time whether a given surface $S(C)$, $C \in \mathcal{C}^+(\underline{D})$, is essential.*

Remark. By definition, this means that the time it takes to check whether $S(C)$, $C \in \mathcal{C}^+(\underline{D})$, is essential is independent from the actual choice of C.

Proof. We want to apply the criterion from prop. 12.20. For this let d, e be constants satisfying the conclusion of prop. 12.20 (e.g. $d = 4 \cdot c(\underline{D})$ and $e = 48 \cdot c(\underline{D})$).

Given $C \in \mathcal{C}^+(\underline{D})$, we first construct $(\Gamma^+)^d C$. According to 12.15, this can be done in constant time.

Next, we check whether $(\Gamma^+)^d C$ has a compression line of length $\leq e$. Now, any such compression line is a disjoint union of arcs in the 2-manifold, F, underlying $\underline{D}$, with the properties (1) - (4) from def. 12.18. There are of course only finitely many embedding types of arcs with properties (1) and (2). In order to find out whether a given collection of at most e such arcs actually forms a compression-line for C, we need to check properties (1)-(4) one by one. Using lemma 12.14, we find that this can be done in constant time. Thus, altogether, it can be checked in constant time whether $(\Gamma^+)^d C$ has a compression-line of length $\leq e$.

Finally, we check whether $\chi(\Gamma^+)^d C = \chi C$. This fails to be the case iff there is some $0 \leq a \leq d-1$ with $\chi(\Gamma^+)(\Gamma^+)^a C \neq \chi(\Gamma^+)^a C$. Now, let C' be any circle-system in $\underline{D}$. Then, according to the way $\Gamma^+ C'$ is defined, $\chi \Gamma^+ C' \neq \chi C'$ if and only if there is a component A of the positive compression region $R^+(C', D')$ and a circle C_0' in C' such that $hA \cap C_0'$ is disconnected. By lemma 12.14, the latter condition can be checked in constant time. $\Diamond$

12.22. *Isotopy of Surfaces.*

We next study the algorithmic problem of deciding whether two given essential surfaces in relative handlebodies are isotopic. It turns out that for this question the concept of compression arcs is relevant. This concept is an adaption of the concept of compression regions for the present purpose.

Given a Heegaard diagram $\underline{D}$, define

$$\mathcal{E}^+(\underline{D}) := \{\, C \in \mathcal{C}^+(\underline{D}) \mid S(C) \text{ is an essential surface in } M(\underline{D}) \,\}.$$

Let $C \in \mathcal{E}^+(\underline{D})$. An arc l in D with $l \cap C = \partial l$ is a *compression arc* (*for* C) if there is a circle C_0 form C such that l joins two different components of $C_0 \cap D$. Given a compression-arc l for the circle-pattern C, define

$$C_l := (C - U(l \cup hl)) \cup (l' \cup l'') \cup h(l' \cup l''),$$

where the regular neighborhood is taken in D and where l', l'' are the copies of l in $\partial U(l)$ (see fig. 12.1 for a schematic illustration of this process). We say that C_l is obtained from C by *compressing along the compression arc* l. Notice that hl is not a compression arc for C if l is. But the arc $l' \subset U(hl)$, joining the two components of $(hl - C_l)^-$, is a compression arc for C_l. In fact, $(C_l)_{l'}$ is pattern-isotopic to C. Notice also that compression regions contain all compression arcs.

The following result follows as in 12.7.

12.23. Lemma. *Let* $\underline{D} = (D, h, \underline{K})$ *be a Heegaard-diagram such that* $M(\underline{D})$ *is* ∂-*irreducible. Suppose* $C \in \mathcal{E}^+(\underline{D})$ *and* l *is a compression-arc for* C. *Then* $C_l \in \mathcal{E}^+(\underline{D})$ *and* $S(C_l)$ *is isotopic to* $S(C)$, *using an isotopy fixing the boundary.* ◊

This lemma tells us that, given an essential surface $S(C)$, $C \in \mathcal{E}^+(\underline{D})$, we obtain other essential surfaces within its admissible isotopy class by successively compressing along compression-arcs. Note that any circle-pattern obtained in this way intersects ∂D in the same number of points as C. Thus this process generates only a finite set of circle-patterns. According to our next result, however, it is this very set which represents the whole admissible isotopy class of $S(C)$.

12.24. Proposition. *Let* $\underline{D}$ *be a Heegaard-diagram such that* $M(\underline{D})$ *is* ∂-*irreducible. Let* $C, C' \in \mathcal{E}^+(\underline{D})$. *Then* $S(C)$ *is admissibly isotopic to* $S(C')$ *in* $M(\underline{D})$ *if and only if there is a finite sequence* $C := C_1, C_2, ..., C_n$ *of oriented circle-patterns in* $\mathcal{E}^+(\underline{D})$ *such that*

(1) C_n *is pattern-isotopic to* C', *and*

(2) C_{i+1}, $i \geq 1$, *is obtained from* C_i *by compressing along some compression-arc for* C_i.

Proof. One direction follows directly from lemma 12.23. To prove the other direction, suppose that $S(C')$ is admissibly isotopic to $S(C)$. Then there is a surface $S \cong S(C)$ and an admissible *homotopy* $f : S \times I \to M(\underline{D})$ with $f \mid S \times 0 = S(C)$ and $f \mid S \times 1 = S(C')$. Let $\mathcal{D} \subset M(\underline{D})$ be the disc-system associated to $\underline{D}$. W.lo.g. we may suppose that f is admissibly deformed so that the above holds and that, in addition, the pre-image $G := f^{-1}\mathcal{D}$ is a codim 1 submanifold in $S \times I$.

Now, fix the boundary-pattern for $S \times I$, as given by the components of $S \times \partial I$ and $\partial S \times I$ (see [Joh 1] about more information on boundary-patterns). Then G carries the induced boundary-pattern and so it may be considered as an admissible surface in the product I-bundle $S \times I$. Our goal is to simplify G by permitted operations, so that in the end it is a system of vertical squares.

First, by the usual surgery-argument, we may suppose that f is admissibly deformed so that the above holds and that, in addition, G is incompressible in $S \times I$.

Next, we verify that this argument carries over to show that $G \cap (\partial S \times I)$ is vertical. To do this assume there is a component l from $G \cap (\partial S \times I)$ which is not vertical, i.e., not an I-fiber.

If l is an arc, then we may suppose l is chosen to be outermost. Then, by definition, there is a disc A in $\partial S \times I$ with $l = (\partial A - (\partial S \times \partial I))^- = A \cap G$. W.l.o.g., A meets $\partial S \times 0$. Consider the arc $l' := A \cap (\partial S \times 0)$. This arc may be identified with a component of $(C - D)^-$. Moreover, l is mapped under f into $\partial D - \partial \underline{K}$. Thus the restriction $f \mid A$ provides a deformation (fixing end-points) of l' in $(F - D)^-$ into $\partial D - \partial \underline{K}$. This, however, is impossible since C is supposed to be a circle-pattern in $\underline{D}$.

If l is a closed curve, observe that no component from $\partial \mathcal{D}$ is contractible in ∂M. So l can be removed by surgery in the usual way, if it were contractible in $S \times I$. Observe also that every component of $(C - D)^-$ is an arc. So every component of $\partial S \times \partial I$ has to meet G. In particular, no component of $G \times (\partial S \times I)$ can be an essential closed curve in $\partial S \times I$. Altogether, our claim is established.

By what we have seen so far, we may suppose that $G = f^{-1}\mathcal{D}$ is an incompressible 2-manifold which intersects $\partial S \times I$ in vertical arcs. To continue the proof, it is now convenient to distinguish between the following two cases:

Case 1. G is essential.

If G is an essential 2-manifold in the I-bundle $S \times I$, observe first that w.l.o.g. no component of G can be an admissible i-faced disc, $1 \leq i \leq 3$. Indeed, any such disc from G has to be disjoint to $\partial S \times I$ since $G \cap (\partial S \times I)$ is vertical. Therefore at least one component of C or C' would be contained in D which is impossible since C and C' are supposed to be circle-patterns.

By [Joh 1, prop. 5.6], it follows that, modulo admissible isotopy, G is either horizontal, or vertical in the I-bundle $S \times I$. But $S(C) \cap \mathcal{D}$ is non-empty, and so also $G \cap (S \times \partial I)$. Thus G has to be vertical. Then, splitting f along G, we obtain a homotopy in the underlying 2-manifold F of $\underline{D}$ which deforms C into C' without changing $\#(C \cap \partial D)$. It follows that C and C' are pattern-isotopic. So we are done in Case 1.

Case 2. G is inessential.

By the very definition of essential 2-manifolds and by our choice of the boundary-pattern for $S \times I$, there is a ∂-compression-disc A in $S \times I$ such that

(1) $A \cap G = (\partial A - \partial(S \times I))^-$ is an essential arc in G, and

(2) $A \cap (\partial S \times \partial I)$ consists of at most one point.

Observe that w.l.o.g. A does not meet $\partial S \times \partial I$ since $G \cap (\partial S \times I)$ consists of vertical arcs (see above), and so $A \cap \partial(S \times I)$ is entirely contained either in $S \times \partial I$, or in $\partial S \times I$.

Suppose first that $A \cap (\partial S \times I)$ is contained in $S \times \partial I$. To fix ideas we may suppose that the arc $\beta := A \cap (S \times \partial I)$ is contained in $S \times 0$. Since $\mathcal{D}$ is a system of discs, it follows that the restriction $f \mid \beta$ is a singular arc in $\mathcal{D}$ which joins two different components of $S(C) \cap \mathcal{D}$. In fact, this singular arc can be straightened (fixing end-points) to some non-singular arc b' in $\mathcal{D}$ with $b' \cap S(C) = \partial b'$. The end-points of b' are joined in $S(C)$ by the singular arc $f \mid (\partial A - S \times 0)^-$. Now, the singular disc $f \mid A$ is contained in one of those 3-balls obtained from $M(\underline{D})$ by splitting along $\mathcal{D}$. It follows that f can be deformed so that afterwards the above holds and that, in addition, $f \mid A$ is an embedding. It follows that b' corresponds to a compression-arc b of C with the additional property that the surface $S(C_b)$ can be obtained by isotoping $S(C)$ across the disc $f(A)$. This isotopy can be extended to an admissible ambient isotopy of $M(\underline{D})$ whose composition with f defines a deformation of f which results in an admissible homotopy g with $g \mid S \times 0 = S(C_b)$ and $g \mid S \times 1 = S(C')$ and such that $g^{-1}\mathcal{D}$ is the surface obtained from G by splitting along A. Now, cutting a surface along a compression-disc, strictly increases its Euler characteristic and so the above procedure applied over and over again has to stop after a finite number of steps.

By what we have seen so far, there are two finite sequences of circle-patterns in $\mathcal{E}^+(\underline{D})$

$$C = C_1, C_2, ..., C_m \quad \text{and} \quad C' = C'_1, C'_2, ..., C'_n.$$

such that

(1) C_{i+1} resp. C'_{i+1} is obtained from C_i resp. C'_i by a compression along some compression-arc of C_i resp. C'_i, and

(2) there is an admissible homotopy $g : S \times I \to M(\underline{D})$ with $g \mid S \times 0 = S(C_m)$, $g \mid S \times 1 = S(C'_n)$, and $g^{-1}(\mathcal{D})$ is a "pseudo-essential" 2-manifold in $S \times I$.

Here an incompressible surface G in $S \times I$ is *pseudo-essential*, if all ∂-compression discs A of G with $A \cap (\partial S \cap \partial I) = \emptyset$ are disjoint to $S \times \partial I$.

In order to continue the argument, let G_0 be any inessential component from G, and let A be a ∂-compression disc for G_0. Then $A \cap \partial(S \times I) \subset \partial S \times I$. Consider the arc $l' := A \cap (\partial S \times I)$. The end-points of l' are contained in the vertical arcs from $G \cap (\partial S \times I)$. Thus either $\partial l'$ is contained in one arc from $G \cap (\partial S \times I)$ or l' joins two different arcs from $G \cap (\partial S \times I)$. If $\partial l'$ is contained in one arc b from $G \cap (\partial S \times I)$, then l' together with a sub-arc from b bounds a disc in $\partial S \times I$. The union of this disc with A is a disc whose boundary lies in G. But G is incompressible, and so it follows that A cannot be a ∂-compression-disc for G in the first place. Thus l' has to join two different arcs, say b_1, b_2, from $G \cap (\partial S \times I)$. But b_1 and b_2 are both vertical and so, pushing l' down to $S \times 0$ or up to $S \times 1$ (and a little bit further), we obtain two discs from A which do not meet $\partial S \times I$ anymore. Since G is pseudo-essential, we thus may conclude that the arc $A \cap G$ is contained in some square from G. This square can certainly be pushed to a vertical square (e.g. that one near one of the components separated from $S \times I$ by $b_1 \cup b_2$).

By what has been demonstrated so far, we may suppose that all inessential components of G are vertical and so, by an argument of Case 1, we may suppose that G is vertical and that therefore the circle-patterns C_m and C_n' are pattern-isotopic in $\underline{\underline{D}}$. Furthermore, it is easily seen that the converse of a compression along a compression-arc is also a compression along a compression-arc. Thus, altogether, $C = C_1, ..., C_m = C_n', ..., C_1' = C'$ is the required sequence of compressions joining C with C'.

This completes the proof of the proposition. $\Diamond$

12.25. Note that the set, $\mathcal{Q}(C)$, of all circle-patterns in $\underline{\underline{D}}$, obtained from C by successively compressing along compression arcs, must be finite. A (very) crude upper bound for $\#\mathcal{Q}(C)$ is given by $2 \cdot 3^{m(C)}$, where $m(C) := \frac{1}{2}\#(C \cap \partial D) - 2$. To verify this estimate observe that $C_l \cap \partial D = C \cap \partial D$, for every $C \in \mathcal{E}^+(\underline{\underline{D}})$ and every compression arc l for C. Thus $\#\mathcal{Q}(C)$ is estimated from above by the size of the set of all arc-systems in D whose boundary equals $C \cap \partial D$. An explicit formula for this size can be derived as follows. Let D be a disc. Denote by $\mathcal{P}_{2n} = \{ p_1, p_2, ..., p_{2n} \} \subset \partial D$ a collection of $2n$ pairwise different points (in clockwise order) and denote by $\mathcal{K}(\mathcal{P}_{2n})$ the set of all arc-systems $\mathcal{K} \subset D$ with $\partial \mathcal{K} = \mathcal{P}_{2n}$, modulo isotopy fixing $\partial \mathcal{K}$. Let α_{2n} denote the total number of arcs from $\mathcal{K}(\mathcal{P}_{2n})$. Now fix $p_1 \in \mathcal{P}_{2n}$. Then every $\mathcal{K} \in \mathcal{K}_{2n}$ contains exactly one arc which joins p_1 with some point from $\mathcal{P}_{2n}$ with even index. Given any arc, k, joining p_1 with $p_{2m} \in \mathcal{P}_{2n}$, $2 \leq 2m \leq 2n$, let $\mathcal{K}_{2m} \subset \mathcal{K}(\mathcal{P}_{2n})$ denote the set of all arc-systems containing k. Then $\#\mathcal{K}_{2m} = \alpha_{2(m-1)} + \alpha_{2(n-m)}$. Thus, altogether, $\alpha_{2n} = \#\mathcal{K}(\mathcal{P}_{2n}) = \sum_{1 \leq m \leq n} \#\mathcal{K}_{2m} = 2 \cdot (\alpha_2 + \alpha_4 + ... + \alpha_{2n-2}) = 2 \cdot 3^{n-2}$, $n \geq 2$. It follows that $\#\mathcal{Q}(C) \leq 2 \cdot 3^{m(C)}$. Since this estimate is exponential one might think that an algorithm for deciding whether two surfaces are isotopic must be exponential. It turns out, however, that this is not the case. Indeed, in the next section (see 13.6) we will give a much better estimate for $\#\mathcal{Q}(C)$ which in turn will yield a polynomial algorithm for this decision-problem.

12.26. Remark. We remark that there are two necessary (and easily verified) conditions for the isotopy of surfaces which often can be used to decide very quickly whether two surfaces are isotopic. To formulate these conditions let $C_1, C_2 \in \mathcal{E}^+(\underline{D})$ be two oriented circle-patterns in the Heegaard-diagram $\underline{D} = (D, h, \underline{K})$. Suppose $S(C_1)$ and $S(C_2)$ are admissibly isotopic in $M(\underline{D})$, then the following holds:

(1) $\#(C_1 \cap d) = \#(C_2 \cap d)$, for every component d of $\partial D - \partial \underline{K}$,

(2) $a(C_1, l) = a(C_2, l)$, for every arc in l in ∂D with $\partial l \subset \partial \underline{K}$. (Recall that $a(.,.)$ denotes the algebraic intersection-number.)

12.27. Boundary-Parallelity.

We wish to introduce a criterion for the existence of essential and not boundary parallel surfaces in relative handlebodies. We begin with the definition of ∂-parallel circle-patterns.

12.28. Definition. *Let $\underline{D} = (D, h, \underline{K})$ be a Heegaard diagram and let F be its underlying 2-manifold. A circle-pattern $C \in \mathcal{C}^+(\underline{D})$ is called* boundary-*parallel if there is a disc-system $A \subset F$ with $\partial A = C$ and such that*

(1) $h(A \cap D) = A \cap D$, and

(2) $(\partial B - \partial D)^-$ is connected, for every component B from $A \cap D$.

If $C \in \mathcal{C}^+(\underline{D})$ is boundary-parallel, then the surface $S(C)$ is boundary-parallel in the handlebody $M(\underline{D})$, but the converse is not necessarily true. Here we say that a surface in $M(\underline{D})$ is *boundary-parallel* if it is parallel to some surface in $\partial M(\underline{D})$ (possibly containing some relator curves from $M(\underline{D})$). We mention the following lemma which makes it easy to check whether a circle-pattern is not boundary-parallel.

12.29. Lemma. *Let $\underline{D}$ be a Heegaard diagram. Let $C \in \mathcal{C}^+(\underline{D})$ be a boundary-parallel circle-pattern and let C_0 be a component of C. Then C_0 bounds a disc A_0 in F such that the intersection of A_0 with every disc from D is connected.*

Proof. Since C is boundary-parallel, there is a disc-system $\mathcal{A}$ in F with properties (1) and (2) from def. 12.28. Let A_0' be that component from $\mathcal{A}$ with $\partial A_0' = C_0$. Set $A_0 := (F_0 - A_0')^-$, where F_0 denotes that component from F containing C_0. Then A_0 is a disc. Moreover, the intersection of A_0 with every disc from D is connected since, by property (2) of $\mathcal{A}$, every component from $(D - A_0')^-$ is a 2-faced disc. $\Diamond$

12.30. Proposition. *Let $\underline{D}$ be a Heegaard diagram such that $M(\underline{D})$ is ∂-irreducible. Suppose $S(C)$ is incompressible in $M(\underline{D})$. Then $S(C)$ is boundary-parallel in $M(\underline{D})$ if and only if either $(\Gamma^+)^d C$ or $(\Gamma^-)^d C$ is a boundary-parallel circle-pattern, where $d := 4 \cdot c(\underline{D})$.*

Proof. One direction follows easily from the definition of boundary-parallel circle-patterns.

For the other direction we use the approach from 12.24. Assume that $S(C)$, $C \in \mathcal{C}^+(\underline{D})$, is boundary-parallel.

Since $S := S(C)$ is supposed to be boundary-parallel, there is a surface $S' \subset \partial M$ such that S is isotopic to S', using an isotopy $\alpha : S \times I \to M$ with $\alpha | S \times 0 = S$, and which is constant on ∂S. In the following we will consider α as a homotopy. In particular, we will allow deformations of α into maps which are no isotopies anymore.

Set $G := \alpha^{-1}\mathcal{D}$, where $\mathcal{D}$ is the disc-system in $M(\underline{D})$ associated to $\underline{D}$.

If G is compressible, then, by the usual surgery argument, there is a deformation of α, constant on $\partial(S \times I)$, so that afterwards the pre-image $\alpha^{-1}\mathcal{D}$ is incompressible.

If G is incompressible, but inessential, then there is a compression disc $A \subset S \times I$ for G. Suppose that the normal orientation of S is chosen so that the normal vector of $S \times 0$ points into $S \times I$. If $A \cap S \times 0 = \emptyset$, there must be a complementary compression disc, A', for G with $A' \cap \partial(S \times I) \subset S \times 0$. The existence of the latter disc however shows that the positive compression region of C is non-empty. Now the result of a compression process along a square is the ∂-compression of G. More precisely, it is a deformation of the map α after which G is compressed along a ∂-compression disc. In other words, the compression process removes ∂-compression disc. An adaption of the argument for prop. 12.12 shows that after $d = 4 \cdot c(\underline{D})$ applications of the compression-process *all* ∂-compression discs for G are vanished, i.e., all compressions along these discs are carried out. Thus, replacing C by $(\Gamma^+)^d C$ if necessary, we may suppose that G is essential in $S \times I$.

If G is essential in the product I-bundle $S \times I$, then G is a system of vertical squares in $S \times I$ (see e.g. [Joh 1, prop. 5.6]). In this case, it is easily seen that C is boundary-parallel. $\Diamond$

12.31. Proposition. *Let $\underline{D}$ be a Heegaard diagram such that $M(\underline{D})$ is ∂-irreducible. Let $C \in \mathcal{C}^+(\underline{\underline{D}})$. Suppose $S := S(C)$ is incompressible and $(\Gamma^+)^d C$ as well as $(\Gamma^-)^d C$, $d := 4 \cdot c(\underline{D})$, is not boundary-parallel. Then there exists an essential surface in $(M, k) = M(\underline{D})$ which is disjoint to k and not boundary-parallel in M.*

Proof. If S is essential, we are done. If not, then there are ∂-compressions for S. Let S' denote the 2-manifold obtained from S by carrying out all ∂-compressions. Then we reobtain S from S' by a sequence of "tunnelings". We here say that a 2-manifold S is obtained from S' by *tunneling* if there is an arc α in ∂M such that $\alpha \cap S' = \partial \alpha$ and S is one component of $(\partial U(S' \cup \alpha) - \partial M)^-$. If S' is ∂-parallel, then either S is compressible or ∂-parallel according whether or not the tunnel-arc α is contained in the parallelity region. But S is supposed to be incompressible. Therefore S' is the desired essential and not ∂-parallel surface. So prop. 12.31 follows from prop. 12.30. $\Diamond$

12.32. For an illustration of the last results we refer to the appendix. We finally also remark that many of the intricacies of this section vanish for full and simple diagrams $\underline{D}$ since in that case $\#(C \cap D) \leq l(\chi C)$, $C \in C^+(\underline{D})$, for some linear function $l(x)$.

§13. The Haken Spectrum.

In this section we discuss the problem of generating and classifying iso-topy classes of essential surfaces in relative handlebodies. In [Ha 1] it has been shown how to obtain all incompressible surfaces in Haken 3-manifolds as solutions of certain systems of linear equations associated to these manifolds. In [F-O] Haken's approach has been extended and the existence of finitely many branched surfaces has been established which carry all incompressible surfaces. It turns out that from a theoretical point of view this is a very useful way of representing surfaces. Numerically, however, this approach is still not very efficient. One of the problems being that the relevant systems of linear equations depend on the combinatorial structure of the underlying triangulations and can be quite involved (and so can the set of all branched surfaces). Here we use the piping procedure in order to obtain a more coherent way for deducing such (or closely related) systems of linear equations in the case of relative handlebodies. As a result we will be able to read off the relevant system of linear equations directly and in polynomial time from a presentation of the fundamental group. We then use this, together with the algorithms from the last section, in order to get a process for enumerating essential surfaces without repetition. Finally, we introduce the "Haken spectrum", a function which counts surfaces, and determine its growth rate to be polynomial.

13.1. *Matrices for Labeled Discs.*

We begin with an observation concerning parametrizations of essential arc-systems in discs with boundary-patterns.

For this consider the following matrices

$$\begin{pmatrix} 1 & 1 & 0 \\ 0 & 1 & 1 \\ 1 & 0 & 1 \end{pmatrix} \text{ and } \begin{pmatrix} -1 & -1 & 0 \\ 0 & 1 & 1 \\ 1 & 0 & 1 \end{pmatrix}.$$

Given these building blocks, set

$$C_1 := \begin{pmatrix} 1 & 1 & 0 \\ 0 & 1 & 1 \\ 1 & 0 & 1 \end{pmatrix}, \quad C_2 := \begin{pmatrix} 1 & 1 & 0 & 0 & 0 & 0 \\ 0 & 1 & 1 & 0 & 0 & 0 \\ 1 & 0 & 1 & -1 & -1 & 0 \\ 0 & 0 & 0 & 0 & 1 & 1 \\ 0 & 0 & 0 & 1 & 0 & 1 \end{pmatrix}$$

and, generally,

$$C_{i+1} := \begin{pmatrix} & & & & 0 & \\ & C_i & & & & \\ & & & -1 & -1 & 0 \\ & 0 & & 0 & 1 & 1 \\ & & & 1 & 0 & 1 \end{pmatrix}.$$

Given a disc D and a system $\underline{d}$ of pairwise disjoint arcs $d_0, d_1, ..., d_{n+1}$, $n \geq 0$, in ∂D (indices in clockwise order say), we associate to $(D, \underline{d})$ the matrix

$A_{(D,\underline{d})} := C_n$. Given a label l_i, $l_i \in \mathbb{N} \cup \{0\}$, for the arc d_i, $0 \leq i \leq n+1$, we associate to the *labeling-vector* $L := (l_0, l_1, ..., l_{n+1})$ for $(D, \underline{d})$ the vector b_L whose transpose is given by

$$b_L := (l_0, l_1, 0, l_2, 0, l_3, 0, ..., 0, l_n, l_{n+1}).$$

In this way, we associate to any labeled disc $(D, \underline{d}, L)$ its *inhomogeneous spectral equation*

$$A_{(D,\underline{d})} \cdot x = b_L.$$

13.2. *Matrices for Presentations.*

Let $\mathcal{P} = < g_1, g_2, ..., g_m \mid r_1, ..., r_n >$ be a finite presentation whose relators have word length $l(r_i) \geq 3$, for all $1 \leq i \leq n$. To any such presentation we associate a *complexity* $c(\mathcal{P})$, by setting

$$c(\mathcal{P}) := m + \sum_{1 \leq i \leq n} (l(r_i) - 2).$$

To every relator, r, from $\mathcal{P}$ we associate the matrix

$$C(r) = C_{l(r)-2},$$

where C_n denotes the recursively defined matrix from 13.1. Furthermore, associate to r the matrix $B(r)$ whose column-vectors b_i, $1 \leq i \leq m$, are given as follows. If the relator r is the word $r = c_1 c_2 ... c_q$ of length $q = l(r)$ with $c_i \in \{g_1, g_1^{-1}, ..., g_m, g_m^{-1}\}$, then define

$$b_i := \begin{pmatrix} \epsilon_{i1} \\ \epsilon_{i2} \\ 0 \\ \epsilon_{i3} \\ 0 \\ \epsilon_{i4} \\ ... \\ 0 \\ \epsilon_{i(q-1)} \\ \epsilon_{iq} \end{pmatrix} \quad \text{by setting} \quad \epsilon_{ij} := \begin{cases} -1, & \text{if } c_j = g_i, \\ -1, & \text{if } c_j = g_i^{-1}, \\ 0, & \text{otherwise.} \end{cases}$$

Given the matrices $B(r_i)$ and $C(r_i)$, $1 \leq i \leq n$, form the matrix

$$A_{\mathcal{P}} := \begin{pmatrix} C(r_1) & 0 & & 0 & B(r_1) \\ 0 & C(r_2) & & 0 & B(r_2) \\ & & ... & & ... \\ 0 & & & C(r_n) & B(r_n) \end{pmatrix}.$$

In this way we associate to every finite presentation $\mathcal{P}$ its *homogeneous spectral equation* $A_{\mathcal{P}} \cdot x = 0$.

13.3. *Spectral Equations and Arc-Systems.*

Let D be a disc and let $\underline{d} = d_0 \cup d_1 \cup \ldots \cup d_{n+1}$, $n \geq 0$, be a collection of pairwise disjoint arcs in ∂D (indices in clockwise direction say). Given a labeling-vector $L = (l_0, l_1, \ldots, l_{n+1}) \in \mathbf{N}^{n+2}$ for $(D, \underline{d})$, define $\mathcal{K}(L)$ to be the set of all arc-systems $\mathcal{K} \subset D$ with $\mathcal{K} \cap \partial D = \partial \mathcal{K}$ and $\#(\mathcal{K} \cap d_i) = l_i$, $0 \leq i \leq n+1$. Of course, $\mathcal{K}(L)$ may or may not be empty.

Given a labeling-vector $L \in \mathbf{N}^{n+1}$ for $(D, \underline{d})$, we associate, to every solution $x \in \mathbf{N}^{3n}$ of the inhomogeneous spectral equation $A_{(D,\underline{d})} \cdot x = b_L$, an essential arc-system $K_x \in \mathcal{K}(L)$ as follows. First, embed the graph Δ_n in the disc D in such a way that the free vertices of Δ_n become the midpoints of the faces of $(D, \underline{d})$. Here the graph Δ_n is given by the following chain of triangles.

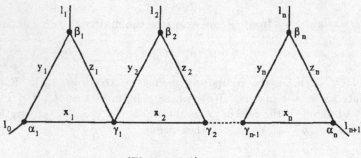

(Figure 13.1)

Next, given a non-negative integer solution-vector $x = (x_1, y_1, z_1, x_2, y_2, z_2, \ldots, x_n, y_n, z_n) \in \mathbf{N}^{3n}$ for $A_{(D,\underline{d})} \cdot x = 0$, take x_i, y_i resp. z_i copies of the edges from Δ_n labeled by x_i, y_i, z_i, respectively. Notice that to every vertex of the graph Δ_n, there is associated a coherence condition as follows.

$$\left\{ \begin{array}{r} x_1 + y_1 = l_0 \\ y_i + z_i = l_i \\ x_i + z_i - x_{i+1} - y_{i+1} = 0 \\ x_n + z_n = l_{n+1} \end{array} \right\} \quad \text{associated to the vertex} \quad \left\{ \begin{array}{l} \alpha_1 \\ \beta_i \\ \gamma_i \\ \alpha_n \end{array} \right\}$$

The set of all coherence conditions yields a system of linear equations in the variables $x_1, y_1, z_1, \ldots, x_n, y_n, z_n$. The matrix associated to this system equals the matrix $A_{(D,\underline{d})}$. It follows that the above copies of arcs must fit together to form an essential arc-system, K_x, in $(D, \underline{d})$ with $\#(K_x \cap d_i) = l_i$. The assignment $x \mapsto K_x$ defines a map θ_L from the set $\mathcal{A}(L)$ of all non-negative integer solution-vectors $x \in \mathbf{N}^{3n}$ for $A_{(D,\underline{d})} \cdot x = b_L$ to the set $\mathcal{K}(L)$.

13.4. Proposition. *The map* $\theta_L : \mathcal{A}(L) \to \mathcal{K}(L)$ *is a surjection, for every* $L \in \mathbf{N}^{n+2}$.

Proof. Let $K \in \mathcal{K}(L)$. Then K is an essential arc-system in $(D, \underline{d})$. It remains to show that it is carried by the graph $\Delta_n \subset D$. To see this let $T \subset D$ be a system of arcs with $T \cap \partial D = \partial T \subset \partial D - \underline{d}$ and $T \cap \Delta_n = \bigcup(\gamma_1 \cup \gamma_2 \cup ... \gamma_n)$ (dotted arcs in fig. 13.2).

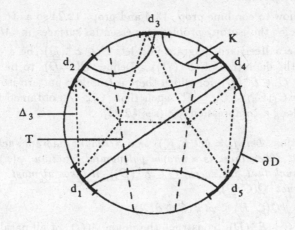

(Figure 13.2.)

Set $D' := (D - U(T))^-$ and set $\underline{d}' := \underline{d} \cup (D' \cap U(T))$. Then the intersection of $\underline{d}'$ with every component of D' consists of exactly three intervals. Moreover, the intersection of K with every component from D' consists of arcs joining these intervals. Thus the latter arcs are all carried by $\Delta_n \cap D'$. So K is carried by Δ_n. This proves the proposition. $\Diamond$

In a similar way, the homogeneous spectral equation for presentations is also related to arc-systems in discs. To explain this, associate to every presentation $\mathcal{P} = <\ g_1, ..., g_m \mid r_1, ..., r_n\ >$ a system of labeled discs as follows. First, pick for every relator r_i a disc $(D_i, \underline{d}_i)$ where $\underline{d}_i$ is a system of $l(r_i)$ pairwise disjoint arcs in ∂D_i. The relator r_i is a word $r_i = c_1 c_2 ... c_q$ of length $q = l(r_i)$ with $c_j \in \{g_1, g_1^{-1}, ..., g_m, g_m^{-1}\}$. Label the faces of $(D_i, \underline{d}_i)$, in clockwise direction of ∂D_i, according to the word $r_i = c_1 c_2 ... c_q$ (i.e., the labels are letters and not integers as before). Given this labeling define $\mathcal{K}(\mathcal{P})$ to be the set of all essential arc-systems, K, in $\bigcup(D_i, \underline{d}_i)$ with $K \cap \bigcup \partial D_i \subset \underline{d}_i$ and the property that K intersects any two arcs from $\underline{D}_i$ which are labeled by the same letter g_i (or g^{-1}) in the same number of points. Let $\mathcal{A}(\mathcal{P})$ denote the set of all solution vectors $x \in \mathbf{N}^{c(\mathcal{P})}$ of the homogeneous spectral equation $A_{\mathcal{P}} \cdot x = 0$.

As above, associate to any given $x \in \mathcal{A}(\mathcal{P})$ the essential arc-system K_x in $\bigcup(D_i, \underline{d_i})$. The assignment $x \mapsto K_x$ defines a map $\theta_\mathcal{P} : \mathcal{A}(\mathcal{P}) \to \mathcal{K}(\mathcal{P})$.

The following proposition follows in exactly the same way as prop. 13.4.

13.5. Proposition. *The map $\theta_\mathcal{P} : \mathcal{A}(\mathcal{P}) \to \mathcal{K}(\mathcal{P})$ is a surjection.* ◊

We next utilize the above results both for the isotopy problem for surfaces and for the problem of generating essential surfaces in relative handlebodies. We begin with the isotopy problem.

13.6. *Isotopy-Algorithm.*

We show how to combine prop. 13.4 and prop. 12.24 so as to get a *polynomial* algorithm for the isotopy problem for essential surfaces in $M(\underline{D})$.

Let $\underline{D}$ be a Heegaard diagram and let $C \in \mathcal{E}^+(\underline{D})$ be a circle-pattern (see 12.22 for the definition of $\mathcal{E}^+(\underline{D})$). Define $\mathcal{Q}(C, \underline{D})$ to be the set of all circle-patterns $C' \in \mathcal{E}^+(\underline{D})$ for which there is (modulo pattern-isotopy) a finite sequence $C = C_1, C_2, ..., C_n = C'$ such that C_{i+1} is obtained from C_i by compression along a compression-arc (see 12.22).

13.7. Proposition. *Let $\underline{D} = (D, h, \underline{K})$ be a Heegaard diagram such that $M(\underline{D})$ is ∂-irreducible. Then there is a known polynomial function $\varphi(x)$ (depending only on $\underline{D}$) such that, for every $C \in \mathcal{E}^+(\underline{D})$, it takes at most $\varphi(\#(C \cap D))$ steps to construct $\mathcal{Q}(C, \underline{D})$.*

In particular, $\#\mathcal{Q}(C, \underline{D}) \le \varphi(\#(C \cap \partial D))$.

Proof. Given $C \in \mathcal{E}^+(\underline{D})$, construct the union $A(C)$ of all parallelity regions in $(F - D)^-$ of arcs from $(C - D)^- \subset F$ (recall F denotes the underlying 2-manifold of $\underline{D}$). This can certainly be accomplished in a number of steps which depends linearly on $\#(C \cap \partial D)$. Define $\underline{d} := A(C) \cap \partial D$ and denote by $\mathcal{K}(C)$ the set of all essential arc-systems K in $(D, \underline{d})$ with $K \cap \partial D = C \cap \partial D$. By prop. 13.4, $\mathcal{K}(C)$ is parameterized by the non-negative integer solution-vectors of a linear equation $Ax = b_L$, where A is a matrix whose size depends on $\#\underline{d}$ alone and where b_L is a vector whose entries is given by the numbers $\#(a \cap C)$, taken over all components a of $A(C) \cap \partial D$. The set of all solutions of $Ax = b_L$ depends polynomially on the norm of b_L. Thus $\#\mathcal{K}(C)$ depends polynomially on $\#(C \cap \partial D)$, i.e., there is a polynomial $\varphi(x)$ such that $\#\mathcal{K}(C) \le \varphi(\#(C \cap \partial D))$.

Set $\mathcal{L}(C) := \{ C \in \mathcal{C}(\underline{D}) \mid C \cap D \in \mathcal{K}(C) \}$. We know that $\mathcal{Q}(C, \underline{D}) \subset \mathcal{L}(C)$. It remains to sort out $\mathcal{Q}(C, \underline{D})$ from $\mathcal{L}(C)$ and to do so in polynomial time. Now, for every $K \in \mathcal{K}(C)$, the number of all non-square regions of K in D is bounded by some constant (depending only on the diagram $\underline{D}$ but not on the choice of K). The same is true for the number of faces of those regions. Moreover, all compression-arcs lie in non-square regions since $C \in \mathcal{E}^+(C)$. It follows that, for every $C' \in \mathcal{L}(C)$, there is only a constant number of choices for compression-arcs for C'. It also takes only constant time to find out whether a given essential arc in a non-square region is actually a compression-arc, for no circle in $\mathcal{E}^+(\underline{D})$ can intersect ∂D in more than $\#\partial \underline{K}$ points. Thus, given a subset $\mathcal{L}_i(\overline{C}) \subset \mathcal{L}(C)$, it takes at most $const \cdot \#\mathcal{L}_i(\overline{C}) \leq const \cdot \#\mathcal{K}(C) \leq const \cdot \varphi(\#(C \cap \partial D))$ number of steps to construct the set $\mathcal{L}_{i+1}(C)$ of all circle-patterns which is obtained from circle-patterns of $\mathcal{L}_i(C)$ by a compression along a single compression-arc. Starting with $\mathcal{L}_0(C) := \{C\}$, we get in this way a sequence of sets $\mathcal{L}_0(C), \mathcal{L}_1(C), \ldots \subset \mathcal{L}(C)$ each one of which can be constructed in polynomial time from the set before it. Moreover, $\mathcal{L}_n(C) = \mathcal{L}_{n+1}(C)$, for some $n \leq \#\mathcal{K}(C) \leq \varphi(\#(C \cap \partial D))$. Taken these estimates together, it follows that $\mathcal{L}_n(C) = \mathcal{Q}(C, \underline{D})$ can be constructed in polynomial time. $\Diamond$

13.8. Corollary. *Let $\underline{D}$ be a Heegaard diagram such that $M(\underline{D})$ is ∂-irreducible. Then there is a computable polynomial function $\varphi(x)$ and an algorithm $\mathcal{A}$, such that, for any two given circle-patterns $C, C' \in \mathcal{E}^+(\underline{D})$, the algorithm $\mathcal{A}$ decides in less than $\varphi(\#(C \cap D))$ steps whether or not $S(C)$ and $S(C')$ are admissibly isotopic in the relative handlebody $M(\underline{D})$.*

Proof. Let $C, C' \in \mathcal{E}^+(\underline{D})$. Construct $\mathcal{Q}(C) := \mathcal{Q}(C, \underline{D})$. According to the previous proposition this can be done in polynomial time (w.r.t. $\#(C \cap \partial D)$). By prop. 12.24, $S(C)$ is admissibly isotopic to $S(C')$ if and only if C' is pattern-isotopic to a circle-pattern from $\mathcal{Q}(C)$. We therefore finish the proof by verifying that it can be determined in linear time (depending on $\#(C \cap \partial D)$) whether or not two given circle-patterns C, C' are pattern-isotopic. To do this simply collapse all parallelity regions of the arc-systems $(C - D)^-$ and $C \cap D$ to arcs. The same with C'. In this way we get graphs $\mathcal{G}(C), \mathcal{G}(C')$ whose vertices lie in ∂D. In fact, $\mathcal{G}(C), \mathcal{G}(C')$ are labeled graphs whose edges are labeled by the total number of arcs from $C \cap D$ resp. $(C - D)^-$ parallel to them. Note that these graphs can be constructed in linear time. Now C, C' are pattern-isotopic if and only if this holds for the labeled graphs $\mathcal{G}(C), \mathcal{G}(C')$. We already know that it can be decided in polynomial time whether or not $\mathcal{G}(C) \cap D$ is isotopic to $\mathcal{G}(C') \cap D$ mod ∂D. Moreover, $(\mathcal{G}(C) - D)^-/h$ and $(\mathcal{G}(C') - D)^-/h$ are both systems of curves in $\partial M(\underline{D})$. The corollary therefore follows from the fact that the isotopy problem for curve-systems in surfaces can be solved in polynomial time since those curve-systems are also parameterized by solutions of a linear equation (namely the one associated to branching complexes say). $\Diamond$

13.9. Remark. (1) The algorithm $\mathcal{A}$ is polynomial in $\#(C \cap D)$ but not necessarily in $|\chi S(C)|$. However, there is a constant $e(\underline{D})$ such that, for every circle-pattern $C \in \mathcal{E}^+(\underline{D})$, there is a circle-pattern $C' \in \mathcal{E}^+(\underline{D})$ such that $\#(C' \cap D) \leq e(\underline{D}) \cdot |\chi C'|$ and that $S(C)$ and $S(C')$ differ by Dehn-twists along essential annuli. Thus the algorithm $\mathcal{A}$ will be polynomial in $|\chi C|$ when we consider only circle-patterns C with $\#(C \cap D) \leq e(\underline{D}) \cdot |\chi C|$ (as will be the case later).

(2) It is tempting to try to solve the isotopy problem for surfaces by first defining a circle-pattern to be in positive (negative) "normal form" if its positive (negative) compression region is empty, and then to compare normal forms. Now, it is

easy to transform a circle-pattern into a normal form by using arc-compressions say. However, examples of non pattern-isotopic normal forms of circle-patterns are easy to construct which represent isotopic surfaces. So this approach appears to be of little use.

13.10. *Generating Surfaces.*

We next show how to use the matrices introduced in the beginning in order to generate surfaces in relative handlebodies. For this let $\underline{D}$ be a Heegaard diagram and let $\mathcal{D}$ be a meridian-system for $M(\underline{D}) = (M, k)$ contained in the disc-system associated to $\underline{D}$. Suppose $\mathcal{D}$ intersects every component from k in at least three points (this is no big restriction, for otherwise either $M^+(\underline{D})$ is reducible or a component from $\mathcal{D}$ meets a component from k in a single point).

13.11. *The pipe-equivalence for surfaces.* Relative handlebodies $M(\underline{D})$ usually have many more essential surfaces than the corresponding 3-manifold $M^+(\underline{D})$. One of the reasons being that surfaces in $M^+(\underline{D})$ can be pushed across the attached 2-handles. This operation often creates a superficial variety of surfaces which do not really reflect any new properties of $M^+(\underline{D})$. To accommodate for this we introduce an equivalence relation for essential surfaces in $M(\underline{D})$.

Suppose S is a surface in a ∂-irreducible relative handlebody $M(\underline{D})$, supported by a basis $\mathcal{B}$, and let $(\mathcal{B}, \mathcal{P}, \mathcal{A})$ be its piping decomposition (see 11.5). Let $\mathcal{B}' \subset \mathcal{B}$ be the disc-system characterized by the properties that $\mathcal{P} \cap \mathcal{B}' = (\partial \mathcal{B}' - \partial \mathcal{B})^-$ and every component from $\mathcal{B} - \mathcal{B}'$ contains exactly one point from $k \cap \mathcal{B}$. Then $\mathcal{B}' \cup \mathcal{P}$ is a surface in $M(\underline{D})$ (i.e., it does not meet the relator curves) and S is obtained from this surface by pruning (along $\mathcal{A}$). Given a surface T in $M(\underline{D})$, we denote by T^+ the surface obtained from T by pruning all compressible boundary curves. So

$$S = (\mathcal{B}' \cup \mathcal{P})^+.$$

For lack of a better name, we say that two surfaces S_i, $i = 1, 2$, in $M(\underline{D})$ with piping decomposition $(\mathcal{B}_i, \mathcal{P}_i, \mathcal{A}_i)$ are *pipe-equivalent* (notation: $S_1 \sim S_2$), if $\mathcal{B}_1 = \mathcal{B}_2$ and if, for every pipe P from $\mathcal{P}_1$, there is a pipe P' from $\mathcal{P}_2$ such that $P \cap \mathcal{B}'_1 = P' \cap \mathcal{B}_2$. Note that, after pruning, pipe-equivalent surfaces in $M(\underline{D})$ yield isotopic surfaces in $M^+(\underline{D})$.

13.12. *The Haken matrix.* We now associate a matrix to the fundamental group of any relative handlebody, by making use of the matrices associated to presentations as introduced in 13.2. We call this matrix the "Haken matrix". Since Haken matrices are associated to fundamental groups rather than homology groups, as in the case of Seifert matrices, they can be expected to reflect deeper properties of relative handlebodies.

Here is the formal construction. First read off, as usual, a presentation, $\mathcal{P}(\underline{D})$, for the fundamental group of $M(\underline{D})$, i.e., for $\pi_1 M^+(\underline{D})$. Specifically, let labels and normal directions for all discs from the meridian-system $\mathcal{D}$ be chosen. Then the generators, $g_1, ..., g_n$, of $\mathcal{P}(\underline{D})$ are given by all these labels. The relators, r_i, $i \geq 1$, are obtained by traveling along the relator curves k_i and recording the directed intersections with the respective discs from $\mathcal{D}$. The matrix $A_{\mathcal{P}(\underline{D})}$ is then defined (see 13.2) since, by hypothesis, $\#(k_i \cap \mathcal{D}) \geq 3$ for all $i \geq 1$.

13.13. Definition. *The matrix* $A_{\underline{D}} := A_{\mathcal{P}(\underline{D})}$ *is called the* <u>Haken matrix</u> *for* $\underline{D}$. *The equation* $A_{\underline{D}} \cdot x = 0$ *is the* <u>(homogeneous) spectral equation</u> *for* $\underline{D}$.

13.14. Proposition. *Let* $\underline{D}$ *be a full pant-diagram. Then the dimension of the null-space of the Haken matrix* $A_{\underline{D}}$ *is bounded from above by the complexity* $c(\mathcal{P}(\underline{D}))$ *(13.2).*

Proof. Take a look at fig. 13.1. Observe that the variables y_i, z_i in Δ_n are already determined by the variables a_i and x_i. The number of different a_i's equals the total number of generators of $\mathcal{P}(\underline{D})$. The number of different x_i's equals the total number of triangles of Δ_n. But, for every relator, r, from $\mathcal{P}(\underline{D})$, the number $l(r) - 2$ equals the total number of triangles in the disc of E_r associated to the relator r. So the estimate follows. $\Diamond$

13.15. *The map* $\theta_{\underline{D}}$. The map $\theta_{\underline{D}}$ assigns to every non-negative integer solution-vector for $A_{\underline{D}} \cdot x = 0$ (the pipe-equivalence class of) a surface in $M(\underline{D})$. To define it, pick for every relator $r_i = c_1 c_2 ... c_q$, $c_j \in \{g_1, ..., g_n\}$, from $\mathcal{P}(\underline{D})$ a labeled disc $(E_i, \underline{e}_i)$ labeled by $c_1, c_2, ..., c_q$ (in clockwise direction). View E_i as being attached along the relator curve k_i of $M(\underline{D})$ in such a way that the meridian-disc D_j from $\mathcal{D} \subset M(\underline{D})$ intersects ∂E_i in the midpoints of those arcs from $\underline{e}_i$ labeled g_j.

Now let x be a non-negative integer solution-vector for $0 = A_{\underline{D}} \cdot x = A_{\mathcal{P}(\underline{D})} \cdot x$. Recall from 13.3 that $K_x := \theta_{\mathcal{P}}(x)$ is an essential arc-system in $(E, \underline{e}) = \bigcup (E_i, \underline{e}_i)$ with the property that any two arcs form $\underline{d}$ with the same label contain the same number of end-points of K_x. Thus there is a (unique) basis $\mathcal{B}_x \subset M(\underline{D})$, carried by the meridian-system $\mathcal{D}$, such that $\mathcal{B}_x \cap \partial E = \partial K_x$. The arcs from K_x can be pushed into $M(\underline{D})$ to form a system $\mathcal{P}_x$ of pipes for $\mathcal{B}_x$ (unique modulo pipe-equivalence). Let S_x be the 2-manifold obtained from $\mathcal{B}_x$ by piping along the piping system $\mathcal{P}_x$, i.e., let $S_x := \mathcal{B}'_x \cup \mathcal{P}_x$. Finally, define

$$\theta_{\underline{D}}(x) := S_x^+ = (\mathcal{B}'_x \cup \mathcal{P}_x)^+.$$

In this way we assign to x a surface S_x^+ in $M(\underline{D})$. Note that S_x^+ may or may not be essential in $M(\underline{D})$ and it may or may not be orientable. In fact, for a generic x chances are high that S_x is non-orientable, for only one of the usually many pipes of $\mathcal{P}_x$ must be twisted in order for S_x to be non-orientable.

So far S_x^+ stands for a pipe-equivalence *class* of surfaces. Since it is more convenient to deal with representatives for these classes, we will use the following convention for the construction above (which yields canonical representatives; again written S_x). Firstly, suppose the graphs $\Delta_{n_i} \subset E_i$, used for defining $\theta_{\mathcal{P}}(x)$, are embedded in such a way that they miss the midpoint of the discs E_i. Secondly, suppose the arcs from K_x are all pushed away from the midpoints of the E_i's.

The next theorem tells us that the pipe-equivalence classes of virtually *all* essential surfaces in relative handlebodies are governed by the Haken matrix. It therefore may be viewed as an extension of prop. 9.20 which says that the homology classes of *non-separating* surfaces are governed by the Seifert matrix. In particular, it points out the special relevance of the piping concept for the study of essential surfaces in relative handlebodies.

13.16. Theorem. *Let $\underline{D}$ be a full and simple pant-diagram. Then, for every ∂-reduced surface S in $M(\underline{D})$ (i.e., not meeting relator curves), there is a non-negative integer solution-vector x for $A_{\mathcal{P}(\underline{D})} \cdot x = 0$ with $\theta_{\underline{D}}(x) \sim S$.*

Remark. The proof shows that the same conclusion holds if $\underline{D}$ is a full pant-diagram, not necessarily simple, and if S is not a disc with one or two punctures.

Proof. Let $S \subset M(\underline{D})$ be a ∂-reduced surface. Then S is essential. Thus, by cor. 9.13, S is obtained from some basis $\mathcal{B} \subset U(\mathcal{D})$ by pseudo-piping. In particular, $S = (\mathcal{B}' \cup \mathcal{P})^+$ for some appropriate system $\mathcal{P}$ of pipes. Since S is ∂-reduced, prop. 10.16 applies. Thus we may suppose $\mathcal{B}$ and $\mathcal{P}$ are chosen so that the above holds and that, in addition, no pipe from $\mathcal{P}$ joins two parallel discs from $\mathcal{B}$. Pushing the pipes out of $M(\underline{D})$ and into $E \times I$ (fixing $\mathcal{D} \cap \mathcal{P}$), yields an essential arc-system $\mathcal{K}$ in the disc-system $(E, \underline{e}) = \bigcup (E_i, \underline{e}_i)$. The claim then follows from prop. 13.5. $\Diamond$

13.17. *Basic Surfaces.*

In [Ha 1] (see also [Schu 1]) it has been observed that the set of non-negative integer solution-vectors of a system of linear equations is linearly spanned by a finite and constructable set of so called "fundamental vectors". This fact has been used in [Ha 1] to show the existence of a finite and constructable set of fundamental surfaces generating, by cut-and-paste alone, the set of all incompressible surfaces in a 3-manifold. Our construction is related to Haken's approach but differs from it, largely because it allows pruning in addition to cut-and-paste. In this way, we obtain a set of basic surfaces which generates the set of all essential surfaces too, but whose number is in general far smaller than the total number of Haken's fundamental surfaces (see the Appendix).

Let $\underline{D}$ be a full pant-diagram. Consider the spectral equation $A_{\underline{D}} \cdot x = 0$. Let

$$\mathcal{F}(\underline{D}) = \{ x_1, x_2, ..., x_m \}$$

be some constructable and finite set of linear independent, non-negative integer solution-vectors for $A_{\underline{D}} \cdot x = 0$, generating the whole solution-set. As mentioned above, such a set of *fundamental solutions* exists (see the Appendix for an example). For every linear combination, $\sum \alpha_i x_i$, of fundamental solutions, set $l(\sum \alpha_i x_i) := \sum |\alpha_i|$. Furthermore, define the *complete set of basic solutions* to be

$$\mathcal{B}(\underline{D}) := \{ \sum \beta_i x_i \mid x_i \in \mathcal{F}(\underline{D}) \text{ and } \beta_i = 0 \text{ or } 1 \}.$$

A surface S_y^+, $y \in \mathcal{B}(\underline{D})$, is called a *basic surface*. Notice that the set $\{ S_y^+ \mid y \in \mathcal{B}(\underline{D}) \}$ of all basic surfaces in $M(\underline{D})$ is finite and constructable since the set $\mathcal{F}(\underline{D})$ of all fundamental solutions is. But notice also that, in contrast to $\mathcal{F}(\underline{D})$, the set $\mathcal{B}(\underline{D})$, and so the set of all basic surfaces, grows exponentially in the complexity of the relative handlebody $M(\underline{D})$.

Observe that, for every linear combination $x = \sum \alpha_i x_i$, $\alpha_i \in \mathbf{N}$ and $x_i \in \mathcal{F}(\underline{D})$, there is a *unique* $y \in \mathcal{B}(\underline{D})$ and a non-negative integer solution-vector, z, for $A_{\underline{D}} \cdot x = 0$ with

(1) $x = y + z$, and

(2) $l(y) \geq l(y')$, for every $y' \in \mathcal{B}(\underline{D})$ with property (1).

We call y the *maximal basic element* for $x = \sum \alpha_i x_i$, and we call S_y^+ the *maximal basic surface* for S_x^+.

The following property of maximal basic surfaces is crucial for our purpose.

13.18. Lemma. *Let y, z be non-negative integer solution-vectors for $A_{\underline{D}} \cdot x = 0$ such that y is the maximal basic element for $y + z$. Then S_{y+z}^+ is obtained from S_y^+ and S_z^+ by cut-and-paste alone.*

Notation: $S_{y+z}^+ = S_y^+ + S_z^+$.

Proof. Assume the converse. Then there is a disc B in $\partial M - k$ whose boundary is a union of arcs $a_1, b_1, ..., a_n, b_n$ which alternately lie in ∂S_y^+ and ∂S_z^+, and so in ∂S_y and ∂S_z. Moreover, all angles between these arcs which lie in B are obtuse. Let the letters be chosen so that $b_i \subset \partial S_z$, and consider b_1. There is a fundamental solution x_j such that $b_1 \subset S_{x_j}$. It follows that $y = \sum \beta_i x_i$ with $\beta_j \neq 0$. Otherwise $y + z = y + x_j + (z - x_j)$ and $l(y + x_j) > l(y)$, contradicting the fact that y is the maximal basic element for $y + z$. It follows that there is a copy of S_{x_j} which is part of S_y. But b_1 was arbitrary and so this property holds for all $b_1, ..., b_n$. In particular, ∂B can be deformed into ∂S_y. So it cannot lie in the *boundary* of the surface $S_y^+ + S_z^+$ obtained from S_y^+ and S_z^+ by cut-and-paste. $\Diamond$

As indicated before, our approach differs from Haken's theory in that it allows pruning in addition to cut-and-paste. The next proposition provides the link between the two approaches. It tells us that pruning is needed for constructing basic surfaces but that the set of all ∂-reduced surfaces can be generated from basic surfaces by cut-and-paste alone.

13.19. Proposition. *Let $\underline{D}$ be a full and simple pant-diagram. Then every ∂-reduced surface in $M(\underline{D})$ can be obtained from basic surfaces by cut-and-paste alone (modulo pipe-equivalence and isotopy).*

Remark. (1) We do not claim that basic surfaces are essential or ∂-reduced. Moreover, surfaces obtained from essential (or ∂-reduced) basic surfaces need not be essential (or ∂-reduced).

(2) The same conclusion holds for all ∂-reduced surfaces different from once or twice punctured disc if $\underline{D}$ is a full, but not necessarily simple, pant diagram.

Proof. Let S be a ∂-reduced surface in $M(\underline{D})$. Then, by thm. 13.16, there is a non-negative integer solution-vector, x, of the spectral equation $A_{\underline{D}} \cdot x = 0$ such that S_x^+ is pipe-equivalent to S. Let y be a maximal basic element for x and let z be the non-negative integer solution-vector with $x = y + z$. Then, by lemma 13.18, the surface S_x^+ is obtained from S_y^+ and S_z^+ by cut-and-paste alone. Since S_y^+ is a basic surface, the proposition follows by an induction on the length $l(x)$. $\diamond$

13.20. *Searching for Essential Surfaces.*

Let S be a surface obtained from surfaces S_1, S_2 in a 3-manifold by cut-and-paste. Then S_1 and S_2 need not be essential when S is. For incompressible surfaces in closed 3-manifolds, it has been analyzed in [J-O] when this is the case, using Haken's theory of fundamental surfaces. The next result is basically a translation of the main theorem of [J-O] into our context. To formulate it let $d(S_x^+) := \#\mathcal{B}_x$ (see 13.15).

13.21. Theorem. *Let $\underline{D}$ be a simple Heegaard diagram. Let y, z be a pair of non-trivial, non-negative integer solution-vectors for the spectral equation $A_{\underline{D}} \cdot x = 0$ such that $S_{y+z}^+ = S_y^+ + S_z^+$ and S_{y+z}^+ is an ∂-reduced surface (possibly non-orientable) in $M(\underline{D})$. Then one of the following holds:*

(1) S_y^+ or S_z^+ is essential in $M(\underline{D})$.

(2) There is a pair of non-trivial, non-negative integer solution-vectors, y', z', for $A_{\underline{D}} \cdot x = 0$ with $S_{y'+z'}^+ = S_{y'} + S_{z'}^+$, $\#(S_{y'}^+ \cap S_{z'}^+) < \#(S_y^+ \cap S_z^+)$, and $S_{y'+z'}^+$ is admissibly isotopic to S_{y+z}^+.

(3) There is a non-negative integer solution-vector, x, with $d(S_x) < d(S_{y+z})$ and S_x^+ admissibly isotopic to S_{y+z}^+.

Proof. By hypothesis, $S_{y+z}^+ = S_y^+ + S_z^+$. By definition, this means that we get S_{y+z}^+ from S_y^+ and S_z^+ by cut-and-paste. More precisely, the cut-and-paste procedure produces first a singular surface, S' say, which touches itself along its singularity curves, i.e., along $S_y^+ \cap S_z^+$, and which yields S_{y+z}^+ after a small general position deformation (constant outside the regular neighborhood of the singularity curves). So, in particular, every curve l from $S_y^+ \cap S_z^+$ gives rise to a "thin" square or annulus in $U(l)$ joining the two components of the intersection of $S_{y+z}^+ \cap U(l)$. Let C denote the union of all these squares and annuli.

13.22. Lemma. *Suppose there is an annulus C from $\mathcal{C}$ and two disjoint discs $D_1, D_2 \subset S_{y+z}^+$ such that $(D_1 \cup D_2) \cap C = \partial C$. Then (2) or (3) of thm. 13.21 holds.*

Proof of lemma. W.l.o.g. let the indices be chosen so that $D_1 \subset S_y^+$ and $D_2 \subset S_z^+$. Define $S_1 := (S_y^+ - D_1) \cup D_2$ and $S_2 := (S_z^+ - D_2) \cup D_1$. Let $E \subset M(\underline{D})$ denote the 3-ball bound by the 2-sphere $D_1 \cup C \cup D_2$ (recall $M(\underline{D})$ is irreducible). Since S_{y+z}^+ is connected, it follows that S_{y+z}^+ cannot meet the interior of this 3-ball. Thus the pipes from $\mathcal{P}_y$ and $\mathcal{P}_z$ (see 13.15) are cut-and-pasted to form pipes for S_1 and S_2. Let S'' be the surface obtained from S_1, S_2 by cut-and-paste. Then $S'' = S_1 + S_2$, $\#(S_1 \cap S_2) < \#(S_y^+ \cap S_z^+)$ and S'' is isotopic to S_{y+z}^+, by some isotopy which is the identity outside of E. Thus (2) of thm. 13.21 follows, if there are non-negative integer solution-vectors y', z' for $A_{\underline{D}} \cdot x = 0$ with $S_1 = S_{y'}$ and $S_2 = S_{z'}$. Now S_1 (and so S_2) can only fail to have this property if one of the pipes from S_1 lies in the parallelity region between two discs from its basis. In this case it is easy to see that S'' must have this very property too. But S'' is ∂-reduced since S_{y+z}^+ is. So, by prop. 10.16, there is then a ∂-reduced surface S^* with $d(S^*) < d(S^+)$ (it is a straightforward matter to verify that the proof of prop. 10.16 also works if S'' is non-orientable). Thus (3) of thm. 13.21 follows, inductively. This finishes the proof of the lemma.

To continue the proof of the theorem, let us assume that S_y^+, say, is inessential in $M(\underline{D})$ (we are done if both S_y^+, S_z^+ are essential). We here treat the case that S_y^+ is compressible; the argument is similar if S_y^+ is ∂-compressible.

The situation is now analogous to thm. 1.1 of [J-O], and we closely follow the argument given there.

According to our assumption, there is a disc, B, in M with $B \cap S_y^+ = \partial B$ and whose boundary is not contractible in S_y^+. W.l.o.g. we may suppose that $B \cap S_z^+$ is a system of arcs. Moreover, we suppose B is chosen so that the above holds and that, in addition, the number, $\#(B \cap S_z^+)$, of arcs from $B \cap S_z^+$ is as small as possible.

We now distinguish between two cases:

Case 1. $B \cap S_z^+ = \emptyset$.

In this Case we have $\partial B \subset S_{y+z}^+$. Moreover, it follows that ∂B bounds a disc, say G_1, in S_{y+z}^+, for S_{y+z}^+ is essential in $M(\underline{D})$. $G_1 \cap C$ is a *non-empty* system of simple closed curves (B is a compression-disc for S_y^+). Thus there is at least one disc, say e_1, in G_1, different from G_1, with $e_1 \cap C = \partial e_1$. Let C_1 be the annulus from C with $\partial e_1 \subset \partial C_1$. Then $\partial C_1 - \partial e_1$ is a curve in S_{y+z}^+ which bounds a disc, say G_2, in S_{y+z}^+ since it bounds the disc $C_1 \cup e_1$ and since S_{y+z}^+ is essential. As before, there is at least one disc $e_2 \subset G_2$, different from G_2, with $e_2 \cap C = \partial e_2$. Let C_2 be the annulus from C with $\partial e_2 \subset \partial C_2$. Continuing this process, we get a sequence of discs and annuli, $e_1, C_1, e_2, C_2, \ldots$. Since C has finitely many components, there must be an integer $n > 1$ and an integer $1 \leq i \leq n$ such that $\partial C_n - \partial e_n$ is a curve in the annulus $(G_i - e_i)^-$; w.l.o.g. $i = 1$ (recall C_n denotes the annulus from C with $\partial e_n \subset C_n$). The curve $\partial C_n - \partial e_n$ may or may not be contractible in $(G_1 - e_1)^-$.

Assume $\partial C_n - \partial e_n$ is contractible in $(G_1 - e_1)^-$. Let $D \subset G_1$ be the disc bound by $\partial C_n - \partial e_n$. Notice that both $B \cup G_1$ and $D \cup C_n \cup e_n$ are 2-spheres in $M(\underline{D})$. So they bound 3-balls $E_1, E_2 \subset M(\underline{D})$, respectively, since $M(\underline{D})$ is irreducible. Moreover, $E_1 \cap E_2 = D$ since S_{y+z}^+ is connected (it is a surface). But, by lemma 13.22, D contains at least one curve from ∂C other than ∂D (recall our minimal choice of the e_i's). Let C_0 be the annulus from C containing this curve. Let T be the component from S_{y+z}^+ containing $\partial C_0 - D$. Then T lies in the 3-ball E_1, or E_2, without meeting its boundary. In particular, $T \neq S_{y+z}^+$. But S_{y+z}^+ is supposed to be connected. So we have a contradiction to our assumption.

Thus $\partial C_n - \partial e_n$ is not contractible in $(G_1 - e_1)^-$. Then the annuli $C_i \cup (G_i - e_i)^-$, $1 \leq i \leq n$, form a torus T (it cannot be a Klein bottle since all closed surfaces in a handlebody are orientable). Moreover, we can write $S_{y+z}^+ = S_0 + T$, where S_0 is a surface isotopic to S_{y+z}^+. This implies (3) of thm. 13.21. So the theorem is established in Case 1.

Case 2. $B \cap S_z^+ \neq \emptyset$.

Recall from 13.15 that the surface S_x is the union $S_x = B'_x \cup \mathcal{P}_x$ of the basis $\mathcal{B}_x$ and the pipe-system $\mathcal{P}_x$. The pipe-system $\mathcal{P}_x$ is carried by $\Delta \times I$ contained in the attached 2-handles, where Δ is the graph shown in fig. 13.1. In particular, every pipe P from $\mathcal{P}_x$ is the finite union of squares above edges of Δ. Adjacent squares form certain angles, namely the angle of the underlying edges. We distinguish between two types of angles. We say that two adjacent edges form an *acute* or an *obtuse angle* according whether or not they lie in the same triangle of Δ.

Each end-point of every arc from $B \cap S_z^+$ has one acute and one obtuse angle with ∂B. Thus the total number of acute angles in B is twice the number of arcs. But the arc-system $B \cap S_z^+$ splits B into discs whose total number is one bigger than the number of arcs in $B \cap S_z^+$. It follows that at least one of those discs, say B_0, contains at most one acute angle.

If B_0 contains no acute angle at all, then $B_0 \cap S_{y+z}^+$ is a simple closed curve. This curve bounds a disc, G, in S_{y+z}^+ since S_{y+z}^+ is essential. ∂C is a system of arcs and closed curves in S_{y+z}^+ and so $G \cap C$ is a system of arcs in G (see our minimal choice of B). Let $G_0 \subset G$ be a disc such that $\ell := G_0 \cap C = (\partial G_0 - C)^-$ is an arc from $G \cap C$. Note furthermore that G_0 lies either in S_y^+ or in S_z^+. If G_0 lies in S_y^+, then, pushing ∂B across G_0, yields a compression disc for S_y^+ which intersects S_z^+ in strictly fewer arcs. This is a contradiction to our minimal choice of B. If G_0 lies in S_z^+, then let B_1, B_2 be the two discs obtained by first splitting G along ℓ and then taking the union of the resulting discs with two copies of G_0. Then $\partial B_1, \partial B_2 \subset S_y^+$. It follows that B_1 (or B_2) must be a compression-disc for S_y^+. But, after a small general position deformation, B_1 intersects S_z^+ in fewer arcs than B. Thus we get again a contradiction to our minimal choice of B.

Therefore B_0 contains exactly one acute angle. The vertex of this angle is contained in a component, l, from $S_y^+ \cap S_z^+$. This component is either an arc or a closed curve. In any case, let C denote the component from $\mathcal{C}$ associated to l. Then C is a square or annulus according whether l is an arc or closed curve.

If ∂C is inessential in S_{y+z}^+, there are two discs in S_{y+z}^+ which touch each other along l (S_{y+z}^+ is essential). In this case we argue as in Case 1 if l is a closed curve (and similarly if l is an arc).

Thus we may suppose that both components of ∂C are essential in S_{y+z}^+ ($S_{y+z}^!$ is essential). Note that B_0 may be considered as a 2-faced disc where one face is an essential arc in C and where the other face lies in S_{y+z}^+. Set $C' := (C - U(B_0))^- \cup B_0' \cup B_0''$, where B_0', B_0'' are the two copies of B_0 in $\partial U(B_0)$. Since S_{y+z}^+ is essential, it follows that every component from $C' \cap S_{y+z}^+$ separates a disc from S_{y+z}^+. None of these discs contains a component of ∂C since every component from ∂C is essential in S_{y+z}^+. It follows that ∂C separates a surface, C', from S_{y+z}^+ which is either a square or an annulus. Set $S^* := (S_{y+z}^+ - C')^- \cup C$. Then S^* is admissibly isotopic to S_{y+z}^+ since C' is parallel to C (recall $M(\underline{D})$ is ∂-irreducible and B_0 is a 2-faced disc). Straightening bands to pipes and applying prop. 10.16 if necessary, it now follows easily that there is a non-negative integer solution-vector, x, for $A_{\underline{D}} \cdot x = 0$ with S_x^+ is isotopic to S_{y+z}^+ and $d(S_x^+) < d(S_{y+z}^+)$. This establishes (3) of thm. 13.21.

So the theorem is proved in Case 2 as well. $\Diamond$

13.23. Corollary. *Let $\underline{D}$ be a simple Heegaard diagram. Then there is a surface in $(M, k) = M(\underline{D})$ (i.e., disjoint to k) which is orientable and ∂-reduced if and only if the set $\{ S_y^+ \mid y \in \mathcal{B}(\underline{D}) \}$ of basic surfaces contains at least one ∂-reduced surface (which may or may not be orientable).*

Proof. Suppose $\mathcal{B}(\underline{D})$ contains an ∂-reduced surface, say S. Then we are done if S is orientable. Thus suppose the converse. Consider $S' := (\partial U(S) - \partial M(\underline{D}))^-$. Note that S' is orientable. Assume S' can be turned into a disc-system, by successively splitting along compression discs, ∂-compression discs and compression annuli. Then note that this disc-system must lie in the complement of S (since S is ∂-reduced) and that its boundary does not meet k. Hence it must be parallel to a disc-system in $\partial M - k$ since $M(\underline{D})$ is ∂-irreducible. It follows that $k = \emptyset$. This contradiction shows that we get the desired surface by applying the above compressions to S'.

For the other direction suppose there is a surface S in $M(\underline{D})$ which is orientable and ∂-reduced. Then, according to prop. 13.16, there is a non-negative integer solution-vector x for $A_{\underline{D}} \cdot x = 0$ such that $S = S_x^+$. If $x \in \mathcal{B}(\underline{D})$, we are done. If not, let S_y^+ be the maximal basic surface for S_x^+. Set $z = x - y$. Then, by lemma 13.18, S_x^+ is obtained from S_y^+ and S_z^+ by cut-and-paste, i.e., $S_x^+ = S_y^+ + S_z^+$. Hence S_y^+ and S_z^+ cannot be both ∂-parallel since S_x^+ is not ∂-parallel. Moreover, it follows that S_y^+ (resp. S_z^+) is ∂-reduced if S_z^+ (resp. S_y^+) is ∂-parallel (see the argument from lemma 10.14). Thus, replacing S_x^+ by S_y^+ or S_z^+ if necessary, we may suppose that neither S_y^+ nor S_z^+ is ∂-parallel. If S_y^+ is ∂-reduced, we are done. If S_y^+ is essential but not ∂-reduced, then the argument from prop. 10.15 provides us with a ∂-reduced surface whose complexity is strictly smaller than that of $S = S_x$. If S_y^+ is not essential, then thm. 13.20, applied to $S_y^+ + S_z^+$, shows that S_z^+ must be essential (or we get a reduction). Moreover, it is not ∂-parallel. Thus either S_z^+ is ∂-reduced, or prop. 10.15 provides us again with a ∂-reduced surface whose complexity is strictly smaller than that of $S = S_x$. Thus in any case we are either done, or we get a reduction. Hence the existence of the desired basic surface follows by induction. $\Diamond$

13.24. Remark. The previous corollary stresses the importance of the set $\{\ S_y^+ \mid y \in \mathcal{B}(\underline{D})\ \}$ of all basic surfaces for decision problems. (We will later see how decision problems concerning 3-manifolds reduce to decision problems for relative handlebodies). But recall that, in contrast to $\mathcal{F}(\underline{D})$, the set $\mathcal{B}(\underline{D})$ is exponentially growing. For numerical purposes the question therefore arises whether already the set $\{\ S_y^+ \mid y \in \mathcal{F}(\underline{D})\ \}$ has the property expressed in cor. 13.23. Specifically, it may be asked whether a surface obtained from two totally compressible surfaces S_1, S_2 has to be totally compressible itself. This is indeed the case if S_1 or S_2 is a system of discs and 2-spheres. In other cases, however, this may well not be true. To give a concrete example, consider a 3-manifold which contains a Seifert fiber space X over the disc with two exceptional fibers such that ∂X is incompressible. Then there is an annulus in X which splits X into two solid tori V_1 and V_2, say. The boundaries $T_1 := \partial V_1$ and $T_2 := \partial U(V_2)$ are two tori which (after a small general position isotopy) intersect transversally in two curves. It is easy to see that the incompressible torus ∂X is isotopic to a surface obtained from the two totally compressible surfaces T_1, T_2 by some appropriate cut-and-paste. This example illustrates the fact that cut-and-paste is a delicate operation whose results are, in particular, not invariant under isotopy.

13.25. Example. As an illustration of cor. 13.23, consider the presentation $\mathcal{P} = \langle\ g_1, g_2 \mid g_1 g_2 g_1 g_2^{-1} g_1^{-1} g_2^{-1},\ g_1 g_2 g_1^{-2} g_2 g_1 g_2^{-1}\ \rangle$. This presentation comes from a

relative handlebody (M, k) (see [He, p.19] for a picture). In fact, the completed
n-relator 3-manifold $M^*(k)$ is Poincaré's homology 3-sphere. The complexity
of $\mathcal{P}$ is $2 + (l(r_1) - 2) + (l(r_2) - 2) = 11$. Thus, by prop. 13.14, there are
at most 11 fundamental solutions for the associated spectral equation and so
at most $2^{11} = 2048$ basic surfaces. None of these surfaces can be extended
to an essential surface in $M^*(k)$, for Poincare's homology sphere has a finite
fundamental group and so contains no essential surface whatsoever. This raises
the question whether (M, k) itself has any essential surface at all. Because of
cor. 13.23, this is the case if one of the maximal 2048 basic surfaces is essential,
and this in turn can be decided by means of the compression algorithm from
section 12. If the answer is yes, it would not only show that $M^*(k)$ is a non-
Haken homology sphere in a very strong sense, but also this fact would have
been established by a systematic procedure rather than an ad hoc argument.
For an answer see the Appendix.

13.26. *Enumerating Essential Surfaces.*

Let (M, k) be a ∂-irreducible, relative handlebody. Set $\mathcal{F}_n(M, k) :=$
{ admissible isotopy classes of essential surfaces S in (M, k) with $S \cap k = \emptyset$
and $| \chi(S) | = n$ }. Denote by $\mathcal{A}(M, k)$ the subgroup of the mapping class
group of (M, k) generated by all Dehn-twists along essential annuli in (M, k).
Then, by 7.18, $\mathcal{F}_n(M, k)/\mathcal{A}(M, k)$ is a finite set since $\mathcal{F}_n(M, k) \subset \mathcal{S}_n(M, k)$.
$\#(\mathcal{F}_n(M, k)/\mathcal{A}(M, k))$ Denotes the cardinality of this set.

13.27. Definition. *Let (M, k) be a ∂-irreducible, relative handlebody. Then
the* <u>Haken</u> <u>spectrum</u> *of (M, k) is the arithmetic function*

$$\Theta_{(M,k)} : \mathbf{N} \to \mathbf{N}$$

defined by the assignment $n \mapsto \#(\mathcal{F}_n(M, k)/\mathcal{A}(M, k))$.

13.28. Theorem. *Let $\underline{D}$ be a full pant-diagram. Let $(M, k) = M(\underline{D})$ be the
associated relative handlebody. Then the Haken spectrum $\Theta_{(M,k)}$ of (M, k) is
an arithmetic function which can be computed in polynomial time (and therefore
also grows polynomially).*

Remark. *By definition, a function $\theta : \mathbf{N} \to \mathbf{N}$ can be computed in polynomial
time, if there is a polynomial $p : \mathbf{N} \to \mathbf{N}$ such that, for every $n \in \mathbf{N}$, it takes
at most $p(n)$ steps to calculate $\theta(n)$.*

Proof. Let $\mathcal{R}_n(M, k)$, $n \geq 0$, denote the set of all ∂-reduced surfaces, S,
in (M, k) with $|\chi S| = n$. Define $\Theta^* : \mathbf{N} \to \mathbf{N}$ by setting $\Theta^*(n) =$
$\#(\mathcal{R}(M, k)/\mathcal{A}(M, k))$.

Theorem 13.28 follows when we show that Θ^* can be computed in polynomial time. This is due to the special way essential surfaces can be obtained from
∂-reduced ones. To describe this way, let us say that a surface S' is obtained
from an admissible surface S in (M, k) by *adding a surface G*, provided
(1) $G \subset \partial M$ and $\partial G \subset k \cup \partial S$ and provided (2) $S' = (S - U(G))^- \cup G'$,

where G' is the copy of G in $(\partial U(G) - \partial M)^-$. It is easily seen that every essential surface in (M, k) is obtained from a ∂-reduced surface by successively adding surfaces (property (1) follows from the ∂-incompressibility of essential surfaces). Our claim then follows from the fact that only a limited number of essential surfaces can be obtained from a ∂-reduced one by adding surfaces in ∂M (limited by some linear function in the Euler characteristic of the ∂-reduced surfaces involved).

By what we have just seen, it remains to show that Θ^* can be computed in polynomial time. Let us first assume that (M, k) is simple; in the sense that it contains no essential annulus (the other case will be discussed below).

Let $\mathcal{P} = \mathcal{P}(\underline{D}) = <g_1, ..., g_m \mid r_1, ..., r_n>$ be the presentation associated to $\underline{D}$. Let $\mathcal{B}(\underline{D})$ be a complete set of basic solutions of the spectral equation $A_{\underline{D}} \cdot x = 0$. By prop. 13.19, every ∂-reduced surface S in (M, k) can be obtained from basic surfaces by cut-and-paste. Thus there is a linear combination $y = \sum \gamma_i y_i$, $\gamma_i \geq 1$, $y_i \in \mathcal{B}(\underline{D})$, such that S_y^+ is admissibly isotopic to S. Moreover, we have $\chi S = \sum \gamma_i \chi S_{y_i}^+$ (which incidentally gives an alternative proof for the finiteness of essential surfaces in simple relative handlebodies). The set $\mathcal{G}_n$ of all linear combinations $y = \sum \gamma_i \chi S_{y_i}^+$ with $|\chi S_y^+| \leq n$ can clearly be enumerated in polynomial time (in n).

By what has been said before, the set $\mathcal{R}_n(M, k)$ is a subset of the set $\{ S_y^+ \mid y \in \mathcal{G}_n \}$. By cor. 12.21, it can be decided in polynomial time whether a surface S_y^+, $y \in \mathcal{G}_n$, is essential in (M, k). Moreover, by cor. 13.8, it can be decided in polynomial time whether two surfaces S_y^+, S_z^+, $y, z \in \mathcal{G}_n$, are admissibly isotopic in (M, k). Using the presentation of surfaces by circle-patterns as in sections 12, it is finally not difficult to construct a polynomial algorithm for deciding whether a surface S_y, $y \in \mathcal{B}(\underline{D})$, is orientable and connected (see Appendix). It follows that the set $\mathcal{F}_n(M, k)$ can be sorted out in polynomial time from the set $\{ S_y^+ \mid y \in \mathcal{G}_n \}$. Since $\#\mathcal{G}_n$, and so $\#\mathcal{R}_n(M, k)$, grows polynomially, it follows that $\#\mathcal{R}_n(M, k)$ can be determined in polynomial time when $\mathcal{R}_n(M, k)$ is constructed. Thus, altogether, the function $\Theta^* : \mathbf{N} \to \mathbf{N}$ can be computed in polynomial time. This proves the theorem under the hypothesis that (M, k) is simple.

If (M, k) is not simple, then proceed as follows. First construct the characteristic submanifold V in $(M, U(k))$ (see [Joh 1] and note that $(M, U(k))$ is a Haken 3-manifold with useful boundary-patterns). Then form the relative handlebody (M', k') by setting

$$M' = (M - V)^- \text{ and } k' := k \cup \partial(\partial V - \partial M)^-.$$

It is not difficult to see that the theorem holds for (M, k) if it holds for (M', k'). But, by what we have seen before, the theorem is indeed true for (M', k') since (M', k') is a simple relative handlebody. $\diamond$

III. GENERALIZED ONE-RELATOR 3-MANIFOLDS

Having studied relative handlebodies to some extent, we now turn our attention to 3-manifolds obtained by adding 2-handles to the relator-curves of relative handlebodies. We call these 3-manifolds n-relator 3-manifolds, where n denotes the number of 2-handles attached. Notice that n-relator 3-manifolds are not just 3-manifolds. Indeed, the notion of an n-relator 3-manifold does not only fix the homeomorphism type of a 3-manifold but also an internal combinatorial structure for it.

There is a considerable difference between one- and higher relator 3-manifolds. One-relator 3-manifolds are well-behaved and much easier to deal with. Very much like the similar situation in combinatorial group theory, where one-relator groups are known to be much more accessible than higher relator groups. In this chapter we concentrate on one-relator 3-manifolds. However, it will turn out that some important properties of one-relator 3-manifolds carry over to n-relator 3-manifolds as well. Indeed, we view the class of one-relator 3-manifolds as a testing ground for finding results which may be valid for general n-relator 3-manifolds. In this sense the present chapter is a preparation for later chapters.

By its very definition, a one-relator 3-manifold is obtained by attaching a single 2-handle to a handlebody M. It turns out, however, that many results concerning one-relator 3-manifolds remain valid under the weaker hypothesis that M is only ∂-reducible. Thus, in order to treat the subject in its proper generality, we decided to consider "generalized one-relator 3-manifolds" as much as possible. Here a *generalized one-relator 3-manifold* is defined to be an (orientable) 3-manifold obtained from some ∂-reducible 3-manifold by attaching a 2-handle along some simple closed curve in its boundary. Of course, every one-relator 3-manifold is also a generalized one-relator 3-manifold. Notice further that, dually, a 3-manifold N is a (generalized) one-relator 3-manifold if and only if it contains at least one proper arc whose complement is a handlebody (resp. a ∂-reducible 3-manifold). Such an arc is called a *Heegaard string* for N. The study of one-relator 3-manifolds is therefore also the study of certain (knotted) arcs in 3-manifolds.

This chapter begins with a study of the interaction between incompressible surfaces and Heegaard strings. Specifically, we establish in section 15 the fundamental technical result that non-separating surfaces with minimal complexity avoid Heegaard strings. In section 16 we generalize this result to essential surfaces in 3-manifolds with useful boundary-patterns. This generalization makes induction over hierarchies of surfaces possible. In section 17 we use such an induction in order to show the remarkable fact that homotopy implies isotopy for Heegaard strings; a fact which is well-known and basic for incompressible surfaces in Haken 3-manifolds. As a consequence of this property we establish and study in section 18 the close relationship between the mapping class group of one-relator 3-manifolds and relative handlebodies. In section 19 and 20 we reduce the incompressibility problem for one-relator 3-manifolds to that for relative handlebodies which in turn has been solved in section 12. Section 21 is concerned with the calculation of the Haken spectrum for one-relator 3-

manifolds. In section 22 we illustrate our discussion by means of a concrete example.

§14. Good Pencils.

The purpose of this section is to set up the basic notion of a "good pencil". The concept of "good pencils" helps to clarify the difference between (generalized) one-relator 3-manifolds with *non-separating* and those with *separating* attaching curve. It turns out that the latter are much simpler to deal with and enjoy stronger properties, mainly because their good pencils are easier to handle. Higher relator 3-manifolds have good pencils too. But good pencils for higher relator 3-manifolds are somewhat hidden and have to be recovered with some care. This will create some extra complications. But, in any case, good pencils will play a crucial part in our study. In this section we introduce "good pencils" and establish some of their properties. We do this general enough so that we can use the material in our later study of surfaces in n-relator 3-manifolds as well.

14.1. Throughout this book $M^+ = M^+(k)$ denotes a (generalized) n-relator 3-manifold, i.e., by definition, a 3-manifold obtained from some handlebody (resp. ∂-reducible 3-manifold) M by attaching a system E of 2-handles along a system $k \subset \partial M$ of n simple closed curves. Let $\mathcal{D}$ be a system of essential, i.e. non ∂-parallel, discs in M. Often $\mathcal{D}$ will be just a single disc or it will be a meridian-system, i.e., a system of discs which splits M into a single 3-ball. Given $\mathcal{D}$, set

$$I = I(\mathcal{D}) := U(k) \cap \partial \mathcal{D},$$

where the regular neighborhood $U(k)$ is taken in ∂M (it is to be understood that $U(k)$ equals $E \cap \partial M = (\partial M - \partial M^+)^-$). We call I the *interval-system* in $\partial \mathcal{D}$.

A 2-manifold S^+ in $M^+(k)$ (not necessarily essential or connected) is called a *normal 2-manifold* (w.r.t. $\mathcal{D}$), if the following holds:

(1) $S^+ \cap E$ consists of discs which are parallel in E to $(\partial E - \partial M)^-$.

(2) $S := M \cap S^+$ is incompressible in M.

(3) $S \cap \mathcal{D} = S^+ \cap \mathcal{D}$ consists of arcs alone, none of which has both its end-points in one component of $I(\mathcal{D})$.

A normal 2-manifold, S^+, is *strongly normal*, if, in addition,

(4) For every component, U_0, from $U(k)$, the closures of any two components from $\partial \mathcal{D} - \partial U_0$, joined by some arc from $S \cap \mathcal{D}$, are disjoint.

Remark. Normal 2-manifolds in the above sense are very similar to Haken's normal surfaces [Ha 1] (see also [Schu]). Observe further that every essential 2-manifold S^+ in a (generalized) *one-relator* 3-manifold $M^+(k)$ can be isotoped into a strongly normal 2-manifold, provided $M^+(k)$ is irreducible (recall discs are not excluded whenever they are essential in $M^+(k)$). In this case, $S \cap \mathcal{D}$ is a system of essential arcs in $(\mathcal{D}, I)$.

14.2. Definition. *Let $M^+(k)$ and $\mathcal{D}$ be given as above, and let S^+ be a normal 2-manifold in $M^+(k)$. Then a sub-system, P, of arcs from $S \cap \mathcal{D}$ is called a $\mathcal{D}$-_pencil_ (or simply a _pencil_) _for_ S (or S^+), if there is a component, I_0, of I such that the following holds:*

(1) every arc from P has one end-point in I_0, and

(2) every arc from $S \cap \mathcal{D}$ which has one end-point in I_0 belongs to P.

The interval I_0 is called a *root* for P (a root for P is unique whenever P is not a system of pairwise parallel arcs). If k_0 is a component from k, then a pencil P is *good for* k_0, if, in addition, I_0 is a component from $U(k_0) \cap \partial\mathcal{D}$ and every arc from P joins I_0 with one of those components from $U(k_0) \cap \partial\mathcal{D}$ which neighbor I_0 in $\partial\mathcal{D}$. Moreover, a good pencil P is called *very good for* k_0, if every arc from P is essential and recurrent in the 2-manifold S (an arc in S is *recurrent*, if both its end-points lie in one component of ∂S).

14.3. Example.

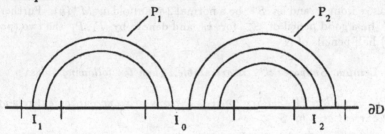

(Figure 14.1.)

The next lemma establishes the existence of good pencils for (generalized) one-relator 3-manifolds.

14.4. Lemma. *Let $M^+(k)$ be a generalized one-relator 3-manifold with a fixed system $\mathcal{D}$ of discs in M. Suppose $\mathcal{D}$ is not disjoint to k. Then every strongly normal 2-manifold in $M^+(k)$, not contained in M, admits at least one (non-trivial) good $\mathcal{D}$-pencil.*

Proof. Either lemma 14.4 is true, or there is at least one arc t from $S \cap D$ which has one end-point in I and which does not join two neighboring intervals from I. Let t be chosen so that the latter holds and that, in addition, it separates a disc D_1 from $\mathcal{D}$ with the property that any arc from $D_1 \cap S$, different from t, joins neighboring intervals. Of course, this choice is always possible since $\mathcal{D}$ is a disc-system. Then either $D_1 \cap S = t$, or $D_1 - t$ contains at least one pencil entirely. The first case is impossible since S^+ is strongly normal, and in the second case note that the pencil contained in $D_1 - t$ has to be good (recall our minimal choice of t). This proves the lemma. $\diamond$

14.5. We next establish some useful properties of good pencils which will be needed later. For this we first introduce some more notation.

A proper arc t from a pencil P for the 2-manifold $S\ (= M \cap S^+)$ is called

(1) *recurrent,* if both its end-points are contained in one component of $E \cap \partial S$ (otherwise: non-recurrent), and

(2) *essential* in S^+, if ∂t is contained in $E \cap \partial S$ and if either it joins two different components of $E \cap S^+$ or its union with some arc in $E \cap S^+$ defines a non-contractible closed curve in S^+.

If P is a good pencil for S^+, note that P falls into at most two sub-systems (notation: *half-pencils*) each of which containing all arcs from P joining the same two components of the interval-system I.

We use this fact to formulate our next result concerning pencils. Let $M^+(k)$ denote an n-relator 3-manifold or a generalized one-relator 3-manifold. Let k_0 be a curve from k and let S^+ be a normal 2-manifold in $M^+(k)$. Furthermore, let P be a good pencil of S^+ for k_0 and denote by P_1, P_2 the two (possibly equal) half-pencils of P.

14.6. Lemma. *Suppose S^+ is orientable. Then the following holds:*

(1) P_1 as well as P_2 either consists entirely of recurrent, or entirely of non-recurrent arcs in S.

(2) If $\#P$ is odd and $\#P_1 \geq \#P_2$, then P_1 consists entirely of recurrent arcs.

Proof. We follow a line of argument from [Joh 3]. Let I_0 be a root for P. Denote by A the annulus from $E \cap \partial M$ with $I_0 \subset A$.

(1) It suffices to show that P_1 consists of recurrent arcs, provided it contains at least one. To this end let b_0 be some recurrent arc from P_1. Let b_1 be an arc from P_1 next to it. Assume b_1 is not recurrent. Then the two end-points of b_1 lie in two different curves from $A \cap (\partial S - \partial_0 S)$, where $\partial_0 S$ denotes that component of ∂S containing ∂b_0. Both these curves lie next to $\partial_0 S$ since b_1 has been chosen next to b_0. Traveling the parallelity region between b_0 and b_1, we find that S must be one-sided. But S^+, and so S, is supposed

to be orientable. So we have a contradiction to our assumption. Hence b_1 is recurrent as well and our claim follows inductively.

(2) By hypothesis, $P \cap I_0$ consists of an odd number of points. In particular, $P \cap I_0$ has a mid-point. In fact, the collection $\partial S \cap A$ has a middle-curve (containing the mid-point). Let b be the arc from P which starts at the mid-point. Then, more precisely, b must be an arc from P_1 since $\#P_1 \geq \#P_2$. Now, P_1 consists of parallel arcs. So, traveling along P_1, we find that b must end at the middle curve (the two end-points of b have the same "distance" from ∂A). In particular, b is recurrent. So, by (1), P_1 consists of recurrent arcs.
◊

§15. The Handle-Addition Lemma (Absolute Form).

This section is concerned with a first version of the handle-addition lemma. Indeed, we here discuss the handle-addition lemma for generalized one-relator 3-manifolds. Other, more general versions, will be given later. Generally speaking, the purpose of any handle-addition lemma is to provide numerical estimates and other information on how essential surfaces in n-relator 3-manifolds do intersect 2-handles. This information is crucial for a variety of applications. The present section deals with the simplest situation. Nevertheless, it contains in a nutshell some of the most basic ideas underlying this book, which in later sections will be developed in various directions. The content of this section is basically a refined version of [Joh 3].

15.1. *Preliminaries.*

Let us begin with the following observation, which we formulate for arbitrary n-relator 3-manifolds.

15.2. Lemma. *Let $M^+(k)$ be an n-relator 3-manifold. Let $\mathcal{D}$ be a meridian-system in the handlebody M and let S^+ be a surface in $M^+(k)$. Suppose that S^+ is normal with respect to $\mathcal{D}$ and that no meridian-system with this property intersects S^+ in a strictly smaller number of curves.*

Then $S^+ \cap \mathcal{D}$ consists of arcs which are essential in $S = S^+ \cap M$.

Remark. Recall a meridian-system splits a handlebody into a single 3-ball.

Proof. Assume the converse. Then there is a component t from $S^+ \cap \mathcal{D} = S \cap \mathcal{D}$ which is an inessential arc in S. Thus t separates a disc D_0 from S, and w.l.o.g. we suppose D_0 is chosen so that $D_0 \cap \mathcal{D} = t$. Now, let D_1 be the component from $\mathcal{D}$ containing the arc t. Then $\mathcal{D} - D_1$ is a system of discs which splits M into a solid torus, say M^*. The 2-manifold $(\partial U(D_1 \cup D_0) - \partial M^*)^-$ consists of three proper discs in M^*. Two of them are non-separating discs in M^* since D_1 is non-separating in M^*. One of the latter discs is parallel in $U(D_1 \cup D_0)$ to D_1; let D_1' be the other one. Then, replacing D_1 by D_1', we clearly obtain from $\mathcal{D}$ a new meridian-system for M, and S^+ is certainly normal w.r.t. this new system as well. However, this new system intersects S^+ in strictly less arcs than $\mathcal{D}$, and so we get a contradiction to our choice of $\mathcal{D}$. This proves the lemma. $\Diamond$

We are now ready to give the first version of the handle-addition lemma. As indicated before, we may distinguish between two cases according whether or not the attaching curve k is separating.

15.3. *The Separating Case.*

We first treat *discs* in one-relator 3-manifolds.

15.4. Proposition. *(1) If $M^+(k)$ is a generalized one-relator 3-manifold such that k is separating in ∂M, then $M^+(k)$ is irreducible and ∂-irreducible if and only if $\partial M - k$ is incompressible in M.*

(2) If $M^+(k)$ is an irreducible one-relator 3-manifold such that k is separating in ∂M, then every disc in $M^+(k)$ can be isotoped into the handlebody M.

Remark. Discs in (generalized) one-relator 3-manifolds have first been studied by Haken, and then by [Pr] and [Ja].

Proof. We only prove (2); the proof for (1) is similar. $M^+(k)$ is supposed to be irreducible. So every disc in $M^+(k)$ can be isotoped into a disc which is strongly normal w.r.t. some meridian-system for M. Thus (2) is an immediate consequence of the next lemma:

15.5. Lemma. *Let $M^+(k)$ be a one-relator 3-manifold such that $k \subset \partial M$ is separating. Suppose $S^+ \subset M^+(k)$ is a disc or 2-sphere which is strongly normal with respect to some meridian-system for M. Then S^+ is not a 2-sphere if k does not bound a disc, and S^+ is contained in M if S^+ is a disc.*

Proof. Assume the converse. Let $\mathcal{D}$ be a meridian-system for M such that S^+ is strongly normal w.r.t. $\mathcal{D}$. Recall from lemma 15.2 that the meridian-system $\mathcal{D}$ can be chosen so that, in addition, $S^+ \cap \mathcal{D}$ consists of arcs which are essential in $S := S^+ \cap M$.

Suppose k bounds a disc in M and S^+ is a disc. Then recall from our assumption, that at least one boundary curve of S lies in $U(k)$ ($= E \cap \partial M$). This boundary curve bounds a disc in M. Thus S is a disc itself since S^+ is normal. This however contradicts the hypothesis that S^+ is a disc.

Suppose k does not bound a disc. Then, in particular, $k \cap \mathcal{D} \neq \emptyset$. Thus the hypothesis of lemma 14.4 is satisfied, and so S^+ admits at least one good $\mathcal{D}$-pencil, say P. Let t be any arc from P which separates a disc D_1 from $\mathcal{D}$ with $D_1 \cap P = t$. Then note that the arc $D_1 \cap \partial M$ intersects only one of the two components of $\partial U(k)$ since k is supposed to be separating. So t has to be recurrent. It then follows from lemma 14.6 that P consists entirely of recurrent arcs. All these arcs are components of $S \cap \mathcal{D}$ and so essential in S (see above). Hence every component from $\partial S - \partial S^+$ contains both end-points of at least one essential arc in S. But this is impossible since ∂S^+ is connected and since $S = S^+ \cap M$ is a planar surface.

Thus, in any case, we get a contradiction to our assumption. The proof of lemma 15.5 is therefore finished. $\Diamond$

In order to study other essential surfaces in (generalized) one-relator 3-manifolds, we need the following notion.

15.6. Definition. *Let S, S' be two 2-manifolds in a 3-manifold N. We say S' is obtained from S by an __annulus-modification__, if*

(1) $(S - S')^-$ is a non-contractible annulus in S, and

(2) $(S - S')^- \cup (S' - S)^-$ is a boundary-parallel annulus in N.

We further say that S' is obtained from S by *annulus-modifications* if there is a finite sequence of surfaces

$$S = S_1, S_2, ..., S_n = S'$$

such that S_{i+1}, $1 \leq i \leq$ n-1, is obtained from S_i, by some annulus-modification.

Given this notation, our previous result on discs may now be generalized to arbitrary surfaces as follows.

15.7. Proposition. *Let $M^+(k)$ be a generalized one-relator 3-manifold which is irreducible and ∂-irreducible. Suppose k is separating in ∂M. Furthermore, let S^+ be an essential surface in $M^+(k)$. Then there is a surface, S^*, in $M^+(k)$ with*

(1) S^* can be obtained from S^+ by annulus-modifications, and

(2) S^* can be (properly) isotoped in $M^+(k)$ into the sub-manifold $M \subset M^+(k)$.

As an immediate corollary we obtain

15.8. Corollary. *Let $M^+(k)$, k separating in ∂M, be a generalized one-relator 3-manifold which is simple. Then every non-separating surface S^+ in $M^+(k)$ whose Betti number is minimal, can be isotoped into the sub-manifold $M \subset M^+(k)$.* ◊

Remark. Notice that splitting a surface along a non-separating, simple closed curve reduces its Betti number (but not its Euler-characteristic).

Proof of Proposition. Fix an essential disc, D, in M with $D \cap \partial M = \partial D$. Such a disc exists since $M^+(k)$ is a generalized one-relator 3-manifold and so, by hypothesis, M is ∂-reducible. Let D be chosen so that the above holds and that, in addition, the intersection $D \cap S^+$ consists of curves whose number is as small as possible.

Suppose $D \cap k = \emptyset$. Then $\partial D \subset \partial M^+(k)$. Hence ∂D bounds a disc $D' \subset \partial M^+(k)$ since $M^+(k)$ is ∂-irreducible. The union $D \cup D'$ is a 2-sphere which bounds a 3-ball $W \subset M^+(k)$ since $M^+(k)$ is irreducible. Obviously, the surface S^+ can be isotoped out of W. But D' cannot be contained in ∂M since D is an essential disc in M. Thus and since $D \subset M$, it follows that $M^+(k) - W \subset M$. So the conclusion of prop. 15.7 follows.

By what we have just seen, we may suppose $D \cap k \neq \emptyset$. Let S^+ be isotoped into a strongly normal surface (see 14.1). Then, by lemma 14.4, we may suppose that the disc D admits at least one non-trivial, good pencil, say P. Let P_1, P_2 denote the two (possibly equal) half-pencils of P. By an argument from 15.5, P consist of recurrent arcs.

We further claim the arcs from P are essential in S^+ (see 14.5 for definition). Assume the converse. Then there is an arc, t, from P_1, say, which is inessential in S^+. Let E denote the 2-handle attached along k. Then, by definition, there is disc E_0 from $E \cap S^+$ and an arc t' in ∂E_0 such that $\partial t = \partial t'$ and $t \cup t'$ forms a closed curve which bounds a disc $A \subset S^+$ with $E_0 \not\subset A$. Observe that, by our minimal choice of $D \cap S^+$, no arc from P is inessential in the surface $(S^+ - E)^-$. Thus, in particular, the disc A has to contain at least one component from $E \cap S^+$. Now every component of $\partial(E \cap S^+)$ is an essential, simple closed curve in the annulus $U(k)$. In particular, every component from $\partial(E \cap S^+)$ must contain at least one end-point from P. But all arcs from P with an end-point in $E \cap A$ are recurrent arcs contained in the disc A. By induction, it easily follows that this is inconsistent with the facts that (1) A is a disc and that (2) all arcs from P are essential in the surface $(S^+ - E)^-$. This contradiction proves our claim.

We are now ready to construct the desired sequence of annulus-modifications of S^+ which yields a surface not meeting the 2-handle E at all. For this let t be again an outermost arc from the pencil P and let D_1 be the disc separated from D by t with $D_1 \cap P = t$. Recall that $D_1 \cap U(k)$ consists of two disjoint arcs. So D_1 may be viewed as a square. Let E' be that component in which the 2-handle E is split by S^+ and which contains the arcs $D_1 \cap U(k)$. Set $U := U(D_1 \cup E')$, where the regular neighborhood is taken in $M^+(k)$. Observe that S^+ splits $(\partial U - \partial M^+(k))^-$ into three annuli. Let A_1 and A_2 be those two annuli of them which meet $\partial M^+(k)$. Define

$$S_1^+ := (S^+ - U)^- \cup A_1 \cup A_2.$$

Then S_1^+ is obtained from S^+ by some annulus-modification. Moreover, it intersects the 2-handle E in strictly fewer discs than S^+. Applying the same procedure to the other arcs from the pencil P as well, we eventually get a sequence of surfaces

$$S^+ = S_0^+, S_1^+, S_2^+, ..., S_n^+$$

such that $S_n^+ \subset M$ and that S_{i+1}^+, $0 \leq i \leq$ n-1 is obtained from S_i^+ by some annulus-modification. This finishes the proof. $\Diamond$

15.9. *The Non-Separating Case.*

We next treat the non-separating case, i.e., the case when the curve k is non-separating in ∂M (actually, we will not really insist in this condition). The idea will be to reduce the non-separating to the separating case. It turns out that we need "torus-modifications" in addition to the annulus-modifications familiar from the separating case.

15.10. Definition. *Let S, S' be two 2-manifolds in a 3-manifold N. We say S' is obtained from S by a torus-modification, if*

(1) $(S - S')^-$ is a non-contractible annulus in S, and

(2) $(S - S')^- \cup (S' - S)^-$ *is a torus which bounds a solid torus in* N *neither containing* S *nor* S'.

We further say that S' is obtained from S by *torus-modifications* if there is a finite sequence of 2-manifolds

$$S = S_1, S_2, ..., S_n = S'$$

such that S_{i+1} is obtained from S_i, $1 \leq i \leq$ n-1, by some torus-modification. Notice that, in contrast to annulus-modifications, a torus-modification does not change the homeomorphism-type of a 2-manifold, but it may very well change the isotopy class of its embedding-type. The next result tells us that surfaces and 2-handles of generalized one-relator 3-manifolds can be freed from each other, using annulus- and torus-modifications. This is a remarkable phenomenon. It occurs in some form for higher-relator 3-manifolds as well (see chapter IV).

15.11. Proposition. *Let* $M^+(k)$ *be a generalized one-relator 3-manifold which is irreducible and* ∂-*irreducible. Let* S^+ *be any essential 2-manifold in* $M^+(k)$. *Then there is a 2-manifold,* S^*, *in* $M^+(k)$ *with*

(1) S^* *can be obtained from* S^+ *by annulus- and torus-modifications, and*

(2) S^* *can be isotoped in* $M^+(k)$ *into the sub-manifold* $M \subset M^+(k)$.

Remark. If S^+ has minimal Betti number and is different from an annulus or torus, then torus-modifications alone suffice.

As an immediate consequence of prop. 15.11 we obtain the following corollary.

15.12. Corollary. *Let* $M^+(k)$ *be a generalized one-relator 3-manifold which is irreducible and* ∂-*irreducible. If* S^+ *is a non-separating surface in* $M^+(k)$, *then there is a non-separating, essential surface,* S, *in* (M, k) *with* $S \cap k = \emptyset$ *and* $\chi S \geq \chi S^+$. $\diamond$

The following observation will be needed to reduce prop. 15.11 to the separating case.

15.13. Lemma. *Let* N *be any 3-manifold with non-empty boundary. Let* $\partial_0 N$ *be a component from* ∂N. *Let* $T_0 \subset \partial_0 N$ *be a torus with one hole such that* $(\partial_0 N - T_0)^-$ *is neither a disc nor compressible in* N. *Then* T_0 *is incompressible in* N.

Proof. Assume the converse. Then there is a disc, D, in N with $D \cap \partial N = \partial D \subset T_0$ such that ∂D is not contractible in T_0. Since $(\partial_0 N - T_0)^-$ is neither a disc nor compressible in N, ∂D cannot be boundary-parallel in T_0. Since T_0 is a punctured torus, it follows that ∂D has to be non-separating in T_0. In this case, however, there is some simple closed curve, t, in T_0 which intersects

∂D in precisely one point. Using this curve, define $D^* := (\partial U(t \cup D) - \partial N)^-$, where the regular neighborhood is taken in N. Then D^* is a disc in N with $D^* \cap \partial N = \partial D^* \subset T_0$. Furthermore, ∂D^* is separating in T_0 since it bounds the regular neighborhood of $t \cup \partial D$ in T_0. Thus ∂D^* is ∂-parallel in T_0. Hence, replacing D by D^*, we obtain a contradiction as above. $\Diamond$

Proof of Proposition 15.11. We use the same set-up as in the proof of prop. 15.7. Specifically, we fix an essential disc, D, in M, $D \cap \partial M = \partial D$, chosen in such a way that the intersection $D \cap S^+$ consists of curves whose number is as small as possible. By an argument from the proof of prop. 15.7, we may suppose that $D \cap k \neq \emptyset$. Let S^+ be isotoped so that it is a strongly normal surface and that, in addition, the intersection with the 2-handle E is as small as possible. Then, again by lemma 14.4, we may suppose that the disc D contains at least one non-trivial good pencil, P, for S^+. Let P_1, P_2 denote the two (possibly equal) half-pencils of P.

We may suppose that there is no recurrent arc in P which is essential in S^+, for any such arc gives rise to an annulus-modification which reduces the intersection of S^+ with the 2-handle E (see the argument from the proof of prop. 15.7). It then follows that at least one arc from P is non-recurrent. To see this, assume all arcs from P are recurrent and inessential in S^+. Then a disc, B, in S^+ is easily constructed with $\partial B \subset S^+ - E$ and which contains P. It then follows, by an induction, that at least one arc from P has to be inessential in the surface $(S^+ - E)^-$. So we get a contradiction to our minimal condition on $D \cap S^+$.

15.14. Lemma. *If* $\#P_1 \geq \#P_2$, *then no arc from* P_1 *is recurrent.*

Proof of lemma. Assume the converse. Then, by lemma 14.6, the half-pencil P_1 consists entirely of recurrent arcs. By what has been seen above, at least one arc from P is non-recurrent, and so, by 14.6 again, the entire half-pencil P_2 consists of non-recurrent arcs.

Now, denote by E' the union of all components from $E \cap S^+$ containing some end-point of P_1. Of course, $\#P_1$ equals the total number of components of E'. Moreover, P_2 has a very special configuration (in the surface $S := (S^+ - E)^-$). Indeed, recalling the hypothesis $\#P_1 \geq \#P_2$, a moments reflection shows that

(1) each component of E' (resp. $(S^+ \cap E) - E'$) contains at most (resp. precisely) one end-point of P_2, and

(2) no two components neither from E', nor from $(S^+ \cap E) - E'$, are joined by an arc from P_1.

Set $U := U((S^+ \cap E) \cup P_2)$, where the regular neighborhood is taken in S^+. Then U is a system of discs and every component from U contains the end-points of at least one arc from P_1. But no arc from P_1 is essential in S^+ (see above). So $U \cup P_1$ is contained in a disc in S^+. It follows, inductively, that at least one arc from P_1 has to intersect $(S - U)^-$ in an inessential arc, where $S = (S^+ - E)^-$. This arc is either inessential in S too, or separates an annulus from S containing an arc from P which is inessential in S. To see the latter recall the properties of P_2 described above and observe that every component of $\partial(E \cap S^+)$ contains the same number of end-points of P. However, the existence of an arc from P which is inessential in S contradicts our minimality condition on $D \cap S^+$. This proves lemma 15.14.

With this lemma we are now ready to construct the desired sequence of annulus- and torus-modifications of S^+ which yields a 2-manifold not meeting the 2-handle E.

To this end let P_1 be a half-pencil from P with $\#P_1 \geq \#P_2$. Then, by the previous lemma, no arc from P_1 is recurrent in S^+. Let I_0 be a root of the pencil P (see 14.2). $\#(P \cap I_0)$ cannot be odd, for otherwise the set $P \cap I_0$ had a mid-point and the arc from P, starting at this mid-point, would be recurrent. Since $\#P$ is even and since $\#P_1 \geq \#P_2$, we may suppose $\#P_1 = \frac{1}{2} \cdot \#(P \cap I_0)$ (otherwise replace P_1 by an appropriate sub-system of P_1). In particular, we suppose the components of $E \cap S^+$ are paired by the arcs from P_1.

Since P_1 is a half-pencil (or part of it), there is an arc, t, from P_1 which separates a disc, D_0, from D with $P \cap D_0 = P_1$. Let $t' \subset \partial M$ be the arc $t' := D_0 \cap \partial D$. Set

$$W := E \cup U(t'), \quad N := (M - U(t'))^-, \quad \text{and} \quad T := (\partial W - \partial M^+(k))^- = W \cap N,$$

where the regular neighborhood is taken in M. Then W is a solid torus and T is a punctured torus in ∂N.

Pushing all arcs from P_1 across the disc D_0 and into $U(t')$, we see that S^+ is isotopic in $M^+(k)$ to a surface S' with

(1) $S' \cap W$ consists of discs (recall no arc from P_1 is recurrent), and

(2) $\partial(S' \cap W) \subset U(\partial T)$, where the regular neighborhood is taken in ∂N.

Thus there is a 2-handle, E', embedded in W and attached to the curve ∂T, such that $S' \subset N \cup E'$. Now, ∂T is clearly separating in ∂N and $N^+(\partial T) := N \cup E'$ is a generalized one-relator 3-manifold. Thus we are in the situation of prop. 15.7. Therefore the intersection of S' with the 2-handle E' can be removed by successive annulus-modifications.

More precisely, there is a sequence, $S' = S_1, S_2, ..., S_n$, of 2-manifolds such that $S_n \cap E' = \emptyset$ and that S_{i+1} is obtained from S_i by an annulus-modification. Let us consider the annulus-modification which transforms S_1 into S_2. At this point we have to refer to the end of the proof of prop. 15.7 for the special way these annulus-modifications are actually constructed. We find that there is a square, $\hat{A}$, in M', crucial for this construction, with two faces in one component $\hat{E}$ (obtained from E by splitting along S'), one face in S'

and the remaining face in ∂N. Indeed, $(S_2 - S_1)^- \cup (S_1 - S_2)$ is the frontier of $U := \hat{E} \cup U(\hat{A})$ (where the regular neighborhood is taken in the closure of the complement of S^+).

Recall one disc from $E' \cap \partial N^+ (\partial T)$ lies in a torus component, say T', from $\partial N^+ (\partial T)$ and the other lies in $\partial N^+ (\partial T) - T' \subset \partial M^+(k)$. In particular, $U \cap \partial N^+ (\partial T)$ lies either in T', or in $\partial N^+ (\partial T) - T'$. If $U \cap \partial N^+ (\partial T) \subset \partial N^+ (\partial T) - T'$, then the above annulus-modification is actually an annulus-modification for S', and so for S^+, in the original generalized one-relator 3-manifold $M^+(k)$, reducing the intersection of S' with the original 2-handle E. If $U \cap \partial N^+ (\partial T) \subset T'$, then recall T' bounds a solid torus in $W \subset M^+(k)$. The union of U with this solid torus is again a solid torus, say W'. Moreover, $W' \cap S_1 = U \cap S_1 \subset S_1 \cap \partial W'$ is an essential annulus in S_1. Thus

$$S_2' := (S_1 - W')^- \cup (\partial W' - S_1)^-$$

is obtained from S_1 by a torus-modification (in $M^+(k)$ and along W'). Moreover, S_2' intersects W, and so the 2-handle E, in strictly fewer discs than S_1. Replacing S_1 by S_2' and repeating the previous procedure over and over again, we get a sequence of annulus- and torus-modifications which turns S', and so S^+, into a 2-manifold S^* with $S^* \cap E = \emptyset$.

This finishes the proof of prop. 15.11. $\diamond$

15.15. By an obvious modification (or better: simplification) of the above method we obtain the original form of the *handle-addition lemma* (due to [Ja]).

15.16. Proposition. *Let $M^+(k)$ be a generalized one-relator 3-manifold whose boundary is different from the 2-sphere. Then $M^+(k)$ is irreducible and ∂-irreducible if and only if $\partial M - k$ is incompressible in M.* $\diamond$

Remark. This particular result has attracted some attention in recent years. For other proofs and applications see also [Sch 1][CG][CGLS].

15.17. Corollary. *Let $\underline{D}$ be a useful, one-relator diagram. Then $M^+(\underline{D})$ is irreducible and ∂-irreducible if and only if $\underline{D}$ is irreducible. (See lemma 7.4).* $\diamond$

Other interesting variations of prop. 15.16 can be deduced along the same line of reasoning. Here is an example.

15.18. Proposition. *Let $M^+(k)$ be an irreducible one-relator 3-manifold. Then every proper disc in $M^+(k)$ can be properly isotoped into M.* $\diamond$

Remark. To see that irreducibility is really necessary for this proposition, take a knotted arc, l, in a handlebody M and let k be one curve from $\partial(U(l) \cap \partial M)$. Attach a 2-handle, E, to k to obtain $M^+(k)$. Then $M^+(k)$ is clearly reducible. The union of the annulus $(\partial U(l) - \partial M)^-$ with some disc in E forms a disc in $M^+(k)$ which *cannot* be isotoped into M.

15.19. *Heegaard-Strings.*

One-relator structures give rise to Heegaard-strings. In fact, Heegaard-strings provide a dual way of looking at one-relator structures. In this book we use both points of view depending on which one is more convenient. For this reason we now translate the results of this section into the language of Heegaard-strings.

A *Heegaard-string* in a 3-manifold N is defined to be an arc, t, in N, $t \cap \partial N = \partial t$, such that $(N - U(t))^-$ is a handlebody. An arc t in a ∂-irreducible 3-manifold N is called a *generalized Heegaard-string* if $(N - U(t))^-$ is ∂-reducible. Closed 3-manifolds have no Heegaard-strings. Heegaard-strings are also known as "tunnels", especially in the context of knot spaces.

Closely related to Heegaard-strings are Heegaard-wands. Here a *wand* in a 3-manifold N is defined to be the union, s, of a simple closed curve, ring(s), and an arc, stem(s), in N such that stem(s) joins ring(s) with ∂N and that stem$(s) \cap (\text{ring}(s) \cup \partial N) = \partial$ stem(s). A wand s is a *Heegaard-wand* (resp. a *generalized Heegaard-wand*), provided $(N - U(s))^-$ is a handlebody (resp. a reducible 3-manifold and N ∂-irreducible). Notice that, given a (generalized) Heegaard-wand s in a 3-manifold N, then stem(s) is a (generalized) Heegaard-string in the 3-manifold $N(s) := (N - U(\text{ring}(s))^-$ which, in addition, joins two *different* components of $\partial N(s)$. In particular, $N(s)$ is again a (generalized) one-relator 3-manifold.

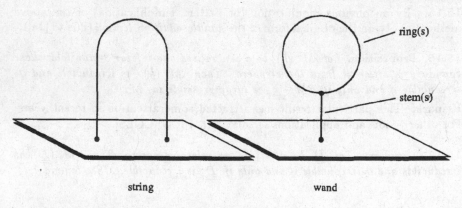

string　　　　　　　　　wand

(Figure 15.1.)

There is a simple process which turns (generalized) Heegaard strings into (generalized) Heegaard-wands. To describe it, let t be a (generalized) Heegaard-string in the 3-manifold N. Suppose there is an arc $l \subset \partial N$ with $l \cap t = \partial l = \partial t$, and let l' be an arc in $(\partial U(l) - \partial N)^-$ with $l' \cap t = \partial l'$ and joining the two components, t_1, t_2, from $t \cap U(l)$. Then $t_i(l) := ((t \cup l') - t_i)^-$, $i = 1, 2$, is a (generalized) Heegaard-ward. We say that $t_i(l)$ is obtained by *sliding* t along the *sliding-arc* l. Note that, for any fixed sliding-arc $l \subset \partial N$ the Heegaard-wands $t_1(l)$, $t_2(l)$ are (ambient) isotopic and so homeomorphic. But many different Heegaard-wands may be obtained from t by choosing different sliding-arcs.

Now, an essential surface in N cannot always be isotoped out of a (generalized) Heegaard-string. However, the obstruction for this is easy to describe. Indeed, the following two propositions taken together describe what happens. These propositions can be proved by simply translating the proofs of propositions 15.7 and 15.11 into the context of (generalized) Heegaard-strings.

15.20. Proposition. *Let N be an irreducible (generalized) one-relator 3-manifold. Suppose t is a (generalized) Heegaard-string in N which joins two different components of ∂N. Let S be an essential 2-manifold in N. Then S can be isotoped so that afterwards either $S \cap t = \emptyset$, or there are two annuli $C_1, C_2 \subset N$, disjoint or equal, such that, for $i = 1, 2$:*

(1) $\partial C_i \subset S \cup \partial N$ and $C_i \cap \partial N \neq \emptyset$.

(2) $C_i \cap t$ is an arc in C_i joining both components of ∂C_i.

(3) $S \cap C_i$ consists of simple closed curves which are essential in both C_i and S.

(4) $S \cap t = (C_1 \cap t) \cup (C_2 \cap t)$. ◊

15.21. Proposition. *Let N be as before. Suppose t is a (generalized) Heegaard-string in N. Let S be an essential 2-manifold in M. Then either t joins different components of ∂N, or t can be slid into a (generalized) Heegaard-wand, s, so that $S \cap U(\mathrm{ring}(s))$ is empty. ◊*

Remark. It is immediate from the proof that the existence of the annuli, C_1, C_2, with properties (1)-(4) from 15.20 follows for all strongly normal surfaces, S, in M (without applying any isotopy).

§16. The Handle-Addition Lemma (Relative Form).

From a technical point of view, the handle-addition lemma in its absolute form as given in the form of prop. 15.11 constitutes a first, important step in our general attempt to understand n-relator structures for 3-manifolds. Other steps will be taken in later chapters in order to get a more general picture. In this chapter however we wish to concentrate on applications of the handle-addition lemma to one-relator 3-manifolds alone. In this section we extend the handle-addition lemma to the category of 3-manifolds with boundary-patterns and admissible maps (as introduced in [Joh 1]). To this end we first generalize the concept of generalized Heegaard strings in 3-manifolds to the concept of admissible Heegaard strings in 3-manifolds with boundary-patterns. Then we prove (thm. 16.23) the handle-addition lemma for admissible Heegaard strings in irreducible 3-manifolds with useful boundary-pattern, i.e., we prove that an admissible Heegaard strings "usually" has a non-separating, essential surface in its complement. This is the relative form of the handle-addition lemma. As a consequence of this result we get that admissible Heegaard strings behave well under splittings along essential surfaces. This property makes admissible Heegaard strings the right concept for us to work with. The results of this section are technical, but they are crucial for the next section in which we study homotopies between generalized Heegaard strings. In this sense the present section is a preparation for the next.

16.1. *Admissible Heegaard Strings.*

As indicated above, this section deals with 3-manifolds with boundary-patterns as introduced in [Joh 1]. We here assume some rudimentary familiarity with 3-manifolds with boundary-patterns. For basic definitions and results we refer to [Joh 1].

Recall from [Joh 1] that, given a boundary-pattern $\underline{m}$ of a 3-manifold M, we distinguish between *bound faces* (= elements of $\underline{m}$) and *free faces* (= components of $(\partial M - \bigcup \underline{m})^-$). The set of all bound and free faces of $(M, \underline{m})$ is called the *completed boundary-pattern*, denoted by $\bar{\underline{m}}$. A boundary-pattern may or may not be useful. But useful boundary-patterns stay useful under splittings along essential surfaces [Joh 1, 4.8].

Now let s be a proper arc in M whose end-points lie in free faces of $(M, \underline{m})$. Define

$$M_s := (M - U(s))^- \quad \text{and} \quad \underline{m}_s := \underline{m} \cup \{ (\partial U(s) - \partial M)^- \},$$

where $U(s)$ denotes the regular neighborhood of s. Given a disc, D, in M_s, $D \cap \partial M_s = \partial D$, we denote by $\underline{d}$ and $\underline{d}_s$ the boundary-patterns of D induced by $\underline{m}$ and $\underline{m}_s$, respectively. A *good disc* for s in $(M, \underline{m})$ is defined to be a disc, D, in M_s, $D \cap \partial M_s = \partial D$, with the following properties:

(1) $D \cap U(s)$ is non-empty,

(2) $(D, \underline{d})$ is an i-faced disc, $1 \le i \le 3$, and

(3) $(D,\bar{\underline{d}}_s)$ is an essential disc in $(M_s,\bar{\underline{m}}_s)$.

In particular, the intersection of D with every face, G, of $(M_s,\bar{\underline{m}}_s)$ is either empty or an essential arc in G (w.r.t. the boundary-pattern induced by $\bar{\underline{m}}_s$).

16.2. Definition. *An* <u>admissible Heegaard string</u> *in* $(M,\underline{m})$ *is an arc in* M *whose end-points lie in free faces of* $(M,\underline{m})$ *and which has a good disc in* $(M,\underline{m})$.

An admissible Heegaard string s in $(M,\underline{m})$ is *trivial* if it has a good disc D in $(M,\underline{m})$ with $D\cap U(s)$ connected. Trivial Heegaard strings will have special relevance for us.

In this section we study admissible Heegaard strings in irreducible 3-manifolds $(M,\underline{m})$ with boundary-patterns. Of course, any generalized Heegaard string in $(\bar{M},\underline{m})$ is admissible if $\underline{m}$ is empty. So the concept of admissible Heegaard strings is in an extension of the notion of *generalized Heegaard strings* for 3-manifolds to the more general context of 3-manifolds with boundary-patterns. In fact, the following proposition tells us that admissible Heegaard strings come up in a natural way when splitting along essential surfaces.

16.3. Proposition. *Let* $(M,\underline{m})$ *be an irreducible 3-manifold whose completed boundary-pattern is useful. Let* s *be an admissible Heegaard string in* $(M,\underline{m})$. *Then the following holds:*

(1) $(M_s,\underline{m}_s)$ *is an irreducible 3-manifold whose completed boundary-pattern is useful if and only if there is no disc,* C, *in* M *with* $s=(\partial C-\partial M)^-$ *and* $C\cap\partial M\subset(\partial M-\bigcup\underline{m})^-$.

(2) Suppose S *is an essential surface in* $(M,\bar{\underline{m}})$ *with* $s\cap S=\emptyset$. *Then* s *is an admissible Heegaard string in* $(M',\underline{m}')$, *where* $(M',\underline{m}')$ *denotes the 3-manifold obtained from* $(M,\underline{m})$ *by splitting along* S.

Proof. Clearly M_s is irreducible since M is. If there is a disc C in M as in (1) of the proposition, then $(\partial U(C)-\partial M)^-$ is a 1-faced disc in $(M_s,\bar{\underline{m}}_s)$ whose boundary is essential in one of the faces from $\bar{\underline{m}}_s$. Thus $\bar{\underline{m}}_s$ is not useful for M_s. For the other direction of (1) let D be any admissible i-faced disc, $1\le i\le 3$, in $(M_s,\bar{\underline{m}}_s)$. If $s=(\partial D-\partial M)^-$ and $D\cap\partial M\subset(\partial M-\bigcup\underline{m})^-$, then we are done. If not, then $D\cap U(s)=\emptyset$ since ∂s lies in free faces of $(M,\underline{m})$. Thus D is an i-faced disc in $(M,\bar{\underline{m}})$. $\bar{\underline{m}}$ is useful for M. Therefore, by definition, ∂D bounds a disc $D'\subset\partial M$ such that $J\cap D'$ is the cone over $J\cap\partial D'$. Here $J:=\bigcup\partial G$, $G\in\underline{m}$. $D\cup D'$ is a 2-sphere in M. So it bounds a 3-ball, E, in M since M is irreducible. If $s\subset M-E$, then $D'\subset\partial M_s$, and we are done. If $s\subset E$, then note that s is an unknotted arc in E since $(E-U(s))^-$ is ∂-reducible. Thus there is a disc, C, in M (actually in E) as in (1) since D' meets only one free face of $(M,\underline{m})$. This proves (1).

To prove (2) let D be a good disc for s in $(M, \underline{m})$. Let D be chosen so that, in addition, $D \cap S$ is a system of arcs and curves whose number is as small as possible. Now S is essential in $(M, \underline{\tilde{m}})$, $S \cap s = \emptyset$, $\underline{\tilde{m}}$ is useful for M and M is irreducible. Thus it follows, by an innermost-disc-argument, that $D \cap S$ is a system of essential arcs in $(D, \underline{\bar{d}}_s)$, not meeting $D \cap U(s)$. Any one of these arcs separates a j-faced disc, $1 \leq j \leq 3$, from the i-faced disc $(D, \underline{\bar{d}})$, $1 \leq i \leq 3$. So some outermost arc of them separates a disc from D which is good for s in $(M', \underline{m}')$ since D is a good disc for s in $(M, \underline{m})$. $\Diamond$

We now discuss the interaction of admissible Heegaard strings and essential surfaces.

16.4. *Displacements.*

We will be concerned, almost exclusively, with 3-manifolds with useful boundary-patterns. In fact, some of the following discussion becomes invalid for non-useful boundary-patterns (see 17.1 for a striking counter-example). This in turn makes essential surfaces important for us (e.g. for any proofs involving hierarchies of surfaces). Now, given an admissible Heegaard string in a 3-manifold $(M, \underline{m})$, the handle-addition lemma tells us that there is always at least one non-separating and *incompressible* surface in M disjoint to this admissible Heegaard string. The question arises naturally whether this surface can be taken to be *essential* for admissible Heegaard strings. Theorem 16.19 tells us that this is indeed the case, provided the admissible Heegaard string is non-trivial. This is the envisioned relative handle-addition lemma.

To be more specific, let S be an admissible surface in $(M, \underline{m})$ and for the moment let $(M', \underline{m}')$ denote the manifold obtained from $(M, \underline{m})$ by splitting along S. An i-faced disc D, $1 \leq i \leq 3$, in $(M', \underline{m}')$ is called an *i-faced compression disc* for S in $(M, \underline{m})$, provided $D \cap \overline{S}$ is an essential arc in S (w.r.t. the boundary-pattern induced by $\underline{m}$). Recall a surface in $(M, \underline{m})$ is *incompressible* if it has no 1-faced compression disc. It is *essential* if it has no i-faced compression disc, $1 \leq i \leq 3$, in $(M, \underline{m})$ whatsoever (see [Joh 1]). Throughout this section let the *complexity* $e(S)$ of S defined to be

$$e(S) := (\chi(S), \alpha(S), -\beta(S), \gamma(S))$$

(w.r.t. the lexicographical order). Here $\alpha(S)$ denotes the number of free boundary curves, $\beta(S)$ denotes the number of all arc-faces and $\gamma(S)$ denotes the number of all free arc-faces of S.

16.5. Lemma. *Let $(M, \underline{m})$ be an irreducible 3-manifold whose completed boundary-pattern is non-empty and useful. Then there is an admissible, non-separating surface S in $(M, \underline{m})$ whose complexity $e(S)$ is maximal and any such surface is essential in $(\overline{M}, \underline{\tilde{m}})$.*

Proof. Since $\partial M \neq \emptyset$, there is a non-separating surface in M. Note that χS is bounded from above. Notice further that $\beta(S)$ is bounded from below, that $\alpha(S)$ is bounded from above by some function in $\chi(S)$ and that $\gamma(S)$ is bounded from above by some function in $\beta(S)$. Thus it follows that a non-separating surface S exists in $(M, \underline{m})$ with $e(S)$ being maximal (w.r.t. to all non-separating surfaces in M). It is easy to verify that such a maximal surface has to be essential. ◊

In particular, note that essential surfaces do exist in all irreducible 3-manifolds whose completed boundary-pattern is non-empty and useful. For later use we further single out "good" essential annuli. Here an essential annulus S in $(M, \underline{m})$ will be called *good* if

(1) every boundary curve of S lies in a bound face of $(M, \underline{m})$,

(2) every annulus A in M with $\partial A = \partial S$ is isotopic (rel ∂A) to S, and

(3) there is no incompressible annulus A in M, $A \cap (S \cup \partial M) = \partial A$, with one boundary curve in S and the other one in a free face of $(M, \underline{m})$.

(Recall from [Joh 1] that, as part of their definition, admissible and so essential annuli have no arc-faces).

The basic operation for us is the "displacement" of essential surfaces (especially those non-separating surfaces with maximal complexity). To introduce the concept in its proper generality (cf. fig. 16.1), let S be any admissible surface in the 3-manifold $(M, \underline{m})$. Set $J := \bigcup \partial G$, $G \in \underline{m}$. Let b be a proper arc in the surface $\bigcup \underline{m}$. Let $U(b)$ be the regular neighborhood of b in M. Let H_0, H_1 be the two components of $U(b) \cap (\partial M - \bigcup \underline{m})^-$. Let S_i, $i = 0, 1$, be the proper surface given by

(1) $(S_i - U(b))^- = (S - U(b))^-$ and $\chi S_i = \chi S$,

(2) $(b \cup H_{i+1}) \cap \partial S_i = \emptyset$ (indices mod 2), and

(3) the number of points of $J \cap \partial S_i$ is as small as possible.

Note that S_i is unique up to admissible isotopy in $(M, \tilde{m})$ (for $i = 0$ as well as $i = 1$). S_i is called a *displacement* of S along b. Thus S has exactly two displacements (along b) if $b \cap S \neq \emptyset$ and none otherwise. Note further that displacements of essential surfaces in $(M, \tilde{m})$ are no longer essential. They will however serve us well as (temporary) substitutes for essential surfaces. Displaced surfaces arise naturally when relative positions of essential surfaces and admissible Heegaard strings are considered.

Indeed, let s be an admissible Heegaard string in the 3-manifold $(M, \underline{m})$. Fix a good disc D for s in $(M, \underline{m})$. Then $b := D \cap \bigcup \underline{m}$ is connected. So it is a proper arc in $\bigcup \underline{m}$. The two displacements of a surface S in $(M, \tilde{m})$ along b are called the *Heegaard displacements* of S along s (or better: along the good disc D).

16.6. Example.

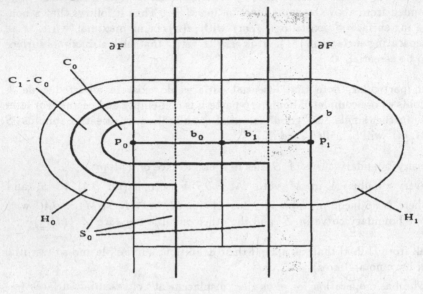

(Figure 16.1.)

We now turn to the following specific situation (which is the setting for 16.7-16.18):

(*) *Let $(M, \underline{m})$ be an irreducible 3-manifold whose completed boundary-pattern is useful. Suppose there is no essential annulus in $(M, \tilde{\underline{m}})$ which has a boundary curve in a free face of $(M, \underline{m})$. Let S be either a good annulus, or a non-separating surface with maximal complexity, $e(S)$. Let s be an admissible Heegaard string in $(M, \underline{m})$ and let D be a good disc for s in $(M, \underline{m})$. Set $b := D \cap \bigcup \underline{m}$. Let D be chosen so that $(\#(D \cap U(s)), \#(b \cap \partial S))$ is as small as possible (w.r.t. the lexicographical order). Suppose $S \cap b \neq \emptyset$. Denote by S_0, S_1 the two Heegaard displacements, i.e., the two displacements of S along b.*

From time to time we will deform s. This will always be done through admissible ambient isotopies in $(M, \underline{m})$ which fix the regular neighborhood of $\bigcup \underline{m}$ in M. Such an isotopy deforms the good disc for s through good discs, and note that the above complexity, $(\#(D \cap U(s)), \#(b \cap \partial S))$, of these discs never changes.

Given the previous setting, we define the *distance*, $d(s)$, for the admissible Heegaard string s as follows. First define a distance for points in $F - \partial S_0$, where $F := (\partial M - \bigcup \underline{m})^-$. To do this note that $H_0 \cap \partial S_0$ consists of inessential and pairwise admissibly parallel arcs in $F = (\partial M - \bigcup \underline{m})^-$ (H_0, H_1 are the two components of $F \cap U(b)$). More precisely, every arc from $H_0 \cap \partial S_0$ separates a 2-faced disc from F contained in H_0. Let $C_0 \subset C_1 \subset ... \subset C_n$, $n \geq 0$, be the entire collection of all these discs. Furthermore, set $C_{-1} = \emptyset$. Then, for every point $z \in F - \partial S_0$, define

$$d(z) := \begin{cases} i, & \text{if } z \in C_i - C_{i-1} \text{ with } i \leq n, \\ \infty, & \text{if } z \notin C_n. \end{cases}$$

Let x_1, x_2 be the two points from ∂s. W.l.o.g. we will suppose the indices are chosen so that $d(x_1) \leq d(x_2)$. Finally, set

$$d(s) := d(x_1) = \min\{ d(x_1), d(x_2) \}.$$

The following lemma is a straightforward consequence of the definitions.

16.7. Lemma. *Let $(M, \underline{m})$, s, S and S_0 be given as in (*). Suppose $s \cap S_0 = \emptyset$ and $d(s) = \infty$. Then $s \cap S = \emptyset$.* ◇

We will argue in prop. 16.20 that there basically always is a non-separating surface $S \subset M$ with $e(S)$ maximal and $s \cap S_0 = \emptyset$. But $d(s)$ may or may not be infinite. In view of 16.7, the hard part is to localize the geometric obstruction for increasing the distance $d(s)$. For this we need to analyze the intersection-pattern $D \cap S_0$ in D. In the following we carry out this analysis under the additional assumption that s is non-trivial (trivial admissible Heegaard strings will have to be treated separately later).

We first show that this intersection-pattern is non-trivial.

16.8. Lemma. *Let $(M, \underline{m})$, s, S and S_0 be given as in (*). Suppose s is non-trivial and $s \cap S_0 = \emptyset$. Then $D \cap S_0$ is non-empty, where D is the good disc for s used for the displacement S_0.*

Proof. Assume $D \cap S_0 = \emptyset$. One arc, say l_0, from $D \cap F$ (recall $F = (\partial M - \bigcup \underline{m})^-$) has the boundary-point p_0 of b as an end-point. By assumption, this arc joins p_0 with a component, say U_0, from $U(s) \cap \partial M$ without meeting ∂S_0. Thus $d(s) = 0$, i.e., an end-point x_1 of s lies in C_0.

Assume $\partial s \subset C_0$. Then note that $(C_0 - U(l_0 \cup (U(s) \cap C_0)))^-$ is an annulus. Thus no free face from $(D, \underline{d}_s)$ can have both end-points in one component from $U(s) \cap C_0$ (see definition of good discs). Then it is easy to see that there must be a pair of arcs from $D \cap C_0$ joining the pair of components from $U(s) \cap C_0$ with ∂C_0. But $D \cap (C_0 \cap \partial F) = p_0 = b \cap H_0$ (recall $F = (\partial M - \bigcup \underline{m})^-$). So at least one arc from $D \cap C_0$ must join $U(s) \cap C_0$ with the arc $(\partial C_0 - \partial F)^-$. Then $D \cap \partial S_0 \neq \emptyset$ since $(\partial C_0 - \partial F)^- \subset \partial S_0$. Contradiction.

Assume C_0 contains exactly one end-point of s (i.e. only x_1). Then $(C_0 - U(s))^-$ is an annulus. So again no free face of $(D, \underline{d}_s)$ can have both end-points in $U(s) \cap C_0$. Therefore every free face of $(D, \underline{d}_s)$, contained in C_0, joins $U(s) \cap C_0$ with ∂C_0. So it has $p_0 = b \cap H_0$ as one of its end-points $(D \cap S_0 = \emptyset$ and $D \cap \bigcup \underline{m} = b)$. Thus there is exactly one free face of $(D, \underline{d}_s)$

contained in C_0. Since every component of $D \cap U(s)$ has to meet this free face, it follows that $D \cap U(s)$ consists of one arc. Therefore s is trivial in $(M, \underline{m})$. Contradiction.

Thus, in any case, our assumption leads to a contradiction. ◊

16.9. *Displacements are Almost Essential.*

By the previous lemma, we know the intersection-pattern $D \cap S_0$ is non-empty (and so we have something to start with). But *a priori* this intersection-pattern could be rather complicated since S_0 is inessential. Indeed, S_0 has a 2-faced compression disc (unless of course $S \cap b = \emptyset$). It may even have 3-faced compression discs. But on the other hand it has no 1-faced compression discs. Moreover, we have the following property concerning 2-faced compression discs, which in turn causes the intersection-pattern $D \cap S_0$ to be in reality comparatively simple.

16.10. Proposition. *Let $(M, \underline{m})$, s, S and S_0 be given as in (*). Suppose s is non-trivial. Then no 2-faced compression disc for S_0 in $(M, \underline{m})$ meets a free face.*

In addition: if A is an incompressible annulus in M, $A \cap (S_0 \cup \partial M) = \partial A$, which has one boundary curve in S_0 and the other in a free face of $(M, \underline{m})$, then $A \cap S_0 \subset S_0$ is parallel to a free face of S_0.

Remark. Note we do not require $s \cap S_0$ to be empty in this proposition.

Proof. We first verify the additional remark. For this let A be an annulus as given there. Then S cannot be a good annulus. So S must be a non-separating surface in M with $e(S)$ maximal. But S_0 is a displacement of S. Hence w.l.o.g. $A \cap (S \cup \partial M) = \partial A$. Then $A \cap S$ is parallel in S to some free face of S since $e(S)$ is maximal. But free faces of S do not meet $b = D \cap \bigcup \underline{m}$. Thus the free face in question is also a free face of S_0. This proves the additional remark.

Now, assume there is a 2-faced compression disc, say A', for S_0 which meets a free face of $(M, \underline{m})$. Let D be the good disc for s used for the displacement S_0. Then we will get the desired contradiction by constructing a good disc, D', for s in $(M, \underline{m})$ with $(\#(D' \cap U(s)), \#(b' \cap \partial S)) < (\#(D \cap U(s)), \#(b \cap \partial S))$, where $b' := D' \cap \bigcup \underline{m}$.

Before starting the actual argument, we wish to emphasize that D is a good disc. In particular, the arc $b = D \cap \bigcup \underline{m}$, is the union of one or two arcs, say b_0, b_1, each of which is entirely contained in a bound face of $(M, \underline{m})$. Let the indices be chosen so that b_0 meets H_0 and ∂S (recall H_0, H_1 are the components of $F \cap U(b)$). Then ∂S meets either only b_0, or both b_0 and b_1. We will see that the situation is somewhat easier to handle if ∂S meets only b_0.

To begin with set $k := A' \cap \partial M$. Then, by our choice of A', k lies in $F = (\partial M - \bigcup \underline{m})^-$. In particular, the points of ∂k lie in components, say l_1, l_2, (possibly equal) of $F \cap \partial S_0$.

A' cannot be a 2-faced compression disc for S since S is supposed to be essential in $(M, \bar{\bar{m}})$. Thus l_1, l_2 cannot be both components of $F \cap \partial S$; at least one of them must be a component of $H_0 \cap \partial S_0$. This leaves us with two cases.

Case 1. l_1 *and* l_2 *are both components of* $H_0 \cap \partial S_0$.

If $l_1 \neq l_2$, then recall l_1 and l_2 are admissibly parallel in F (we are in Case 1). Moreover, k must lie in the corresponding parallelity region since $k \cap \partial S_0 = \partial k$. Thus the ambient isotopy which pushes S_0 back to S pushes, at the same time, A' into a 2- or 3-faced compression disc, A, for S with $A \cap \partial M \subset b \subset \bigcup \underline{m}$ (it is a 2- or 3-faced compression disc according whether or not $b_1 \cap \partial S$ is empty). But S is essential in $(M, \bar{\bar{m}})$. So $A \cap S$ separates a 2- resp. 3-faced disc, B, from S. The union $A \cup B$ is an admissible 1- or 2-faced disc in $(M, \underline{m})$. $\underline{m}$ is useful for M. Thus, by definition, $\partial(A \cup B)$ bounds a disc $C \subset \bigcup \underline{m}$ such that $C \cap J$ is either empty or an arc (recall $J = \bigcup \partial G$, $G \in \underline{m}$). Note $B \cap \partial C \subset \partial S$ and $(A \cap \partial C) \cap \partial S = \partial(A \cap \partial C)$. In particular, $b \cap C$ is a system of (proper) arcs in C with end-points in $B \cap \partial C$. Thus $b \cap C$ separates a disc, C', from C (possibly $C' = C$) with $C' \cap \partial C \subset B \cap \partial C$ and $C' \cap D = (\partial C' - \partial C)^-$. Set $D' := (D - U(C'))^- \cup C''$, where C'' is the copy of C' in $(\partial U(C') - \partial M)^-$. Set $b' := D' \cap \bigcup \underline{m}$. Then, by construction, D' is a good disc for s in $(M, \underline{m})$ with $D' \cap U(s) = D \cap U(s)$, but $\#(b' \cap \partial S) < \#(b \cap \partial S)$. This however contradicts our minimal condition on $(\#(D \cap U(s)), \#(b \cap \partial S))$.

If $l_1 = l_2$, then there is an arc, J, in l_1 with $\partial l = \partial k$. The surface S_0 splits the regular neighborhood $U(l)$ (taken in M) into two 3-balls. Let U_0 be that one of them meeting A'. Then there is a 3-faced disc, say B_0, in U_0 with $B_0 \cap \partial U_0 = \partial B_0$ and $B_0 \cap A' = \emptyset$ such that the three faces of B_0 lie in S_0, $(\partial U_0 - (S_0 \cup \partial M))^-$, and $F = (\partial M - \bigcup \underline{m})^-$, respectively. Now A' splits $(\partial U_0 - (S_0 \cup \partial M))^-$ into three discs. One of them meets A' in two arcs. The union of this disc with $(A' - U_0)^-$ is an annulus $C \subset M$ with $C \cap (S_0 \cap \partial M)^- = \partial C$ which has one boundary curve in S_0 and the other one in a free face of $(M, \underline{m})$. This annulus is incompressible in M since $A' \cap S_0$ is an essential arc. Thus S cannot be a good annulus. Therefore S is a non-separating surface with $e(S)$ maximal. It follows that the curve $C \cap S_0$ is admissibly parallel to a free boundary curve of S_0 (see the additional remark verified above). Let $C' \subset S_0$ be the corresponding parallelity region. Then the union $C \cup C'$ is an incompressible annulus in M whose boundary curves lie in free faces of $(M, \underline{m})$. Thus, by hypothesis on $(M, \underline{m})$, this annulus is inessential in $(M, \bar{\bar{m}})$. Since M is irreducible and $\bar{\bar{m}}$ is useful for M, it follows that $C \cup C'$ is parallel in M to an annulus in $\bar{F}$. Let W be the corresponding parallelity region. Notice that the interiors of W and U_0 do not meet. Notice further that there is a 3-faced disc, B_1, in W with $B_1 \cap \partial W = \partial B_1$ such that the three faces lie in C, C' and F, respectively. Now both B_0 and B_1 intersect C in an essential arc (but from different sides of C. Thus w.l.o.g. $B_0 \cap C = B_1 \cap C$. So the union $B_0 \cup B_1$ is a 2-faced compression disc for S_0.

But, by construction, the arc $(B_0 \cup B_1) \cap \partial M$ joins a free boundary curve of S_0 with a free arc face of S_0. It follows that, splitting S_0 along $B_0 \cup B_1$, we obtain from S_0 a non-separating surface in M whose Euler characteristic is strictly bigger than that of S_0, and so of S. This contradicts our maximality condition on $e(S)$.

Thus Case 1 leads to contradictions.

The next case cannot occur if S is a good annulus, for good annuli have no free faces. (Thus the proof of the proposition is already complete for good annuli S).

Case 2. l_1, *but not* l_2, *is a component of* $H_0 \cap \partial S_0$.

The ambient isotopy which pushes S_0 back to S, makes l_1 disappear but leaves l_2 untouched. Thus it pushes A' to a disc A which is a 3- or 4-faced compression disc for S in $(M, \underline{\bar{m}})$ with $A \cap (\partial M - F)^- = A \cap \bigcup \underline{m} \subset b$ (it is a 3- or 4-faced disc according whether or not $b_1 \cap \partial S$ is empty).

If A is a 3-faced compression disc for S, then we may argue as in Case 1. Indeed, $A \cap S$ separates again a 3-faced disc, B, from S. The union $A \cup B$ is a 2-faced disc in $(M, \underline{\bar{m}})$. $\underline{\bar{m}}$ is useful for M. Thus $\partial(A \cup B)$ bounds a disc, C, in ∂M such that $C \cap J$ is an arc. This arc splits C into two discs. One of them lies in a free and the other one in a bound face of $(M, \underline{m})$. Let C_1 be that one of them which lies in a bound face of $(M, \underline{m})$. Let C'_1 be an outermost disc in C_1 separated by b, i.e., either $C'_1 = C_1$ (if $b \cap C_1 = A \cap C_1$) or $C'_1 \cap \partial C_1 \subset (\partial C_1 - A)^-$ and $C'_1 \cap b = (\partial C'_1 - \partial C_1)^-$. Then $D \cap C'_1 = (\partial C'_1 - \partial C_1)^-$. Set $D' := (D - U(C'_1))^- \cup C''_1$, where C''_1 is the copy of C'_1 in $(\partial U(C'_1) - \partial M)^-$. Let $b' := D' \cap \bigcup \underline{m}$. Then, by construction, $D' \cap U(s) = D \cap U(s)$, but $\#(b' \cap \partial S) < \#(b \cap \partial S)$. So we have again a contradiction.

Thus we may suppose A is a 4-faced compression disc for S in $(M, \underline{\bar{m}})$. In this case the construction of D' is more delicate. We proceed as follows.

First, $A \cap S$ separates a 4-faced disc, B, from S since $e(S)$ is maximal. The union $C := A \cup B$ is an admissible 4-faced disc in $(M, \underline{\bar{m}})$ with $A \cap \bigcup \underline{m} \subset \partial C$.

16.11. Lemma. *s can be admissibly (ambient) isotoped in* $(M, \underline{\bar{m}})$ *so that afterwards* $s \cap C = \emptyset$.

Proof of Lemma. Note that C is an admissible square in $(M, \underline{\bar{m}})$ of a very special form. Indeed, C has exactly one free face. $b \cap \partial C$ equals $b_1 \cap \partial C$ and is a collection of points, all contained in the bound face, c, of C opposite to the free face (we here assume $A \cap \partial C$ is slightly moved away from b). Now, let C_1 be the displacement of C, obtained by displacing C along b without meeting b_0. Then, in particular, $C_1 \cap b = \emptyset$. Let $\underline{c}_1$ be the boundary-pattern of C_1, induced by $\underline{m}$. Then the collection of free faces of $(C_1, \underline{c}_1)$ consists of the free face of C and the components of $U(c \cap b)$. We next study the intersection $D \cap C_1$.

By the argument from Case 1, it follows that C_1 has no 2-faced compression disc which meets a free face (note that $C \cap b_0 = \emptyset$ and that the displacement of C takes place along b_1 alone). Thus we may suppose s and D are isotoped, using an ambient isotopy constant on $\bigcup \underline{m}$, so that afterwards $D \cap C_1$ consists of arcs which are essential in both $(D, \underline{d}_s)$ as well as $(C_1, \underline{c}_1)$ (strictly speaking, we first collapse $U(s)$ to the arc s to turn D into a 2-complex, and then move this 2-complex with the ambient isotopy). After this isotopy, one or two end-points of s may now lie in H_1.

Actually, the lemma follows, if we can show that $s \cap H_1 = \emptyset$. To see this note that, in the situation at hand, we have $s \cap C_1 = \emptyset$, for otherwise there would be a good pencil of $D \cap C_1$ in D (see the proof of 14.4) but there are no good pencils for discs (see proof of 15.11). Thus, if $s \cap H_1 = \emptyset$, we can push C_1 back to C without meeting s, and we were done.

Thus assume $s \cap H_1 \neq \emptyset$. We will arrive at a contradiction by constructing a good disc for s which intersects $U(s)$ in strictly fewer arcs than D. To do this fix an end-point of s, say x_1, contained in H_1. W.l.o.g. we suppose that C_1 is deformed so that as many arcs from $C \cap H_1$ as possible are pushed out of H_1, using an ambient isotopy of M which fixes ∂s and $(\partial M - U(H_1))^-$. Moreover, we suppose x_1 is chosen so that it lies in that component from $H_1 - \partial C$ whose intersection with $\partial F = \partial \bigcup \underline{m}$ is connected. Then, by an argument from lemma 16.8, there must be at least one free face of $(D, \underline{d}_s)$ which intersects all arcs from $C \cap H_1$. This arc therefore intersects every free face of $(C_1, \underline{c}_1)$ except possibly that one which C_1 has in common with C. Let $\mathcal{L}$ be the collection of all arcs from $C_1 \cap D$ having at least one of these intersection points as end-points. By our choice of $\mathcal{L}$, it follows that at least one of the outermost arcs, say l, from $\mathcal{L}$ must separate a disc C_1' from C' such that $(C_1', \underline{c}_1')$ is a 2- resp. 3-faced disc. Now, l is also an arc in D. More precisely, it joins two different free faces of $(D, \underline{d}_s)$ (recall C_1 is displaced and $D \cap C_1$ is essential in $(D, \underline{d}_s)$) but not the two free faces meeting b. The arc l splits D into two discs and let D_1 be that one of them which does not contain b. The union $D_1 \cup C_1'$ has all properties of a good disc except that it might have self-singularities (for the latter we have to remember that l has been chosen outermost only w.r.t. $\mathcal{L}$ and not to all of $C_1 \cap D$). But all self-singularities can be removed by using cut-and-paste and the result is a good disc D' for s. By construction, D' meets $U(s)$ in strictly fewer arcs than D. This, however, contradicts our minimal choice of D (see (*)). The lemma is therefore established.

It is now a simple matter to finish the proof. Indeed, the following lemma establishes the desired contradiction:

16.12. Lemma. *There is a good disc, D', for s in $(M, \underline{m})$ with $(\#(D' \cap U(s), \#(b' \cap \partial S)) < (\#(D' \cap U(s)), \#(b \cap \partial S))$, where $b' := D' \cap \bigcup \underline{m}$.*

Proof of Lemma. C is a square. One face of C, say c_0, lies in $b_0 \subset b$. Set $D' := (D - U(c_0))^- \cup C'$, where C' is a copy of C in $\partial U(C)$. Then D' is a possibly singular disc. By lemma 16.11, we may suppose that s is admissibly isotoped so that it does not meet C. Thus D' is actually a singular disc in $(M - U(s))^-$. But all self-singularities can be removed by using cut-and-paste. Of course, these cut-and-paste operations can be chosen so that D' becomes a single non-singular disc - which we denote by D' again. It is easily verified that D' is then a good disc for S with $D' \cap U(s) = D \cap U(s)$. Set $b' = D' \cap \bigcup \underline{m}$. Notice that $b' \cap \partial S = (b \cap \partial S) - (b_0 \cap \partial S)$, and so $\#(b' \cap \partial S) < \#(b \cap \partial S)$. Thus $(\#(D' \cap U(s), \#(b' \cap \partial S)) < (\#(D' \cap U(s)), \#(b \cap \partial S))$ and this proves the lemma.

With this lemma the proof of the proposition is finished. $\Diamond$

Note that the previous proposition tells us, in particular, that both displacements S_0, S_1 of S are incompressible and have no 2-faced compression discs which meet free faces of $(M, \underline{m})$. As a first application of this proposition we obtain the following two corollaries.

16.13. Corollary. *Let* $(M, \underline{m})$, s, S *and* S_0 *be given as in* (*). *Suppose* s *is non-trivial and* $s \cap S_0 = \emptyset$. *Let* D *be the good disc used for the displacement* S_0. *Suppose* D *is deformed, using an admissible isotopy in* $(M, \underline{\tilde{m}})$ *constant on* $D \cap (U(s) \cup \bigcup \underline{m})$, *so that afterwards* $D \cap S_0$ *is a system of curves whose number is as small as possible.*

Then $D \cap S_0$ *is a non-empty system of essential arcs in* $(D, \underline{\bar{d}}_s)$ *whose end-points lie in free faces of* $(D, \underline{d}_s)$.

Remark. This corollary holds for all good discs, E, for s with $(E \cap \bigcup \underline{m}) \cap \partial S_0 = \emptyset$. **Proof.** M is irreducible and S_0 is incompressible. Thus, by the usual innermost-disc argument, $D \cap S_0$ is a system of arcs. None of them meets $D \cap U(s)$ since $S_0 \cap s = \emptyset$, and none of them meets $b = D \cap \bigcup \underline{m}$ since $S_0 \cap b = \emptyset$ (S_0 is a displacement). Thus all arcs from $D \cap S_0$ have end-points in free faces of $(D, \underline{d}_s)$. So every arc from $D \cap S$ which is inessential in $(D, \underline{\bar{d}}_s)$ separates a 2-faced disc from D. But, by prop. 16.10, there is no 2-faced compression disc for S_0 in $(M, \underline{\tilde{m}})$ meeting a free face of $(M, \underline{m})$. Moreover, M is irreducible and $\underline{\tilde{m}}$ is useful for M. Thus, by another application of the innermost-disc-argument, it follows that $D \cap S_0$ consists of essential arcs in $(D, \underline{\bar{d}}_s)$ (recall our minimality condition on $D \cap S_0$). Moreover, $D \cap S_0 \neq \emptyset$, by lemma 16.8. $\Diamond$

16.14. Corollary. *Let* $(M, \underline{m})$, s, S *and* S_0 *be given as in* (*). *Suppose* s *is non-trivial and* $s \cap S_0 = \emptyset$. *Then* s *is an admissible Heegaard string in* $(M', \underline{m}')$, *where* $(M', \underline{m}')$ *denotes the 3-manifold obtained from* $(M, \underline{m})$ *by splitting along* S_0.

Proof. It remains to find a good disc for s in $(M', \underline{m}')$. Consider the good disc, D, for s in $(M, \underline{m})$, used for the displacement S_0. By cor. 16.13,

$D \cap S_0$ consists of essential arcs in $(D, \bar{\underline{d}}_s)$. If $D \cap S_0$ is empty, then D itself is a good disc for s in $(M', \underline{m}')$. If not, then an outermost arc from $D \cap S_0$ separates a 2-faced disc from $(\overline{D}, \underline{d})$ which contains a component of $U(s) \cap \partial D$. This disc is clearly a good disc for s in $(M', \underline{m}')$. $\Diamond$

16.15. Tubular Cages.

We next introduce "tubular cages". With this notion we are in the position to state our main result concerning displacements. It tells us that admissible Heegaard strings can be pushed out of displacements, except when they are hidden in "tubular cages".

A *tubular cage* for an admissible Heegaard string, s, in a 3-manifold $(M, \underline{m})$ is a solid torus, $W = W(s)$, in M with $s \subset W$ and the following additional properties:

(1) $W \cap \partial M$ is an annulus whose winding number with respect to W is bigger than one.

(2) There is a bound face, B_W, of $(M, \underline{m})$ such that $W \cap B_W = W \cap \bigcup \underline{m}$ and that $B_W \cap (\partial W - \partial M)^-$ is essential in B_W.

(3) There is a free face, F_W, of $(M, \underline{m})$ such that $D_W := W \cap F_W = W \cap (\partial M - \bigcup \underline{m})^-$ is a disc with $D_W \cap \partial F_W$ connected and contained in ∂B_W.

(4) $s \cup l$ is isotopic to the core of W, where l is an arc in D_W with $\partial l = \partial s$.

16.16. Proposition. *Let* $(M, \underline{m})$, s, S *and* S_0 *be given as in* (*). *Suppose* s *is non-trivial and* $s \cap S_0 = \emptyset$. *Then* s *is admissibly ambient isotopic in* $(M, \bar{\underline{m}})$ *to an admissible Heegaard string,* s', *so that*

(1) $s' \cap S = \emptyset$, *or*

(2) $s' \cap S_0 = \emptyset$ *and* s' *has a tubular cage,* W, *in* $(M_0', \underline{m}_0')$ *which is not a tubular cage in* $(M, \underline{m})$.

In addition, $D_W \subset C_m$ *and* $D_W \cap C_{m-1} = \emptyset$, *where* $m = d(s)$ *(see 16.4 for definition of* C_m *and* $d(s)$).

Remark. $(M_0', \underline{m}_0')$ denotes the manifold obtained from $(M, \underline{m})$ by splitting along S_0.

Proof. We begin by introducing some notation.

Consider the good disc, D, of s, used for the displacement of S. $U(s) \cap \partial D$ is a system of arcs in ∂D. Let the end-points of $U(s) \cap \partial D$ be labeled z_1 or z_2 according whether they lie next to x_1 or x_2, respectively (i.e. in that component of $U(s) \cap \partial M$ containing the point x_1 resp. x_2 of ∂s). Then $d(z_1) \leq d(z_2)$ since, by convention from 16.4, $d(x_1) \leq d(x_2)$.

W.l.o.g. $b = D \cap \bigcup \underline{m}$ is non-empty (otherwise there was no displacement in the first place and we are done). Let p_0, p_1 be the two points from ∂b. It is important to remember that $b \cap H_0$ is a single point, say p_0. Thus $d(p_0) = 0$ and $d(p_1) = \infty$. Let l_0, l_n be the components from $(\partial D - (b \cup U(s)))^-$ which contain p_0, p_1, respectively.

Given a system, $\mathcal{K}$, of essential arcs in $(D, \underline{\bar{d}}_s)$, we say that an arc from $\mathcal{K}$ is *outermost* if it separates an "outermost disc". A disc $D' \subset D$ is *outermost* (for $\mathcal{K}$) if $D' \cap \mathcal{K} = (\partial D' - \partial D)^-$ and $D' \cap b = \emptyset$. An arc from $\mathcal{K}$ is called *admissibly ∂-parallel* if it is admissibly parallel to some component of $U(s) \cap \partial D$. Finally, we say $\mathcal{K}$ is a *good arc-system* w.r.t. l_0 (or l_n) if every component from $\mathcal{K}$ is an arc which has an end-point in l_0 (or l_n).

The proof of the proposition is in two steps. In the first step we show how to turn $D \cap S_0$ into a good arc-system without increasing $\#(D \cap S_0)$. In the second step we show that a good arc-system can be changed into some smaller good arc-system whenever $d(s) < \infty$. The following lemma is a preparation for this strategy.

16.17. Lemma. *Suppose $s \cap S_0 = \emptyset$. Then there is an admissible ambient isotopy α_t, $t \in I$, of $(M, \underline{m})$ constant on $\bigcup \underline{m}$, with the following property:*

(1) $d(\alpha_1 s) = \infty$ or every admissibly ∂-parallel arc from $\alpha_1 D \cap S_0$ meets l_0.

(2) $\alpha_1 D \cap S_0$ is a good system w.r.t. l_0, provided $d(s) < \infty$ and $D \cap S_0$ is a good arc-system w.r.t. l_0.

(3) $\alpha_1 s \cap S_0 = \emptyset$, $d(\alpha_1 s) \geq d(s)$ and $\#(\alpha_1 D \cap S_0) \leq \#(D \cap S_0)$.

(4) $\alpha_1 D \cap S_0$ is a system of essential arcs in $(D, \underline{\bar{d}}_s)$.

Proof of Lemma. Suppose (1) does not hold for s and D; otherwise we are done since, by cor. 16.13, we may suppose that (4) holds. Then $d(s) < \infty$ and there is an admissibly ∂-parallel arc from $D \cap S_0$ which does not meet l_0. Set $m := d(s)$. Then, by definition, $x_1 \subset C_m - C_{m-1}$. Moreover, recall $b := D \cap \bigcup \underline{m}$ and $p_i := b \cap H_i$, $i = 0, 1$.

Let t be an admissibly ∂-parallel arc from $D \cap S_0$ which does not meet l_0. Denote by A its corresponding parallelity region. Then w.l.o.g. $A \cap S_0 = t$. Let α_t, $t \in I$, be the ambient isotopy which is constant outside of the regular neighborhood of A and which pushes s across A.

To describe the effect of this isotopy, note that the end-points of every component of $U(s) \cap \partial D$ are labeled z_1, z_2. Thus at least one component of $((A \cap \partial D) - U(s))^-$ is contained in $(C_m - C_{m-1})^-$ (namely that one containing the end-point of $U(s) \cap \partial D$ labeled z_1). More precisely, every arc from $A \cap (C_m - C_{m-1})^-$ is an arc in $(C_m - C_{m-1})^-$ joining $U(s) \cap C_m$ with $(\partial C_m - \partial F)^-$. Thus $\alpha_1 s \cap S_0 = \emptyset$ and $d(\alpha_1 s) > d(s)$. Moreover, $\alpha_1(D) \cap S_0 = ((D \cap S_0) - t) \cup ((\partial C - \partial D)^- - t'))^-$. Here C denotes the regular neighborhood of $U(s) \cap \partial D$ in D and t' is the component of $(\partial C - \partial D)^-$ admissibly parallel in $(D, \underline{\bar{d}}_s)$ to t.

We now view C as part of $\alpha_1 D$. Let t'' be that component of $(\partial C - \partial \alpha_1 D)^-$ which meets l_0. Let c be a component of $(\partial C - \partial \alpha_1 D)$ different from t', t''. Then, by construction, c has an end-point in $(\partial C_m - \partial F)^-$. Let l_c be the component of $\partial \alpha_1 D \cap C_m$ containing this end-point. Then $\partial l_c \subset (\partial C_m - \partial F)^-$. So $\partial l_c \subset S_0$ and l_c is inessential in $(C_m - C_{m-1})^-$. In particular, l_c can be pushed out of C_m. Thus $\alpha_1 D$ can be deformed into D', using an admissible isotopy which is constant on $\alpha_1 D \cap (U(s) \cup \bigcup \underline{m})$, so that $D' \cap S_0 = ((\alpha_1 D \cap S_0) - U(\bigcup l_c))^- \cup \bigcup l'_c)$, where l'_c denotes the copy of l_c in $(\partial U(l_c) - \partial \alpha_1 D)^-$ and where the unions are taken over all components c from $(\partial C - \partial \alpha_1 D)^-$ different from t', t''. Thus, altogether, $d(s') > d(s)$ and D' is a good disc for s' with $s' \cap S_0 = \emptyset$ and $\#(D' \cap S_0) = \#(D \cap S_0)$. The next picture illustrates a typical situation:

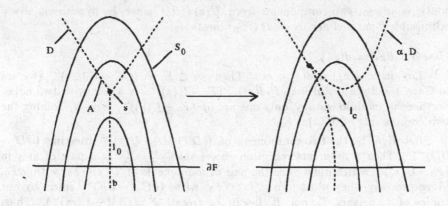

(Figure 16.2)

Notice that none of the arcs l_c above can have an end-point in $l_0 \cap S_0$, for no component c of $\partial C - t' \cup t''$ is admissibly parallel to the arc from $U(s) \cap \partial D$ next to p_0. Thus every arc from $D' \cap S_0$ has an end-point in l_0, provided all the arcs from $D \cap S_0$ have an end-point in l_0. Hence $D' \cap S_0$ is a good arc-system if $D \cap S_0$ is.

Now we apply the above procedure to s' and so on. After at most $d(s)$ steps the process stops. The composition of all isotopies involved is the desired ambient isotopy α_t.

This proves the lemma.

The next lemma tells us that, using the previous lemma, the intersection-pattern $D \cap S_0$ can always be turned into a good arc-system.

16.18. Lemma. *Suppose that $D \cap S_0$ is a system of essential arcs and that every admissibly ∂-parallel arc from $D \cap S_0$ meets l_0. Suppose further that*

$d(s) < \infty$ and that $D \cap S_0$ is not a good arc-system. Then either one of the following holds:

(1) $D \cap S_0$ is a good arc-system (w.r.t. l_0).

(2) s has a tubular cage as in prop. 16.16.

(3) There is an admissible ambient isotopy α_t, $t \in I$, of $(M, \underline{m})$, constant on $\bigcup \underline{m}$, such that $\alpha_1 s \cap S_0 = \emptyset$, $d(\alpha_1 s) > d(s)$ and $\#(\alpha_1 D \cap \overline{S}_0) \leq \#(D \cap S_0)$.

Proof of Lemma. Suppose $D \cap S_0$ is not a good arc-system (otherwise we are done). Then there is an arc from $D \cap S_0$ which does not meet l_0. In fact, there must be such an arc, say t, which is outermost for $D \cap S_0$. t separates a disc, D', form D with $D' \cap b = \emptyset$ and $D' \cap S_0 = (\partial D' - \partial D)^- = t$. Note that D' contains at least two components from $U(s) \cap \partial D$ since, by hypothesis, every admissible ∂-parallel arc from $D \cap S_0$ meets l_0.

Case 1. $d(x_1) = d(x_2)$.

Let $m := d(x_1) = d(s) < \infty$. Then $\partial s \subset F_0 := (C_m - C_{m-1})^-$ (we are in Case 1 and see definition of $d(s)$). $(F_0 - U(s))^-$ is a disc with two holes. So there is, modulo isotopy, only one arc in $(F_0 - U(s))^-$, say α, joining the two components of $U(s) \cap F_0$.

Let l'_1, l'_2 be the two components of $((D' \cap \partial D) - U(s))^-$ meeting $(\partial D' - \partial D)^-$. Then a moments reflection shows that $l'_1 \cup l'_2$ is a pair of arcs in $(F_0 - U(s))^-$ which must joins the pair of components of $U(s) \cap F_0$ with ∂F_0. More precisely, they must join $U(s) \cap F_0$ with $(\partial C_m - \partial F)^-$ since, by our choice of t, neither l'_1 nor l'_2 lies in l_0 (recall $F = (\partial M - \bigcup \underline{m})^-$). Thus, in particular, $l'_1 \cup (U(s) \cap F_0) \cup \alpha \cup l'_2$ separates a disc A from F_0 with $A \cap \partial F_0 \subset (\partial C_m - \partial F)^-$. Thus the regular neighborhood $U(A \cup D' \cup U(s))$ forms a tubular cage for s in $(M'_0, \underline{m_0}')$ which is not a tubular cage in $(M, \underline{m})$. In fact, by construction, this tubular cage has the properties of the additional remark of prop. 16.16 since $A \cap F_0 \subset (\partial C_m - \partial F)^-$. Thus (2) of the lemma follows in Case 1.

Case 2. $d(x_1) \neq d(x_2)$, i.e., $d(x_1) < d(x_2)$.

We first claim D' contains exactly two components from $U(s) \cap \partial D$. We already know it contains at least two components. Assume for a moment it contains more than two. Then there must be at least one component, l, from $(\partial D - (b \cup U(s)))^-$ with $l \subset D'$, $\partial l \subset U(s)$ and such that the two points from ∂l are either labeled z_1, z_1 or z_1, z_2 (since the end-points of every component of $U(s) \cap \partial D$ are labeled z_1, z_2). In both cases we find that $l \cap S_0 \neq \emptyset$ since $d(x_1) = d(s) < \infty$ and $d(x_1) < d(x_2)$. But this is impossible since, by our choice of D', the disc D' contains no arc from $D \cap S_0$ other than t. The claim is therefore established.

Let k_1, k_2 denote the two components from $U(s) \cap \partial D$ contained in D'. Let the indices be chosen so that k_1 lies closer (in $(\partial D - b)^-$) to the point p_0 from ∂b than k_2. Let l_1, l_2 be the two components from $(\partial D - (b \cup U(s)))^-$ meeting k_1. Let the indices be chosen so that $\partial l_2 \subset k_1 \cup k_2$. Then $l_2 \cap \partial S_0 = \emptyset$ and so ∂l_2 must be labeled z_2, z_2. (Incidentally, it also follows that $d(z_2) = d(x_2) = \infty$).

l_1 does not meet the arc b; i.e. $l_1 \neq l_0$ (see our choice of t). Thus $\partial l_1 \subset U(s)$. The points from $\partial(k_1 \cup l_2 \cup k_2)$ are labeled z_1, z_1 since the points from ∂l_2 are labeled z_2, z_2. Set $m := d(z_1) = d(x_1)$. Then $x_1 \in C_m - C_{m-1}$, and so $\partial(k_1 \cup l_2 \cup k_2) \subset C_m - C_{m-1}$. Hence the two arcs of $((D' \cap \partial D) - (k_1 \cup l_2 \cup k_2))^-$, say $l_1,', l_2'$, are contained in $(C_m - C_{m-1})^-$. In fact, because $l_1 \cap b = \emptyset$, it follows that both l_1' and l_2' are arcs which join $U(x_1)$ with the arc $(\partial C_m - F)^-$. Attaching k_1, k_2 via $U(s)$, we obtain from D' an annulus A'. Since l_1', l_2' both join $U(x_1)$ with $(\partial C_m - F)^-$, it follows further that A' can be pushed into an annulus A, $A \cap (S_0 \cup \partial M) = \partial A$, with one boundary curve in S_0 and the other one in a free face of $(M, \underline{m})$. This annulus is incompressible, for all faces of $(D, \underline{d}_s)$ are essential arcs in faces from $(M_s, \underline{\tilde{m}}_s)$. Thus S is not a good annulus. Therefore, by prop. 16.10, $A \cap S_0$ is parallel to a free boundary curve of S_0. Let $A' \subset S_0$ be the corresponding parallelity region. Then $A \cup A'$ is an admissible annulus in $(M, \underline{\tilde{m}})$ which is incompressible and whose boundary lies in free faces of $(M, \underline{m})$. But $(M, \underline{m})$ is supposed to have no essential annulus with a free boundary curve. It follows that A is admissibly isotopic (rel $A \cap S_0$) to A' (use that M is irreducible and that $\underline{\tilde{m}}$ is useful for M). The same isotopy pushes s into an admissible Heegaard string, s', with $s' \cap S_0 = \emptyset$ and $d(s') > d(s)$ (note that no boundary curve of A lies in $C_m - C_{m-1}$ since A is incompressible).

We claim this isotopy can in fact be chosen so that it satisfies (3) of the lemma. We verify this claim by a careful analysis of the special situation at hand. First note that there is a square, B, in M with $B \cap (S_0 \cup s \cup \partial M) = \partial C$ and $B \cap D = \emptyset$ such that s is one of its faces, $B \cap S_0$ is a face and the remaining faces lie in free faces of $(M, \underline{m})$. Let β_t, $t \in I$, be the ambient isotopy of M, constant outside of $U(B)$, which pushes s across B. Note that

$$\beta_1 s \cap S_0 = \emptyset, \ d(\beta_1 s) > d(s) \text{ and } \beta_1 D \cap S_0 = (D \cap S_0) \cup (\partial C - \partial D)^-,$$

where C is the regular neighborhood of $U(s) \cap \partial D$ in D.

We now proceed in a similar way as in the proof of lemma 16.17. First we view C as part of of $\beta_1 D$. Let c_0 be the component of $(\partial C - \partial \beta_1 D)^-$ with $c_0 \cap l_0 \neq \emptyset$. Every component, c, from $(\partial C - \partial \beta_1 D)^-$ has an end-point, x_c, in $(\partial C_m - \partial F)^-$, $m = d(s)$. Let l_c be the component of $\beta_1 D \cap C_m$ containing x_c. Then $\partial l_c \subset (\partial C_m - \partial F)^-$, if $c \neq c_0$. Thus l_c, $c \neq c_0$, is inessential in C_m and so it can be pushed out of C_m. Thus there is an admissible ambient isotopy, γ_t, $t \in I$, constant on $U(s) \cap \bigcup \underline{m}$, so that

$$(\gamma_1 \beta_1 D) \cap S_0 = ((\beta_1 D \cap S_0) - U(\bigcup_{c \neq c_0} l_c))^- \cup \bigcup_{c \neq c_0} l_c'.$$

Here the regular neighborhood is taken in D and l'_c denotes the copy of l_c in $(\partial U(l_c) - \partial D)^-$. Thus

$$\#((\gamma_1\beta_1 D) \cap S_0) = \#(D \cap S_0) + 1.$$

(Note that c_0 is not affected by γ_t).

Now consider D' again. Recall $D' \cap U(s) = k_1 \cup k_2$. Let c_i, $i = 1, 2$, be the component of $(\partial C - \partial D)^-$ which is admissibly parallel to k_i. Then l_{c_i}, $i = 1, 2$, joins c_i with the arc $t = (\partial D' - \partial D)^-$. Thus the arc

$$t' := ((c_1 \cup t \cup c_2) - U(l_{c_1} \cup l_{c_2}))^- \cup (l'_{c_1} \cup l'_{c_2})$$

is a component of $(\gamma_1\beta_1 D) \cap S_0$. On the other hand, this arc has both end-points in a single component of $(\partial D - (b \cup U(s))^-$ and so it is inessential in $(D, \underline{\tilde{d}}_s)$. Thus, applying prop. 16.10, we find an additional admissible ambient isotopy γ_t, $t \in I$, constant on $U(s) \cup \bigcup \underline{m}$ so that

$$(\delta_1\gamma_1\beta_1 D) \cap S_0 = ((\gamma_1\beta_1 D) \cap S_0) - t'.$$

Hence, altogether,

$$\#((\delta_1\gamma_1\beta_1 D) \cap S_0) = \#(D \cap S_0).$$

Of course, $(\delta_1\gamma_1\beta_1 s) \cap S_0 = \emptyset$ and $d(\delta_1\gamma_1\beta_1 s) > d(s)$. Thus with $\alpha_t := \delta_t\gamma_t\beta_t$ we have found the desired ambient isotopy satisfying (3) of the lemma.

This proves the lemma.

We finally show that good arc-systems can be reduced whenever $d(s) < \infty$.

16.19. Lemma. *Suppose $D \cap S_0$ is a good arc-system and $d(s) < \infty$. Then there is an admissible ambient isotopy α_t, $t \in I$, such that $\alpha_1 s \cap S_1 = \emptyset$, $\#(\alpha_1 D \cap S_1) < \#(D \cap S_0)$ and $\alpha_1 D \cap S_1$ is a good arc-system.*

Remark. Note that in the statement of this proposition we switched from one displacement of S, namely S_0, to the other, namely S_1.

Proof of Lemma. Let l_n be the component of $(\partial D - (b \cup U(s))^-$ containing the point p_1 of ∂b. Let k_n be the component of $U(s) \cap \partial D$ meeting l_n. Let l_{n-1} be the component of $(\partial D - (b \cup U(s))^-$, different from l_n, which meets k_n.

Our starting point is the claim $(l_n \cup l_{n-1}) \cap S_0 \neq \emptyset$. To see this claim recall the points from ∂k_n are labeled z_1, z_2. So either $k_n \cap l_n$ or $k_n \cap l_{n-1}$ is labeled z_1. If $k_n \cap l_n$ is labeled z_1, then clearly $l_n \cap S_0 \neq \emptyset$ since $d(z_1) = d(x_1) = d(s) < \infty$ and $l_n \neq l_0$. If $l_{n-1} \cap k_n$ is labeled z_1, then the points from ∂l_{n-1} are labeled z_1, z_1 or z_1, z_2. If $d(z_1) = d(z_2)$, then $d(z_2) < \infty$ and so $l_n \cap S_0 \neq \emptyset$. Thus we may suppose $d(z_1) < d(z_2)$. Then $l_{n-1} \cap S_0 \neq \emptyset$, if ∂l_{n-1} is labeled z_1, z_2. We are left with the case that l_{n-1} is labeled z_1, z_1. Let $m = d(s)$. Then $x_1 \in (C_m - C_{m-1})^-$. More precisely, $((C_m - C_{m-1})^- - U(s))^-$ is an annulus since $d(x_1) < d(x_2)$. Hence l_{n-1}

cannot be contained in C_m since l_{n-1} is an essential arc in $(F-U(s))^-$ (see definition of good discs). Therefore $l_{n-1} \cap S_0 \neq \emptyset$ if ∂l_{n-1} is labeled z_1, z_1. This establishes the claim.

Now consider the other displacement of S, i.e. consider the surface S_1. It is easy to describe the intersection-pattern $\mathcal{K}_1 := D \cap S_1$ in terms of the old intersection-pattern $\mathcal{K}_0 := D \cap S_0$. To do this denote $h_i := H_i \cap \partial D$ (recall H_0, H_1 are the two components of $F \cap U(b)$). Then h_1 is a neighborhood of the point p_1 of b in $(\partial D - (b \cup U(s)))^-$. h_0 is the union of l_0 with the neighborhood (in $(\partial D - (b \cup U(s))^-)$ of either (1) all end-points from $U(s) \cap \partial D$, or (2) all end-points from $U(s) \cap \partial D$ labeled z_1. (1) or (2) holds according whether or not $d(x_2) = \infty$. Let α_t, $t \in I$, denote the isotopy which contracts $(U(s) \cap \partial D) \cup b \cup h_0 \cup h_1$ into $(U(s) \cap \partial D) \cup h_1$. Let α_t^+ be the extension of α_t to an ambient isotopy of D which is the identity outside of the regular neighborhood of $(U(s) \cap \partial D) \cup b \cup h_0 \cup h_1$ in D. Then it is not difficult to see that $\mathcal{K}_1 = \alpha_1^+(\mathcal{K}_0)$ (modulo admissible ambient isotopy of $(D, \underline{d}_s)$). In particular, $\#\mathcal{K}_1 = \#\mathcal{K}_0$.

Let h_0' be the union of all components of h_0 meeting k_n. Let $\mathcal{T}_0$ be the collection of all arcs from $D \cap S_0$ having an end-point in h_0'. Then, by the argument for the above claim, $\mathcal{T}_0$ is non-empty. Moreover, by lemma 16.17, we may suppose $\mathcal{T}_0$ consists of arcs joining h_0' with l_0. Set $\mathcal{T}_1 = \alpha_1^+(\mathcal{T}_0)$. Then $\#\mathcal{T}_1 = \#\mathcal{T}_0 = \#(s \cap S_1) > 0$. But every arc from $\mathcal{T}_1$ is inessential in $(D, \underline{d}_s)$. More precisely, there is a disc A in D with $\mathcal{T}_1 \subset A$ which is a neighborhood of the point $k_n \cap l_n$. Since $\#\mathcal{T} = \#(s \cap S_1)$, there is an admissible ambient isotopy β_t, $t \in I$, of $(M, \underline{m})$, constant outside of a regular neighborhood of A, which deforms S_1 into a surface S_1' with $s \cap S_1' = \emptyset$ and $S_1' \cap D = (S_1 \cap D) - \mathcal{T}_1$.

Finally, $S_1' \cap D$ is a good arc-system w.r.t. l_n. Indeed, this is a straight-forward consequence of the fact that $S_1 \cap D$ is a good arc-system w.r.t. l_n since $S_0 \cap D$ is a good arc-system w.r.t. l_0.

It follows that $\alpha_t := \beta_{1-t}$ is the desired ambient isotopy.

Given the above lemmas it is now easy to finish the proof. First set $p := \#(D \cap S_0)$. Then, by lemma 16.17 and 16.18, we may suppose s (and so D) is admissibly isotoped in $(M, \underline{m})$ so that $D \cap S_0$ is a good arc-system. If $d(s) = \infty$, then (1) of the proposition follows from lemma 16.7. If not, then, by lemma 16.19, there is an admissible ambient isotopy α_t, $t \in I$, such that $\alpha_1 s \cap S_1 = \emptyset$, $\#(\alpha_1 D \cap S_1) < \#(D \cap S_0)$ and $\alpha_1 D \cap S_1$ is a good arc-system (w.r.t. l_n). Then we substitute the displacement S_1 for S_0 and apply the above process again, and so on. The procedure has to stop after finitely many steps. Thus we eventually end up with an admissible Heegaard string, s', with $d(s') = \infty$ (measured w.r.t. S_0 or S_1). So again (1) of the proposition follows from lemma 16.7. ◊

The next lemma gives a few useful facts concerning tubular cages. In particular, it implies that s (or better $d(s)$) cannot be pushed all the way to infinity if it is contained in a tubular cage.

16.20. Lemma. *Let $(M, \underline{m})$, s, S and S_0 be given as in (*). Suppose that $s \cap S_0 = \emptyset$ and that $\partial s \subset C_m$, for some $m \geq 0$ (see 16.4). Let W be a tubular cage for s in $(M_0', \underline{m}_0')$. Then the following holds:*

(1) There is no 1-faced compression disc for $(\partial W - \partial M_0')^-$ in either M or M_0'.

(2) There is no 2-faced compression disc for $(\partial W - \partial M_0')^-$ in $(M_0', \underline{m}_0')$ meeting a free face of $(M_0', \underline{m}_0')$.

(3) s cannot be deformed into $W \cap \partial M_0'$, using a homotopy in W which fixes ∂s.

(4) s cannot be deformed out of W, using an admissible homotopy in $(M_0', \bar{\underline{m}}_0')$.

Remark. $(M_0', \underline{m}_0')$ denotes the manifold obtained from $(M, \underline{m})$ by splitting along S_0.

Proof. (1) Assume there is a 1-faced compression disc, D, for $A := (\partial W - \partial M_0')^-$. Then $U(W \cup D)$ is punctured lens space in M. By the irreducibility of M, it then follows that A has to have winding number one (w.r.t. W). This however contradicts property (1) of tubular cages. This proves (1)

(2) Assume there is a 2-faced compression disc, D, for A in $(M_0', \underline{m}_0')$ meeting a free face of $(M_0', \underline{m}_0')$. Then $e := D \cap \partial M' \subset F_0$, where F_0 is a free face of $(M_0', \underline{m}_0')$. Moreover, $\partial e \subset (\partial D_W - \partial F_0)^-$ (see definition of tubular cages for the definition of D_W). Note that $F_0 = (C_m - C_{m-1})^-$ and so it is a disc. e splits the disc D_W or the disc $(F_0 - D_W)^-$ into two discs. Let D' be that one of them whose intersection with $(\partial D_W - \partial F_0)^-$ is connected. Then $\partial(D \cup D')$ is a closed curve in A. More precisely, it is a contractible curve in A since A has no 1-faced compression discs (see (1)). Thus $D \cap A$ is inessential in A and so D is no 2-faced compression disc for A. This is the required contradiction.

(3) By property (1) of tubular cages, the core of W cannot be deformed into $W \cap \partial M_0'$. By property (4) of tubular cages, it therefore follows that s cannot be deformed (rel ∂s) into $W \cap \partial M$.

(4) Assume there is an admissible homotopy $f : s \times I \to M$ in $(M_0', \bar{\underline{m}}_0')$ which pushes s out of W. Let f be admissibly deformed in $(M_0', \bar{\underline{m}}_0')$ so that $f^{-1}(\partial W - \partial M_0')^-$ consists of curves (general position) whose number is as small as possible. Let k be a component of $f^{-1}(\partial W - \partial M_0')^-$ (this exists since $f(s \times 0) \subset W$ and $f(s \times 1) \subset M_0' - W$. M_0' is aspherical (since it is irreducible) and $(\partial W - \partial M_0')^-$ is incompressible in M_0' (see (1)). Thus

it follows, by the usual surgery argument involving the loop-theorem, that k cannot be a closed curve. A similar argument shows that ∂k cannot lie in one component of $\partial s \times I$ (use [Joh 1, prop. 2.1] and the fact, see (1) above, that $(\partial W - \partial M_0')^-$ has no 2-faced compression discs in $(M_0', \underline{m}_0')$ meeting free faces). Thus $f^{-1}(\partial W - \partial M_0')^-$ consists of arcs in $s \times I$ joining the two components of $\partial s \times I$. In particular, $f^{-1}(\partial W - \partial M_0')^-$ splits $s \times I$ into 4-faced discs. Let Q be that one of them containing $s \times 0$. The existence of $f|Q$ then shows that s can be deformed in W into $(\partial W - \partial M_0')^-$, using an admissible homotopy in $(M_0', \underline{\tilde{m}}_0')$. More precisely, it can be deformed into a (singular) arc whose end-points lie in $F_W \cap (\partial W - \partial M_0')^-$. But $F_W \cap (\partial W - \partial M_0')^-$ is connected. It therefore follows that s can be deformed in W into $(\partial W - \partial M_0')^-$ and then (in the annulus $(\partial W - \partial M_0')^-$ and fixing end-points) into $W \cap \partial M_0'$. This contradicts (3) above.

16.21. *Relative Handle-Addition Lemma.*

In order to state our next result in its proper generality we need a little preparation. Start with a collection $\mathcal{A} = A_1, A_2, ..., A_n, \ n \geq 1,$ of pairwise disjoint and pairwise parallel, incompressible annuli in a solid torus W. Any incompressible annulus in a solid torus is boundary parallel [Wa 1]. It follows that any incompressible annulus in W has a well-defined winding number and this is bigger than zero. Winding numbers of parallel annuli are the same. So we may define the winding number of the collection $\mathcal{A}$ to be the winding number of A_1, say. Suppose the winding number of $\mathcal{A}$ is bigger than one. Then there is another collection $\mathcal{B} = B_1, B_2, ..., B_n$ of incompressible annuli in W such that, for all $1 \leq i \leq n$, $\partial A_i = \partial B_i$ and A_i is not isotopic (rel ∂A_i) in W to B_i. $\mathcal{B}$ is unique, up to isotopies fixing boundaries. We call $\mathcal{B}$ the *collection corresponding to* $\mathcal{A}$. Given two surfaces S and S' in a 3-manifold M, we say that S' is obtained from S by a *torus-modification along a solid torus*, W, if W is a solid torus with the following properties:

(1) $W \cap S$ is a collection of pairwise parallel, incompressible annuli in W with winding number bigger than one,

(2) $(S' - W)^- = (S - W)^-$, and

(3) $S' \cap W$ is the collection corresponding to $S \cap W$.

Let s be an admissible Heegaard string in an irreducible 3-manifold $(M, \underline{m})$ whose completed boundary-pattern is useful. Let S be either a good annulus or an essential, non-separating surface in $(M, \underline{\tilde{m}})$ with $e(S)$ maximal. Let S_0 be a Heegaard displacement (along a good disc D for s in $(M, \underline{m})$). Then we are in situation (*) from 16.4.

16.22. Proposition. *Suppose there is no essential annulus in $(M, \underline{m})$ which has a boundary curve in a free face of $(M, \underline{m})$. Then there is a surface, S', obtained from S by the torus modification along some appropriate solid torus,*

*and there is an admissible (ambient) isotopy of s in $(M, \bar{\underline{m}})$ which deformes
s so that afterwards $s \cap S$, or $s \cap S'$, or $s \cap S_0$, or $s \cap S_0'$ is empty.*

Remark. Here S_0' denotes a displacement of S'.

Proof. The idea is to translate the proof of lemma 15.11 into the context of 3-manifolds with boundary-patterns. We suppose $(D \cap \bigcup \underline{m}) \cap \partial S \neq \emptyset$ (the treatment of other case is exactly the same; use only S instead of S_0).

Let s be admissibly isotoped in $(M, \bar{\underline{m}})$ so that $\#(s \cap S_0)$ is as small as possible. Let D be the good disc for s in $(M, \underline{m})$ used for the displacement of S (see definition of displacements). Set $b := D \cap \bigcup \underline{m}$. By general position, we may suppose that $D \cap S_0$ is a system of admissible arcs and curves in $(D, \underline{d}_s)$. More precisely, by an innermost-disc-argument, we may suppose $D \cap S_0$ is a system of arcs (S_0 is incompressible since S is) none of which has both end-points in one free face of $(D, \underline{d}_s)$ (see prop. 16.10). In fact, it is a system of arcs in D not meeting b since $\bar{b} \cap S_0 = \emptyset$ (by definition of displacement). Every inessential arc from $D \cap S_0$ separates a 3-faced disc from $(D, \bar{\underline{d}}_s)$ not meeting b. This 3-faced disc defines an admissible (ambient) isotopy for s in $(M, \bar{\underline{m}})$ which reduces $\#(s \cap S_0)$. Thus we may suppose s is admissibly isotoped so that $D \cap S_0$ is a system of essential arcs in $(D, \bar{\underline{d}}_s)$ not meeting b.

If $s \cap S_0 = \emptyset$, we are done. So suppose the converse. Then there must be a good pencil for S_0 (see the proof of 15.4), i.e., there is a component, k_0, from $U(s) \cap \partial D$ so that every arc from $D \cap S_0$ meeting k_0 is admissibly parallel to one of the two a components, l_1, l_2, from $(\partial D - (b \cup U(s)))^-$ meeting k_0.

As shown in the proof of prop. 15.11, any good pencil gives rise to a sequence of annulus-modifications and/or one torus modification along a solid torus which turns S_0 into a surface, S_0', with $s \cap S_0' = \emptyset$.

It remains to show that (in our situation) no annulus-modification actually occurs. To see this recall from the proof of 15.11 how annulus-modifications arise from pencils. It turns out that in our situation (see our construction of pencils) an annulus-modification would be along an incompressible annulus A with the following properties:

(1) $s \cap A$ is an arc in A joining the two components of ∂A,

(2) one component of ∂A lies in S_0, and

(3) the other component of ∂A lies in a free face of $(M, \underline{m})$.

By the very definition of "good annuli", this is impossible if S is a good annulus. Thus w.l.o.g. S is not. Then recall $e(S)$ is supposed to be maximal. It therefore follows that $A \cap S_0$ is parallel to a free face of S_0 (see prop. 16.10). Let A' be the corresponding parallelity region. Then, in particular, $A' \subset S$ (S_0 is a displacement of S along D). $A \cup A'$ is an annulus whose boundary lies in free faces of $(M, \underline{m})$. By hypothesis, this annulus has to be inessential in $(M, \bar{\underline{m}})$. M is irreducible, $\bar{\underline{m}}$ is useful for M and $A \cup A'$ is incompressible. Thus it follows that $A \cup A'$ is parallel to some annulus in a free face of $(M, \underline{m})$. Therefore A' is isotopic to A, using an admissible isotopy in

$(M, \underline{m})$ fixing $A \cap A'$. This isotopy can be extended to an admissible isotopy of S_0 which reduces the intersection $s \cap S_0$ (see property (1) of A). Thus we get a contradiction to our minimality condition on $\#(s \cap S_0)$. This establishes the claim. $\Diamond$

From the previous proposition we now finally deduce the main result of this section. It is the "relative handle-addition lemma". Note that it does not deal with trivial, admissible Heegaard strings.

16.23. Theorem. *Let $(M, \underline{m})$ be an irreducible 3-manifold whose completed boundary-pattern is useful. Suppose that no essential annulus in $(M, \underline{\bar{m}})$ has a free face. Let s be a non-trivial, admissible Heegaard string in $(M, \underline{m})$. Let S be a non-separating surface in $(M, \underline{m})$ with $e(S)$ maximal. Then s can be deformed, using an admissible ambient isotopy in $(M, \underline{\bar{m}})$, so that afterwards either $s \cap S$ or $s \cap S'$ is empty, where S' is a obtained from S by the torus modification along some appropriate solid torus.*

Proof. By prop. 16.22, there is a surface S', obtained from S by the torus modification along some appropriate solid torus, such that $s \cap S$, $s \cap S_0$, or $s \cap S'$, or $s \cap S_0'$ is empty (modulo admissible isotopy of s).

If $s \cap S$ or $s \cap S'$ is empty, then we are done. So we suppose $s \cap S_0$ is empty (if $s \cap S_0' = \emptyset$, we argue exactly the same). Moreover, by prop. 16.16, we may suppose that $s \cap S_0 = \emptyset$ and that s has a tubular cage W in $(M_0', \underline{m_0}')$ which is not a tubular cage in $(M, \underline{m})$ (here $(M_0', \underline{m_0}')$ denotes the manifold obtained from $(M, \underline{m})$ by splitting along S_0). Finally, $D_W \subset C_m$ and $D_W \cap C_{m-1} = \emptyset$.

Form $W' := (W - U(D_W))^-$ and define

$$S_0'' := (S_0' - W')^- \cup (\partial W' - \partial M_0')^-.$$

Then $\partial S_0'' = \partial S' = \partial S_0''$ and S_0'' is obtained from S_0' by a torus modification along the solid torus W'. Moreover, s can be admissibly isotoped in $(M, \underline{\bar{m}})$ into an admissible Heegaard string $s' \subset S'' = \emptyset$ and $d(s') > d(s)$. Thus, replacing S_0' by S_0'', we may now repeat the above process and so on. In the end we obtain a non-separating surface, S_0^*, from S_0' with $\partial S_0^* = \partial S'$ and an admissible isotopy which pushes s to s^* with $s^* \cap S^* = \emptyset$ and $d(s^*) = \infty$.

It remains to verify that S_0^* is obtained from S_0 by a torus modification along a single solid torus, for then the proposition follows from lemma 16.7. Here is a brief indication of the relevant argument. First, by induction, it suffices to show the claim is true for S_0''. To see the claim for S_0'' recall S_0' is obtained from S_0 by a torus modification along some solid torus, say V. Since $s \cap V \neq \emptyset$, it follows that $V \cap W' \neq \phi$. It is then not hard to see that $V \cup W'$ has to be a solid torus ($V \cap W'$ is a solid torus) and that S_0'' is obtained from S_0 by a torus modification along that solid torus. $\Diamond$

§17. Homotopy and Isotopy.

In this section we use the relative handle-addition lemma from the previous section in order to control homotopies of generalized Heegaard strings (see 17.5, 17.10 and 17.21). Our main result of this section is the property that homotopies of generalized Heegaard strings in simple 3-manifolds can be deformed into embeddings (see thms. 17.7, 17.12 and 17.22 for the precise statement). In particular, it follows that generalized Heegaard strings in simple 3-manifolds are homotopic if and only if they are isotopic (see cor. 17.32). In fact, the isotopy in question can be taken to be an ambient isotopy in the 3-manifold. This feature of generalized Heegaard strings is quite remarkable and somewhat unexpected. It is of course far from being true for arbitrary arcs in 3-manifolds (otherwise there would be no knot theory) and so (generalized) Heegaard strings must be really very special. Moreover, it is a well-known and important fact that incompressible surfaces in Haken 3-manifolds are homotopic if and only if they are isotopic [Wa 2]. Thus, although codim 2 objects, generalized Heegaard strings share an important feature with codim 1 objects in 3-manifolds.

17.1. *Deformations of Generalized Heegaard Strings.*

Throughout this section we deal with homotopies and isotopies of codim two objects in 3-manifolds. Recall that, in contrast to codim one objects, isotopies between codim two objects do not necessarily extend to ambient isotopies of the underlying 3-manifold (otherwise every knot would be trivial). But all isotopies encountered in this section will have this additional property. So, throughout this section, the term "isotopy" will always mean "ambient isotopy". Recall from [Joh 1] that a homotopy in a 3-manifold with boundary-pattern is *admissible* if it preserves the boundary-pattern. Here is the main result of this section (see thm. 17.22 for a slightly more general statement).

17.2. Theorem. *Let $(M, \underline{m})$ be a simple 3-manifold such that no essential annulus in $(M, \underline{m})$ has a free face. Let $f : I \times I \to M$ be an admissible homotopy in $(M, \underline{\tilde{m}})$ such that $f|I \times 0$ and $f|I \times 1$ are admissible Heegaard strings in $(M, \underline{m})$. Then f can be deformed into an embedding, using an admissible homotopy whose restriction to $I \times 0$ and $I \times 1$ is an isotopy.*

Remark. A Haken 3-manifold is simple if its characteristic submanifold is trivial (see [Joh 1, p.159]).

17.3. Corollary. *Generalized Heegaard strings in simple 3-manifolds are homotopic if and only if they are isotopic.*

Recall from 16.1 that the concept of admissible Heegaard strings generalizes the concept of generalized Heegaard strings which in turn generalizes the concept of Heegaard strings. In particular, cor. 17.3 applies to Heegaard strings as well.

At this point recall that the completed boundary-pattern of a simple 3-manifold is useful (in the sense of [Joh 1, def. 2.2]). This fact is absolutely crucial. Not only are non-useful boundary-patterns difficult to handle, but the above theorem is definitely false for them. Figure 17.1 gives a counter-example. Indeed, it describes two admissible Heegaard strings (in a solid torus with boundary-pattern) which are admissibly homotopic, but not admissibly isotopic. But notice that the boundary-pattern of the underlying 3-manifold is not useful. Theorem 17.2 tells us that such an example is not possible for useful boundary-patterns. This illustrates once more the special relevance of useful boundary-patterns (for more features of useful boundary-patterns, see [Joh 1]).

17.4. Example.

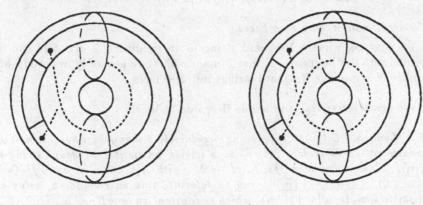

(Figure 17.1.)

Here is our strategy for proving thm. 17.2. The idea is to generalize first the problem at hand to a problem which is invariant under splittings along essential surfaces and then to use induction on a hierarchy of essential surfaces to solve the generalized problem. Thus, instead of only studying admissible homotopies between admissible Heegaard strings, we decide to study a variety of homotopies. Specifically, we study admissible homotopies $f : I \times I \to M$ in $(M, \underline{\tilde m})$ such that $f|I \times 0$ is an admissible Heegaard string and that $f(I \times 1)$ is an arc which is either an admissible Heegaard string, or entirely contained in a bound face, or entirely contained in a free face of $(M, \underline{m})$. Our goal is to show that all these homotopies can be admissibly deformed into an embedding. Our starting point is the observation that this is true for 3-balls with useful boundary-patterns. This is the induction beginning. To carry out the induction step, we use the handle-addition lemma in its relative form as established in the previous section. According to this lemma there is "usually", to any given admissible Heegaard string, an essential surface disjoint to it. In particular, the admissible Heegaard string remains unaffected by splitting along this essential surface. Moreover, we show in this section that this essential surface splits all admissible homotopies of this admissible Heegaard string into admissible homotopies. In particular, we show that, for any two admissibly homotopic, admissible Heegaard strings, there is always an essential surface which is disjoint to *both* of them. The problem of studying the original admissible homotopies in $(M, \underline{m})$ is then reduced to study admissible homotopies in the splitted manifolds. But, by induction, these splitted homotopies are admissibly homotopic to embeddings. To complete the induction step it then remains to show that all these deformations, fitted together, deform the original homotopy into an embedding.

As indicated in this outline we will study a variety of admissible homotopies of admissible Heegaard strings. Specifically, we study deformations into free faces (17.5), deformations into bound faces (17.10) and deformations into another admissible Heegaard string (17.19). We begin with the simplest one.

17.5. Deformations into Free Faces.

We first ask when a Heegaard string in an irreducible 3-manifold can be deformed into the boundary of that 3-manifold. More generally, we study homotopies of admissible Heegaard strings into free faces.

We first treat trivial, admissible Heegaard strings.

17.6. Lemma. *Let* $(M, \underline{m})$ *be an irreducible 3-manifold whose completed boundary-pattern is useful. Let* s *be a trivial, admissible Heegaard string in* $(M, \underline{m})$. *Then every map* $f : s \times I \to M$ *with* $f|s \times 0 = s$ *and* $f(\partial(s \times I) - s \times 0)^{-} \subset (\partial M - \bigcup \underline{m})^{-}$ *can be deformed into an embedding, using an admissible homotopy in* $(M, \bar{\underline{m}})$ *which is constant on* $s \times 0$.

Proof. Since s is trivial in $(M, \underline{m})$, there is a good disc, D, for s in $(M, \underline{m})$ such that $D \cap U(s)$ is connected.

If $D \cap \bigcup \underline{m}$ is non-empty, then, identifying D with f along s, we get a singular 2- or 3-faced disc in $(M, \bar{\underline{m}})$. But $\bar{\underline{m}}$ is a useful boundary-pattern. So, using [Joh 1, prop 2.1], we deduce that D does not intersect bound faces of $(M, \underline{m})$ in essential arcs. This is a contradiction since D is supposed to be a good disc for s (see 16.1).

If $D \cap \bigcup \underline{m}$ is empty, then, identifying D with f along s, we get a singular 1-faced disc in $(M, \bar{\underline{m}})$. Now, every free face of $(M, \underline{m})$ is incompressible since $\bar{\underline{m}}$ is supposed to be useful. Moreover, M is aspherical since M is irreducible (see [Wa 2]). It therefore follows that f can be admissibly deformed (rel $s \times 0$) into D. $\Diamond$

The following theorem deals with the general case.

17.7. Theorem. *Let* $(M, \underline{m})$ *be an irreducible 3-manifold whose completed boundary-pattern is useful. Suppose no essential annulus in* $(M, \underline{m})$ *has a free face. Let* s *be an admissible Heegaard string in* $(M, \underline{m})$. *Let* $\bar{f} : s \times I \to M$ *be a homotopy with* $f|s \times 0 = s$ *and such that* $f(\partial(s \times I) - s \times 0)^{-}$ *lies in some free face of* $(M, \underline{m})$.

Then f can be deformed into an embedding, using an admissible homotopy in
$(M, \underline{m})$ *which is constant on* $s \times 0$.

Proof. The proof is by induction on an appropriate hierarchy of essential surfaces.

Suppose M is a 3-ball. Then $\partial(M - U(s))^-$ is a torus and $(M - U(s))^-$ is
∂-reducible. It follows that s is an unknotted arc in M. Moreover, all faces of
$(M, \bar{\underline{m}})$ are discs since $\bar{\underline{m}}$ is supposed to be useful for M. So the conclusion of
the theorem follows easily in this case. This establishes the induction beginning.

It remains to carry out the induction step.

W.l.o.g. we may suppose that s is non-trivial, for otherwise the theorem
follows from lemma 17.7. Thus, by thm. 16.23, there is an essential, non-
separating surface, say S, in $(M, \bar{\underline{m}})$ with $s \cap S = \emptyset$ and $e(S)$ maximal. Let
$(M', \underline{m}')$ be the manifold obtained from $(M, \underline{m})$ by splitting along S. Then
$(M', \underline{m}')$ is again an irreducible 3-manifold whose completed boundary-pattern
is useful [Joh 1, prop. 4.8]. Moreover, we claim there is no essential annulus A
in $(M', \bar{\underline{m}}')$ which has a free face in $(M', \underline{m}')$. Assume for a moment A is
such an annulus. Then one face of A must lie in S (or better: in a bound face
of $(M', \underline{m}')$ which is a copy of S). Since $e(S)$ is maximal, it follows that
$A \cap S$ is parallel in S to a free face of S. Let $A' \subset S$ be the corresponding
parallelity region. Then $A \cup A'$ is an incompressible annulus whose boundary
lies in free faces of $(M, \underline{m})$. By our hypothesis on $(M, \underline{m})$, this annulus has
to be inessential and so it has to have a 2-faced ∂-compression disc since it is
incompressible. But a 2-faced ∂-compression disc for $A \cup A'$ is a 3-faced ∂-
compression disc for A in $(M', \bar{\underline{m}}')$. So A is inessential in $(M', \bar{\underline{m}}')$, and
the claim is established. By prop. 16.3, s is an admissible Heegaard string
in $(M', \underline{m}')$. M is aspherical (it is irreducible), $\bar{\underline{m}}$ is useful for M and S
is essential in $(M, \bar{\underline{m}})$. Thus, by a well-known surgery argument, f can be
admissibly deformed in $(M, \bar{\underline{m}})$ so that afterwards $f^{-1}S$ is empty (use [Joh
1, prop. 2.1]). We therefore have verified that $(M', \underline{m}')$, s and f satisfy the
hypothesis of the theorem. By [Ha 2] (see [He, thm 13.3]), we obtain a 3-ball by
successive splittings of M along incompressible and non-separating surfaces.
The theorem therefore follows by induction on a hierarchy. $\Diamond$

17.8. Corollary. *Let M be an irreducible and ∂-irreducible one-relator 3-
manifold, different from the 3-ball. Then no Heegaard string in M can be
deformed (relative end-points) into ∂M.*

Proof. We distinguish between two cases:

Case 1. M is a simple 3-manifold.

Assume the corollary does not hold. Then there is a Heegaard string, s,
in M and a homotopy $f : s \times I \to M$ such that $f|s \times 0 = s$ and $f(\partial(s \times
I) - s \times 0)^- \subset \partial M$. Set $\underline{m} = \emptyset$. Then, in the terminology of 16.1, s is an
admissible Heegaard string in $(M, \underline{m})$. Moreover, M, s and f satisfy the
hypothesis of the previous theorem. Thus, by this theorem, f can be deformed

(relative $s \times 0$) into an embedding. It then follows that M is ∂-reducible since $(M - U(s))^-$ is a handlebody and since M is not a 3-ball. This contradicts the hypothesis.

Case 2. M is not a simple 3-manifold.

In this Case there is an essential annulus S in M. Let S be chosen so that this holds and that, in addition, $s \cap S$ is a collection of points whose number is as small as possible. We claim $s \cap S = \emptyset$. If not, then it follows from lemma 17.20 and 21 the existence of an annulus C which joins S with ∂M and contains one of the components of $s - S$. Set $S' := (S - U(C))^- \cup (C' \cup C'')$, where C', C'' are the two copies of C in $\partial U(C)$. Notice that S' consists of two annuli. Both of them intersect s in strictly fewer points than S and at least one of them is essential. Thus we have a contradiction to our minimal choice of S and the claim is established.

Now, consider the homotopy $f : I \times I \to M$ which deforms s into ∂M, i.e., $f | I \times 0 = s$ and $f(\partial(I \times I) - I \times 0)^- \subset \partial M$. Since $s \cap S = \emptyset$ and since S is essential, it follows that f can be deformed (rel s) out of S. Moreover, every homotopy of f can be deformed (rel s) out of S. Thus the conclusion of the corollary holds if we can show it for the manifold $M_1 = (M - U(S))^-$. Hence we are finished if M_1 is ∂-irreducible. Because M_1 then satisfies the hypothesis of the corollary and has smaller complexity than M.

Thus suppose M_1 is ∂-reducible. Then let $(M_1, \underline{m}_1)$ be the manifold obtained from $(M, \emptyset)$ by splitting along S_1. Let $E \subset M_1$ be a system of proper discs with the property that every disc in $(M_1 - U(E))^-$ is ∂-parallel (of course such a system always exists because $\chi \partial M_1$ is finite). In fact, E can be chosen so that this holds and that, in addition, it is a system of essential discs in $(M_1, \underline{\underline{m}}_1)$. We claim E can be admissibly isotoped so that afterwards $s \cap E = \emptyset$. To see this recall from prop. 16.3, that s is an admissible Heegaard string in $(M_1, \underline{m}_1)$. Let D be a good disc for s (see 16.1) and let D be chosen so that (*) from section 16.4 holds. Now let E_0 be a displacement of E (see 16.4 for definition). Then $D \cap E$ is a system of arcs whose end-points lie in free faces of $(D, \underline{d}_s)$. It then follows that $s \cap E_0 = \emptyset$, for otherwise there would be a good pencil of $D \cap E$ in D (see the proof of 14.4) but there are no good pencils for discs (see the proof of 15.11). Thus prop. 16.16 applies and so either (1) or (2) of prop. 16.16 holds for s. But (2) is impossible. Indeed, s cannot lie in a tubular cage, for otherwise we could deform f admissibly into the tubular cage and we would get an immediate contradiction to the definition of tubular cages. Hence (1) holds, i.e., s can be admissibly isotoped so that afterwards $s \cap E = \emptyset$. Then f as well as every homotopy of f (rel s) can be admissibly deformed (rel s) in $(M_1, \underline{m}_1)$ out of E (use [Joh 1, prop. 2.1] and the fact that $(M_1, \underline{\underline{m}}_1)$ has a useful boundary-pattern). Therefore the conclusion of the corollary holds if it holds for $M_2 = (M_1 - U(E))^-$ (without boundary-pattern). But, by construction, M_2 is ∂-irreducible, and so it satisfies the hypothesis of the corollary. Moreover, it has smaller complexity than M, and so, by induction (on a hierarchy) the corollary holds for M_2. This proves the corollary in Case 2. ◊

17.9. Corollary. *Let* M *be any one-relator 3-manifold and let* s *be a Heegaard-string in* M. *Then* s *can be deformed (relative* ∂s) *into* ∂M *if and only if* M *is a handlebody and* s *is an unknotted arc in* M.

Remark. Recall that an arc, s, in M is *unknotted*, if there is a disc, D, in M with $(\partial D - \partial M)^- = s$.

Proof. One direction is obvious. For the other direction, let $\mathcal{D}$ be a maximal system of essential discs in M. It follows from the handle-addition lemma (prop. 15.16), that $\mathcal{D}$ can be chosen in the complement of the Heegaard-string s. In particular, there is a component, N, of $(M - U(\mathcal{D}))^-$ which contains s. Every component of $(M - U(\mathcal{D}))^-$, different from N, is a 3-ball since $\mathcal{D}$ splits $(M - U(s))^-$ into handlebodies. Moreover, ∂N has to be the 2-sphere. Otherwise it follows from the handle-addition lemma that N is a Haken 3-manifold and that s is a Heegaard-string in N which can be deformed into ∂N. This is impossible, by cor. 17.8. It follows that N is a punctured lens space since $(N - U(s))^-$ is a handlebody and so a solid torus. To continue, let N^* be the lens space obtained from N by attaching a 3-ball, E, to ∂N. Let s' be an arc in ∂N with $\partial s' = \partial s$. The union $s \cup s'$ is homotopic to the core of the solid torus $E \cup U(s)$. So it represents (the conjugacy class of) the generator of $\pi_1 N^*$. Now $s \cup s'$ is a contractible curve in N^* since s can be deformed into the 2-sphere ∂N. It follows that N^* is a lens space with trivial fundamental group. Thus N^* is the 3-sphere and so N is the 3-ball. Altogether, M is a handlebody. Moreover, $(N - U(s))^-$ is a solid torus. Thus s is an unknotted arc in the 3-ball N. By definition, there is therefore a disc, D, in N with $s = (\partial D - \partial N)^-$. Now, $U(\mathcal{D}) \cap \partial N$ consists of discs and so D can be chosen disjoint to $U(\mathcal{D})$. Then D is a disc in M with $s = (\partial D - \partial M)^-$. Thus s is unknotted in the handlebody M. This proves the corollary. ◊

17.10. *Deformations into Bound Faces.*

Having studied deformations of admissible Heegaard strings into free faces, we are now interested in deformations into *bound* faces. The study of such deformations is more involved than the previous study of deformations into free faces; mainly because trivial, admissible Heegaard strings are harder to handle in this setting. As a result, we will be able to prove the corresponding theorem (thm. 17.12) only for simple 3-manifolds, i.e., for Haken 3-manifolds whose characteristic submanifold is trivial. The idea for its proof is again to use hierarchies of essential surfaces. However, the set of simple 3-manifolds is *not* closed under splittings along essential surfaces. We are therefore forced to weaken the concept of simple 3-manifolds. It emerges the concept of "almost simple 3-manifolds". At first sight this concept is somewhat technical but it turns out that it is just about right to serve our purpose well. Here is the definition.

An irreducible 3-manifold $(M, \underline{m})$ whose completed boundary-pattern is useful is *almost simple* if the following holds:

(1) Every incompressible torus in M is parallel to a free face of $(M, \underline{m})$.

(2) Suppose X is an essential I-bundle in $(M, \bar{\underline{m}})$ which meets free faces of $(M, \underline{m})$, and whose lids lie either in one bound face or in two adjacent bound faces of $(M, \underline{m})$. Then X is an I-bundle over the square.

(3) Every essential I-bundle in $(M, \bar{\underline{m}})$ with at least one lid in a free face of $(M, \underline{m})$ is an I-bundle over the disc.

(4) Every essential I-bundle, X, in $(M, \bar{\underline{m}})$ with at least one lid in a free face of $(M, \underline{m})$ and $(\partial X - \partial M)^-$ connected, is the neighborhood of a face from $(M, \bar{\underline{m}})$.

(Recall the convention that the completed boundary-pattern of an annulus or Möbius band consists of all boundary components).

Of course, all simple 3-manifolds $(M, \emptyset)$ with empty boundary-patterns are almost simple. In addition, they have no essential annuli (meeting free faces). This is our starting point. We next have to verify the crucial statement that these properties are preserved under splittings along essential surfaces. To make this statement precise, we need the notion of a "push" for essential surfaces.

Let S be an essential surface in $(M, \bar{\underline{m}})$. Let $(B, \underline{b})$ be a bound face of $(M, \underline{m}$ with the boundary-pattern induced by $(M, \bar{\underline{m}})$. Suppose $\mathcal{A}$ is a collection of pairwise disjoint squares, A, in bound faces, B_A, of $(M, \underline{m})$ such that $(\partial A - \partial B_A)^-$ is a face of S. Define $S' := (S - U(\mathcal{A}))^- \cup \mathcal{A}'$, where $\mathcal{A}'$ is the copy of $\mathcal{A}$ in $(\partial U(\mathcal{A}) - \partial M)^-$. We say S' is obtained from S by a *push (across $\mathcal{A}$)*; or simply: S' is a *push* of S (we need this notion only for 17.11).

17.11. Proposition. *Let $(M, \underline{m})$ be an almost simple 3-manifold such that no essential annulus in $(M, \bar{\underline{m}})$ meets a free face of $(M, \underline{m})$. Let S be either an essential square in $(M, \bar{\underline{m}})$ (separating or not), or a non-separating surface in $(M, \bar{\underline{m}})$ such that $e(S)$ is as large as possible. Suppose every bound face of S that lies next to a square-face of $(M, \underline{m})$ is contained in a square-face of $(M, \underline{m})$.*

Then the manifold $(M', \underline{m}')$ obtained from $(M, \underline{m})$ by splitting along S, or one of its pushs, is an almost simple 3-manifold and no essential annulus in $(M', \bar{\underline{m}}')$ meets a free face of $(M', \underline{m}')$.

Remark. See 16.4 for the definition of $e(S)$. We say that a face of S *lies next to* a face, B', of $(M, \bar{\underline{m}})$ if it is admissibly parallel, in the face of $(M, \underline{m})$ containing it, to a face of B' (w.r.t. the boundary-pattern of B induced by $\underline{m}$). A face of $(M, \underline{m})$ is a *square-face* if it is a disc whose induced completed boundary-pattern consists of four arcs.

Proof. S is essential in $(M, \bar{\bar{m}})$ (this is clear if S is a square and see 16.5 if S is not a square). So, by [Joh 1, prop. 4.8], $(M', \underline{m}')$ is an irreducible 3-manifold whose completed boundary-pattern is useful. In the following, B_1, B_2 will denote the two bound faces of $(M', \underline{m}')$ which are copies of S. View B_1, B_2 as being components of $(\partial U(S) - \partial M)^-$.

We first show that there is no essential annulus, A, in $(M', \bar{\bar{m}}')$ which has at least one boundary curve in a free face of $(M', \underline{m}')$. We have seen this before but let us repeat the argument for completeness. It is easy to see that $A \cap (B_1 \cup B_2) \neq \emptyset$ since $(M, \underline{m})$ has no essential annulus which meets a free face. Thus w.l.o.g. A joins $\overline{B}_1$ with a free face of $(M', \underline{m}')$. Splitting S along A, we get a non-separating surface whose complexity is larger or equal to $e(S)$. By our maximality condition on S, it has to be equal to $e(S)$. This however is only possible if the curve $A \cap B_1$ is parallel in S to a free face of S. Let A' be the corresponding parallelity region in S. Then $A \cup A'$ is an admissible annulus in $(M, \bar{\bar{m}})$ whose boundary lies in free faces of $(M, \underline{m})$. It has to be inessential in $(M, \bar{\bar{m}})$ since $(M, \underline{m})$ has no essential annulus meeting a free face. But, of course, $A \cup A'$ is incompressible in M since A is in M'. It follows that $A \cup A'$ is parallel (but not admissibly parallel) to some annulus in a free face of $(M, \underline{m})$, since $(M, \bar{\bar{m}})$ is an irreducible 3-manifold with useful boundary-pattern. The parallelity region avoids S since $S \cap (A \cup A') = \emptyset$ and since S is non-separating. Therefore it follows that A is inessential in $(M', \bar{\bar{m}}')$. This contradicts our choice of A.

We next show that $(M', \underline{m}')$ has the four properties of an almost simple 3-manifold.

(1) Let T be an incompressible torus in M'. Then T is also incompressible in M since S is incompressible. Thus T is parallel to a free face of $(M, \underline{m})$ since $(M, \underline{m})$ is almost simple. S cannot lie in the corresponding parallelity region, for otherwise S would be a separating torus. Thus T is admissibly parallel to a free face of $(M', \underline{m}')$.

(2) Let X' be an essential I-bundle in $(M', \bar{\bar{m}}')$ which meets at least one free face of $(M, \underline{m})$ and whose lids lie in one, or two adjacent, bound faces of $(M', \underline{m}')$.

Suppose one lid of X' lies in B_1, say. Then $X' \cap B_2 = \emptyset$ since the lids of X' lie in the same or in adjacent bound faces of $(M, \underline{m})$. It follows that $(B_1 - X')^- \cup (\partial X' - \partial M')^-$ is a non-separating 2-manifold in $(M, \underline{m})$. By our maximality condition on $e(S)$, it follows that $X' \cap B_1$ must be a square. So X' is the I-bundle over the square.

Thus we may suppose that no lid of X' lies in $B_1 \cup B_2$.

If $X' \cap (B_1 \cup B_2) = \emptyset$, then X' is also an essential I-bundle in $(M, \bar{\bar{m}})$. Hence it is the I-bundle over the square since $(M, \underline{m})$ is almost simple.

If $X' \cap (B_1 \cup B_2) \neq \emptyset$, then recall at least one face of X' lies in a free face of $(M', \underline{m}')$. Thus, for every component, A, of $(B_1 \cup B_2) \cap \partial X'$, there is an essential square in X' which joins A with a free face of $(M, \underline{m})$. Splitting S along A, we get a non-separating surface whose complexity is not smaller than $e(S)$. So, by our maximality condition on $e(S)$, it must be equal to $e(S)$. This in turn is only possible if the arcs of $(\partial A - \partial(B_1 \cup B_2))^-$ are admissibly parallel in B_1 to a free face of $(M, \underline{m})$. Thus every component, A, of $(B_1 \cup B_2) \cap X'$ separates at least one square, say A', from $B_1 \cup B_2$. It follows that $X'' := U(X' \cup \bigcup A')$ is an admissible I-bundle in $(M, \underline{\tilde{m}})$, where the union is taken over all components, A, of $X' \cap (B_1 \cup B_2)$. A moments reflection shows that this I-bundle must be contained in an essential I-bundle, say X, in $(M, \underline{\tilde{m}})$. Of course, X meets free faces of $(M, \underline{m})$ and the lids of X are contained in one or two adjacent bound faces of $(M, \underline{m})$ (see our choice of X'). Thus X is the I-bundle over the square since $(M, \underline{m})$ is almost simple. It is then easy to see that $X'' \subset X$ and so $X' \subset X''$ must also be such an I-bundle.

(3) Let X' be an essential I-bundle in $(M', \underline{m}')$ with at least one lid in a free face of $(M', \underline{m}')$. If no lid of X' lies in $B_1 \cup B_2$, then X' is also an admissible I-bundle in $(M, \underline{m})$. So it has to be an I-bundle over the disc, for otherwise it contains an annulus which is essential in $(M, \underline{\tilde{m}})$ and which meets a free face of $(M, \underline{m})$. If a lid of X' lies in $B_1 \cup B_2$, then X' must be the product I-bundle over the square or annulus since $e(S)$ is maximal. By the same reason, there is at least one component, A', from $(B_1 - X')^-$ which is a square or annulus with one face in a free face of $(M, \underline{m})$. It follows that A' cannot be an annulus, for otherwise the union of A' with a component of $(\partial X' - \partial M')^-$ would be an essential annulus in $(M, \underline{m})$ with a free face. Thus X' is an I-bundle over the square, and (3) is established.

(4) Let X' be an essential I-bundle in $(M', \underline{m}')$ with $(\partial X' - \partial M')^-$ connected and at least one lid in a free face of $(M', \underline{m}')$. We claim X' is the regular neighborhood of some face of $(M', \underline{m}')$.

Suppose $X' \cap (B_1 \cup B_2) = \emptyset$. Then X' is also an essential I-bundle in $(M, \underline{m})$, and the claim follows since $(M, \underline{m})$ is supposed to be almost simple.

Assume no lid of X' lies in $B_1 \cup B_2$. Then X' is an admissible I-bundle in $(M, \underline{m})$. If $X' \cap (B_1 \cup B_2)$ is connected and $(\partial X' - \partial M)^-$ is inessential in $(\overline{M}, \underline{\tilde{m}})$, then X' is the neighborhood in $(M', \underline{m}')$ of some face of $(M', \underline{m}')$, for $(M, \underline{\tilde{m}})$ is an irreducible 3-manifold with useful boundary-pattern. If $X' \cap (B_1 \cup B_2)$ is disconnected and $(\partial X' - \partial M)^-$ is inessential in $(M, \underline{\tilde{m}})$, then notice first that $X' \cap (B_1 \cap B_2)$ consists of two squares. But these two squares are not the same across $U(S)$ since $(M, \underline{m})$ has no essential annulus or Möbius band meeting a free face of $(M, \underline{m})$ (incidentally, S is not a square by the same argument). It is easily seen that this situation can be avoided by pushing S. Thus we may suppose w.l.o.g. that $(\partial X' - \partial M)^-$ is essential in $(M, \underline{\tilde{m}})$. Then, by definition, X' is an essential I-bundle in $(M, \underline{\tilde{m}})$. Moreover, $(\partial X' - \partial M)^-$ is connected (unless X' is a neighborhood of B_1) and a lid of X' lies in a free face of $(M, \underline{m})$. Hence X' is a neighborhood in

$(M, \underline{m})$ of some face in $(M, \underline{m})$ since $(M, \underline{m})$ is almost simple. In particular, the latter face is a square. Therefore at least one bound face of S lies next to a square-face, say C_1, of $(M, \underline{m})$. Hence, by our hypothesis on S, the latter face of S must itself be contained in a square-face, say C_0, of $(M, \underline{m})$. Then the regular neighborhood $U(C_0 \cup C_1)$ is an essential I-bundle in $(M, \bar{m})$ which cannot be a regular neighborhood of a face of $(M, \bar{m})$. This, however, is impossible since $(M, \underline{m})$ is supposed to be almost simple.

Thus we may suppose that at least one lid of X' lies in $B_1 \cup B_2$. Then X' is a product I-bundle with one lid in B_1, say, and the other one in a free face of $(M', \underline{m'})$. Set $G_1 := B_1 \cap X'$ and let $\underline{g}_1$ denote the boundary-pattern of G_1 induced by $\underline{m}$. $(\partial G_1 - \partial B_1)^-$ is an arc since $(\partial X' - \partial M')^-$ is connected. Thus it suffices to show that $(G_1, \bar{g}_1)$ is a square. This is clear if S is a square. Thus suppose the converse. Recall every face of $(G_1, \underline{g}_1)$, contained in $G_1 \cap \partial B_1$, must lie in a bound face of $(M, \underline{m})$ since $e(S)$ is maximal and one lid of X' lies in a free face of $(M', \underline{m'})$. Hence and since $(\partial G_1 - \partial B_1)^-$ is connected it suffices to show that every essential arc in $(G_1, \underline{g}_1)$ is admissibly parallel to a free face of B_1. But this follows immediately from our maximality condition on $e(S)$ since every essential arc in $(G_1, \underline{g}_1)$ is the face of some essential square in X' joining B_1 with a free face of $(M, \underline{m})$. So we are done.

This finishes the proof of the proposition. $\Diamond$

17.12. Theorem. *Let $(M, \underline{m})$ be an almost simple 3-manifold such that no essential annulus in $(M, \bar{m})$ meets a free face of $(M, \underline{m})$. Let s be an admissible Heegaard string in $(M, \underline{m})$. Let $f : s \times I \to M$ be a map with $f | s \times 0 = s$ which maps $s \times 1$ into a bound face of $(M, \underline{m})$ and which maps the components of $\partial s \times I$ into free faces of $(M, \underline{m})$.*

Then f can be deformed into an embedding, using an admissible homotopy in $(M, \bar{m})$ which is an isotopy on s.

Proof. The proof is again by induction on a hierarchy.

Suppose M is a 3-ball. Then all faces of $(M, \bar{m})$ are discs since $\bar{m}$ is supposed to be useful for M. Thus we may suppose f is admissibly deformed in $(M, \bar{m})$ (fixing $s \times 0$) so that afterwards $f | s \times 1$ is an embedding. In fact we may suppose $f | (\partial(s \times I) - s \times 0)^-$ is an embedding. Now s is an unknotted arc in M since it is an admissible Heegaard string. So $f(\partial(s \times I))$ is a simple closed curve in M which bounds a disc in M (M is a 3-ball). Clearly f can then be deformed (rel $\partial(s \times I)$ into this disc. This establishes the induction beginning.

We claim the induction step can be carried out whenever there is a non-separating, essential surface, S, in $(M, \bar{m})$ (as in prop. 17.11) disjoint to s.

To see the claim consider the manifold $(M', \underline{m'})$ obtained from $(M, \underline{m})$ by splitting along S. By prop. 17.11, $(M', \underline{m'})$ is almost simple and no essential annulus in $(M', \tilde{m}')$ meets a free face of (M', m'). Moreover, by prop. 16.3, s is an admissible Heegaard string in (M', m'). Finally consider the map f. $\tilde{m}$ is useful for M, M is aspherical since it is irreducible (see e.g. [Wa 2]) and $\overline{S}$ is essential in $(M, \tilde{m})$. Therefore we may suppose f is deformed, using an admissible homotopy in $(M, \tilde{m})$ fixing $s \times 0$, so that afterwards $f^{-1}S$ consists of essential arcs in the square $s \times I$. Then $f^{-1}S$ splits $s \times I$ into a system of squares. One of them, say Q, contains $s \times 0$ (recall $s \cap S = \emptyset$). By [Ha 2], we obtain a 3-ball by successive splittings of M along non-separating incompressible surfaces. Thus, by induction, we may suppose $f|Q$ is deformed into an embedding, using an admissible homotopy in $(M', \tilde{m}')$ fixing $s \times 0$. Pushing S across the square $f(Q)$, we may reduce the number of arcs from $f^{-1}S$ without changing the intersection of S with s. Applying this operation a finite number of times if necessary, we may suppose $f^{-1}S = \emptyset$. Then, by induction again, f can be deformed into an embedding, using an admissible homotopy of f in $(M', \tilde{m}')$ which is an isotopy on $s \times 0$. But this homotopy is also an admissible homotopy in $(M, \tilde{m})$. So the theorem would follow.

Now, by thm. 16.23, there is a non-separating, essential surface in $(M, \tilde{m})$ as required as long as s is non-trivial in (M, m). Thus, by what we have just seen, thm. 17.12 holds true, as long as it holds for all admissible Heegaard strings which are trivial. This requirement will be established in the next proposition. ◇

The next proposition deals with trivial, admissible Heegaard strings. To appreciate the significance of this special case one only needs to realize that a non-trivial admissible Heegaard string may well become trivial after splitting along a disjoint, essential surface. Actually, this special case is the hard part of proving thm. 17.12. It is important to note that for prop. 17.13 we allow essential annuli in $(M, \tilde{m})$ to meet free face of (M, m).

17.13. Proposition. *Let (M, m) be an almost simple 3-manifold. Suppose that s is an arc in (M, m) whose end-points lie in free faces and which has a good disc D in $(M, \tilde{m})$ with $D \cap U(s)$ non-empty and connected. Let $f : s \times I \to M$ be a map with $f|s \times 0 = s$ which maps $s \times 1$ into a bound face of (M, m) and which maps the components of $\partial s \times I$ into free faces of (M, m).*
Then f can be deformed into an embedding, using an admissible homotopy in $(M, \tilde{m})$ which is an isotopy on s.

Remark. See 16.1 for definition of "good discs".

Proof. In the course of this proof we will have to consider a number of special situations. For clarity we will formulate these special situations in separate lemmas. Here is the easiest one.

17.14. Lemma. *The proposition holds if $D \cap \bigcup m = \emptyset$.*

Proof of Lemma. Push $f|s \times 0$ across D and into the free face of $(M, \underline{m})$ containing ∂s. Let f' be the resulting map. f' is an admissible, singular 2-faced disc in $(M, \underline{\tilde{m}})$. $\underline{\tilde{m}}$ is useful for M. Thus, by [Joh 1, prop. 2.1], $f'|\partial(s \times I)$ extends to a map $h : s \times I \to \partial M$ such that $h^{-1}J$ is an arc, say ℓ, where $J := \bigcup \partial B$, $B \in \underline{m}$. It follows that f' can deformed into $(\partial U(\ell) - \partial M)^-$, using an admissible homotopy in $(M, \underline{\tilde{m}})$ which is an isotopy on s. The desired homotopy for f is then easily constructed. This proves lemma 17.14.

Thus for the following we may suppose that $D \cap \bigcup \underline{m} \neq \emptyset$. Then, given f and D as in the proposition, we form their "union" by identifying f and D along s. Here is a more formal definition of this operation (plus some extra notation needed later). Let $(C, \underline{c})$ be a disc with two free faces and so that $(C, \underline{\tilde{c}})$ is a 4- or 5-faced disc. Let c_1, c_2, c_3 (possibly $c_1 = c_2$) denote the bound faces and let γ_1, γ_2 denote the free faces of $(C, \underline{c})$. Let the indices be chosen so that $(\partial C - \gamma_1 \cup \gamma_2)^-$ is the disjoint union of $c_1 \cup c_2$ and c_3. Let ℓ be an essential arc in $(C, \underline{c})$, joining γ_1 with γ_2. Then ℓ separates a 4-faced disc, say C_1, from $(C, \underline{\tilde{c}})$ containing c_3. Of course, $C_2 := (C - C_1)^-$ is a 4- resp. 5-faced disc again. Define an admissible map $g : (C, \underline{c}) \to (M, \underline{\tilde{m}})$ by the requirement that (1) $g|C_1 = f$, (2) $g|C_2$ is the embedding $D \hookrightarrow M$ and (3) $g|(\partial C_1 - \partial C)^- = s = g|(\partial C_2 - \partial C)^-$. Then g is called the *union of f and D*. Note that to prove the proposition it suffices to show that g can be admissibly deformed into an embedding since, by Nielsen's theorem [Wa 2], homotopies between curves in surfaces can be deformed into isotopies.

g is an admissible, singular 4- or 5-faced disc in $(M, \underline{\tilde{m}})$. The next lemma tells us that w.l.o.g. we may suppose it is a singular pentagon.

17.15. Lemma. *The proposition holds if $D \cap \bigcup \underline{m}$ is contained in one bound face of $(M, \underline{m})$.*

Proof of Lemma. Let B denote the bound face of $(M, \underline{m})$ containing $D \cap \bigcup \underline{m}$. Then the union $g : (C, \underline{\tilde{c}}) \to (M, \underline{\tilde{m}})$ of f and D is an admissible, singular square in $(M, \underline{\tilde{m}})$ such that at at least one of its faces is non-singular and contained in B.

Suppose g is essential in $(M, \underline{\tilde{m}})$. Then, by [Joh 1, thm. 12.5], g can be admissibly deformed into a component, say X, of the characteristic submanifold of $(M, \underline{\tilde{m}})$, using an admissible homotopy in $(M, \underline{\tilde{m}})$. Moreover, by [Joh 1, prop. 5.10], g can be admissibly deformed in X into a fiber preserving map, using an admissible homotopy in $(M, \underline{\tilde{m}})$. Thus we may suppose f is deformed into a fiber preserving map, using an admissibly homotopy which is an isotopy on $s \times 0$. Then, pushing all self-intersection arcs along parts of $f(s \times I)$ and across s, we see that f can be admissibly deformed into an embedding, using an admissible homotopy in $(M, \underline{\tilde{m}})$ which is an isotopy on $s \times 0$.

Suppose g is inessential in $(M, \underline{\tilde{m}})$. Then, by definition, there is an essential arc, k, in $(C, \underline{\tilde{c}})$ and an admissible map $\alpha : (E, \underline{e}) \to (M, \underline{\tilde{m}})$ such that $(E, \underline{\tilde{e}})$ is a 2- or 3-faced disc and $\alpha|(\partial E - \bigcup \underline{e})^- = g|k$. Now let C_1, C_2

denote the two components into which C is split by k. Then $\alpha \cup g|C_1$ and $\alpha \cup g|C_2$ are both either an admissible, singular 2-faced disc, or an admissible singular 3-faced disc.

If both $\alpha \cup g|C_1$ and $\alpha \cup g|C_2$ are admissible, singular 3-faced discs, then, using [Joh 1, prop. 2.1], the asphericity of M and the fact that $\bar{m}$ is useful for M, we find an edge, l, of the graph $J = \bigcup \partial B$, $B \in \underline{m}$, such that g can be admissibly deformed into the square $(\partial U(l) - \partial M)^-$.

If both $\alpha \cup g|C_1$ and $\alpha \cup g|C_2$ are admissible, singular 2-faced discs, we may argue similarly. Indeed, we first find, using [Joh 1, prop. 2.1], that $g(\underline{e})$ must be contained in the bound face, B, for otherwise it would follow that D intersects B in an inessential arc (which is impossible since D is supposed to be a good disc). Then, by the above argument, we conclude that f can be admissibly deformed into the non-singular square $(\partial U(D \cap B) - \partial M)^-$.

Thus, in any case, g can be admissibly deformed into an embedding. This finishes the proof of lemma 17.15.

Set $b := D \cap \bigcup \underline{m}$. Let B_1, B_2 be the two bound faces of $(M, \underline{m})$ meeting b. Set $b_i := b \cap B_i$, i=1,2. Recall b_i, i=1,2, is an essential arc in B_i (w.r.t. the boundary-pattern induced by $\underline{m}$). Finally, let B be the bound face of $(M, \underline{m})$ containing $f(s \times 1)$.

17.16. Lemma. *The proposition holds if B is neither B_1 nor B_2.*

Proof of Lemma. If $B \neq B_1, B_2$, then $f(s \times 1) \cap b = \emptyset$. It follows that f can be pushed out of D. More precisely, f can be deformed so that afterwards $f^{-1}D = s \times 0$, using an admissible homotopy in $(M, \bar{m})$ fixing $s \times 0$. Then f is an essential, singular square in $(M_s, \bar{m}_s)$ (see 16.1 for the definition of $(M_s, \underline{m}_s)$). By prop. 16.3, M_s is irreducible and $\bar{m}_s$ is useful for M_s. Thus, by [Joh 1, thm. 12.5], f can be admissibly deformed in $(M_s, \bar{m}_s)$ into a component, say X, of the characteristic submanifold of $(M_s, \bar{m}_s)$. Moreover, by [Joh 1, prop. 5.10], f can be admissibly deformed in X into a fiber preserving map. Pushing the self-intersections of f along parts of $f(s \times I)$ and across s, we eventually see that f can be admissibly deformed in $(M, \bar{m})$ into an embedding. Lemma 17.16 therefore follows.

We next have to establish a few properties of "very good" surfaces in $(M, \underline{m})$. Throughout this proof a *good surface* is an essential surface in $(M, \bar{m})$ which is either non-separating in M with $e(S)$ maximal, or a square which is not admissibly parallel to a face of $(M, \bar{m})$. A good surface, S, in $(M, \underline{m})$ is *very good* if there is no good surface, S', in $(M, \underline{m})$ with $\#(b \cap \partial S') < \#(b \cap \partial S)$.

For the next two lemmas let S be a very good surface in $(M, \underline{m})$. Moreover, let the union $g : (C, \underline{c}) \to (M, \bar{m})$ of f and D be admissibly deformed so that $g^{-1}S$ is a system of arcs and curves whose number is as small as possible. M is aspherical (since it is irreducible), $\bar{m}$ is useful for M, S is essential in

$(M, \underline{\tilde{m}})$ and $\#(b \cap \partial S)$ is minimal. Thus it follows that $g^{-1}S$ is a system of essential arcs in $(C, \underline{\tilde{c}})$ (apply [Joh 1, prop. 2.1]). Recall c_1, c_2 and c_3 are the bound faces of $(\tilde{C}, \underline{c})$ and w.l.o.g. $g(c_i) \subset B_i$, i=1,2 and $g(c_3) \subset B$.

17.17. Lemma. *Suppose there is an arc, k, arc from $g^{-1}S$ such that $k \cap (c_1 \cup c_2) \neq \emptyset$ and that $g|k$ is inessential in S. Then the proposition holds.*

Proof of Lemma. Observe that every essential arc in the pentagon $(C, \underline{\tilde{c}})$ is admissibly parallel to some face. In particular, k separates a 4-faced disc, Q, form $(C, \underline{\tilde{c}})$.

Now w.l.o.g. $k \cap (c_1 \cup c_2) = k \cap c_1$. Moreover, $g|k$ is inessential in S. Thus, by definition, there is an admissible map $\alpha : (E, \underline{\tilde{e}}) \to (M, \underline{\tilde{m}})$ such that $(E, \underline{\tilde{e}})$ is a 2- or 3-faced disc and that $\alpha|(\partial E - \bigcup \underline{e})^- = g|k$.

Assume $(E, \underline{\tilde{e}})$ is a 2-faced disc. Then the union of α with $g|Q$ and $g|(C - Q)^-$ is a singular 2- and 3-faced disc in $(M, \underline{\tilde{m}})$, respectively. But $\underline{\tilde{m}}$ is useful for M. Thus, by [Joh 1, prop. 2.1], all the faces of these singular discs are inessential in the faces of $(M, \underline{\tilde{m}})$ containing them. In particular, three different faces of the singular pentagon g are inessential in the faces of $(M, \underline{\tilde{m}})$ containing them. One of them has to be b_1 or b_2. Contradiction.

Assume $(E, \underline{\tilde{e}})$ is a 3-faced disc. Then $\alpha \cup g|Q$ is an admissible, singular 3-faced disc in $(M, \underline{\tilde{m}})$. Thus, by [Joh 1, prop. 2.1], all its faces are inessential (singular) arcs in the faces of $(M, \underline{\tilde{m}})$ containing them. It follows that k must be parallel to a free face, say γ_1, of $(C, \underline{c})$. The union $\alpha \cup g|(C - Q)^-$ is an admissible, singular square in $(M, \underline{\tilde{m}})$. If this singular square is inessential in $(M, \underline{\tilde{m}})$, then two opposite faces of it are inessential in the faces of $(M, \underline{\tilde{m}})$ containing them (apply [Joh 1, prop. 2.1] again). One of them is also a face of g. In fact, it has to be the free face, γ_2, of $(C, \underline{c})$ different from γ_1. Pushing this free face in the right direction, we get from $\alpha \cup g|(C - Q)^-$ an admissible singular 2- or 3-faced disc in $(M, \underline{\tilde{m}})$. If it is an admissible, singular 3-faced disc, then $B \neq B_1, B_2$ and lemma 17.17 follows from lemma 17.16. If it is an admissible, singular 2-faced disc, then all faces of it can be deformed into edges of $J := \bigcup \partial B$, $B \in \underline{m}$, in the faces of $(M, \underline{\tilde{m}})$ containing it. It follows that b_1 must be inessential in the face containing it. Contradiction. So $\alpha \cup g|(C-Q)^-$ has to be an essential, singular square in $(M, \underline{\tilde{m}})$. Thus, by [Joh 1, 12.5 and 5.10], it can be admissibly deformed into a vertical, singular square in some essential I-bundle of $(M, \underline{\tilde{m}})$ (which is a component of the characteristic submanifold of $(M, \underline{\tilde{m}})$). But $g|c_1 \cup c_2$ is an embedding. Thus, pushing the self-singularities of f along parts of $f(s \times I)$ and across s, we obtain an embedding. In this case the proposition is proved.

This finishes the proof of the lemma 17.17.

For the next lemma suppose k is an arc from $g^{-1}S$ with $k \cap (c_1 \cup c_2) \neq \emptyset$ which is outermost. Then it separates a 4-faced disc, Q, from $(C, \underline{\tilde{c}})$ with $Q \cap g^{-1}S = k$. Moreover, let $(M', \underline{m}')$ denote the manifold obtained from $(M, \underline{m})$ by splitting along S. Set $h := g|Q$.

17.18. Lemma. *If h cannot be admissibly deformed in $(M', \underline{\bar{m}}')$ into an embedding, then the proposition holds.*

Proof of Lemma. Assume h cannot be admissibly deformed into an embedding.

By [Joh 1, prop. 4.8], $(M', \underline{\bar{m}}')$ is an irreducible 3-manifold whose completed boundary-pattern is useful. Moreover, note that h is an admissible, singular square in $(M', \underline{m}')$. It may or may not be essential in $(M', \underline{\bar{m}}')$.

Suppose h is inessential in $(M', \underline{\bar{m}}')$. Then, by definition, there is an arc, say l, in Q, joining opposite faces, such that $h|l$ is inessential in $(M', \underline{\bar{m}}')$. Thus if l is admissibly parallel to k, lemma 17.18 follows from lemma 17.17. So we may suppose that l meets k. Then it joins k with the face of $(C, \underline{\bar{c}})$ contained in Q. M' is aspherical and $\underline{\bar{m}}'$ is useful for M'. Thus, using [Joh 1, prop. 2.1], we find that h can be deformed into S, using an admissible homotopy in $(M, \underline{\bar{m}})$ fixing $(\partial Q - \partial C)^-$. So, clearly, g can be admissibly deformed in $(M', \underline{\bar{m}}')$ into an embedding, and the proposition follows.

Suppose h is essential in $(M', \underline{\bar{m}}')$. Then, by [Joh 1,thm. 12.5], h can be admissibly deformed in $(M', \underline{\bar{m}}')$ into a component, say X', of the characteristic submanifold of $(M', \underline{\bar{m}}')$. By definition, the characteristic submanifolds consists of essential I-bundles and Seifert fiber spaces. So, by [Joh 1, prop. 5.10], we conclude that X' has to be an I-bundle since h is an essential singular square. But X' cannot be the I-bundle over the disc since h is not admissibly homotopic to an embedding.

Assume one lid of X' lies in a bound face of $(M', \underline{m}')$ which is a copy of S. For simplicity, we view this lid as being contained in S. By our choice of h, X' must be a product bundle which joins S with some face of $(M, \underline{m})$. Then it follows from our maximality condition on $e(S)$ that every component from $(\partial(S \cap X') - \partial S)^-$ is admissibly parallel to a face of S (note that S is no square since X' is not the I-bundle over the disc). Thus $(S - X')^-$ consists of squares since X' is not the I-bundle over the disc. Hence one of the components of $(S - X')^- \cup (\partial X' - \partial M')^-$ is non-separating. But all these components are admissible squares in $(M, \underline{m})$ (in fact, they must be essential squares since X' is a component of the characteristic submanifold). Thus, by our maximality condition on $e(S)$, S must be a square. So X' is the I-bundle over the disc. Contradiction.

No lid of X' lies in a copy of S and X' is not the I-bundle over the disc. Thus it follows that X' is also an essential I-bundle in $(M, \underline{\bar{m}})$. The lids of X' lie in those faces of $(M, \underline{\bar{m}})$ containing the end-points of $h\overline{|k}$. Now $h|k$ joins B_1 (or B_2) with either B or with some free face of $(M, \underline{m})$. X' is not the I-bundle over the disc. Moreover, it cannot be the neighborhood of an essential annulus in $(M, \underline{\bar{m}})$ (X' contains the singular square h). Hence no lid of X can lie in a free face since $(M, \underline{m})$ is supposed to be almost simple (see property (3) of almost simple 3-manifolds). Thus the lids of X are contained in $B_1 \cup B$ (resp. $B_2 \cup B$). If $B = B_1$ or $B = B_2$, recall B_1, B_2 are adjacent bound faces of $(M, \underline{m})$ ($B_1 \cup B_2$ contains b). Then X' is the I-bundle over the disc since $(M, \underline{m})$ is supposed to be almost simple (see property (2) of almost simple

3-manifolds). But this is impossible since h cannot be admissibly deformed into an embedding. So B is neither B_1 nor B_2. Hence lemma 17.18 follows from lemma 17.16.

17.19. Lemma. *Suppose there is a very good surface, S, in $(M, \underline{m})$ with $b \cap \partial S \neq \emptyset$. Then the proposition holds provided (1) or (2) is true:*

(1) S is a square.

(2) Every essential square in $(M, \bar{\underline{m}})$ is admissibly parallel to a face of $(M, \bar{\underline{m}})$.

Proof of Lemma. As above, we may suppose the union $g : (C, \bar{\underline{c}}) \to (M, \bar{\underline{m}})$ is admissibly deformed so that $g^{-1}S$ is a system of essential arcs in the pentagon $(C, \bar{\underline{c}})$ whose number is as small as possible. Since $b \cap \partial S \neq \emptyset$, there is at least one arc, say k, from $g^{-1}S$ with $k \cap (c_1 \cup c_2) \neq \emptyset$. W.l.o.g. we suppose k is chosen so that it separates a 4-faced disc, Q, from $(C, \bar{\underline{c}})$ with $Q \cap g^{-1}S = k$. Let $(M', \underline{m}')$ denote the manifold obtained from $(M, \underline{m})$ by splitting along S.

By lemma 17.18, we may suppose $h := g|Q$ is admissibly deformed in $(M', \bar{\underline{m}}')$ into an embedding. Then $Q' := h(Q)$ is an admissible square in $(M', \bar{\underline{m}}')$.

Suppose S is a square (i.e., (1) of lemma 17.19 is true). Then $X := U(S \cup Q')$ is an admissible I-bundle in $(M, \bar{\underline{m}})$. One component, say S', of $(\partial X - \partial M)^-$ is admissibly parallel to S. Let A_1, A_2 be the other two components. By construction, $\#(b \cap \partial A_i) < \#(B \cap \partial S)$, i=1,2. So A_1 as well as A_2 has to be either inessential or admissibly parallel to a face of $(M, \underline{m})$ since S is a very good surface. It follows that X is contained in an essential I-bundle, X', in $(M, \bar{\underline{m}})$ with $(\partial X' - \partial M)^- = S'$. But the lids of X' are joined by $g|k$. So either a lid of X' is free or every lid of X' is contained in $B \cup B_1 \cup B_2$. Since $(M, \underline{m})$ is almost simple and since S is not admissibly parallel to a face of $(M, \bar{\underline{m}})$, it follows that $B \neq B_1$ and $B \neq B_2$ (see property (2) and (4) of almost simple 3-manifolds). So the proposition follows from lemma 17.16.

Thus we may suppose every essential square in $(M, \bar{\underline{m}})$ is admissibly parallel to a face of $(M, \underline{m})$ (i.e., (2) of lemma 17.19 is true). In particular, S is then a non-separating surface different from a square.

$Q' = h(Q)$ is a square. So, by our maximality condition on $e(S)$, the arc $h(k)$ has to be admissibly parallel in S to some face of S. Let A be the corresponding parallelity region. Then $Q^* := A \cup Q'$ is an admissible square in $(M, \bar{\underline{m}})$.

Let k^* be that face of Q^* which is a face of S and let l^* be the face of Q^* opposite to k^*. Let t_1^*, t_2^* be the two components of $(\partial Q^* - (k^* \cup l^*))^-$. Recall from the definition of Q that one face of $Q' = h(Q)$ lies in b, say in b_1. Thus we may suppose the indices are chosen so that t_1^* meets b_1. Then $t_1^* \subset B_1$.

Case 1. Q^* *is inessential in* $(M, \tilde{\underline{m}})$.

Recall $\tilde{\underline{m}}$ is useful for M. It therefore follows that Q' is admissibly parallel in $(\overline{M'}, \tilde{\underline{m}}')$ to a face of $(M', \tilde{\underline{m}}')$. Let E be the corresponding parallelity region in M'.

Note that l^* cannot lie in a free face of $(M, \underline{m})$. Otherwise k^* must lie in a bound face (we are in Case 1). So $S^* := (S - \overline{E})^- \cup Q'$ is a non-separating surface with $e(S^*) > e(S)$. But this contradicts our maximality condition of $e(S)$ (S is a very good surface).

Thus l^* lies in a bound face of $(M, \underline{m})$. But then l^* is part of the arc b. In particular, one end-point of l^* lies in a free face of $(M, \underline{m})$ since both end-points of b do. Thus k^* lies in a bound face since free faces are disjoint. Then $S^* := (S - E)^- \cup Q'$ is a non-separating surface in M with $e(S^*) = e(S)$. Moreover, $b \cap Q' = \emptyset$. Thus $\#(b \cap S^*) < \#(b \cap \partial S)$. This however contradicts our minimality condition on S.

Therefore Case 1 holds not true.

Case 2. Q^* *is essential in* $(M, \tilde{\underline{m}})$.

By our assumption on $(M, \tilde{\underline{m}})$, The essential square Q^* is admissibly parallel in $(M, \tilde{\underline{m}})$ to a face, say B_0, of $(M, \tilde{\underline{m}})$. Let E be the corresponding parallelity region. Then $E \cap \partial M$ is a disc which in turn is a neighborhood of B_0. Define

$$S^* := (S - E)^- \cup Q^*.$$

Sub-Case 1. l^* *lies in a free face of* $(M, \underline{m})$.

In this Sub-Case, l^* joins b_1 with $f(s \times 1)$. So it joins the face B_1 with B since $b_1 \subset B_1$ and $f(s \times 1) \subset B$. Moreover, by our maximality condition on $e(S)$, it follows that k^* lies in a free face of $(M, \underline{m})$ as well.

Let B_1^*, B_2^* be the components of $B_1 \cap (E \cap \partial M)$, $B \cap (E \cap \partial M)$ containing t_1^*, t_2^*, respectively. Let F_1^*, F_2^* be the components of $((E \cap \partial M) - (B_1^* \cup B_0 \cup B^*))^-$ containing k^*, l^*, respectively.

By our maximality condition on the complexity of S, it follows that $\#(b \cap \partial S^*) \geq \#(b \cap \partial S)$.

Suppose $b \cap B_0 \neq \emptyset$. Then $b \cap B_0$ is an essential arc in B_0. But B_0 is a square and at least one end-point of $b \cap B_0$ has to lie in a free face of $(M, \underline{m})$ (recall $b = b_1 \cup b_2$). Thus $b \cap B_0$ has to join the two free faces of B_0. In particular, $b = D \cap B_0$. So the proposition follows from lemma 17.15.

Thus we may suppose $b \cap B_0 = \emptyset$. Then $b \cap (E \cap \partial M) \subset B_1^* \cup B_2^*$. It follows that $b \cap B_1^*$ and $b \cap B_2^*$ is non-empty since $\#(b \cap \partial S)$ is minimal. In particular, $B_2^* \subset B_2$ since $b \subset B_1 \cup B_2$ and $B_1 \neq B_2$ (see lemma 17.15). Thus the original string s has to have one end-point, say x_1, in F_1^* and the other, say x_2, in F_2^*. Let e_1, e_2 be the two components of $\partial s \times I$. Let the indices be chosen so that $f|e_i$, $i = 1, 2$, is a (singular) arc which starts at x_i. Then notice that w.l.o.g. $f|e_1$ does not meet k^* (otherwise S^* would be an essential, non-separating surface in $(M, \underline{\tilde{m}})$ with $e(S^*) = e(S)$ and $\#(b \cap S^*) = \#(b \cap S)$, which intersects the boundary of the singular pentagon g in strictly fewer points than S; and we have a reduction). Thus $f|e_1$ has to join x_1 either with $B_1^* \subset B_1$ or with $B_2^* \subset B_2$. The following picture illustrates the first situation.

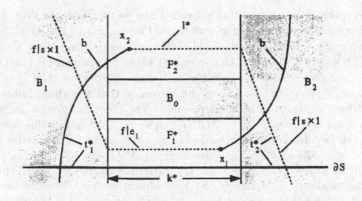

(Figure 17.2.)

If $f|e_1$ joins x_1 with B_1, then recall that $f|e_2$ joins x_2 with B_2 (since $t_2^* \cap B_2^* \neq \emptyset$) But $f(s \times 1) \subset B$. Thus $B_1 = B = B_2$. Hence the lemma follows from lemma 17.15.

If $f|e_1$ joins x_1 with B_2, then g can be pushed into an admissible, singular 3-faced disc in $(M, \underline{\tilde{m}})$. An application of [Joh 1, prop. 2.1], shows that g can be deformed so that afterwards $f|\partial C$ is an embedding, using an admissible homotopy in $(M, \underline{\tilde{m}})$ fixing $b = c_1 \cup c_2$. By Dehn's lemma, $g(\partial C)$ bounds a disc. g can be deformed into this disc. So clearly f can be deformed into an embedding, using an admissible homotopy in $(M, \underline{\tilde{m}})$ fixing s.

This proves prop. 17.13 in Sub-Case 1.

Sub-Case 2. l^* lies in a bound face of $(M, \underline{m})$.

In this Sub-Case, l^* is contained in b (see our choice of Q). Thus one end-point of l^* lies in the bound face B_1 and the other end-point lies in a free face. Then B_0 must be a bound face and k^* must lie in a bound face, for free faces are disjoint. In particular, $e(S^*) = e(S)$. But $U(E \cap \partial M)$ contains only one end-point of s. More precisely, $b \cap U(E \cap \partial M)$ is connected. Thus, applying a small additional isotopy, S^* intersects b in strictly fewer points than S. This contradicts our minimality condition on $\#(b \cap \partial S)$. Thus Sub-Case 2 is not possible.

After having shown that prop. 17.13 holds under various conditions, we are now finally in the position to show that it holds without any qualification whatsoever.

17.20. Lemma. *The proposition holds.*

Proof of Lemma. $\partial M \neq \emptyset$. It follows that there is at least one very good surface, S, in $(M, \underline{m})$.

Suppose S is a separating square. Then we may suppose that $b \cap \partial S = \emptyset$, for otherwise the proposition follows from (1) of lemma 17.19. So $s \cap S = \emptyset$. Let $(M(S), \underline{m}(S))$ denote the manifold obtained from splitting along S. Denote by $(M_0(S), \underline{m}_0(S))$, resp. $(M_1(S), \underline{m}_1(S))$ the component of $(M(S), \underline{m}(S))$ which does resp. does not contain s. Set $c(S) := c(M_0(S), \underline{m}_0(S)) := (-\chi \partial M_0(S), \# \underline{m}_0(S))$. Let S be chosen so that the above holds and that, in addition, $c(S)$ is as small as possible. The good disc for s in $(M, \underline{m})$ is also a good disc for s in $(M_0(S), \underline{m}_0(S))$. The admissible homotopy f of s can be admissibly deformed (fixing $s \times 0$) into an admissible homotopy for s in $(M_0(s), \underline{m}_0(s))$. Moreover, $(M_0(S), \underline{m}_0(S))$ is an irreducible 3-manifold whose completed boundary-pattern is useful (see [Joh 1, prop. 4.8]). In fact, we claim $(M_0(S), \underline{m}_0(S))$ is almost simple. Assume the converse. Then it follows that a bound face, a, of S is next to a bound square-face, A_1, of $(M_0(S), \underline{m}_0(S))$. By definition (see the remark following prop. 17.11), the arc a is admissibly parallel, in the face of $(M, \underline{m})$ containing it, to a face of A_1. Let A_0 denote the corresponding parallelity-region. Then $S' := (\partial(U(A_0 \cup A_1) \cup M_1) - \partial M)^-$ is a square. Since A_0 and A_1 lie in bound faces of $(M, \underline{m})$ and since $b \cap \partial S = \emptyset$, it follows that $b \cap \partial S' = \emptyset$ (recall ∂s lies in free faces and b intersects bound faces in essential arcs). Thus S' is a very good, separating square. But $c(S') < c(S)$. So we get a contradiction to our minimal choice of $c(S)$. Thus $(M_0(S), \underline{m}_0(S))$ is indeed almost simple. Hence the hypothesis of prop. 17.12 is reproduced. Now every essential square in $(M, \tilde{\underline{m}})$, disjoint to b and not admissibly parallel to a face of $(M, \tilde{\underline{m}})$, is very good. Thus, splitting along appropriate squares if necessary, we may suppose that no such essential square in $(M, \tilde{\underline{m}})$ is separating.

Suppose S is a non-separating square. Then we may suppose again that $b \cap \partial S = \emptyset$ and so $s \cap S = \emptyset$ (s is trivial). Suppose that there is a bound face, say a_1, of S which lies next to a square-face, A_1, of $(M, \underline{m})$. Then, by definition (see remark of 17.11), a_1 is admissibly parallel, in the face A_0 of $(M, \underline{m})$ containing it, to a face of A_1. Pushing a_1 across the corresponding parallelity region, we obtain from S an admissible square, S_1, in $(M, \tilde{\underline{m}})$, with one face, a_1, in A_1. We still have $s \cap S_1 = \emptyset$, for A_0 is a bound face of $(M, \underline{m})$ and so $s \cap A_0 = \emptyset$. S_1 is again an essential square since S is not admissibly parallel to a face from $(M, \tilde{\underline{m}})$. If a face from S_1 lies next to a bound square-face, A_2, of $(M, \underline{m})$, then A_1, A_2 are adjacent squares. $S_2 := \partial U(A_1 \cup A_2) - \partial M)^-$ is an essential square which does not meet b. But S_2 is separating and so excluded. Hence it follows that splitting $(M, \underline{m})$ along S_0 or S_1, we get an almost simple 3-manifold (see propositions 17.11 and

12). But this splitting process increases the Euler characteristic of the boundary of the 3-manifold. Thus, after finitely many such steps if necessary, we may suppose that every essential square in $(M, \bar{m})$ is admissibly parallel to a face of $(M, \bar{m})$.

By what we have seen so far, we may suppose S is a non-separating surface. By lemma 17.19 and by our additional hypothesis on $(M, \underline{m})$, we may suppose $b \cap \partial S = \emptyset$, and so $s \cap S = \phi$ (s is trivial). By a similar argument as used before, we find such a very good surface with the additional property that every face of S next to a square-face of $(M, \underline{m})$ lies in a square face itself. Thus, by prop. 17.11, the manifold $(M', \underline{m}')$ (obtained from $(M, \underline{m})$ by splitting along S) is almost simple. Moreover, $s \subset M'$ and the good disc for s in $(M, \underline{m})$ is certainly also a good disc for s in $(M', \underline{m}')$. Thus the hypothesis of prop. 17.13 is reproduced. Moreover, this proposition is true when M is a 3-ball. Thus prop. 17.13 follows by induction on a hierarchy of surfaces.

This finishes the proof of prop. 17.13. ◊

The proof of the previous proposition finally completes the proof of thm. 17.13.

17.21. Deformations between Generalized Heegaard Strings.

We next study admissible homotopies between admissible Heegaard strings. The result will again be that basically all such homotopies can be admissibly deformed into embeddings.

17.22. Theorem. *Let $(M, \underline{m})$ be an almost simple 3-manifold such that no essential annulus in $(M, \bar{m})$ meets a free face of $(M, \underline{m})$. Let s_0, s_1 be two admissible Heegaard strings in $(M, \underline{m})$. Let $f : s \times I \rightarrow M$ be a map with $f | s \times i = s_i$, $i = 0, 1$, which maps the components of $\partial s \times I$ into free faces of $(M, \underline{m})$.*

Then f can be deformed into an embedding, using an admissible homotopy in $(M, \bar{m})$ which is an isotopy on $s \times 0$ and $s \times 1$.

Remark. Notice we do not claim that the homotopy is an isotopy on $s \times \partial I$, i.e., we do allow the admissible Heegaard strings to cross each other throughout their isotopic deformation. Notice also that, in particular, the theorem holds for all simple 3-manifolds with empty boundary-patterns.

Proof. Let us begin with the case that s_1, say, is trivial in $(M, \underline{m})$. Then, by definition (see 16.1), there is a good disc, D, for s_1 such that $D \cap U(s_1)$ is connected. f can be pushed out of D. More precisely, f can be deformed so that afterwards $f^{-1}D = s \times 1$, using an admissible homotopy in $(M, \bar{m})$ which is an isotopy on $s \times 0$ and which fixes $s \times 1$. Let $(M', \underline{m}')$ be the manifold obtained from $(M_s, \underline{m}_s)$ by splitting along D. By [Joh 1, prop. 4.8], $(M', \underline{m}')$ is an irreducible 3-manifold whose completed boundary-pattern is useful since, by prop. 16.3, this holds for $(M_{s_1}, \underline{m}_{s_1})$ and since D is essential in $(M_s, \bar{m}_s)$ (see 16.1). Moreover, it is easily seen that $(M', \underline{m}')$ is an almost simple 3-manifold which has no essential annulus with a free face, for $(M, \underline{m})$ has these properties. Now, by what we have verified above, f may be viewed as an admissible homotopy of s_2 in $(M', \underline{m}')$ into a bound face. Hence thm. 17.22 follows from thm. 17.12.

Thus we may suppose that neither s_0 nor s_1 is trivial in $(M, \underline{m})$. Under this additional hypothesis, we will now prove the theorem by induction.

First suppose M is a 3-ball. Then s_0 and s_1 are unknotted arcs in M since they are admissible Heegaard strings. Thus they are admissibly isotopic if and only if they join the same two free faces of $(M, \underline{m})$. But they do join the same two free faces since they are admissibly homotopic. The conclusion of the theorem therefore follows. This establishes the induction beginning.

Now suppose for a moment there is a non-separating surface, S, in $(M, \bar{\underline{m}})$ with $e(S)$ maximal as in prop. 17.11 (see 16.4 for the definition of $e(S)$) and such that both s_0 and s_1 can be admissibly isotoped out of S. Then we may argue as in the proof of thm. 17.12. As there consider the manifold $(M', \underline{m}')$ obtained from $(M, \underline{m})$ by splitting along S. By prop. 17.11, $(M', \underline{m}')$ is almost simple and no essential annulus in $(M', \bar{\underline{m}}')$ meets a free face of $(\overline{M'}, \underline{m}')$. By prop. 16.3, s_0 as well as s_1 is an admissible Heegaard string in $(M', \underline{m}')$. Moreover, $f : s \times I \to M$ can be deformed so that afterwards $f^{-1}S$ consists of arcs joining $\partial s \times I$. Then $f^{-1}S$ splits $s \times I$ into squares. One of them, say Q, contains s_0. If $Q \neq s \times I$, then, by thm. 17.12, $f|Q$ can be deformed into an embedding, using an admissible homotopy in $(M', \bar{\underline{m}}')$ which is an isotopy on s_0. Pushing s_0 across the square $f(Q)$, we may reduce $f^{-1}S$. Thus w.l.o.g. we may suppose $f^{-1}S = \emptyset$. Then, by induction, f can be deformed into an embedding, using an admissible homotopy in $(M', \bar{\underline{m}}')$ which is an isotopy on s_0 and s_1. But this homotopy is also an admissible homotopy in $(M, \bar{\underline{m}})$. So the conclusion of the theorem follows.

It remains to establish the existence of the above surface S for non-trivial, admissible Heegaard strings s_0, s_1. This will be accomplished in the next proposition. $\Diamond$

17.23. Proposition. *Let* $(M, \underline{m})$ *be an almost simple 3-manifold such that no essential annulus in* $(M, \bar{\underline{m}})$ *meets a free face of* $(M, \underline{m})$. *Let* s_0, s_1 *be two admissibly homotopic, non-trivial, admissible Heegaard strings in* $(M, \underline{m})$.

Then there is a non-separating surface, S, *in* $(M, \underline{m})$ *with* $e(S)$ *maximal and there are admissible Heegaard strings* s_0', s_1' *in* $(M, \underline{m})$ *such that*

(1) s_i' *is admissibly isotopic in* $(M, \underline{m})$ *to* s_i, $i=1,2$.
(2) $s_i' \cap S = \emptyset$, $i=1,2$.

Remark. See 16.4 for the definition of the complexity $e(S)$.

Proof. Of course, there is no reason to expect, for any pair of admissible Heegaard strings, a surface as predicted in 17.23. So the existence of an admissible homotopy between s_0 and s_1 will be of crucial importance for the argument.

In order to construct the desired surface we proceed as follows. First, by thm. 16.23, there is a non-separating surface, S, in $(M, \underline{\tilde{m}})$ with $S \cap s_0 = \emptyset$ and $e(S)$ maximal (recall s_0 is non-trivial). W.l.o.g. we may also suppose that every bound face of S, next to a square-face of $(M, \underline{m})$, is actually contained in a square-face of $(M, \underline{m})$ (see the remark following prop. 17.11 for definition). It remains to modify S so that afterwards it satisfies the conclusion of the proposition.

Let s_1 be admissibly isotoped in $(M, \underline{\tilde{m}})$ so that $s_1 \cap S$ is a finite collection of points whose total number is as small as possible. Assume $s_1 \cap S \neq \emptyset$ (for otherwise we are done).

s_1 is an admissible Heegaard string in $(M, \underline{m})$ and so there is a good disc, D, for s_1 in $(M, \underline{m})$ (see 16.1 for definition). By prop. 16.22, there is a surface, S', which is either S or obtained from S by the torus modification along some solid torus, and there is an admissible isotopy which pushes s_1 so that afterwards $s_1 \cap S'_0 = \emptyset$, where S'_0 denotes a displacement of S' along $b := D \cap \bigcup \underline{m}$ (see 16.4). Moreover, the isotopy can be chosen so that, in addition, it does not change the number of intersections of s_1 with S throughout its deformation (see proof of 16.22).

Let $d(s_1)$ be the distance of s_1 w.r.t. S'_0 as defined in 16.4. Then we distinguish between the following two cases.

Case 1. $d(s_1) = \infty$.

Note that in this Case, $s_1 \cap S' = \emptyset$ (and not just $s_1 \cap S'_0 = \emptyset$).

Before we proceed we introduce some notation. Recall S' is obtained from S by a torus-modification along some solid torus, say P. Let $(M', \underline{m}')$, resp. P' be the 3-manifolds obtained from $(M, \underline{m})$ resp. P by splitting along S. $S \cap P$ is a system of incompressible and pairwise disjoint annuli in P whose winding number (w.r.t. P) is bigger than one (see definition of torus-modification along solid tori in 16.21). Thus, in particular, P' consists of solid tori, $P'_0, P'_1, ..., P'_m$, $m \geq 1$. Set $A'_i := (\partial P'_i - \partial M')^-$ and $A' := \bigcup A'_i$, $0 \leq i \leq m$ W.l.o.g. we suppose the indices are chosen so that A'_m is connected, P'_m is the parallelity region between $A'_m = (\partial P'_m - \partial M')^-$ and $P'_m \cap \partial M'$ and P'_{i+1}, $0 \leq i \leq m - 1$, is adjacent to P'_i. Then A'_0, A'_m are annuli and A'_i, $1 \leq i \leq m - 1$, is a pair of admissibly parallel annuli. In fact, $A'_0 \subset \partial P'_0$ is an annulus with winding number bigger than one w.r.t. P'_0. It is easily verified that $A' - A'_m$ is a system of essential annuli in $(M', \underline{m}')$ since $\#(s_1 \cap S)$ is supposed to be minimal. Moreover, it follows that $s_1 \cap P'_0$ is an arc in P'_0 which cannot be deformed (in P'_0 and fixing end-points) into the annulus $P'_0 \cap \partial M'$. $s_1 \cap P'_i$, $1 \leq i \leq m$, consists of pairs of unknotted arcs joining the two annuli from $P'_i \cap \partial M'$. Finally, the two components from $(s_1 - P)^-$ are two arcs which lie next to each other, i.e., they are opposite faces of a square. To see the statements concerning s_1 see how P comes up in the proof of 16.22. (Fig. 17.3 illustrates the situation at hand).

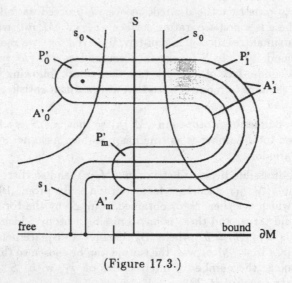

(Figure 17.3.)

17.24. Lemma. *Either prop.* *17.23 holds, or* s_0 *is a trivial, admissible Heegaard string in* $(M', \underline{m}')$.

Proof of Lemma. We first claim A' consists of good annuli in $(M', \underline{m}')$ (see 16.4 for definition of good annuli). Assume some component, say A'_i, of A' is not a good annulus. Then, by definition, there is an incompressible annulus $B \subset M$ with $B \cap (A'_i \cup \partial M') = \partial B$ such that either $\partial B = \partial A$, or one boundary curve lies in A'_i and the other one in a free face of $(M', \underline{m}')$. A moments reflection shows that $U(P \cup B)$ must contain either an essential torus, or an essential annulus in $(M, \underline{\tilde{m}})$ whose boundary lies in a free face of $(M, \underline{m})$. This contradicts our hypothesis on $(M, \underline{m})$, The claim is therefore established.

Now recall S is essential in $(M, \underline{\tilde{m}})$ and $s_0 \cap S = \emptyset$. Thus, by prop. 16.3, s_0 is an admissible Heegaard string in $(M', \underline{m}')$. Since A' is a system of good annuli, it allows no torus modification. Moreover, $e(S)$ is maximal. It follows from prop. 16.22 and 16.16 that either s_0 is trivial in $(M', \underline{m}')$, or that s_0 can be admissibly isotoped in $(M', \underline{\tilde{m}}')$ out of A'. In the latter case $s_0 \cap S' = \emptyset$ since $s_0 \cap S = \emptyset$. Moreover, $s_1 \cap S' = \emptyset$ and $e(S') = e(S)$. So S' satisfies the conclusion of prop. 17.23.

This proves lemma 17.24.

By lemma 17.24, we may suppose s_0 is a trivial, admissible Heegaard string in $(M', \underline{m}')$. It is our goal to use this property in order to construct an admissible isotopy for s_1 in $(M, \underline{m})$ which reduces $\#(s_1 \cap S)$. To construct such an isotopy we start with a "good" singular square or pentagon for s_1.

To define this notion fix a boundary-pattern, $\underline{\sigma}$, for $s \times I$ as follows. Let $s \times 0$ and $s \times 1$ be bound faces. Let one component of $\partial s \times I$ be a free face. Let the other component of $\partial s \times I$ be either a free face, or decomposed into one bound face and one free face. Let this decomposition be chosen so that the bound face meets $s \times 0$. $(s \times I, \underline{\bar{\sigma}})$ is a square or pentagon according whether or not both components of $\partial s \times I$ are free faces. A *good singular square or pentagon* for s_1 is a map $h : s \times I \to M$ with $h(s \times 0) \subset S$ and $h|s \times 1 = s_1$, which maps the bound resp. free faces from $(s \times I, \underline{\sigma})$ contained in $\partial s \times I$ into bound resp. free faces of $(M, \underline{m})$.

17.25. Lemma. *There exists a good singular square or pentagon h for s_1 such that $h^{-1}S$ lies in the union of $s \times 0$ with all free faces of $s \times I, \underline{\sigma})$.*

Proof of Lemma. s_0 is supposed to be a trivial, admissible Heegaard string in $(M', \underline{m}')$. Then, by definition, there is a good disc D for s_0 in $(M', \underline{m}')$ whose intersection with $U(s_0)$ is connected. Identifying f and this good disc along s_0, we obtain a good singular square or pentagon for s_1. It has the desired property since $D \subset M'$. Lemma 17.25 is established.

Let $h : s \times I \to M$ be a good singular square or pentagon for s_1 which is chosen so that $h^{-1}S$ and $h^{-1}\partial P$ are systems of curves and that, in addition,

$$c(h) := (\#h^{-1}S, \#h^{-1}\partial P, \#(h^{-1}S \cap h^{-1}\partial P))$$

is as small as possible (w.r.t. the lexicographical order). We next collect a few properties of h. For the upcoming discussion consult fig. 17.4:

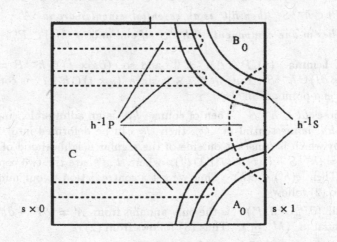

(Figure 17.4)

17.26. Lemma. $h^{-1}S$ *is the union of $s \times 0$ and arcs joining $s \times 1$ with a component of $\partial s \times I$.*

Proof of Lemma. Note that no arc from $h^{-1}S$ joins the two components of $\partial s \times I$. Indeed, this follows from our minimality condition on h and the fact that any arc, joining the two components of $\partial s \times I$, separates a square from $(s \times I, \underline{\bar{a}})$ containing $s \times 0$ or $s \times 1$.

Note further that no arc from $h^{-1}S$ has both its end-points in one component of $\partial s \times I$. This follows easily from the usual surgery argument involving the relative loop-theorem [Joh 1, prop. 2.1].

Next we claim that no arc from $h^{-1}S$ can have both end-points in $s \times 1$. Assume the converse for the moment. Then there is a disc, D_0, in $s \times I$ with $D_0 \cap \partial(s \times I) \subset s \times 1$ and such that $D_0 \cap h^{-1}S = (\partial D_0 - \partial(s \times I))^-$. Set $g := h|D_0$. Then $g : D_0 \to M'$ is an admissible map into $(M', \underline{m}')$. Now recall $s_1 \cap P$ is connected, $s_1 \cap A' = s_1 \cap A'_m$ and $A' - A'_m$ is essential in $(M', \underline{m}')$. It follows that $t := g(D_0 \cap s \times 1) = s_1 \cap P'_0$ and that g can be admissibly deformed (fixing t) so that afterwards $g(D_0) \subset P'_0$. But t is an admissible Heegaard string in $(P'_0, \underline{p_0}')$, where $\underline{p_0}' = \{ (\partial P'_0 - \partial M')^- \}$. So, by thm. 17.7, g can be deformed into an embedding, using an admissible homotopy in $(P'_0, \underline{\bar{p}}_0')$ fixing t. Thus $\#(s_1 \cap S)$ can be diminished by some admissible isotopy of s_1 in $(M, \underline{\bar{m}})$. Contradiction.

This proves lemma 17.26.

17.27. Lemma. *Let* t *be the closure of a component from* $h^{-1}\partial P - h^{-1}S$. *Then the following holds:*

(1) $\partial t \subset (s \times \partial I) \cup h^{-1}S$ *but* $\partial t \not\subset s \times 1$ *(recall* $s \times 0 \subset h^{-1}S$).

(2) If $\partial t \subset h^{-1}S$, *then* $h|t$ *is an essential, singular arc in* A'.

(3) ∂t *lies in one component of* $h^{-1}S$ *if and only if* $h(t) \subset P'_m$.

Proof of Lemma. (1) $P \cap \partial M = \emptyset$ and so $(\partial s \times I) \cap h^{-1}P = \emptyset$. Thus $\partial t \subset (s \times \partial I) \cup h^{-1}S$. But $\partial t \not\subset s \times 1$ since $(s \times 1) \cap h^{-1}P$ is connected and contains end-points of $h^{-1}S$.

(2) Suppose $\partial t \subset h^{-1}S$. Then of course $h|t$ is an admissible, singular arc in A'. If $h|t$ is inessential in A', then h can be deformed into h', using a homotopy which is constant outside of the regular neighborhood of t, so that $(h')^{-1}S = (h^{-1}S - U(t))^- \cup (t' \cup t'')$, where t', t'' are the two copies of t in $\partial U(t)$. Then $c(h') < c(h)$. Thus we get a contradiction to our minimal choice of h. So (2) follows.

(3) Recall $(\partial P'_m - \partial M')^-$ is the only annulus from $A' = (\partial P' - \partial M')^-$ which is inessential in $(M', \underline{m}')$. Thus (3) follows from (2).

Hence lemma 17.27 is established.

A disc $E \subset s \times I$, not contained in $h^{-1}P$, is a *bad square* (for $h^{-1}S \cup h^{-1}\partial P$) if

(1) $E \cap h^{-1}P_0 \neq \emptyset$ and $E \not\subset h^{-1}P$,

(2) $D \cap h^{-1}S$ consists of two disjoint arcs in ∂D, and

(3) every component of $(\partial D - h^{-1}S)^-$ lies either in $h^{-1}P_0'$, or in a bound or free face of $(s \times I, \underline{\sigma})$ different from $s \times 1$.

17.28. Lemma. *There are no bad squares for* $h^{-1}S \cap h^{-1}\partial P$.

Proof of Lemma. Assume E is a bad square for $h^{-1}S \cup h^{-1}\partial P$. We suppose w.l.o.g. that $int(E) \cap h^{-1}P_0' = \emptyset$. S is essential in $(M, \underline{\bar{m}})$ and $c(h)$ is supposed to be minimal. It is therefore easy to verify that the restriction $g := h|E$ must be an essential, singular square in $(M_0', \underline{\bar{m}_0}')$. Here $(M_0', \underline{m_0}')$ denotes the manifold obtained from $(M', \underline{m}')$ by splitting along $(\partial P_0' --\partial M')^-$. Using [Joh 1, prop. 4.8], it follows that $(\overline{M_0'}, \underline{\bar{m}_0}')$ is an irreducible 3-manifold with useful boundary-pattern (recall $(\partial P_0' - \partial M')^-$ is essential in $(M', \underline{m}')$). Thus, by [Joh 1, thm. 12.5], g can be admissibly deformed into a component, say X', of the characteristic submanifold of $(M', \underline{m}')$. X' must be an I-bundle since g is an essential, singular square. Recall $P_0' \cap \partial M'$ lies in a bound face of $(M', \underline{m}')$ which is a copy of S. Thus $X' \cup U(\partial P_0' - \partial M')^-$ is an I-bundle whose lids lie in *one* copy of S. Then this I-bundle must be an I-bundle over the square, annulus or Möbius band since $e(S)$ is supposed to be maximal (see proof of (2) of 17.11). Hence g cannot be essential in $(M_0', \underline{m_0}')$. Contradiction. Lemma 17.28 therefore holds true.

Notice that $\#(s_1 \cap S)$ is even and so is $\#((s \times 1) \cap h^{-1}S) = \#(s_1 \cap S)$. In particular, $h^{-1}S$ splits $s \times 1$ into an odd number of intervals. Thus there is a unique middle interval, say ℓ. Note that $\ell \subset h^{-1}P_0$. Let C be the closure of the component of $(s \times I - h^{-1}S)^-$ containing ℓ.

17.29. Lemma. $(\partial C - \partial(s \times I))^-$ *consists of two components which in turn are not admissibly parallel in* $(s \times I, \underline{\bar{\sigma}})$.

Proof of Lemma. By lemma 17.26, $(\partial C - \partial(s \times I))^-$ consists of two components. If they are admissibly parallel in $(s \times I, \underline{\bar{\sigma}})$, then C_0 is a 4-faced disc. ℓ is one of its faces. Hence it is easily seen that C must contain a bad square. This however contradicts lemma 17.27. Lemma 17.30 is therefore established.

We know that $((s \times I) - C)^-$ consists of two components, say A, B (see lemma 17.25 and 17.26). Let A be that one of them which meets the component of $\partial s \times I$ which is a free face of $(s \times I, \underline{\sigma})$. Let $A_0, A_1, ..., A_\alpha$, $\alpha \geq 1$, be the components in which A is split by $h^{-1}S$. Let the indices be chosen so that A_0 contains an end-point of $s \times 1$ and that, in addition, A_{i+1} is adjacent to A_i, $0 \leq i \leq \alpha - 1$.

17.30. Lemma. $A_0 \cap h^{-1}P$ *is connected.*

Proof of Lemma. Assume the converse. Set $\mathcal{K}_A := A \cap h^{-1}S$ and $\mathcal{K}_B := B \cap h^{-1}S$. Then $h^{-1}S = (s \times 0) \cup \mathcal{K}_A \cup \mathcal{K}_B$ (see lemma 17.26). Note also that $\#\mathcal{K}_A = \#\mathcal{K}_B$, by our choice of C. Set $k_0 := (\partial A_0 - \partial(s \times I))^-$. Then, by assumption, $k_0 \cap h^{-1}P$ is disconnected. In particular, there is a component, a_0, of $(k_0 - h^{-1}P)^-$ which does not meet ∂k_0. If $A_1 \neq \emptyset$, let E_1 be the component of $(A_1 - h^{-1}P)^-$ containing a_0. It follows from lemma 17.27 that E_1 has to be a 4-faced disc. Let a_1 be the face of E_1 opposite a_0. Continuing the construction, we get a sequence $E_1 \subset A_1, E_2 \subset A_2, ..., E_\alpha \subset A_\alpha$ of 4-faced discs and a sequence $a_1, a_2, ..., a_\alpha$ of faces. a_α lies in C. Note that $h(A_0 \cap h^{-1}P) \subset P'_m$ (recall P'_m is the only component of P' which is inessential in $(M', \underline{m}')$). So $\partial a_0 \subset h^{-1}P'_m$. Thus a counting argument shows that the two arcs from $C \cap h^{-1}\partial P$, starting from the end-points of $a_\alpha \subset C \cap A_\alpha$, must in fact lie in $h^{-1}P'_0$. Therefore they cut out a bad square from C, provided they are parallel in C. Hence, by lemma 17.28, they cannot join the same two arcs from $h^{-1}S$. It follows that one of them joins K_A with $s \times 0$. But, by our choice of A, it follows that this particular arc must separate a 4-faced disc from C meeting the component of $\partial s \times I$ which is a free face of $(s \times I, \underline{\sigma})$ (see fig. 17.4 for an illustration). However, this 4-faced disc is a bad square and so we get a contradiction to lemma 17.28. This finishes the proof of lemma 17.30.

According to the previous lemma, $A_0 \cap h^{-1}P$ is connected. Set $A'_0 = (A_0 - h^{-1}P)^-$ and $s'_1 := A'_0 \cap (s \times 1)$. Then A'_0 is a 4-faced disc in a natural way with s'_1 as one of its faces. Define $M_0 := (M - P)^-$ and set $\underline{m}_0 = \underline{m}$. Furthermore, let $(M'_0, \underline{m}_0')$ be the manifold obtained from $(M_0, \underline{m}_0)$ by splitting along $T_0 := S \cap M_0$. Set $g := h(A'_0)$. Then g is an admissible map into $(M'_0, \underline{m}_0')$

17.31. Lemma. *The map g can be deformed into an embedding, using an admissible homotopy in $(M'_0, \bar{\underline{m}}_0')$ which is an isotopy on s'_1.*

Proof of Lemma. Of course, s'_1 is an admissible Heegaard string in $(M_0, \underline{m}_0)$. So, by 16.3, it is an admissible admissible Heegaard string in $(M'_0, \underline{m}_0')$ since $s'_1 \cap T_0 = \emptyset$.

Note that $(M'_0, \underline{m}_0')$ is almost simple. To show this, we have to verify properties (1)-(4) of almost simple 3-manifolds. Now (1) is obvious. (2) follows since, by prop. 17.11, $(M', \underline{m}')$ is almost simple. (3) and (4) follow since $(M', \underline{m}')$ is almost simple and since the only free faces of $(M'_0, \underline{m}_0')$ different from faces of $(M', \underline{m}')$ are annuli. Thus $(M'_0, \underline{m}_0')$ is indeed almost simple. But notice that $(\overline{M'_0}, \underline{m}_0')$ may well have an essential annulus which meets a free face (e.g., an annulus joining S with P). So we cannot simply refer to thm. 17.12 in order to establish lemma 17.31. To overcome this obstacle, consider for a moment the manifold, $(M^*, \underline{m}^*)$, obtained from $(M, \underline{m})$ by splitting along S'. Recall $e(S') \geq e(S)$. So $e(S')$ is maximal since $e(S)$ is. Thus, by prop. 17.11, no essential annulus in $(M^*, \bar{\underline{m}}^*)$ meets a free face of $(M^*, \underline{m}^*)$. Moreover, $S \cap M^*$ is a system of good annuli in $(M^*, \bar{\underline{m}}^*)$ (see the proof of 17.24). Thus, by prop. 16.22 and 16.16, either s_1 is trivial in $(M^*, \underline{m}^*)$ or it can be admissibly isotoped in $(M^*, \underline{m}^*)$ so that afterwards it does not meet S anymore. Hence, by our minimality condition on $s_1 \cap S$, it follows s_1 must be trivial in $(M^*, \underline{m}^*)$. But then s'_1 is a trivial, admissible Heegaard string in $(M'_0, \underline{m}_0')$. So, instead of referring to thm. 17.12, we may refer to prop. 17.13. Thus lemma 17.31 follows.

By what we have just seen, we may suppose that $h|A_0$ is an embedding. Now $B_0 \cap (s \times 1)$ lies next to $A_0 \cap (s \times 1)$ (in M). So w.o.l.g. $s \cap g(A_0) = A_0 \cap s \times 1$. Thus, pushing $A_0 \cap (s \times 1)$ across $g(A_0)$, we reduce the intersection $s_1 \cap S$. So, by our minimality condition on $\#(s_1 \cap S)$, it follows that $s_1 \cap S = \emptyset$. Thus S is the desired surface. This finishes the proof of prop. 17.23 in Case 1.

Case 2. $d(s_1) < \infty$.

In this Case, S_0' separates a 2-faced disc, F_0, from the free part $F := (\partial M - \bigcup \underline{m})^-$ with $F_0 \cap S_0' = (\partial F_0 - \partial F)^-$ (see definition of displacements). W.l.o.g. $s_0 \cap F_0 = \emptyset$. If $s_1 \cap F_0 = \emptyset$, then reduce $\#(\partial S_0' \cap \partial F)^-$ without $\#(s_i \cap S_0')$, i=1,2, by using an isotopy of S_0' which is constant outside of $U(F_0)$ (the regular neighborhood is taken in M). If the new surface separates a 2-faced disc from F, then repeat the above reduction process, and so on. After finitely many of such steps the process stops. Let G denote the resulting surface.

Since $d(s_1) < \infty$, G separates a 2-faced disc, F_0, from F and F_0 contains at least one end-point of s_1 but none of s_0. Let $(M', \underline{m}')$ be the manifold obtained from $(M, \underline{m})$ by splitting along G. Recall s_1 is a non-trivial, admissible Heegaard string in $(M, \underline{m})$ with $s_1 \cap G = \emptyset$. Thus, by prop. 16.16, we may suppose that s_1 is admissibly isotoped in $(M, \underline{m})$ so that afterwards $s_1 \cap G = \emptyset$ and s_1 is contained in a tubular cage, W, in $(M, \underline{m})$ which is not a tubular cage in $(M', \underline{m}')$. Define $W' := U(W - U(F_0))^-$, where the regular neighborhood is taken in M. Let G' be the torus modification along W'. Pushing G' across F_0, we obtain a surface, G'', with $s_1 \cap G'' = \emptyset$ again. Continuing this process, we eventually get a surface, G^*, with $s_1 \cap G^* = \emptyset$ and $\partial G^* = \partial S$. In fact, it is not hard to see that G^* is obtained from S by the torus modification along some solid torus (see proof of 16.23). Thus setting $S' = G^*$, we are in Case 1 again.

This finishes the proof of prop. 17.23. $\Diamond$

As an immediate consequence we get the main result of this section.

17.32. Corollary. *Two generalized Heegaard strings in a simple 3-manifold are homotopic if and only if they are (ambient) isotopic.* $\Diamond$

17.33. Remarks. (1) It follows from the previous theorem that any two homotopic generalized Heegaard strings s, s' in a 3-manifold M are *strongly homeomorphic* in the sense that there is a homeomorphism $h : M \to M$ isotopic to the identity with $h(s) = s'$.

(2) It is important to realize that thm. 17.22 does *not* claim that, for any homotopic generalized Heegaard-strings s and s', there exists a square A in M with $(\partial A - \partial M)^- = s \cup s'$. This should be contrasted with Waldhausen's theorem [Wa 2] which tells us that isotopies between incompressible surfaces in 3-manifolds are composed out of parallelity regions. Indeed, a similar theorem for Heegaard-strings seems very unlikely.

§18. The Mapping Class Group.

In this section we address the problem of calculating, or better estimating, the mapping class groups of one-relator 3-manifolds.

18.1. *Preliminaries.*

The relator structure for one-relator 3-manifolds adds a new feature to the general study of the mapping class group of Haken 3-manifolds (as, e.g., given in [Joh 1, cor. 27.6]). This extra feature is of interest for an actual computation of such groups. To be more specific, let $M^+(k)$ be a one-relator 3-manifold which is irreducible and ∂-irreducible. Recall from prop. 15.16 and lemma 7.4 that it is a simple matter to check whether $M^+(k)$ has this property. Recall from section 5 that it is easy to produce various examples of such 3-manifolds. Observe further that any homeomorphism

$$h : (M, k) \to (M, k)$$

has a canonical extension to a homeomorphism

$$h^+ : M^+(k) \to M^+(k).$$

The assignment $h \mapsto h^+$ defines a map

$$\mathrm{Diff}(M, k) \to \mathrm{Diff}\, M^+(k),$$

which in turn induces a homomorphism of mapping class groups

$$\varphi : \pi_0 \mathrm{Diff}(M, k) \to \pi_0 \mathrm{Diff}\, M^+(k).$$

In general, this homomorphism is neither an injection nor a surjection. Here is a concrete example.

18.2. Example.
Let M^+ be an I-bundle over some surface, S. Let t be a fiber of this bundle. Then $M := (M^+ - U(t))^-$ is a handlebody. Hence $M^+ = M^+(k)$ is a one-relator 3-manifold, where k denotes the core-curve of the annulus $M \cap U(t)$.

Let $p \in S$ be the point corresponding to the fiber t. Fix a closed path $w : I \to S$ based at p. Note that w may also be considered as an isotopy of the point p. This isotopy can be extended to an ambient isotopy, $\alpha : S \times I \to S$, which in turn can be lifted to an ambient isotopy $\alpha^+ : M^+ \times I \to M^+$. Then $\alpha^+ \mid M^+ \times 1$ is a homeomorphism of M^+ which maps t to itself and which therefore induces an element, h_w^+, of $\pi_0 \mathrm{Diff}(M, k)$.

18.3. Proposition.
Let M^+ be an I-bundle over some surface, S, with $\chi S < 0$. Then the assignment $w \mapsto h_w^+$ induces an isomorphism $\psi : \pi_1 S \to$ kern $(\varphi : \pi_0 \mathrm{Diff}(M, k) \to \pi_0 \mathrm{Diff}\, M^+(k))$.

Proof. We first verify that ψ is well-defined. To this end recall that every isotopy between fiber-preserving homeomorphisms of M^+ can be isotoped into a fiber-preserving isotopy. Thus h_w^+ is *isotopic* to the identity if and only if $h_w := \alpha | S \times 1$ is. Moreover, w is null-homotopic if and only if h_w is *homotopic* to the identity rel base-point, p. Thus the claim follows when we show that every homeomorphism of S, homotopic to the identity rel base-point, is isotopic to the identity rel base-point. This in turn is a well-known fact in surface-topology. We recall the relevant argument for the convenience of the reader. Suppose h_w is homotopic to the identity rel base-point, p. Then let $k : I \to S$ be some non-singular and non-trivial loop based in p. Then w.l.o.g. $\ell := h_w k$ is a non-singular loop based in p which is in general position to k and which is homotopic (rel p) to k. Let $\tilde{p}$ be a point above p in the universal cover, $\tilde{S}$, of S. Let $\tilde{k}$, $\tilde{\ell}$ be the lifts of k, ℓ, respectively, starting at $\tilde{p}$. Then $\tilde{k}$ and $\tilde{l}$ are non-singular arcs with the same end-points. Since $\tilde{S} = \mathbf{R}^2$, there must be two different points $x, y \in \tilde{k} \cap \tilde{\ell}$ bounding arcs $\tilde{k}_1 \subset \tilde{k}$ and $\tilde{\ell} \subset \tilde{l}$ with $\tilde{k}_1 \cap \tilde{\ell}_1 = \partial \tilde{k}_1 = \partial \tilde{l}_1$. Moreover, there must be a disc $\tilde{D} \subset \tilde{S}$ with $\tilde{D} = \tilde{k}_1 \cup \tilde{\ell}_1$. This disc projects to a disc $D \subset S$. Pushing l across this disc, we either reduce $\#(\tilde{k} \cap \tilde{l})$ or we isotope l into k. Thus, inductively, we may suppose h_w is isotoped so that $h_w | U(k) = \text{id} | U(k)$. Set $S' := (S - U(k))^-$. Since S is not the torus and since $h_w \cong \text{id}$, it follows from a well-known theorem of Nielsen [Wa 2] that $h_w | S'$ is homotopic to the identity in S'. Then, by Nielsen [Wa 2] again, it is also *isotopic* to the identity. Altogether, h_w is isotopic to the identity rel base-point.

Suppose h_w is isotopic to the identity rel base-point. Then w.l.o.g. we may suppose $\alpha \mid S \times 1$ is deformed into the identity, using an isotopy fixing p. Choose an essential, simple closed curve, k, in S containing the point p. Then $\alpha(k \times 0) = \alpha(k \times 1)$. Using Nielsen's theorem once more, it follows that $f := \alpha \mid k \times I$ can be deformed into k, using a homotopy fixing $k \times \partial I$ (recall S is not a torus or a Klein bottle). So, in particular, w can be deformed (rel end-points) into k. By the same argument, w can be deformed (rel end-points) into any simple closed curve, $\ell \subset S$, with $\ell \cap k = p$. It follows that w can be contracted into p since k and ℓ are curves which are neither homotopic nor null-homotopic. This shows that ψ is injective.

It remains to verify surjectivity of ψ. To this end we define an inverse $\theta : \ker n\varphi \to \pi_1 S$ as follows. Let $h : (M, k) \to (M, k)$ be a homeomorphism whose extension $h^+ : M^+ \to M^+$ is isotopic to the identity. The restriction of such an isotopy to the fiber t defines a singular essential annulus in M^+. Since M^+ is a product I-bundle, we may suppose that this singular annulus is vertical (see [Joh 1, prop. 5.10]). Therefore its projection to the base-surface S defines a path, say $(w_h)^{-1}$. The assignment $h \mapsto w_h$ is the desired inverse. $\Diamond$

This ends the discussion of example 18.2.

18.4. *Heegaard Strings and Mapping Class Groups.*

Although the notion of one-relator 3-manifolds is much more general than that of a product I-bundle the example from 18.2 describes nevertheless a typical feature with respect to the mapping class group. This is the content of our next result.

18.5. Proposition. *Let* $M^+(k)$ *be a one-relator 3-manifold which is irreducible and ∂-irreducible. Let*

$$\varphi : \pi_0 \mathrm{Diff}(M, k) \to \pi_0 \mathrm{Diff}\ M^+(k)$$

be the homomorphism as introduced earlier. Then there is a, possibly non-orientable, surface S with

$$\pi_1(S) \cong \mathrm{kern}\ \varphi.$$

In addition, ∂S is non-empty, if M^+ is not the I-bundle over a surface (orientable or not), and S is the disc, if the 3-manifold M^+ is simple.

Remark. Recall from [Joh 1] the definition of *simple* 3-manifolds.

Proof. Recall $(M^+(k) - M)^-$ is the regular neighborhood of some proper arc, t, in $M^+ := M^+(k)$ (the co-core of the attached 2-handle of $M^+(k)$). In order to describe the kernel of φ in the spirit of example 18.2, we wish to construct an essential I-bundle, W, in M^+ which carries the kernel of φ. This can be done as follows.

First observe that by [Joh 1] (see also [JS]) the characteristic submanifold exists for M^+ since M^+ is a Haken 3-manifold. Let W denote the union of all those components of this characteristic submanifold into which the arc t can be pushed, using a (proper) homotopy in M^+. By the properties of the characteristic submanifold (use [Joh 1, 8.2, 5.10 and 12.5]), it follows that W cannot have more than two components. More precisely, one of its components, say W_1 (possibly empty), must be an essential I-bundle and the other one, say W_2 (possibly empty), must be an essential Seifert fiber space (which has no I-bundle structure whatsoever) since t can be deformed into both W_1 and W_2 (use again [Joh 1, 8.2, 5.10 and 12.5]).

Now let us consider the non-trivial elements in kern φ. Any such element is represented by some homeomorphism $h : (M, k) \to (M, k)$ such that its extension $h^+ : M^+ \to M^+$ is isotopic to the identity. In this case there is an isotopy $\alpha : M^+ \times I \to M^+$ with $\alpha \mid M^+ \times 0 = h^+$ and $\alpha \mid M^+ \times 1 = \mathrm{id}$. Consider $\alpha \mid t \times I$. The two components from $t \times \partial I$ are both mapped to the arc t, and identifying them, we obtain from $\alpha \mid t \times I$ a singular annulus or Möbius band $f : A \to M^+$.

We claim f cannot be deformed into t. Assume the converse. Note that any deformation of f can be extended to a deformation of the map α. Thus, for the sake of argument, we may suppose that α is deformed (rel $M^+ \times \partial I$) into a *homotopy* with $\alpha(t \times I) \subset t$. Identify the handlebody M with $(M^+ - U(t))^-$. Then α restricts to a homotopy of M as well, and, by assumption, $\alpha(\partial M \times I) \subset \partial M$. Let $\mathcal{D}$ be a meridian-system in the handlebody M and consider $g := \alpha \mid \mathcal{D} \times I$. The restriction $g \mid \partial(\mathcal{D} \times I)$ consists of singular 2-spheres in M. But handlebodies are aspherical. So $g \mid \partial \mathcal{D} \times I$ can be contracted in M. Since $(M - U(\mathcal{D}))^-$ is a 3-ball, it follows, by asphericity of M again, that g can be deformed (relative $\partial \mathcal{D} \times I$) into M. By the same argument, $\alpha \mid M \times I$ can be deformed into M, by a homotopy which fixes $\partial M \times I$. The result is a homotopy, which pushes $h : (M, k) \to (M, k)$ to the identity. By [Wa 1], it then follows that h is isotopic in (M, k) to the identity. But this contradicts the hypothesis that h is non-trivial.

By what we have just seen, $f : A \to M^+$ cannot be deformed into t. By the asphericity of M^+ (see [Wa 1]), it follows that the restriction of f to at least one boundary curve of A is a non-contractible curve in ∂M^+. Since M^+ is supposed to be ∂-irreducible, it then follows from the loop-theorem (see [St][Joh 9]) that $f : A \to M^+$ induces an injection of fundamental groups. Moreover, t is an essential arc in A which, by cor. 17.9, is essential in M^+ as well. Thus f is essential, i.e., either an essential singular annulus or Möbius band in M^+.

Since f is an essential, singular annulus or Möbius band, it follows from [Joh 1, thm. 12.5] that it can be deformed into W. Thus we may suppose f is deformed so that $f(A)$ lies either in W_1 or in W_2. In particular, t lies either in W_1 or W_2. In fact, we may suppose that every homotopy of t into itself can be deformed (rel t) into the component of W containing t.

Suppose $t \subset W_1$. Recall W_1 is an essential I-bundle. Then an argument from thm. 18.3 may be combined with the previous observation, in order to show that there is an isomorphism $\pi_1 S \to \ker \varphi$, where S denotes the base-surface of W_1, and we are done.

Suppose $t \subset W_2$. Recall W_2 is an essential Seifert fiber space. Moreover, every essential singular annulus, which comes from a homotopy of t to itself as above, can be deformed into W_2 (rel t). Moreover, by [Joh 1, prop. 5.10], it is homotopic to a fiber-preserving map, and so to a rational multiple of f. Thus $\ker \varphi$ is either trivial or free cyclic. Hence it is isomorphic to the fundamental group of $S :=$ disc or annulus.

The proof of thm. 18.5 is therefore finished. $\Diamond$

18.6. Remark. Note that thm. 18.5 generalizes the well-known fact that the kernel of the natural map from any point stabilizer of the mapping class group of a closed orientable surface to that mapping class group is isomorphic to the fundamental group of that surface. To see this relationship, apply thm. 18.5 to the product I-bundle over the surface in question. This is a one-relator 3-manifold which equals its own characteristic submanifold.

18.7. Remark. In view of the above result, the mapping class group of the relative handlebody (M, k) becomes important for a computation of the mapping class group of simple one-relator 3-manifolds $M^+(k)$ (or even non-simple ones). The mapping class group of (M, k) is almost cyclic (see prop. 8.11) and the determination of its order may be feasible for specific examples (with the help of computers). It remains to determine the cokernel $\pi_0 \operatorname{Diff} M^+(k)/\pi_0 \operatorname{Diff}(M, k)$. We will later see that, in general, this cokernel is non-trivial since there are non-homeomorphic relative handlebodies (M, k) and (M, l) with $M^+(k) \cong M^+(l)$ (see section 26). However, the size of the cokernel is limited from above by the total number of non-homeomorphic Heegaard strings in $M^+(k)$ and the Rigidity Theorem (see section 31) tells us that in "most" cases the latter number is finite. This means, in particular, that the relative handlebody (M, k) captures the infinite part of the mapping class group of $M^+(k)$.

§19. Incompressibility.

The handle-addition lemma in its simplest form deals with discs in (generalized) one-relator 3-manifolds (see prop. 15.16). In prop. 15.11 the relevant statement has been extended from discs to arbitrary surfaces in one-relator 3-manifolds. Here we are interested in yet another extension. For this note that the handle-addition lemma in its original form can also be read as a (very convenient) criterion for the incompressibility of a very special type of surfaces in (generalized) one-relator 3-manifolds, namely those closed surfaces parallel to boundary components. This interpretation raises the question whether there is a similar criterion for other closed surfaces in those 3-manifolds as well. It turns out that to a large extent this is the case. The next two sections are concerned with this type of problem. The present section deals with the case of separating relator curves.

19.1. Recall from section 6 the convenient way of representing surfaces by circle-patterns in Heegaard-diagrams. Recall that to every circle-pattern, C, in a Heegaard-diagram, $\underline{D}$, there are associated 2-manifolds, $S(C) \subset M(\underline{D})$ and $S^+(C) \subset M^+(\underline{D})$, and that every incompressible surface in $M(\underline{D})$ and $M^+(\underline{D})$ can be obtained in this way (modulo isotopy). By cor. 12.21, there is a fast algorithm for deciding whether $S(C)$ is essential in $M(\underline{D})$. One may therefore wonder to which extent incompressibility of $S^+(C)$ is already determined by that of $S(C)$. In general, there is of course no reason to expect that incompressibility of $S(C) \subset M(\underline{D})$ implies incompressibility of $S^+(C) \subset M^+(\underline{D})$. It turns out however that for one-relator 3-manifolds this is almost true (see 19.2 and 20.2). To explain this phenomenon, let $\underline{D}$ be a one-relator diagram and suppose $S(C)$ is incompressible in the handlebody $M(\underline{D})$. Then, in the terminology of section 14, the surface $S^+(C)$ is normal. But any normal surface can be isotoped into a strongly normal surface, by a constructable and efficient isotopy in $M^+(\underline{D})$. It is therefore sufficient to consider only strongly normal surfaces in $M^+(\underline{D})$.

19.2. Theorem. *Let $M^+(k)$ be a generalized one-relator 3-manifold which is irreducible and ∂-irreducible. Suppose k is separating in ∂M. Then every closed and (strongly) normal surface in $M^+(k)$ is incompressible in $M^+(k)$.*

Proof. Assume the converse. Then there is a surface, S^+, in $M^+(k)$ which is strongly normal and compressible. In particular, the surface

$$S := S^+ \cap M$$

is incompressible in M and there is an essential disc, D, in M such that $D \cap S$ is a system of arcs. In fact, $D \cap S$ has to be non-empty since the surface S^+ is not contained in M (every closed surface in a handlebody is compressible) and since $\partial M - k$ is incompressible in M (handle-addition lemma).

Let $E := (M^+(k) - M)^-$ denote the 2-handle attached to k. Set

$$I := E \cap \partial D.$$

Then, in the terminology of section 14, I is the interval-system for ∂D.

Since S^+ is compressible, there is a compression-disc for S^+, i.e., a disc A in $M^+(k)$ such that $A \cap S^+ = \partial A$ and that ∂A is not contractible in S^+. W.l.o.g. we suppose A is chosen so that, in addition, (1) all components of $A \cap E$ are discs parallel in E to $(\partial E - M)^-$ (general position) and that (2) $A \cap D$ consists of arcs whose total number is as small as possible.

It follows from our minimality condition on $A \cap D$ that (1) no arc from $A \cap D$ has both its end-points in one component of the interval-system I and that (2) no arc from $A \cap D$ has both its end-points in one component of $S \cap D$. Thus there are actually only three different types of arcs, t, from $A \cap D$. Here is a complete list of them:

Type I. t joins one component of $D \cap S$ with one of I,

Type II. t joins two different components of $D \cap S$, and

Type III. t joins two different components of I.

19.3. Lemma. *If no type I arc occurs, then A lies in M.*

Proof of Lemma. By hypothesis, $D \cap A$ consists of type II and type III arcs alone. By an argument given in the proof of lemma 14.4, it then follows the existence of at least one component, I^*, from I with the property that every arc from $A \cap D$ which has one end-point in I^* has its other end-point in a component of I neighboring the component I^* in ∂D. Let P denote the sub-system of all arcs from $A \cap D$ with one end-point in I^*. W.l.o.g. P consists of essential arcs in $(A - E)^-$. Moreover, P consists of recurrent arcs in $(A - E)^-$ since k is separating in ∂M. Thus every component from $\partial(A - E)^-$ different from ∂A contains both the end-points of at least one essential arc in $(A - E)^-$. But this is impossible since $(A - E)^-$ is a planar surface. This finishes the proof of the lemma.

Of course, A cannot lie in M, for S is incompressible and ∂A does not bound a disc in S. Thus there must be at least one type I arc, say t. Recall t joins a component, l, of $D \cap S$ with some component, I_1, of I. Let I_0, I_2 denote the two components of I containing ∂l. These exists since S^+ is supposed to be closed and since $S^+ \cap E \neq \emptyset$. We now may distinguish between three cases. We get the desired contradiction to our assumption by showing that none of these cases can actually occur.

Case 1. $I_0 = I_1$ or $I_2 = I_1$.

In this Case, the number of components of $A \cap E$ can be diminished by an isotopy of A, and this contradicts our choice of A.

Case 2. I_0 *neighbors* I_1, i.e., at least one component of $\partial D - (I_0 \cup I_1)$ is disjoint to I.

In this Case, the arc $l \cup t$ separates a disc, D_0, from D such that $D_0 \cap I$ equals $D_0 \cap (I_0 \cup I_1)$. We may suppose that t is chosen so that, in addition, D_0 contains no type I arcs from $A \cap D$ different from t.

Consider the end-point, x_0, (resp. y_0) of l (resp. t) contained in I_0 (resp. I_1). Note that x_0 lies in a disc, B_0, from $E \cap S^+$, and that y_0 lies in a disc, C_0, from $E \cap A$ with $B_0 \cap C_0 = \emptyset$.

We claim the point $y_1 := B_0 \cap I_1$ cannot lie in D_0. Assume the converse. Then observe that the arc $D_0 \cap \partial D$ intersects only one of the two components of $\partial U(k)$, where $U(k) = E \cap \partial M$, since k is separating in ∂M. It follows that the point $C_0 \cap I_0$ has to lie outside of D_0 (recall our assumption). But y_1 is the end-point of an arc from $D \cap S = D \cap S^+$. This arc is certainly contained in D_0 and has its other end-point, say x_1, in I_0 since S^+ is strongly normal. Now, the point x_1 again lies in some component, B_1 from $E \cap S^+$. It follows that the point $y_2 := B_1 \cap I_1$ lies in that interval, separated by y_1 from I, which does not contain y_0. Thus, repeating the above procedure over and over again, we eventually obtain an infinite set of components from $S^+ \cap D$ which in turn is impossible.

By what we have seen so far, the point y_1 defined above has to lie outside of D_0. But then the point $x_1 := C_0 \cap I_0$ lies in D_0. Now, this point is the end-point of an arc from $A \cap D$, which in turn has to join I_0 with I_1 (according to our choice of t and our minimality condition on $A \cap D$). Thus, considering the arcs from $A \cap D_0$ instead those from $S^+ \cap D_0$, we again get a contradiction as above.

Thus, altogether, Case 2 cannot occur.

Case 3. I_0 does not neighbor I_1.

As a matter of fact, we may even suppose that there is no type I arc, t, from $A \cap D$ such that I_0 neighbors I_1 (where I_0, I_1 are defined as before, for any given type I arc t). In this case, let us again consider one of those discs, say D_0, separated by $l \cup t$ from D. W.l.o.g. t and D_0 are chosen in such a way that D_0 contains no type I arc from $A \cap D$ different from t.

Since we are in Case 3, the arc $D_0 \cap \partial D$ contains at least one component of I entirely. Observe that any arc from $S^+ \cap D$ or $A \cap D$ which has one end-point in such an interval from I has to have its other end-point in I as well. This follows directly from our choice of D_0, our minimality condition on $A \cap D$, and the fact that S^+ is closed. By an argument in the proof of lemma 14.4, it then follows the existence of at least one component, I^*, from I with the following properties:

(1) I^* is entirely contained in $D_0 \cap \partial D$, and

(2) every arc from $A \cap D$ which has one end-point in I^* has its other end-point in a component from I neighboring I^* in ∂D.

By an argument of lemma 19.3, the existence of such an interval I^* yields a contradiction. Thus also Case 3 cannot occur.

This proves thm. 19.2. $\Diamond$

As an application we get the following corollary concerning certain closed surfaces in general.

19.4. Corollary. *Let $M^+ = M^+(k)$ be a generalized one-relator 3-manifold. Let S^+ be an (orientable) closed, strongly normal surface in M^+. Suppose S^+ has no compression-annulus and $S := S^+ \cap M$ is incompressible in M. Then S^+ is incompressible in M^+.*

Remark. A compression-annulus for S^+ is an annulus, A, $A \cap (S^+ \cup \partial M^+) = \partial A$, joining S^+ with ∂M^+ in such a way that $A \cap S^+$ is non-contractible in S^+.

Proof. (Sketch) We may suppose that k is non-separating in ∂M, for otherwise the conclusion follows immediately from thm. 19.2. $(M^+(k) - M)^-$ is the regular neighborhood of some Heegaard-string $t \subset M^+(k)$. Let A be a compression-disc for S^+ in $M^+ := M^+(k)$.

S^+ is strongly normal with respect to some essential disc, D, in M. If $A \cap D = \emptyset$, then $A \subset M$, and we are done. Thus we may suppose that $A \cap D$ is a non-empty system of arcs. Not all of these arcs can be proper, for otherwise we get a contradiction by applying the argument used in the handle-addition lemma to the planar surface $(A - U(t))^-$. In particular, it follows that $S^+ \cap D \neq \emptyset$. But $S^+ \cap D$ is a system of arcs since S^+ is normal with respect to D. Thus $S^+ \cap t \neq \emptyset$.

∂t lies in one component of ∂M^+. Thus t can be slid into a (generalized) Heegaard-ward, s, such that $S^+ \cap U(\text{ring}(s)) = \emptyset$. In fact, S^+ is still normal with respect to some essential disc in $(M^+ - U(\text{ring}(s)))^-$ (the argument is the same as for 15.21 since S^+ is normal). Then there are two annuli C_1, C_2 (possibly empty) in $(M^+ - U(\text{ring}(s)))^-$ with (1) - (4) of 15.20 (the argument is the same as for 15.20). Suppose the indices are chosen so that $C_1 \cap \partial M^+ \neq \emptyset$ and that $C_2 \cap W \neq \emptyset$, where $W := U(\text{ring}(s))$. All curves from $C_i \cap S^+$, $i = 1, 2$, are essential in both C_i and S^+. Thus C_1 is actually empty since S^+ has no compression annulus. Therefore C_2 is non-empty.

Consider $V := U(C_2 \cup W)$. V may be viewed as an essential Seifert fiber space in the complement of S^+. In particular it follows that the disc A can be isotoped out of V. Thus w.l.o.g. $A \subset (M^+ - W)^-$. In other words, S^+ is compressible in $(M^+ - W)^-$. But $(M^+ - W)^-$ is not only a generalized one-relator manifold, but stem(s) is a Heegaard-string in $(M^+ - \text{ring}(s))^-$ joining two different boundary components. It follows from thm. 19.2 again, that $(S^+ - U(s))^-$ is compressible in $(M^+ - W)^-$. Sliding the Heegaard-ward s back to the Heegaard-string t, it is not hard to see that S is compressible in $M = (M^+ - U(t))^-$, too. $\Diamond$

§20. The Compression-Algorithm.

Theorem 19.2. told us that incompressibility for closed, normal surfaces in generalized one-relator 3-manifolds, $M^+(k)$, is automatic, provided the relator curve, k, is separating in ∂M. Here in turn we are interested in the case when k is non-separating. In this case the situation is technically more involved, and we will not get the same strong result. But we will be able to show that compression discs are easy to detect (see thm. 20.2).

20.1. In order to study the way in which compression discs intersect 2-handles, we introduce the concept of "compression pencils".

For this let $M^+(k)$ be a generalized one-relator 3-manifold which is irreducible and ∂-irreducible. Let S^+ be any closed surface in $M^+(k)$ which is normal w.r.t. some essential disc, D, in M. D cannot be disjoint to k, by the handle-addition lemma. Set

$$I := U(k) \cap \partial D = (\partial M - \partial M^+(k))^-.$$

A *compression-pencil (for S^+)* is defined to be a simple arc, T, in D with $T \cap (S^+ \cup I) \subset \partial T$ which joins a component, I_0, from I with some component, l, from $S^+ \cap D$ with $l \cap I_0 = \emptyset$. Fix a normal-direction of the surface S^+ in $M^+(k)$. Then we may distinguish between a positive and a negative side of S^+. A compression-pencil for S^+ is called *positive* (resp. *negative*), if it meets S^+ from its positive (resp. negative) side. Compression-pencils, T, give rise to *compression-shifts*. By this we mean an isotopy of S^+ in $M^+(k)$ whose result is given by

$$\sigma_T(S^+) := (S^+ - U(T \cup B))^- \cup B'.$$

Here B denotes a disc in the 2-handle $E := (M^+(k) - M)^-$ which is disjoint to S^+, parallel to $(\partial E - \partial M)^-$ and contains the point $T \cap I_0$. Furthermore, $U(T \cup B)$ is the regular neighborhood in $M^+(k)$, and B' is that component of $\partial U(T \cup B) - S^+$ whose intersection with the 2-handle E is non-empty. Note that it is a simple matter to describe σ_T in terms of circle-patterns.

Using the operator σ_T, we may now formulate our criterion for incompressibility:

20.2. Theorem. *Let $M^+(k)$ and S^+ be given as above. Then S^+ admits a positive compression-disc in $M^+(k)$ if and only if $S_T := M \cap \sigma_T(S^+)$ admits a positive compression-disc in M, for some positive compression-pencil, T, of S^+. The same with negative compression-discs.*

Remarks. (1) Recall a positive (resp. negative) compression-disc meets S^+ from the positive (resp. negative) side.

(2) Note that S_T admits a negative compression-disc, if T is a positive compression-pencil for S^+. In particular, S_T is compressible in M (i.e., $\sigma_T(S^+)$ is not normal in $M^+(k)$).

Before we enter the actual proof of this theorem, we first point out a consequence of it. Recall that there is an efficient procedure which isotopes any given normal surface into a strongly normal surface. So the algorithm below describes a procedure for deciding whether a given surface in a one-relator 3-manifold is incompressible.

20.3. Corollary. *Let* $\underline{D}$ *be a full one-relator Heegaard-diagram such that* $M^+(\underline{D})$ *is irreducible and* ∂*-irreducible. Then there is a* $\kappa(\underline{D})$ *and an algorithm,* A*, such that, for every closed and strongly normal surface* $S^+(C)$*,* $C \in C^+(\underline{D})$*, the algorithm* A *decides in less than* $\kappa(\underline{D})$ *steps whether* $S^+(C)$ *is incompressible in* $M^+(\underline{D})$*.*

Proof of Corollary. Observe that the compression-pencils for $S^+(C)$ are easy to enumerate. In fact, for every $C \in C^+(\underline{D})$, the total number of them is less than 2^n, where n is the complexity, $c(\underline{D})$, of the diagram $\underline{D}$ (see section 6). Now it can be decided in constant time whether S_T admits a positive compression-disc in $M(\underline{D})$ (see the algorithm from cor. 12.21). Thus thm. 20.2 gives rise to the desired algorithm (for deciding whether a given closed, normal surface in $M^+(\underline{D})$ is incompressible). $\Diamond$

Proof of Theorem. We are asked to show two directions.

First suppose there is a positive compression-pencil, T, for S^+ such that the surface $S_T = M \cap \sigma_T(S^+)$ admits a positive compression-disc, say A, in M. ∂A can be isotoped in $\sigma_T S^+$ into a curve, l, in $S^+ \cap \sigma_T S^+ \subset S^+$ since $(\sigma_T S^+ - S^+)^-$ is a disc. But $(\sigma_T S^+ - S^+)^-$ is parallel to the disc $(S^+ - \sigma_T S^+)^-$. So it follows that l is contractible in S^+ if and only if ∂A is contractible in $\sigma_T S^+$. Since every isotopy of ∂A extends to an isotopy of the compression-disc A, it follows that S^+ has a positive compression-disc. This proves one direction of thm. 20.2.

For the other direction, we extend the proof of thm. 19.2. In particular, we will freely use the notion introduced there; notably the notion of "type I, II or III arcs".

To begin with, let A be any positive compression-disc for S^+ in $M^+(k)$. As in the proof of thm. 19.2, we here may again suppose that $A \cap D$ consists of arcs alone *and* that these arcs fall into three types, as given by the type I, type II and type III arcs. Using the disc A, it is our goal to show that a certain type I arc from $A \cap D$ is indeed a positive compression-pencil for S^+ satisfying the conclusion of thm. 20.2.

By what has been just indicated, the proof naturally falls into two parts. Firstly, it has to be demonstrated that S^+ is already incompressible in $M^+(k)$, if there are no type I arcs of $A \cap D$ at all (a fact which also deserves some interest in its own right), and secondly, a compression-pencil has to found in the presence of type I arcs.

The first part (see the proof under hypothesis (A) below) will be similar to the proof of thm. 19.2. But we have to be careful since it is presently not enough to search for good pencils. Instead we have to search for certain nice components of I $(= U(k) \cap \partial D)$, which we call "special".

More precisely, let I_0 be a component of I and let I_1, I_2 denote the components of I neighboring I_0 in ∂D. Then I_0 is called a *special interval* *(for S^+)*, provided the following three conditions hold:

(1) If t is any arc from $S^+ \cap D$ with one end-point in I_0, then the other one lies in $I_1 \cup I_2 \subset I$.

(2) If t is a type III arc from $A \cap D$ with one end-point in I_0, then the other one lies in $I_1 \cup I_2 \subset I$.

(3) If t is a type I arc from $A \cap D$ with one end-point in I_0, then the other one lies in a component l of $S^+ \cap D$ with $l \cap I_1 \neq \emptyset$, say.

20.4. Example.

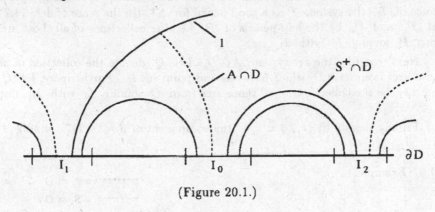

(Figure 20.1.)

The following observation will be crucial for our purpose:

20.5. Lemma. *In the notation above, a special interval always exists.*

Proof of Lemma. The proof is an extension of the argument used for lemma 14.4. Indeed, we may construct a special interval in the following three steps.

First, define a disc, D_1, in D as follows: if every arc from $S^+ \cap D$ is good in the sense that it joins neighboring components of I, then set $D_1 := D$. If not, there is at least one bad arc, t, from $S^+ \cap D$ which separates a disc from D containing no other bad arcs besides t. Let D_1 be that disc.

Next, define a disc, D_2, in D_1 as follows: if every type III arc from $A \cap D_1$ is good in the sense that it joins neighboring components of I, then set $D_2 := D_1$. If not, there is at least one bad type III arc, t, from $A \cap D_1$ which separates a disc from D_1 containing no other bad type III arcs besides t. Let D_2 be that disc.

Finally, let I_2 be a component from I contained in D_2. If I_2 is not already special, then there are type I arcs, t_1, t_2, joining I_1 with two different arcs, s_1, s_2, from $S^+ \cap D_2$. $s_2 \cup t_2$ splits D_2 into three discs. Let D_3 be that one of them containing $s_1 \cup t_1$. Observe that s_1 ends in at least one component, I_3, from I with the property that no type I arc joins I_3 with $\partial D_3 - \partial D_2$. Now consider $I_3 \subset D_3$ instead of $I_2 \subset D_2$ and apply the above process again and so on. This procedure has to stop after finitely many steps since $S^+ \cap D$ is finite. So it results in a special interval.

This proves lemma 20.5.

To continue the proof of thm. 20.2, let I_0 be any special interval from I as established in the previous lemma.

It is next convenient to single out various sub-systems of $S^+ \cap D$ and $A \cap D$ associated to the special interval I_0.

We begin with the arc-system $S^+ \cap D$. Let P denote the sub-system of all those arcs from $S^+ \cap D$ which have one end-point in I_0. Then, by our choice of I_0, the system P is a good pencil for S^+ (in the sense of def. 14.2). Let P_1 and P_2 be the half-pencils of P, i.e., the collections of all those arcs from P joining I_0 with I_1, resp. I_2.

Next, consider the arc-system $A \cap D$. Let Q denote the collection of all those arcs from $A \cap D$ which have one end-point in I_0. Furthermore, let Q_1 and Q_2 be the collections of all those arcs from Q joining I_0 with I_1, resp I_2.

Finally, denote by r_j, $j = 1, 2$, the component of $(\partial D - I)^-$ joining I_0 with I_j.

20.6. Example.

$$\cdots\cdots = A \cap D$$
$$\textemdash = S^+ \cap D$$

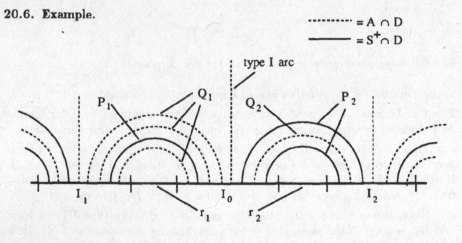

(Figure 20.2.)

The various possible positions of the above sub-system will lead to a number of case-distinctions which we then will have to treat one by one. It will turn out that each one of these cases can be reduced to the following result (in that we deduce consequences contradicting this very result) which in turn technically provides a link between the separating case considered in the last section and the non-separating case considered here.

20.7. Lemma. *Let* $A_1, ..., A_m$ *resp.* $B_1, ..., B_n$ *denote the components of* $A \cap E$ *resp.* $S^+ \cap E$. *Then*

$$A_i \cap (I_0 \cup I_1) \cap \partial Q_1 = \emptyset, \quad \text{for some } 1 \leq i \leq m,$$

provided

$$B_j \cap (I_0 \cup I_1) \cap \partial P_1 \neq \emptyset, \quad \text{for every } 1 \leq j \leq n.$$

Proof of Lemma. Assume the converse. Then $A_i \cap (I_0 \cup I_1) \cap \partial Q_1$ as well as $B_j \cap (I_0 \cup I_1) \cap \partial P_1$ are both non-empty, for every i and j. ∂r_1 cannot be contained in one component of $\partial U(k) = \partial(E \cap \partial M)$, for otherwise Q_1 consists of recurrent and essential arcs in $(A - E)^-$. This in turn, however, is impossible since not every curve from $\partial(A - E)^- - \partial A$ can contain both end-points of some essential arc in $(A - E)^-$ $((A - E)^-$ is planar).

By what we have just seen, r_1 has to join the two components of $\partial U(k)$. It follows that $k^* := \partial U(U(k) \cup r_1)$ defines a simple closed curve in ∂M which is separating in ∂M.

We now argue as in the proof of prop. 15.11. First, observe that the components of ∂S, namely $\partial B_1, ..., \partial B_n$, come in pairs each of which joined by an arc from the half-pencil P_1. In the same way, the components of $\partial(A - E)^- - \partial A$, namely $\partial A_1, ..., \partial A_n$, come in pairs each of which joined by an arc from Q_1. Splitting $(S^+ - E)^- = S$ and $(A - E)^-$ along these arcs and pushing them along D and towards the arc r_1, we obtain surfaces S' and A' from S and A such that $\partial S'$ as well as $\partial A' - \partial A$ is contained in $U(k^*)$, where the regular neighborhood $U(k^*)$ is taken in ∂M. Attaching a 2-handle, E^*, along the curve k^*, we obtain a new (generalized) one-relator 3-manifold, $M^+(k^*)$. Attaching appropriate discs in E^* to all components from $\partial S'$ and $\partial A' - \partial A$ contained in $U(k^*)$, we obtain a surface, S^*, homeomorphic to S^+ and a disc, A^*, with $A^* \cap S^* = \partial A^* = \partial A$. Furthermore, S^* and A^* are contained in $M^+(k^*)$, and it is easily checked that $S^* \cap M$ is incompressible in M since $S^+ \cap M$ is.

Thus we are precisely in the separating case as considered before since k^* is separating in ∂M and since $M^+(k^*)$ is irreducible and boundary-irreducible (see lemma 15.13 and the handle-addition lemma).

It follows that ∂A^* bounds a disc in S^*, and this disc, together with A^*, bounds a 3-ball, W^*, in $M^+(k^*)$ since $M^+(k^*)$ is irreducible. If we now reverse the construction which turned S^+, A into S^*, A^*, respectively, we not only obtain the latter surfaces back, but also obtain a ball, W, from W^* with $A \cap W = A \cap \partial W = A$ and $(\partial W - A)^- = W \cap S^+$. Thus A is not a compression-disc. But this contradicts our choice of A. Lemma 20.7 is therefore established.

As indicated above we will have to consider various cases. We first treat all of them under the following

Hypothesis (A). No type I arc from $A \cap D$ ends in I_0.

Case 1. P equals P_1.

This Case is inconsistent with the following lemma. Indeed, lemma 20.9 contradicts lemma 20.7 since we are in Case 1.

20.8. Lemma. *For every component, A_i, of $A \cap E$, there is a type III arc from $A \cap D$ joining $A_i \cap I_0$ with I_1.*

Proof of lemma. Let A_i be any component of $A \cap E$, and let t_i denote the type III arc from $A \cap D$ which ends in $A_i \cap I_0$ (this arc exists by hypothesis). Assume t_i does not join I_0 with I_1. Then it has to join I_0 with I_2 (see our choice of I_0).

If r_2 joins the two components, a_1, a_2, of $\partial U(k) = \partial(E \cap \partial M)$, it follows from the existence of the arc t_i that ∂A_i can be pushed in $U(k)$ into a_1 as well as into a_2 without meeting P (we are in Case 1). Thus $S^+ \cap U(k) = \emptyset$. So $S^+ \subset M$ which is excluded.

If r_1 joins the two components of $\partial U(k)$, note that ∂A_i can still be pushed in $U(k)$, without meeting P, into that component from $\partial U(k)$ containing $r_2 \cap I_0$. It follows that $A_i \cap I_1$ can be joined in I_1 with $r_1 \cap I_1$. without meeting P. So $A_i \cap I_0$ has to be the end-point of some type III arc of $A \cap D$ joining I_1 with I_0 (recall $P \neq \emptyset$), and we are done.

Thus w.l.o.g. r_1 as well as r_2 have both their end-points in one component of $\partial U(k)$. Then it follows that Q consists of recurrent and essential arcs in $(A - E)^-$. Hence every component of $\partial(A - E)^- - \partial A$ contains both the end-points of at least one essential arc in $(A - E)^-$. But this is impossible since $(A - E)^-$ is a planar surface.

Altogether, lemma 20.8 is established.

Case 2. $\#P_1 > \#P_2$, where as usual $\#P_i$ denotes the number of all components of P_i.

We may distinguish between the following sub-cases:

Sub-Case (a). ∂r_1 as well as ∂r_2 lies in one component of $\partial U(k)$.

In this Sub-Case $Q = Q_1 \cup Q_2$ consists of recurrent and essential arcs in the surface $(A - E)^-$ which in turn is impossible since $(A - E)^-$ is planar.

Sub-Case (b). ∂r_1, but not ∂r_2, lies in one component of $\partial U(k)$.

Let A_i be any component of $A \cap E$, and let t_i be that arc from Q which ends in $A_i \cap I_0$. Then either t_i lies in Q_1 or in Q_2. If it lies in Q_1, then t_i is recurrent in $(A - E)^-$ since ∂r_1 lies in one component of $\partial U(k)$. If it lies in Q_2, then it joins the point $A_i \cap I_0$ with the end-point $A_j \cap I_2$, for some component A_j of $A \cap E$. Notice that the point $A_j \cap I_2$ can be joined in I_2 with $r_2 \cap I_2$ by an arc which intersects P_2, and so S^+, in at most $\#P_2$ points. Since we are in Sub-Case (b), it follows that $r_2 \cap I_2$ lies in the same component of $\partial U(k)$ as ∂r_1. In particular, the point $A_j \cap I_0$ can be joined in I_0 with the point $r_1 \cap I_0$ by an arc which intersects S^+, and so P_1, in at most $\#P_2$ points. But $\#P_2 < \#P_1$ since we are in Case 2, and so there is an arc, t_j, from Q_1 which ends in $A_j \cap I_0$.

By what we have seen so far, there is system $T \subset Q$ of pairwise disjoint, essential arcs in $(A - E)^-$ with the following property: for every component, ∂A_i, of $\partial(A - E)^- - \partial A$, there is an arc, t_i, from T such that either ∂t_i lies in ∂A_i, or t_i joins ∂A_i with a component, ∂A_j, for which there is an arc, t_j, from T with $\partial t_j \subset \partial A_j$.

Since $(A - E)^-$ is a planar surface, there has to be at least one arc, t_0, from T which separates an annulus from $(A - E)^-$. Since every component from $\partial(A - E)^- - \partial A$ intersects D in the same number of points, i.e., contains the same number of points from ∂Q, this annulus contains an arc from Q. Such an arc has to be inessential in $(A - E)^-$ which, however, is impossible.

Sub-Case (c). ∂r_1 is not contained in one component of $\partial U(k)$.

In this Sub-Case we already know that $B_j \cap (I_0 \cup I_1) \cap \partial P_1$ is non-empty, for every component, B_j, of $S^+ \cap E$. Thus, by lemma 20.7, there is a component, A_i, of $A \cap E$ with $A_i \cap (I_0 \cup I_1) \cap \partial Q_1 = \emptyset$.

Let $y_i := A_i \cap I_0$. Then y_i lies in Q_2. So it can be joined in I_0 with the point $r_2 \cap I_0$ by an arc which intersects P, and so S^+, in at most $\#P_2$ points. Now, observe that the points $r_1 \cap I_1$ and $r_2 \cap I_0$ both lie in the same component of $\partial U(k)$ since we are in Sub-Case (c). It follows that the point $A_i \cap I_1$ can be joined in I_1 with the point $r_1 \cap I_1$ by an arc which intersects S^+, and so P_1, in at most $\#P_2$ points. But again $\#P_2 < \#P_1$ since we are in Case 2, and so $A_i \cap I_1$ is contained in ∂Q_1, which contradicts our choice of A_i.

Altogether, we have seen that Case 2 cannot occur.

Case 3. $\#P_1 = \#P_2$ and $P_1 \neq P$.

W.l.o.g. we may suppose the indices have been chosen in such a way that $\#Q_1 \geq \#Q_2$.

Sub-Case (a). ∂r_1 as well as ∂r_2 lies in one component of $\partial U(k)$.

This Sub-Case can be excluded in the same way as Sub-Case (a) from Case 2.

Sub-Case (b). ∂r_1, but not ∂r_2, lies in one component of $\partial U(k)$.

We modify the argument of Sub-Case (b) of Case 2.

As in that Sub-Case, let again A_i be any component of $A \cap E$, and let t_i be that arc from Q which ends in $A_i \cap I_0$. If this arc lies in Q_1, it is recurrent. If it lies in Q_2, it joins the point $A_i \cap I_0$ with the point $A_j \cap I_2$, for some component, A_j, of $A \cap E$. This time note that the point $A_j \cap I_2$ can be joined in I_2 with $r_2 \cap I_2$ by an arc whose interior intersects Q_2, and so A, in strictly less than $\#Q_2$ points. Since we are in Sub-Case (b), it follows that $r_2 \cap I_2$ lies in the same component of $\partial U(k)$ as ∂r_1. In particular, the point $A_j \cap I_0$ can be joined in I_0 with the point $r_1 \cap I_0$ by an arc whose interior intersects A, and so Q_1, in strictly less than $\#Q_2$ points. But $\#Q_2 \leq \#Q_1$, and so there is an arc, t_j, from Q_1 which ends in $A_j \cap I_0$. Now, we may conclude the argument in exactly the same way as in Sub-Case (b) of Case 2.

Sub-Case (c). ∂r_1 is not contained in one component of $\partial U(k)$.

Since $\#P_1 \geq \#P_2$ and $\#Q_1 \geq \#Q_2$, it follows from the counting argument used above that, for every component A_i and B_j from $A \cap E$, resp. $S^+ \cap E$, the sets

$$A_i \cap (I_0 \cup I_1) \cap Q_1, \text{ and } B_j \cap (I_0 \cup I_1) \cap P_1$$

are both non-empty. This, however, contradicts lemma 20.7.

It remains to check the same cases as above, but this time under the new

Hypothesis (B). At least one type I arc from $A \cap D$ ends in I_0.

Let R denote the union of all type I arcs from $A \cap D$ which end in I_0. Since I_0 is a special interval for I, there is a component, l, of $S^+ \cap D$ with $l \cap I_1 \neq \emptyset$ and such that any arc from R has the other end-point in l.

The complexity of the situation at hand can be reduced considerably by simply noting that ∂r_1 cannot lie in one component of $\partial U(k)$. That this is excluded indeed, follows exactly as in Case 2 of the proof of thm. 19.2.

We now distinguish between the following cases.

Case 1. $\#P_2 \leq \#P_1$.

We claim $\#(Q_2 \cup R) \leq \#Q_1$. To see this claim, let t_i be any component from $Q_2 \cup R$, and let A_i be that component of $A \cap E$ containing $t_i \cap I_0$. Then note that the point $A_i \cap I_0$ can be joined in I_0 with the point $r_2 \cap I_0$ by an arc which intersects P_2, and so S^+, in at most $\#P_2$ points. Since ∂r_1 is not contained in one component of $\partial U(k)$, it follows that $r_2 \cap I_0$ and $r_1 \cap I_1$ lie in the same component of $\partial U(k)$. In particular, the point $A_i \cap I_1$ can be joined in I_1 with the point $r_1 \cap I_1$ by an arc which intersects S^+ in at most $\#P_2$ points. But $\#P_2 \leq \#P_1$. Furthermore, l as well as every component of P_1 has one end-point in I_1. It follows that the point $A_i \cap I_1$ is end-point of some component, s_i, of $A \cap D$ which lies in that disc separated from $l \cup R$ containing P_1. Thus s_i is a component of Q_1. The assignment $t_i \mapsto s_i$ defines an injection from the set of components of $Q_2 \cup R$ to that of Q_1, proving the claim.

Since $\#(Q_2 \cup R) \leq \#Q_1$ and since $\#P_2 \leq \#P_1$, it is easily verified that the components of ∂S as well as those from $\partial(A - E)^- - \partial A$ come in pairs each of which joined by an arc from P_1 resp. Q_1. Thus we get a contradiction to lemma 20.7.

Therefore Case 1 cannot occur.

Case 2. $\#P_1 < \#P_2$.

We may distinguish between the following sub-cases:

Sub-Case (a). ∂r_2 is not contained in one component of $\partial U(k)$.

Let A_i be a component of $A \cap E$ which contains a point of $R \cap I_0$. Then note that the point $A_i \cap I_0$ can be joined in I_0 with the point $r_1 \cap I_0$ by an arc which intersects P_1, and so S^+, in precisely $\#P_1$ points. Since neither ∂r_1 nor ∂r_2 is contained in one component of $\partial U(k)$, it follows that $r_1 \cap I_0$ and $r_2 \cap I_2$ lies in the same component of $\partial U(k)$. In particular, the point $A_i \cap I_2$ can be joined in I_2 with the point $r_2 \cap I_2$ by an arc which intersects S^+, and so P_2, in $\#P_1$ points. But $\#P_1 < \#P_2$ since we are in Case 2. Thus there is an arc, t_i, from Q_2 which ends in $A_i \cap I_2$. Let A_j be that component of $A \cap E$ which contains $t_i \cap I_0$, and note that $r_1 \cap I_1$ and $r_2 \cap I_0$ lie in the same component of $\partial U(k)$. It follows that the point $A_j \cap I_1$ can be joined in I_1 with the point $r_1 \cap I_1$ by an arc which intersect S^+ in $\#P_1$ arcs. Thus there is an arc, t_j, from Q_1 which ends in $A_j \cap I_1$. Observe that $t_j \cap I_0$ and $A_i \cap I_0$ can be joined in I_0 without meeting S^+. Thus, if we denote by Q' the union of all those arcs from Q which end in $\partial A_i \cap (I_1 \cup I_0 \cup I_2)$, it follows that $Q' \cap Q$ consists of infinitely many arcs. This is certainly impossible.

Sub-Case (b) ∂r_2 lies in one component of $\partial U(k)$.

It is this Sub-Case in which we finally have to construct the required compression-pencil. This construction in turn is based on the following result.

20.9. Lemma. *Every component of $A \cap E$ is joined in $(A - E)^-$ with ∂A by precisely one component of the arc-system R.*

Remark. Recall that R is defined to be the union of all type I arcs from $A \cap D$ which end in I_0.

Proof of Lemma. We modify the argument from Sub-Case (b) of Case 2 in hypothesis (A).

As there let again A_i be any component of $A \cap E$, and suppose that $(A_i \cap I_0) \cap R = \emptyset$. Let t_i be that arc from $A \cap D$ which ends in $A_i \cap I_0$. Then this arc either lies in Q_1 or in Q_2 (recall I_0 is special). If t_i lies in Q_2, it is recurrent. If it lies in Q_1, it joins the point $A_i \cap I_0$ with the point $A_j \cap I_1$, for some component A_j of $A \cap E$. Note that the point $A_j \cap I_1$ can be joined in I_1 with the point $r_1 \cap I_1$ by an arc which intersects P_1, and so S^+, in at most $\#P_1$ points. Since we are in Sub-Case (b) , it follows that $r_1 \cap I_1$ lies in the same component of $\partial U(k)$ as ∂r_2. In particular, the point $A_j \cap I_0$ can be joined in I_0 with $r_2 \cap I_0$ by an arc which intersects S^+, and so P_2, in at most $\#P_1$ points. But $\#P_1 < \#P_2$, and so there has to be an arc, t_j, from Q_2 which ends in $A_j \cap I_0$. This arc t_j is recurrent since we are in Sub-Case (b) .

By what we have seen so far, we have the following conclusion: if $Q \neq \emptyset$, then, for every component ∂A_i of $\partial (A - E)^- - \partial A$ with $R \cap \partial A_i = \emptyset$, there is an arc, t_i, from Q such that either ∂t_i lies in ∂A_i, or t_i joins ∂A_i in $(A - E)^-$ with some component ∂A_j of $\partial (A - E)^-$ which in turn contains both the end-points of some arc from Q.

We claim $Q = \emptyset$. Assume the converse. Then, by the previous conclusion, there has to be one arc, say t_0, from Q which separates an annulus from $(A - E)^-$, for $(A - E)^-$ is a planar surface. Since every component of $\partial (A - E)^- - \partial A = \bigcup \partial A_i$ intersects the disc D in the same number of points, i.e., contains the same number of points of $\partial (A \cap D)$, this annulus necessarily contains an arc from $A \cap D$ which is inessential in $(A - E)^-$. This, however, is excluded, proving the claim.

Since every component, A_i, of $A \cap E$ intersects I_0 (k is connected), lemma 20.10 now follows from the fact that Q is shown to be empty.

We may now finish the proof of thm. 20.2. For this let T be any arc in D with $T \cap S^+ \subset \partial T$ which joins I_0 with l and which is disjoint to R. We claim T is the required compression-pencil.

Indeed, T is positive since every arc from R joins I_0 with $l \subset S^+ \cap D$ and A is supposed to be a positive compression-disc for S^+. Now, recall the definition of the surface $S^* := \sigma_T(S^+)$. It follows that A is a positive compression-disc for S^* as well and that every arc from R is an arc which joins I_0 with some component of $S^* \cap D$ which too ends in I_0. Thus, by lemma 20.10, an isotopy of A which pushes the arcs from R towards I_0 deforms A into a disc in M which then is the required positive compression-disc for $\sigma_T(S^+)$.

Altogether, this proves thm. 20.2. $\Diamond$

§21. The Haken Spectrum.

In this section we discuss issues centered around the Haken spectrum for
one-relator 3-manifolds. Recall that the Haken spectrum is an arithmetic function which counts essential surfaces. We will show that this function is computable, we will determine its growth rate and we will estimate the occurrence
of its first non-trivial value. We will achieve all this by reducing the related
questions to questions about relative handlebodies which in turn have been an-
swered in chapter II. For this we need to estimate the intersection of essential
surfaces with the attached 2-handle of the one-relator 3-manifold. We already
know from prop. 15.12 that this intersection is empty for non-separating surfaces of maximal Euler characteristic. Here we show that the intersection for all
other surfaces is also limited. The basic technical tool will be the existence of
"Heegaard cylinders". We begin by introducing this concept first.

21.1. *Heegaard-Cylinders.*

As mentioned above we are interested in the intersection of essential surfaces
in one-relator 3-manifolds with the attached 2-handle. Equivalently, we are
interested in the intersection of essential surfaces and Heegaard strings. Now,
in contrast to the intersection of two essential surfaces, the intersection of an
essential surface with a curve can of course be arbitrary large. It turns out
however that the intersection of an essential surface with a Heegaard string
may be interpreted as an intersection of surfaces, namely the intersection of
the surface in question with the Heegaard cylinders associated to the Heegaard
string.

21.2. Definition. *Let M^+ be a one-relator 3-manifold and let t be a Heegaard-string in M^+. Then a <u>Heegaard cylinder</u> (for t) is a non-contractible annulus A in M^+, $A \cap \partial M^+ = \emptyset$, such that $A \cap t$ is a non-empty system of one or two arcs which are essential in A. A disjoint union of finitely many Heegaard cylinders is called a <u>Heegaard cylinder-system</u>.*

21.3. Example.

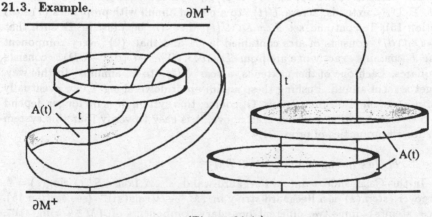

(Figure 21.1.)

21.4. Proposition. *Let M^+ be an irreducible and ∂-irreducible one-relator 3-manifold. Then, to every Heegaard-string, t, in M^+ , there is associated a Heegaard cylinder-system, $\mathcal{A}(t)$, with the property that every orientable and essential 2-manifold, S^+ , in M^+ can be isotoped into a 2-manifold, S^* , so that afterwards*

(1) $S^* \cap \mathcal{A}(t)$ is contained in the union of at most two components of $\mathcal{A}(t)$,

(2) $S^* \cap \mathcal{A}(t)$ consists of closed curves which are essential in both S^* and $\mathcal{A}(t)$, and

(3) $S^* \cap t \subset S^+ \cap t$ (and so $\#(S^* \cap t) \leq \#(S^+ \cap t)$).

In addition, $\mathcal{A}(t)$ with (1)-(3) exists for every strongly normal surfaces, S^+ , in M^+ .

Proof. Without loss of generality we may suppose that S^+ is strongly normal since every essential 2-manifold in an irreducible one-relator 3-manifold can be isotoped into a strongly normal one. Thus the mere existence of a pair of Heegaard cylinders with properties (1) - (3) follows directly from prop. 15.20 and 15.21. But this pair of Heegaard-cylinders depends on the choice of the 2-manifold S^+ . With a little more care the Heegaard cylinder-system can be made independent of this choice. To explain this we distinguish between two cases:

Case 1. t joins two different components of ∂M^+ .

Recall from section 15 the way the above mentioned pair of Heegaard cylinders for S^+ is constructed: S^+ is strongly normal. More precisely, there is an essential disc, D, in $(M^+ - U(t))^-$ such that S^+ is strongly normal w.r.t. D. Since M^+ is ∂-irreducible, it follows that $D \cap U(t) \neq \emptyset$. In particular, there is a good pencil, P, from $S^+ \cap D$ in D whose total number of arcs equals $\#(S^+ \cap t)$. Let P_1, P_2 be the two half-pencils of P. Now, there are squares, B_i , $i = 1, 2$, in D such that $B_i \cap P = P_i$, $(\partial B_i - \partial D)^- \subset P_i$ and $B_i \cap \partial D$ consists of two opposite faces of B_i . Since we are in Case 1, it follows that $B_1 \cup B_2$ extends, across $U(t)$, to a pair of annuli with properties (1) - (3) (section 15). To continue, set $I := D \cap U(t)$. Let D' be a disc in D such that (1) $D' \cap \partial D$ consists of arcs contained in I and that (2) every component from I contains exactly one arc from $D' \cap \partial D$. Then $U(\partial D' - \partial D)^-$ consists of squares. Each one of them extends, across $U(t)$, to an annulus. In this way, we get a *set* of annuli. Pushing these annuli up or down along t, we eventually obtain a *system* of disjoint annuli. Of course, this system does no longer depend on S^+ and, by what we have seen above, it is easy to verify that this system satisfies all properties of prop. 21.4.

Case 2. ∂t lies in one component of ∂M^+ .

In this Case, slide t into a Heegaard-wand s. W.l.o.g. $S^+ \cap ring(s) = \emptyset$. Moreover, stem(s) is a Heegaard-string in $(M^+ - U(ring(s)))^-$ (see section 15). Since stem(s) joins two different boundary components of $(M^+ - ring(s))^-$, it follows from Case 1 the existence of a Heegaard cylinder-system for stem(s). Sliding s back to the Heegaard-string t, we obtain from this Heegaard cylinder-system a Heegaard cylinder-system for t. This finishes the proof of prop. 21.4. $\Diamond$

21.5. *Estimates.*

We begin with estimates concerning the characteristic submanifold in one-relator 3-manifolds.

21.6. Proposition. *Let M^+ be an irreducible and ∂-irreducible one-relator 3-manifold and let t be a Heegaard-string in M^+. Let V be the characteristic submanifold of M^+. Suppose V is isotoped so that $\#(t \cap (\partial V - \partial M^+)^-)$ is minimal. Then $\#(t \cap (\partial V - \partial M^+)^-) \leq 8$.*

Remark. This is not the best possible estimate. The point is rather that this estimate does not depend on the choice of the Heegaard string.

Proof. Let $\mathcal{A}(t)$ be a Heegaard cylinder-system as given by prop. 21.4. Set $S^+ := (\partial V - \partial M^+)^-$. Then S^+ is an essential 2-manifold in M^+. According to prop. 21.4, there is a pair, $\mathcal{A}$, of annuli from $\mathcal{A}(t)$ such that $A \cap S^+$ is a system of curves which are essential both in A and S^+ and which contains $S^+ \cap t$. Let A be one of the components from $\mathcal{A}$. It remains to show that $\#(A \cap S^+) \leq 4$. Assume the converse. Let $A_1, A_2, ..., A_n$, $n \geq 6$, be all the annuli in which A is split by S^+, and let the indices be chosen so that $A_i \cap A_{i+1} \neq \emptyset$, $1 \leq i < n$. Then A_2, A_4 (resp. A_3, A_5) are essential annuli in V. But then A_3 (resp. A_4) is an essential annulus in $(M^+ - V)^-$. This, however, is impossible since characteristic submanifolds are full [Joh 2, def. 8.2]. $\Diamond$

21.7. Proposition. *Let M^+ and t be given as in 21.6. Let V be the characteristic submanifold of M^+. Then*

(1) V has at most three Seifert fiber spaces, different from solid tori, and

(2) if X is a Seifert fiber space from V, then $\chi(F - U)^- \geq -32$, where F is the orbit surface of X and U is the regular neighborhood of all exceptional points in F.

Remark. Recall that an exceptional point is the image of an exceptional fiber under the fiber projection.

Proof. To prove this proposition, we are going to need the full strength of prop. 15.21 and 15.22. Let t be a Heegaard-string and set $S^+ := (\partial V - \partial M^+)^-$. As common by now, we distinguish between the following two cases:

Case 1. t joins two different components of ∂M^+.

Let V be properly isotoped in M^+ so that afterwards $\#(S^+ \cap t)$ is as small as possible. Let C, C' be the two annuli as given by prop. 15.21

(possibly C or $C' = \emptyset$). Let $C_1, C_2, ..., C_m$, $m \geq 1$, be the annuli in which C is split by S^+. Let the indices be chosen so that $C_1 \cap \partial M^+ \neq \emptyset$ and that $C_i \cap C_{i+1} \neq \emptyset$, $1 \leq i \leq m$. In the same way, let $C'_1, C'_2, ..., C'_n$, $n \geq 1$, be the annuli in which C' is split by S^+. C_i, $1 \leq i \leq m$, is an essential annulus either in V or in $(M^+ - V)^-$. Since characteristic submanifolds are full [Joh 2, def. 8.2], it follows that C_1 is contained in V. By the same reason, it then follows that C_3 cannot be contained in V. Thus $m \leq 2$. Similarly, $n \leq 2$. Let X, X' (possibly equal) be the components from V containing $C \cap \partial M^+$, $C' \cap \partial M^+$, respectively. Note that the arc $(t - C \cup C')^-$ is contained either in V or in $(M - V)^-$. Let Y be the (possibly empty) component from V containing $(t - C \cup C')^-$. Now, every Seifert fiber space from V, disjoint to t, must be a solid torus since $(M^+ - U(t))^-$ is a handlebody and since every torus in a handlebody is compressible. Moreover, every Seifert fiber space from V, meeting t, must be either X, X' or Y. This proves (1).

It remains to show property (2), i.e., we have to show that X, X' and Y satisfy (2) provided they are Seifert fiber spaces.

Consider Y. Assume Y is a non-empty Seifert fiber space (possibly equal to X or X') but not one as in (2). Then Y contains an incompressible torus, T, which is disjoint to $C_1 \cup C'_1$ and which is not ∂-parallel in Y. Now, T has to intersect $(t - C \cup C')^-$ since $(M^+ - U(t))^-$ is a handlebody and since T is incompressible. Thus, applying 15.21 to the essential 2-manifold $T \cup (\partial V - \partial M^+)^-$, we see that there must be an annulus, B, in Y which joins T with $C \cap \partial Y$ (resp. $C' \cap \partial Y$). But the annuli C_1 and B are both vertical in V, and so there is a component of $(C - V)^-$ which is an essential annulus in $(M - V^+)^-$ and whose boundary lies in fibers of Seifert fiber spaces from V. This, however, is impossible since V is full.

Consider X. W.l.o.g. $X \neq Y$, for otherwise we are done. Then $Z := (X - U(C_1 \cup C'))^-$ consists of essential Seifert fiber spaces in $(M^+ - U(t))^-$. But $(M^+ - U(t))^-$ is handlebody. Thus Z has to be a union of solid tori. It follows that X is a Seifert fiber space as described in (2). The same for X'.

Case 2. ∂t lies in one component from ∂M^+.

Slide t into a Heegaard-wand s with the properties from prop. 15.22. Set $W := \text{ring}(s)$. Then we may suppose that $S^+ \cap W = \emptyset$. Now, $\text{stem}(s)$ is a Heegaard-string in $(M^+ - W)^-$ which joins ∂M^+ with ∂W, i.e., it joins two different boundary components from $(M^+ - W)^-$. Thus we are in the situation of prop. 15.21 again. Let C, C' be the two annuli in $(M^+ - W)^-$ as given by that proposition. Let C, C' be chosen so that, in addition, $C \cap \partial W \neq \emptyset$ and $C' \cap \partial M^+ \neq \emptyset$.

Suppose $C = \emptyset$. Then the proof may be finished as in Case 1.

Suppose $C \neq \emptyset$. Let $C_1, C_2, ..., C_m$, $m \geq 1$, be the annuli in which C is split by S^+. Suppose the indices are chosen so that, in addition, $C_i \cap C_{i+1} \neq \emptyset$, $1 \leq i < m$, and $C_1 \cap W \neq \emptyset$. $U(S^+ \cup C_1 \cup W)$ is an essential Seifert fiber space in V or $(M^+ - V)^-$. But it cannot lie in $(M^+ - V)^-$ since characteristic submanifolds are full. Thus $U(C_1 \cup W) \subset V$. More precisely, it is a vertical Seifert fiber space in some Seifert fiber space from V. It follows that $\tilde{V} := (V - W)^-$ is the characteristic submanifold of the one-relator 3-manifold $\tilde{M} := (M^+ - W)^-$. Applying the argument from Case 1, we find that $\tilde{V}$ satisfies properties (1) and (2). Thus also V satisfies (1) and (2). ◊

Given a surface S, let us here denote by $c(S)$ the maximal number of pairwise disjoint, pairwise non-parallel, non-contractible and simple closed curves in S. Then

$$c(S) = 3g(S) + 2b(S) - 3 = \begin{cases} 1, & \text{if } \chi S = 0 \\ -\frac{3}{2}\chi(S), & \text{otherwise} \end{cases}$$

where $g(S)$, $b(S)$ denote the genus resp. the total number of boundary curves of S.

21.8. Proposition. *Let M^+ be an irreducible and ∂-irreducible one-relator 3-manifold and let s be a Heegaard-string in M^+. Let S^+ be an essential surface in M^+. Then there is an isotopy, α_t, and a product, h, of Dehn twists along essential tori in $M^+(k)$ such that*

$$\#(h\alpha_1(S^+) \cap s) \leq 32 \cdot c(S^+).$$

Remark. If S^+ is strongly normal in M^+, then the isotopy can be chosen to be constant on ∂S^+.

Proof. Let $\mathcal{A}(s)$ be a Heegaard cylinder-system as given in prop. 21.4. Let S^+ be isotoped so that the conclusion of prop. 21.4 holds and that, in addition, $\#(S^+ \cap \mathcal{A}(s))$ is as small as possible. Let $\mathcal{A}$ be the union of the two annuli from $\mathcal{A}(s)$ which meet S^+ (this union exists because of property (1) of prop. 21.4). Now, recall that there is a system of at most $c(S^+)$ curves in S^+ whose neighborhood $\mathcal{B} \subset S^+$ contains all curves from $\mathcal{A} \cap S^+$ (mod isotopy). Given $\mathcal{A}$ and $\mathcal{B}$, let X denote the union of $U(\mathcal{A} \cup \mathcal{B})$ with all solid tori from $(M^+ - X)^-$. Then X is a system of at most two Seifert fiber spaces, each one of which is either essential or a solid tori. Moreover, $\mathcal{A}$ and $\mathcal{B}$ may be viewed as two systems of proper and vertical annuli in X. More precisely, $\mathcal{A}$ consists of at most two and $\mathcal{B}$ consists of at most $c(S^+)$ annuli. Thus, given components A, B from $\mathcal{A}, \mathcal{B}$, resp., it remains to show that $\#(A \cap B) \leq 16$, mod Dehn twists. This is clear if X is a solid torus.

Suppose X is not a solid torus. Then we argue as follows. First, let $p : X \to F$ be the fiber-projection onto the orbit surface F (w.l.o.g. X is connected). Let $F' := (F - E)^-$, where E is the regular neighborhood of all exceptional points in F. Then $\chi F' \geq -3$. Indeed, X is an essential Seifert fiber space, so it can be isotoped into the characteristic submanifold of M^+ (see [Joh 1, prop. 10.8]) and therefore the claim follows from prop. 21.7. To continue set $\alpha := p(A)$ and $\beta := p(B)$. Then α and β are proper arcs in F'. But F' need not be orientable. To fix this note that at most $-\frac{3}{2} \cdot 2(\chi F'+1)+2 = 8$

arcs from $(\beta - U(\alpha))^-$ are not pairwise parallel in $(F' - U(a))^-$ (this can be seen, e.g., by first taking the double of the surface $(F' - U(\alpha))^-$ and then applying the available estimate for the total number of non-parallel *closed* curves in a *closed* surface; see above). Thus there is a system $p \subset a - \beta$ of at most 8 points such that $G = U(\beta \cup (\alpha - U(p))^-)$ is an orientable surface. Set $\alpha' = (\alpha - U(p))^-$. Then, by lemma 7.16, $\#(\alpha' \cap \beta) \leq 2 \cdot 8 = 16$ mod Dehn twists along two-sided simple closed curves. Thus the proposition follows since all Dehn twists along two-sided curves can be lifted to Dehn twists along vertical tori.

This finishes the proof of prop. 21.8. $\lozenge$

21.9. *The Seifert Characteristic.*

Recall the Seifert characteristic of a 3-manifold, M^+, is defined to be the minimum taken over the negative Euler characteristics of all essential, non-separating surfaces in M^+. The next proposition relates the Seifert characteristic, $\sigma(M^+(k))$, of the one-relator 3-manifold, $M^+(k)$, to the Seifert characteristic, $\sigma(M, k)$, of the associated relative handlebody.

21.10. Proposition. *Let $M^+(k)$ be any irreducible and ∂-irreducible one-relator 3-manifold. Then $\sigma(M^+(k)) = \sigma(M, k)$.*

Proof. This is an immediate consequence of prop. 15.12. $\lozenge$

21.11. Corollary. *Let (M_i, k_i), $i \geq 1$, be a sequence of one-relator relative handlebodies. Then $\sigma(M^+(k_i)) \to \infty$, if $\sigma(M, k_i) \to \infty$.* $\lozenge$

Remark. In section 11 we gave an algebraic estimate for $\sigma(M, k)$. Using this estimate, many examples with $\sigma(M^+(k_i)) \to \infty$, $i \to \infty$, can now be constructed (see, e.g., example 11.13). Compare this result with [Ko 5], where it is shown that $\sigma(M^+(f^i k)) \to \infty$, $i \to \infty$, for many pseudo-Anosov diffeomorphisms $f : \partial M \to \partial M$.

21.12. Remark. The previous proposition may also sometimes help to calculate vice versa the Seifert characteristic of relative handlebodies. To illustrate the latter point, consider the relative handlebodies as indicated by fig. 11.5. They are determined by pairs (p, q) of integers (counting the number of times the relator curve winds around a handle). Thus we may denote them by $M(p, q)$. As in example 11.13, we see that the Euler characteristics $\sigma(M(p, q))$ tend to infinity with p, say (it is to be understood that $(p, q) = 1$). Now, this very fact may also be verified by observing that the one-relator 3-manifold $M^+(p, q) := M^+(k)$ obtained from $M(p, q)$ is actually a Seifert fiber space. It is the Seifert fiber space over the disc with two exceptional fibers (whose monodromies are given by the intersections of k with the two meridian-discs D_1 and D_2, respectively).

To see the latter claim, simply note that the separating disc, D_3, in M can be piped to a separating annulus, A, in M. Let N_1 and N_2 be the two 3-manifolds in which the handlebody M is split by A. Then both N_1 and N_2 are solid tori, and let the indices be chosen so that ∂N_2 contain the curve

k. The separating disc D_3 cuts k into two arcs, one of them contained in the pipe. Thus the regular neighborhood of the pipe is a 1-handle which forms a cancelling pair with the attached 2-handle. It follows that the annulus A splits $M^+(k)$ into two relative solid tori, say (W_1, A_1) and (W_2, A_2). Observe further that $D_1 \cap N_1$ is the meridian-disc of the one solid torus, and that the disc D_2 can be extended across the 2-handle to a meridian-disc for the other solid torus. It follows that (N_1, A) is admissibly homeomorphic to (W_1, A_1), say. In order to describe (W_2, A_2) let M' denote that solid torus in which M is split by D_3 and which does not contain N_1, and let A' be the annulus which is the union of D_3 with the arc $k' := k \cap M'$. Then (W_2, A_2) is admissibly homeomorphic to (M', A') since $k' \cup (k - k')^-$ is the boundary of the attached 2-handle. This proves the claim.

By what has been verified, $M^+(p, q)$ is indeed the Seifert fiber space over the disc with two exceptional fibers. Any non-separating essential surface in this Seifert fiber space is known to be horizontal (see [Wa 1]). In other words, any such surface is the union of a number of copies of the meridian-discs in the solid tori, obtained from $M^+(p, q)$ by splitting along the separating essential annulus A. But, by what has been demonstrated above, those meridian-discs intersect A in p and q arcs, respectively. Thus, for a minimal surface, we have to take q, resp. p copies since $(p, q) = 1$. It follows that

$$\sigma(M(p, q)) = \sigma(M^+(p, q)) = pq.$$

In particular, we see again that $\sigma(M(p, q)) \to \infty$, for $p \to \infty$ say.

21.13. *The Haken Spectrum.*

After the previous preparation we are now ready to discuss the Haken spectrum.

Our estimates for the Thurston-norm and the Haken spectrum are straightforward consequences of prop. 21.8. To formulate them, let $N = M^+(k)$ be some irreducible and ∂-irreducible one-relator 3-manifold. Let us again denote by $\mathcal{S}_n(N)$ the set of isotopy classes of all those (orientable) essential surfaces in N whose negative Euler characteristic equals n. Let us further denote by $\mathcal{A}(N)$ denote that subgroups of the mapping class group $\pi_0 \mathrm{Diff}(N)$ generated by all Dehn twists along essential annuli and tori in N. Moreover, the complexity $c(S)$ for surfaces S (see 21.8) gives rise to an arithmetic function $c_N : \mathbf{N} \to \mathbf{N}$ defined by setting

$$c_N(n) := \max \{ c(S) \},$$

where the maximum is taken over all essential surfaces S in N with $|\chi S| = n$. Finally, for orientable and essential surface S, set

$\Theta_N(n) := \mathcal{S}_n(N)/\mathcal{A}(N)$,

$\tau^+(S) := \min\{-\chi S' \mid S'$ is an orientable surface in N homologous to $S \}$.

Then Θ_N is the *Haken-spectrum* of N and $\tau^+(S)$ is the *Thurston-norm* of S (with respect to $M^+(k)$).

Recall that the first homology group of (M, k) is canonical isomorphic to that of $M^+(k)$. In particular, any element from $H(M, k)$ may be identified with an element of $H_1(M^+(k))$ and vice versa. Moreover, recall that, by duality, there is a relative homology class of surfaces associated to any such torsion-free element. Thus the Thurston-norm is defined on the torsion-free parts of both $H(M, k)$ and $H_1(M^+(k))$. The next result tells us that these two Thurston-norms coincide and that, moreover, the Haken spectrum can be estimated from above by some polynomial.

21.14.   Theorem. *Let $N := M^+(k)$ be a one-relator 3-manifold. Let x be some torsion-free element from $H_1(M^+(k))$ and let $\tau(x), \tau^+(x)$ be the Thurston-norm of x with respect to (M, k) and $M^+(k)$, respectively. Then*

$$\tau^+(x) = \tau(x), \quad and$$
$$\Theta_N(n) \leq \Theta_{(M,k)}(n + 6 \cdot c_N(n)).$$

Remark. Recall from prop. 11.12 that τ may be estimated from below.

Proof. The first equality is a direct consequence of prop. 15.11 since neither annulus- nor torus-modifications alter the Euler characteristic of a surface. The second inequality in turn follows immediately from prop. 21.8.

Recall from thm. 13.28 that the Haken spectrum $\Theta_{(M,k)}$ can be computed in polynomial time. In particular, it is dominated by some polynomial, and so is Θ_N. $\Diamond$

21.15.   Remark. In order to actually compute the Haken spectrum, Θ_N, for a given one-relator 3-manifold $N = M^+(k)$, we need an algorithm which produces, for every $n \in \mathbf{N}$, the value $\Theta_N(n)$. Now, by thm. 13.28, $\Theta_{(M,k)}$ is computable. In particular, there is an algorithm which enumerates all essential surfaces, S, in (M, k). with $\chi S \geq 6 \cdot c_N(n)$. To every such surface, S, there is assigned a unique surface, $S^+ \subset M^+(k)$, obtained from S by capping boundary curves contained in the 2-handle. In order to check whether $S^+ \subset M^+(k)$ is essential, take the double $DS^+ \subset DM^+(k)$. Note DS^+ is closed. Observe that $S^+ \subset M^+(k)$ is essential iff $DS^+ \subset DM^+(k)$ is incompressible. It is known that the latter question can be decided algorithmically (see e.g. [JO] or chapter IV). Thus, in order to compute Θ_N, it remains to decide when two essential, capped surfaces, S_1^+, S_2^+ are isotopic in $M^+(k)$. A (very crude) algorithm for this problem can be constructed by using Waldhausen's characterization [Wa 2] of homotopic incompressible surfaces.

§22. An Example.

In [Th 4] one of the simplest examples of one-relator 3-manifolds is presented which carry the structure of a hyperbolic 3-manifold with totally geodesic boundary (this structure is actually explicitly constructed there). In particular, this example must have finite mapping class group. Here we take this example as one of the simplest examples for briefly illustrating a relevant aspect of the approach taken in this chapter. Specifically, we study the mapping class group of this example in order to point out the potential of surface-enumerations for 3-manifolds for a better understanding of mapping class groups.

We first introduce Thurston's example discussed in this section. Then we establish a result concerning the enumeration of certain of its incompressible surfaces. Finally, we use that result for a determination of its mapping class group (prop. 22.7).

22.1. Thurston's example alluded to above is the one-relator 3-manifold $M^+ = M^+(\underline{D})$ given by the Heegaard-diagram $\underline{D} = (D, h, \underline{K})$ shown in figure 22.1.

In that picture the system of circles defines the system, D, of meridian-discs of $\underline{D}$. The union of all arcs joining meridian-discs defines the arc-system, $\underline{K}$, of $\underline{D}$. Finally, let the involution $h : D \to D$ of $\underline{D}$ be defined by the rule that h interchanges A_i, A_{i+1}, resp. B_i, B_{i+1}, resp. C_i, C_{i+1}, resp. D_i, D_{i+1}, and maps end-points of $\underline{K}$ to end-points with the same label. It is further to be understood that the 2-sphere system, F, underlying the diagram $\underline{D}$, consists of two components such that each one contains precisely one component of $D \cup \underline{K}$. In particular, $\underline{D}$ is a useful and irreducible diagram (in the terminology of def. 6.2). Set $k := \underline{K}/h$. Check that k is connected (the arrows in the picture indicate a fixed orientation for k).

22.2. Proposition. *Let $\underline{D} = (D, h, \underline{K})$ denote the diagram of Thurston's example as presented previously. Let $\mathcal{S}_n(\underline{D})$ denote the set of all admissible isotopy classes of essential surfaces, S, in $M(\underline{D})$ with $S \cap k = \emptyset$ and $\chi S = n$. Then the following holds:*

(1) $\mathcal{S}_{-1}(\underline{D})$ consists of four punctured tori.

(2) If $S_1, S_2 \in \mathcal{S}_{-1}(\underline{D})$, then $S_1^+, S_2^+ \subset M^+(\underline{D})$ are isotopic in $M^+(\underline{D})$.

Remark. Recall that to every essential surface, S, there is associated a unique surface, S^+, by capping off all boundary curves contained in the attached 2-handle.

Proof. Recall from prop. 9.13 that every essential surface in $M(\underline{D})$, not meeting k, is obtained from some disc-system, $\mathcal{D}$, in $M(\underline{D})$ by pseudo-piping along k. More precisely, $\mathcal{D}$ may be taken as a system of discs which are copies of discs from a meridian-system given by $\underline{D}$. We next consider surfaces with Euler characteristic -1 and obtained by piping, rather than pseudo-piping (we will later see that this is sufficient).

To fix ideas let S be any non-separating, essential surface in $M(\underline{D})$ with $\chi(S) = -1$ and obtained by piping from some disc-system, $\mathcal{D}$, as above. Set $B = \partial D$. Then B is a collection of circles, parallel to ∂D, in the surface, F, underlying $\underline{D}$. Let $C \in \mathcal{C}^+(\underline{D})$ be the circle-pattern with $S = S(C)$. Then

22.3. Thurston's Example.

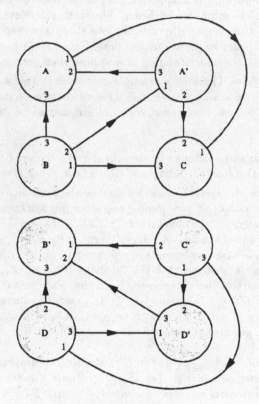

(Figure 22.1.)

$-1 = \chi(S) = \chi(C) = \#C - \frac{1}{2}\#(C \cap D) = \#B - \frac{1}{2}\#(B \cap \underline{K}) = \sum m_i - \frac{1}{2}\sum m_i c_i = \sum m_i(1 - \frac{1}{2}c_i)$, where m_i denotes the total number of components from B parallel to the i-th component $\partial_i D$ of ∂D, and where $c_i := \#(\underline{K} \cap \partial_i D)$. Now, verify that in our example, $c_i = 3$, for all i, and so $-1 = -\frac{1}{2}\sum m_i$ i.e. $\sum m_i = 2$. Thus B has to consist of two components since it generates the surface S of Euler characteristic -1. Moreover, since S is supposed to be non-separating, it follows from lemma 9.3 that the two components of B cannot be both parallel to circles from either $A_1 \cup A_2$, or $B_1 \cup B_2$, or $C_1 \cup C_2$, or $D_1 \cup D_2$. Moreover, the algebraic intersection number, $i(B, k)$, has to be zero since $i(\partial S, k)$ vanishes and piping does not change algebraic intersection numbers. Consequently, if one component of B

is parallel to either one of B_1, B_2, C_1 or C_2, the other component of B has to have this property, too. Thus no component of B can actually have this property, for otherwise B would equal the intersection of two codim 0 submanifolds of $M(\underline{D})$ and S would be separating (see lemma 9.3 again). It follows that one component of B has to be parallel to A_1 or A_2, say, and, by the same reasoning, the other component then has to be parallel to D_1 or D_2. Now, $A_1 \cup A_2$ as well as $D_1 \cup D_2$ are invariant under the attaching map h, and so it makes no difference whether we take as a component of B a circle around A_1 or around A_2. The same with D_1 or D_2. Thus, altogether, we may suppose that B consists of precisely two components, one parallel to A_2 and the other one parallel to D_2. Check that $i(B, k)$ vanishes indeed.

Now, recall the piping-procedure. It is a locally defined process which effects only a neighborhood of k in $M(\underline{D})$. Thus the relevant information is schematically given by figure 22.2. The circle represents the curve $k = \underline{K}/h$ and the little arrows indicate a fixed normal-orientation of B (the argument is independent of the special choice of this orientation). It is immediately checked that the following is a complete list of piping-possibilities:

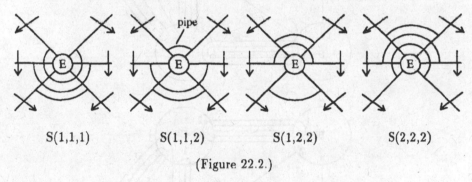

$$S(1,1,1) \qquad S(1,1,2) \qquad S(1,2,2) \qquad S(2,2,2)$$

(Figure 22.2.)

Note that any two neighboring pictures differ by pushing one pipe across the 2-handle E. The resulting surfaces (after piping B along k) differ therefore by an isotopy in $M^+(\underline{D})$ alone which pushes the surface across the 2-handle E in $M^+(\underline{D})$ attached to k, and so they are all pairwise isotopic in $M^+(\underline{D})$.

Figure 22.4 shows the circle-pattern $C(1,1,1)$ obtained from B by piping according to the schematic diagram $S(1,1,1)$. We check that this circle-pattern admits no compression-region and no compression-line at all. Thus, by 12.20, it is essential (alternatively, this property follows of course also from the fact that $\underline{D}$ is simple, i.e. it contains no essential disc or annulus). Since all surfaces above are isotopic in $M^+(\underline{D})$ to $S(C(1,1,1))$, the same holds for all the other surfaces as well. In particular, it is never necessary to extend a piping-process to a pseudo-piping process, and so the surfaces described above are *all* non-separating and essential surfaces in $M(\underline{D})$ with Euler characteristic -1 and there are no other (mod isotopy) with the same property. It still remains to verify that the circle-pattern $C(1,1,1)$ represents a punctured torus indeed. But the little arrows in figure 22.2 indicate a possible choice of a coherent normal-orientation for $C(1,1,1)$. Thus the surface $S(C(1,1,1)$ is 2-sided in $M(\underline{D})$ and so orientable. Furthermore, it is easily checked that $(C-D)^{-}/h$ is connected and so the boundary $\partial S(C(1,1,1))$ is connected. Finally, it has already been computed that $\chi(S) = -1$. This verifies our claim and therefore completes the proof of prop. 22.2. $\Diamond$

22.4. Remark. The previous example has been chosen for its simplicity. See the Appendix for an example which requires more of the machinery from chapter II.

22.5. Example. The circle-pattern $C(1,1,1)$:

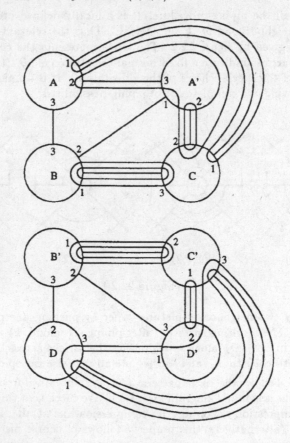

(Figure 22.3.)

22.6. As before let $\underline{D}$ be the Heegaard diagram for Thurston's example. Set $M^{+} := M^{+}(\underline{D})$ and $G := \pi_1 M^{+}$. M^{+} admits a hyperbolic structure with totally geodesic boundary and G appears as a Kleinian group, i.e. as a discrete subgroup of PSL_2C with non-empty region of discontinuity. (As a matter of fact, the regions of discontinuity are interiors of circles). An explicit construction of this hyperbolic structure may be found in [Th 4]. Applying hyperbolic geometry (rigidity etc.), it is then not difficult to show the existence of injections

$$\text{Out } G \hookrightarrow \text{Isom } M \hookrightarrow \text{Isom } \partial M^+.$$

Here Isom ∂M^+ denotes the group of isometries with respect to the induced hyperbolic structure (it is crucial that ∂M^+ is totally geodesic). Thus Out G appears as the subgroup of the moduli group of a closed Riemannien surface of genus three. Hence, by a well-known classical theorem due to Hurwitz [BS], we get

$$|\text{Out } G| \leq 84 \cdot (\text{genus } \partial M^+ - 1) = 168.$$

As an illustration of the method developed so far, we are going to show how to improve this estimate, using a combination of topological and geometric means, and how to determine the structure of the finite group Out G.

Indeed, it is the purpose of this section to sketch an argument for the following result.

22.7. Proposition. *Let* G *be given as above. Then there is a short exact sequence*

$$1 \to \mathbf{Z}_2 \to \text{Out } G \to D_4 *_{D_2} D_6 \to 1.$$

In particular, $|\text{Out } G| \leq 24$.

Proof. $M^+(k)$ contains no essential torus since its fundamental-group is a Kleinian group. Moreover, using lemma 7.4, an inspection of the Heegaard-diagram $\underline{D}$ shows that $M(\underline{D})$ is a simple relative handlebody, i.e. $M(\underline{D})$ contains no essential annulus (apart from the trivial one near the relator curve). It thus follows from an application of prop. 15.11, that either $M^+(\underline{D})$ is annulus-free or that the characteristic submanifold, V, of $M^+(k)$ is a solid torus with $V \cap \partial M^+(k)$ connected (actually, $M^+(k)$ must be annulus-free since $\partial M^+(k)$ is totally geodesic; see [Th 4]). Thus it follows that Out G is isomorphic to the mapping class group $\pi_0\text{Diff } M^+$ (see e.g. [Joh 1, thm. 24.2]). Now, let

$$S(M^+) := \text{set of all isotopy classes of essential surfaces in } M^+,$$
$$S_m(M^+) := \{S^+ \in S(M^+) \mid \chi(S^+) \geq m\}.$$

Then recall that $S_m(M^+)$ is a finite set (see e.g. prop. 20.13 and recall that $M^+(\underline{D})$ is simple). The mapping class group $\pi_0\text{Diff } M^+$ acts on the finite sets $S_m(M^+)$, for every m.

At this point recall from prop. 22.2 that $M(\underline{D})$ contains at least one non-separating punctured torus T. This surface has to be incompressible in $M^+(k)$ since $M^+(k)$ is boundary-irreducible. It further has to be boundary-incompressible in $M^+(k)$ since $M^+(k)$ contains no essential annulus. Thus T is in fact essential in M^+, and this shows, in particular, that $S_m(M^+) \neq \emptyset$

for $m = \chi(T) = -1$. It therefore suggests itself to consider the action of $\pi_0 \text{Diff } M^+$ on this particular set $S_{-1}(M^+)$. Let stab_T and orbit_T denote the stabilizer and the orbit, respectively, of $T \in S_{-1}(M^+)$ under this action. Then

$$| \text{Out } G | = | \pi_0 \text{Diff } M^+ | = | \text{stab}_T | \cdot | \text{orbit}_T | .$$

It remains to estimate $| \text{stab}_T |$ and $| \text{orbit}_T |$.

We begin with stab_T. Set $\text{stab}_T^{\pm} := \text{stab}_T \cap \pi_0 \text{Diff}^+ M^+$, where $\text{Diff}^+ M^+$ denotes the group of all orientation-preserving diffeomorphisms of M^+. We claim there is an inclusion $\text{stab}_T^{\pm} \hookrightarrow \text{Isom } T$. Let $h \in \text{stab}_T^{\pm}$. Then $h(T) = T$. By [Wa 2], $h|T \simeq \text{id}$ in T, provided $h \simeq \text{id}$ in M^+. Thus the assignment $h \mapsto h|T$ gives a well-defined homomorphism $\text{stab}_T^{\pm} \to \pi_0 \text{Diff} T$. If $h|T = \text{id}$, then, in particular, $h|\partial T = \text{id}|\partial T$. But $h|\partial M^+$ is finite modulo isotopy. Thus w.l.o.g. ∂M carries a hyperbolic structure, ∂T is a closed geodesic and h is an isometry preserving ∂T. Then $h|\partial M^+ = \text{id}$ and $h = \text{id}$ since h is orientation preserving. To see this note that every point, p, in ∂M^+ (resp. M^+) determines a unique perpendicular to ∂T (resp. ∂M^+). Since h is an orientation-preserving isometry (and so does not interchange the two sides of ∂T in ∂M^+), it maps such a perpendicular, and so p, to itself. This establishes the claim.

We next claim $|\text{orbit}_T| = 1$. To see this claim let $h \in \pi_0 \text{Diff } M^+$. Then $h(T) =: T'$ is a punctured torus in $M^+(k)$. $E := (M^+ - M(\underline{D})^-)^-$ is the regular neighborhood of some Heegaard-string, t, in M^+. Let T' be isotoped so that the intersection $T' \cap t$ is as small as possible. It follows that T' is strongly normal with respect to any essential disc in $M(\underline{D})$. Let A be the disjoint union of at most two annuli containing $T' \cap t$ as given by prop. 15.21 and 15.22. Since T' is non-separating, it follows that $T' \cap A$ has to be contained in that annulus from A which meets ∂M^+. Moreover, $A \cap T'$ consists of simple closed curves which are essential in both A and T'. In particular, $A \cap T'$ splits T' into a system of surfaces with Euler characteristic bigger than or equal to zero. Pushing the boundary curves of these surfaces out of A and into ∂M^+, results in surfaces which are admissible in $M(\underline{D})$ with the same Euler characteristic. One of these surfaces has to be non-separating in $M(\underline{D})$ since T' is non-separating in M^+. Compressing this particular surface along $(\partial$-$)$ compression-discs, we finally obtain a strongly essential and non-separating surface in $M(\underline{D})$ whose Euler characteristic is bigger than or equal to zero. Since $M(\underline{D})$ is simple, it therefore follows from prop. 22.2 that this surface has to be a punctured torus, and so it is equal to T'. Thus $T' \cap E = \emptyset$. It then follows from prop. 22.2 that T' is isotopic to T. Since T has been chosen arbitrarily, it therefore follows that $| \text{orbit}_T | = 1$.

By what we have seen so far, we have $\text{Out } G \cong \text{stab}_T$ and a short exact sequence

$$1 \to \mathbf{Z}_2 \to \text{stab}_T \to \text{Isom } T \to 1.$$

Now, by [Zie 4, thm. 23.1], Isom $T \cong GL_2\mathbf{Z} \cong D_4 *_{D_2} D_6$. Moreover, in [Zie 4, cor. 23.5] the finite subgroups of $Gl_2\mathbf{Z}$ are classified. They have order at most 12. Thus, taken together, the proof of prop. 22.7 is finished. ◊

IV. N-Relator 3-Manifolds

The purpose of this chapter is to generalize the handle-addition lemma from one-relator (section 15) to arbitrary n-relator 3-manifolds, or, dually, from a statement about Heegaard strings to a statement concerning Heegaard string-systems and Heegaard graphs. As a result, we obtain a finiteness-theorem for essential surfaces in 3-manifolds (extending classical results of Haken).

By definition, a *Heegaard graph* in a 3-manifold, M, is a finite, tri-valent, possibly disconnected graph, Γ, in M, $\Gamma \cap \partial M = \partial \Gamma$, such that $(M - U(\Gamma))^-$ is a handlebody. A *Heegaard string-system* is a Heegaard graph which is a system of arcs. The closed surface $\partial U(\Gamma \cup \partial M) - \partial M$ is the *Heegaard surface* of Γ. Two Heegaard graphs, $\Gamma_1, \Gamma_2 \subset M$, are *slide-equivalent* if the regular neighborhood $U(\Gamma_1 \cup \partial M)$ (taken in M) is proper isotopic to $U(\Gamma_2 \cup \partial M)$. The *relative Euler characteristic* of Γ is defined to be $\chi(\Gamma, \partial \Gamma) := \chi \Gamma - \chi \partial \Gamma$. It is an invariant of the slide-equivalence class of Γ.

In chapter III we considered the case where Γ is a single arc. Recall that the handle-addition lemma for one-relator 3-manifolds estimates the intersection of such strings with essential surfaces. In the case of general Heegaard string-systems, we have to be a bit more careful. In particular, we have to deal with the extra flexibility which comes from slides of one string over others. This extra flexibility is the reason why intersections of essential surfaces with Heegaard string-systems are no longer limited (even not modulo Dehn-twists along tori). To illustrate this point, take e.g. a Heegaard string-system of two strings and consider an essential surface which intersects this Heegaard string-system in a minimal number of points and at least one of its strings non-trivially. Then notice that the surface can be forced to intersect the other string arbitrarily often by sliding it over the first string over and over again. It turns out, however, that this process is essentially the only way (besides Dehn twists along tori) for increasing or decreasing intersections of essential surfaces with Heegaard string systems. In fact, we will show that the intersection of essential surfaces with Heegaard string-systems and Heegaard graphs is still rather rigid.

Here is a formal statement of the main result of this chapter (cor. 23.34).

Theorem. *To any 3-manifold, M, there is associated a known polynomial function, $\sigma_M : \mathbf{N} \times \mathbf{N} \to \mathbf{N}$, such that the following holds.*

For every essential surface, S, and every Heegaard graph, Γ, in M, there is (1) a Heegaard string-system, Γ', slide-equivalent to Γ, (2) an ambient isotopy, α_t, $t \in I$, and (3) a product, h, of Dehn twists along essential tori in M with

$$\#(hS \cap \alpha_1 \Gamma') \le \sigma_M(-\chi S, -\chi(\Gamma, \partial \Gamma)).$$

This theorem has a number of interesting consequences, some of which will be discussed in this chapter. For the unprepared reader, however, the numerical estimate underlying the previous result may seem of no great value. Indeed, it may even be argued that the intersection of any essential surface with Heegaard graphs is already finite anyway. Moreover, by Haken's classical theory, any essential surface can be obtained, using cut-and-paste, from a known, finite set of fundamental surfaces. Thus the intersection with Heegaard graphs is actually determined by those of the finitely many fundamental surfaces. However, it has to be pointed out though that Haken's theory makes no predictions whatsoever on the intersection of fundamental surfaces with Heegaard graphs. Now, the above theorem tells us that the minimal intersection of any essential surface with any Heegaard surface depends only on the *topological type*, but not on the *embedding type* of the surfaces involved. In particular, the theorem can therefore be read as a statement on essential surfaces *as well as* a statement on Heegaard graphs. It is the observation of this duality which makes the theorem so valuable for us. In fact, our later study of Heegaard graphs in Haken 3-manifolds is based on refinements of this observation. It is the basic observation underlying this book.

The crucial idea underlying the proof of the above theorem is the concept of "strictly normal surfaces" as introduced in section 23. This concept generalizes Haken's concept of normal surfaces. Strictly normal surfaces are useful and frequent. As further applications of this concept, we give a simple proof of Waldhausen's theorem concerning Heegaard graphs in S^3 and a quick proof of Haken's 2-sphere result. In fact, we generalize Haken's result to inessential 2-spheres as well (section 24). This in turn enables us to give a new proof of the Reidemeister-Singer theorem (section 25).

§23. The Handle-Addition Lemma (General Case).

In order to deduce the finiteness theorem mentioned above, we introduce the concept of "strictly normal positions" for surfaces and Heegaard graphs. It turns out that the behaviour of surfaces in strictly normal positions is very similar to that of strongly normal surfaces in one-relator 3-manifolds. Specifically, we show that every essential surface can be brought into a strictly normal position (prop. 23.11) and that the intersection of strictly normal surfaces with Heegaard graphs can be estimated (thm. 23.33). We further show that the intersection of essential surfaces and Heegaard graphs is governed by the intersection of essential surfaces and Heegaard cylinders, i.e., by the intersection of codim one objects (prop. 23.37). In fact, for certain special situations this intersection can be calculated. We illustrate this statement with discs and 2-spheres (prop. 23.16). Specifically, we get a generalization of the original handle-addition lemma for one-relator 3-manifolds (prop. 15.18) to arbitrary n-relator 3-manifolds (prop. 23.18). We show that a number of known reducibility criteria for Heegaard graphs follow from this calculation. Specifically, we give new proofs of the respective criteria of Haken and Frohmann concerning the reducibility of Heegaard graphs (prop. 23.21, prop. 23.22). Merging the underlying idea with Gabai's concept of thin position, we get a straightforward proof of Waldhausen's result concerning Heegaard decomposition of the 3-sphere (thm. 23.23). As a by-product of the proof we also obtain the reducibility-criterion of Casson-Gordon for Heegaard graphs (cor. 23.27). For completeness, we generalize cor. 17.9 in that we show that an irreducible 3-manifold, M, is a handlebody if and only if it has a Heegaard graph which can be deformed into ∂M (thm. 23.43).

23.1. *Strictly Normal Surfaces.*

In this section we consider "strictly normal surfaces" in relative handlebodies. We begin with the following definition.

23.2. Definition. *Let $(M, \underline{m})$ be a handlebody whose boundary-pattern is a system of annuli. Let $(\mathcal{D}, \underline{d})$ be a meridian-system of essential discs in $(M, \underline{\bar{m}})$. A 2-manifold, S, in M whose boundary curves are contained in faces of $(M, \underline{\bar{m}})$ is called* <u>strictly normal</u> *(with respect to $\mathcal{D}$) if the following holds:*

(1) S is incompressible in M.

(2) Every arc from $\mathcal{D} \cap S$ is essential in S and essential in $(\mathcal{D}, \underline{\bar{d}})$.

(3) No arc from $\mathcal{D} \cap S$ is bad.

Recall that $\underline{\bar{m}}$, $\underline{\bar{d}}$ denote the completed boundary-patterns. An arc, b, from $\mathcal{D} \cap S$ is called *bad* if there is a component, r_1, of ∂S such that

(1) r_1 is contained in an annulus from $\underline{m}$,

(2) b joins r_1 with a different component from ∂S, and

(3) at least one (open) component of $\partial D - \partial b$ does not intersect r_1, where D denotes the component of $\mathcal{D}$ containing b.

Here are some properties of strictly normal surfaces in handlebodies.

23.3. Proposition. *Let $(M, \underline{m})$, $(\mathcal{D}, \underline{d})$ and S be given as in definition 23.2. Suppose S is strictly normal w.r.t. $\mathcal{D}$. Let r be a component of ∂S which lies in a face of $(M, \underline{m})$, i:e., in a bound face. Then either (1) or (2) holds:*

(1) r is the boundary of a component of S, or

(2) at least one non-recurrent as well as at least one recurrent arc from $\mathcal{D} \cap S$ has an end-point in r.

Remark. An arc in S is *recurrent* if both its end-points lie in one component of ∂S.

Proof. W.l.o.g. S is connected. If $r = \partial S$, we are done. So suppose the converse.

Let $\mathcal{K}_r$ be the collection of all arcs from $S \cap \mathcal{D}$ meeting r. Let U_r be a regular neighborhood of $r \cup \mathcal{K}_r$. Then each component of $\partial U_r - \partial S$ is contained in $\partial (M - U(\mathcal{D}))^-$. Since $(M - U(\mathcal{D}))^-$ is a 3-ball and since S is incompressible in M (property (1) of strictly normal surfaces), it follows that

$\partial U_r - \partial S$ bounds discs in S. The existence of the desired non-recurrent arc from $\mathcal{D} \cap S$ now follows from the hypothesis $r \neq \partial S$.

By what we have just seen, there is at least one arc, say b, from $\mathcal{D} \cap S$ which is a non-recurrent arc in S with one end-point in the boundary curve r. Let D be the disc from $\mathcal{D}$ containing b, and let D_0 be one of the two discs in which D is split by b. Denote $b' := D_0 \cap \partial D$. Since S is a strictly normal surface, it follows that the interior of b' has to intersect r, too. Let x denote one point of this intersection. Then the point x in turn is the end-point of some arc, l, from $\mathcal{D} \cap S$. W.l.o.g., l is recurrent in S, for otherwise recall one end-point of l, namely x itself, lies in r and so we could replace b by l and argue as before. This proves property (2) of prop. 23.3. $\Diamond$

The following corollary is an immediate consequence of the previous proposition.

23.4. Corollary. *Let $(M, \underline{m})$ be a handlebody whose boundary-pattern is a system of annuli. Suppose S is a strictly normal, planar surface in $(M, \underline{m})$ which has at most one free boundary curve. Then S is a disc.*

Remark. Recall a boundary curve of S is *bound* if it lies in an annulus from $\underline{m}$ and *free* if it lies in $\partial M - \bigcup \underline{m}$.

Proof. We may suppose ∂S is disconnected, for otherwise S is a disc and we are done. Thus, by prop. 23.3, every bound component of ∂S contains the end-points of at least one recurrent arc from $\mathcal{D} \cap S$. But S is planar and at most one component of ∂S is free. So it follows by induction that at least one of these recurrent arcs must be inessential in S. This, however, contradicts the fact that S is strictly normal (see (2) of 23.2). $\Diamond$

23.5. *Strictly Normal Positions.*

Using the concept of "strictly normal surfaces" in relative handlebodies, we can now define the notion of "strictly normal positions" for surfaces and Heegaard graphs in arbitrary 3-manifolds. First recall the definition:

23.6. Definition. *Let M be a 3-manifold. A finite, tri-valent, possibly disconnected graph Γ in M, $\Gamma \cap \partial M = \partial \Gamma$, is a Heegaard graph in M if $(M - U(\Gamma))^{-}$ is a handlebody. The closed surface $G_{\Gamma} := \partial U(\Gamma \cup \partial M) - \partial M$ is the Heegaard surface associated to the Heegaard graph Γ.*

Remark. We say Γ is *tri-valent* if every vertex from $\partial \Gamma$, resp. $\Gamma - \partial \Gamma$, is end-point of exactly one, resp. exactly three, edge(s) of Γ.

Two Heegaard graphs, $\Gamma_1, \Gamma_2 \subset M$, are *slide-equivalent* if $U(\Gamma_1 \cup \partial M)$ is proper isotopic to $U(\Gamma_2 \cup \partial M)$. Denote by $\mathcal{G}(M, \Gamma)$ the set of all Heegaard graphs in M, slide-equivalent to Γ.

Given a Heegaard graph, Γ, in M, let $\Gamma^{(0)}, \Gamma^{(1)}$ denote the set of all vertices resp. edges of Γ. Let $U_0 := U(\Gamma^{(0)})$ and denote by U_1 the regular neighborhood of $(\Gamma^{(1)} - U_0)^-$ in M. Set $U(\Gamma) = U_0 \cup U_1$. Let

$$\varphi_\Gamma : M \to M$$

denote the map whose restriction $\varphi_\Gamma \mid U(\Gamma) : U(\Gamma) \to \Gamma$ is the canonical retraction, given by the requirement $\varphi_\Gamma(U_0) = \Gamma^{(0)}$ and $\varphi_\Gamma(U_1 - U_0) = \Gamma^{(1)} - \Gamma^{(0)}$, and whose restriction to $M - U(\Gamma)$ is an embedding. We say φ_Γ is the *canonical contraction associated to* Γ. Set

$$M_\Gamma^+ := (M - U(\Gamma^{(0)}))^-, \quad M_\Gamma := (M - U(\Gamma))^- = (M - U_0 \cup U_1)^-,$$
$$\mathcal{A}_\Gamma := (\partial M_\Gamma - \partial M_\Gamma^+)^- \quad \text{and} \quad E_\Gamma := (U_1 - U_0)^-.$$

Denote by $\underline{m}_\Gamma$ the collection of all annuli from $\mathcal{A}_\Gamma$ and denote by k_Γ the collection of cores of all the annuli from $\mathcal{A}_\Gamma$. Notice that M_Γ is a handlebody since Γ is supposed to be a Heegaard graph. So M_Γ^+ is the n-relator 3-manifold $M_\Gamma^+(k_\Gamma)$. E_Γ is the system of 2-handles attached to M_Γ along k_Γ. Moreover, $\Gamma' := \Gamma \cap M_\Gamma^+$ is a Heegaard string-system in M_Γ^+. We say Γ', resp. $M_\Gamma^+(k)$, is the Heegaard string-system, resp. the n-relator structure, associated to Γ. Note that Γ' and E_Γ are dual to each other.

23.7. Definition. *Let M be a 3-manifold. Let $\Gamma \subset M$ be a Heegaard graph and let $S \subset M$ be a 2-manifold. Then S and Γ are in a <u>strictly normal position</u>, provided there is a meridian-system, $(\mathcal{D}, \underline{d})$, of essential discs in $(M_\Gamma, \underline{m}_\Gamma)$ such that the 2-manifold $S \cap M_\Gamma$ is strictly normal w.r.t. $\mathcal{D}$.*

If Γ and S are in strictly normal position, then we say that S is *strictly normal (w.r.t. Γ)* and that Γ is *strictly normal (w.r.t. S)*.

23.8. For completeness we next verify that our notion of "slide-equivalence" is equivalent to the intuitive notion of "slides" for graphs. To this end let Γ be a Heegaard graph in a 3-manifold M. Let B be a 4-faced disc in M such that the following holds:

(1) $B \cap (\Gamma \cup \partial M)$ equals the union of three faces of B.

(2) One face of B lies in an edge, e, of Γ and two other faces lie either in an edge of Γ or in ∂M.

Notice that $B \cap e$ consists of one or two arcs. Let α be one of them. Define $\Gamma' := (\Gamma - \alpha)^- \cup (\partial B - \Gamma)^-$. Of course, Γ' is a Heegaard graph for M again. We say Γ' is obtained from Γ by a *square-move* across B.

23.9. Proposition. *Let Γ, Γ' be Heegaard graphs in a 3-manifold M. Then Γ and Γ' are slide-equivalent if and only if there is a finite sequence, $\Gamma =$*

$\Gamma_1, \Gamma_2, ..., \Gamma_n = \Gamma'$, *of graphs so that* Γ_{i+1}, $2 \leq i \leq n$, *is obtained from* Γ_i *by some square-move.*

Proof. Suppose Γ' is obtained from Γ by a square-move across a 4-faced disc B. Then observe that $U(\Gamma \cup B \cup \partial M)$ is proper isotopic to both $U(\Gamma \cup \partial M)$ and $U(\Gamma' \cup \partial M)$. So one direction of the proposition follows, inductively.

For the other direction let Γ, Γ' be two slide-equivalent Heegaard graphs in M. Then w.l.o.g. $U := U(\Gamma \cup \partial M) = U(\Gamma' \cup \partial M)$. Let $\mathcal{D}$ be the disc-system in $U(\Gamma \cup \partial M)$ dual to Γ, i.e., every edge of Γ intersects exactly one disc from $\mathcal{D}$ in exactly one point. Let $\mathcal{D}' \subset U$ be the dual disc-system for Γ'.

U may or may not be a handlebody. In any case, however, we can define disc-slides for $\mathcal{D}$ as in 1.2. In fact, it is easily verified that Γ and Γ' differ by a square-move if and only if $\mathcal{D}$ and $\mathcal{D}'$ differ by such a disc-slide. Thus, if U is a handlebody (i.e., if $\partial M = \emptyset$), then prop. 23.9 follows directly from prop. 1.5. If not, then we have to modify the argument from 1.5 slightly as follows. First, using the argument from 1.5, we find that $\mathcal{D}'$ is slide-equivalent in U to a disc-system, $\mathcal{D}^*$, with $\mathcal{D}^* \cap \mathcal{D} = \emptyset$. Now, $(U - U(\mathcal{D}))^-$ equals the collar $U(\partial M)$. Thus every single disc from $\mathcal{D}^* \subset U(\partial M)$ is ∂-parallel in $U(\partial M)$. In particular, we may suppose $\mathcal{D}^*$ is slid so that afterwards the whole system $\mathcal{D}^*$ is parallel in $U(\partial M)$ to a system, say $\mathcal{A}$, in $\partial U(\partial M) - \partial M$. Since no disc from $\mathcal{D}^*$ is separating in U, we may suppose $\mathcal{D}$ is slid so that $\mathcal{A} \cap U(\mathcal{D})$ is connected, for every component, A, of $\mathcal{A}$ and every component D from $\mathcal{D}$. Since $(M - U(\mathcal{D}^*))^-$ is connected and since $\#\mathcal{D}^* = \#\mathcal{D}$, it then follows that $\mathcal{D}$ can be slid into a disc-system, $\mathcal{D}^{**}$, so that afterwards $\mathcal{A} \cap U(\mathcal{D}^{**})$ consists of one disc, for every component, A, of $\mathcal{A}$. Then $\mathcal{D}^*$ and $\mathcal{D}^{**}$ are isotopic in $U(\partial M)$. Thus $\mathcal{D}$ and $\mathcal{D}'$ are slide-equivalent in U. So, dually, Γ and Γ' differ by a sequence of square-moves. $\Diamond$

23.10. *Slides Associated to Bad Arcs.*

Strictly normal surfaces have no bad arcs (see the terminology from def. 23.2). Thus in order to turn a surface into a strictly normal surface, we have to remove bad arcs. This can be done in a reasonable way using certain "controlled" slides and isotopies. But at this point we have to be careful. The reason being that we do not want to exclude compressible surfaces, like inessential 2-spheres, from our consideration. So, in particular, we cannot use arbitrary isotopies. Indeed, a careless chosen isotopy may push a given inessential 2-sphere into the complement of a Heegaard graph and then it is useless.

It turns out, however, that to every bad arc there is associated a certain "slide" of the Heegaard graph which removes this bad arc. In order to describe this operation, let $S \subset M$ be a surface and let $\Gamma \subset M$ be a Heegaard graph. Let $(\mathcal{D}, \underline{d})$ be a meridian-system of essential discs in the handlebody $(M_\Gamma, \underline{m}_\Gamma)$. After a small general position isotopy if necessary, we may suppose $\mathcal{D} \cap S$ is system of arcs and curves in $\mathcal{D}$.

Suppose b is a bad arc from $\mathcal{D} \cap S$. Then, by definition, one end-point of b lies in a component, say r, from $S \cap \partial U(\Gamma)$, and b separates a disc, say D_b, from $\mathcal{D}$ such that $D_b \cap r$ is connected. Let D_b' denote the regular neighborhood of D_b in $\mathcal{D}$. Let $\varphi_\Gamma : M \to M$ be the canonical contraction associated to Γ. Then $\varphi_\Gamma(r)$ is a point, say x_b, in an edge, say e_b, of Γ. $\varphi_\Gamma(b)$ is an arc joining x_b either with ∂M or another point in Γ (possibly in the same edge e_b). $\varphi_\Gamma | D_b \cap \partial \mathcal{D}$ is a, possibly singular, arc in $\Gamma \cup \partial M$. x_b separates e_b into two arcs. Only one of them, say α_b, is contained in $\varphi_\Gamma D_b'$. Define

$$\Gamma' := (\Gamma - \alpha_b) \cup \varphi_\Gamma(\partial D_b' - \partial \mathcal{D})^-.$$

Of course, Γ' is again a Heegaard graph since it differs from Γ by an isotopy of the edge, e_b, in $M - (\Gamma - e_b)$. We say Γ' is obtained from Γ by a *slide associated to* b. Note that this slide depends on the choice of the component r above. In particular, there are at most two slides associated to any given bad arc.

Schematically, the slide associated to b may be illustrated as in fig. 23.1 (but recall that $\varphi_\Gamma | D_b \cap \partial \mathcal{D}$ may well be a singular path in Γ).

23.11. Proposition. *Let M be a 3-manifold. Let $\Gamma \subset M$ be a Heegaard graph and let $S \subset M$ be a 2-manifold. Suppose S is not totally compressible in M. Then there is a Heegaard graph $\Gamma' \subset M$ and a 2-manifold $S' \subset M$ such that S' and Γ' are in strictly normal position.*

In addition, Γ' can be chosen slide-equivalent to Γ, and S' can be obtained from S, using disc-compressions and isotopies.

Remark. Here we say that a 2-manifold S' is obtained from S by a *disc-compression* if it is obtained by a compression along a compression disc or a ∂-compression disc. Recall S' is isotopic to S if M is a Haken 3-manifold and S is essential. A surface is *totally compressible* if it can be turned, using disc-compressions, into a system of inessential discs and 2-spheres.

Proof of Proposition. Since S is not totally compressible, we may suppose it is an essential 2-manifold (possibly a system of discs and 2-spheres). Moreover, we may suppose it is compressed so that afterwards $S \cap M_\Gamma$ is incompressible. Then there is a meridian-system, $(\mathcal{D}, \underline{d})$, of essential discs in M_Γ such that $\mathcal{D} \cap S$ is a system of arcs. Let $(\mathcal{D}, \underline{d})$ be chosen so that this holds and that, in addition, $\#(\mathcal{D} \cap S)$ is as small as possible.

Let b be an outermost arc from $\mathcal{D} \cap S$. Then, by definition, b separates a disc, D_b, from $\mathcal{D}$ with $D_b \cap S = b$.

Suppose b is inessential in $(\mathcal{D}, \underline{d})$. Then notice that the restriction $\varphi_\Gamma | D_b \cap \partial \mathcal{D}$ must be an embedding, i.e, a non-singular arc in $\Gamma \cup \partial M$. W.l.o.g. this arc does not lie in ∂M, for otherwise $\#(\mathcal{D} \cap S)$ can be reduced by the compression of S along D_b. So it is either an arc, c, in some edge of Γ, or the union of such an arc with an arc in ∂M. Thus, the isotopy which pushes c across $\varphi_\Gamma(D_b)$, extends to an isotopy which deforms Γ to a graph Γ' with

$\#(\Gamma' \cap S) < \#(\Gamma \cap S)$. Moreover, $S \cap M_{\Gamma'}$ must be incompressible $M_{\Gamma'}$ since $S \cap M_\Gamma$ is supposed to be incompressible in M_Γ.

Suppose b is a bad arc. Then let Γ' be the result of a slide associated to b. Then $\#(S \cap \Gamma) < \#(S \cap \Gamma')$. Again it is easy to verify that $S \cap M_{\Gamma'}$ must be incompressible in $M_{\Gamma'}$ since $S \cap M_\Gamma$ is supposed to be incompressible in M_Γ.

Thus, by what we have just verified, $\#(\Gamma \cap S)$ can be reduced whenever there is an arc from $\mathcal{D} \cap S$ which is inessential or bad. So, inductively, we may suppose $\mathcal{D} \cap S$ is a system of essential arcs none of which is bad. Then, by definition, S is strictly normal w.r.t. $\mathcal{D}$. This finishes the proof. $\lozenge$

23.12. Example.

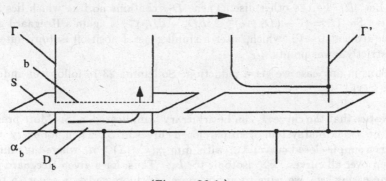

(Figure 23.1.)

23.13. Remark. It is worth noting that the number of slides in the statement of prop. 23.11 can in general not be limited by a function which is independent of the selected surface S. To explain this fact note first that the statement of the previous proposition makes sense for surfaces too. Indeed, let us call a finite and tri-valent graph, Γ, in a closed surface, S, a *Heegaard graph* for S if its complement is an open disc. Then we prove the following statement, using the idea of strictly normal positions.

23.14. Lemma. *Let S be a closed, orientable surface. Let $\Gamma \subset S$ be a Heegaard graph and let $k \subset S$ be a simple closed, non-contractibe curve in S. Then there is a Heegaard graph, Γ', slide-equivalent to Γ, such that $\#(k \cap \Gamma') = 1$.*

Proof. The closure of $S - \Gamma$ is a disc, say D_Γ. Let the boundary-pattern, $\underline{d}_\Gamma$, of D_Γ be given by the edges of Γ. Then $(D_\Gamma, \underline{d}_\Gamma)$ is a polygon. Each face of this polygon is contained in exactly one edge of Γ and every edge from Γ contains exactly two faces of $(D_\Gamma, \underline{d}_\Gamma)$.

Using a small general position isotopy if necessary, we may suppose that k meets no vertices from Γ. Then $k \cap D_\Gamma$ is a system of arcs in D_Γ with end-points in the faces of $(D_\Gamma, \underline{d}_\Gamma)$. Using another isotopy of k if necessary, we may suppose that no arc from $k \cap D_\Gamma$ has both end-points in one face of $(D_\Gamma, \underline{d}_\Gamma)$. Let k_0 be an outermost arc from $k \cap D_\Gamma$ in D_Γ. Let d_1, d_2 be the two faces of $(D_\Gamma, \underline{d}_\Gamma)$ joined by k_0. Let D_1, D_2 denote the two discs in which D_Γ is separated by k_0.

Suppose d_1, d_2 lie in the same edge of Γ. If $d_1 \cap D_1$ is the same arc as $d_2 \cap D_1$ in Γ, then $\#(k \cap \Gamma) = 1$, and we are done. Thus suppose the converse. Then D_1 is embedded in S. Hence $\Gamma' := (\Gamma - (D_1 \cap d_1))^- \cup (\partial D_1 - \partial D_\Gamma)^-$ is a Heegaard graph, slide-equivalent to Γ. And, after a small general position isotopy, Γ' intersects k in strictly fewer points than Γ.

Suppose d_1, d_2 lie in different edges of Γ. Let e be the edge of Γ containing d_1. Note that only one of the two discs D_1, D_2 contains a face of $(D_\Gamma, \underline{d}_\Gamma)$ which lies in e (e contains exactly two faces one of them being d_1). Let D_1 be the other disc. Then D_1 contains no face which lies in the edge e. So $\Gamma' := (\Gamma - (D_1 \cap e)^- \cup (\partial D_1 - \partial D_\Gamma)^-$ is again a Heegaard graph, slide-equivalent to Γ, which, after a small general position isotopy, intersects k in strictly fewer points.

Thus in any case we get a reduction. So lemma 23.14 follows by induction on $\#(k \cap \Gamma)$. $\diamond$

Notice that the curve k can be arbitrary complicated in S. More precisely, let $\mathcal{L}$ be a maximal system of simple closed curves in S. Then, for every $n \geq 1$, there is a simple closed curve, k_n, with min $\#(k'_n \cap \mathcal{L}) \geq n$, where the minimum is taken over all curves, k'_n, isotopic to k_n. Thus, for a given Heegaard graph and increasing n, we have to apply more and more slides in order to turn it into a graph which intersects k_n in one point only.

Taking the product of S with the circle, we see that $S \times S^1$ contains vertical tori for which it takes more and more slides in order to put a given Heegaard graph into a strictly normal position. Thus the number of slides in prop. 23.9 can in general not be limited. However, we do conjecture that this number can be limited for all simple 3-manifolds.

23.15. *The Handle-Addition Lemma.*

As an immediate consequence of the existence result concerning strictly normal positions (prop. 23.11), we get a first generalization of the handle addition lemma from one-relator 3-manifolds to arbitrary n-relator 3-manifolds.

23.16. Proposition. *Let Γ be a Heegaard graph in a 3-manifold M.*

If S is an essential disc and M is irreducible, then there is a Heegaard graph, Γ', in M, slide-equivalent to Γ, such that $S \cap \Gamma' = \emptyset$.

If S is a system of essential 2-spheres in M, there is a Heegaard graph, $\Gamma' \subset M$, slide-equivalent to Γ, and a system of essential 2-sphere, $S' \subset M$, obtained from S by disc-compressions, such that $\#(\Gamma' \cap S'_i) = 1$, for every component, S'_i, from S'.

Proof. By prop. 23.11, there is a Heegaard graph, Γ', and a surface, S', such that (1) S' and Γ' are in strictly normal position and that (2) Γ' is slide-equivalent to Γ and S' can be obtained from S by disc-compressions. By cor. 23.4, $(S_i' - U(\Gamma'))^-$ is a disc, for every component, S_i', from S'. So prop. 23.16 follows if S is a 2-sphere. If S is a disc, then $S' \cap \Gamma' = \emptyset$ (cor. 23.4). Moreover, S' is obtained from S by compressions along ∂-compression discs alone since M is supposed to be irreducible. Vice versa, S is obtained from S' by successive band-connected sums along ∂-parallel bands. The first part of prop. 23.16 now follows since all these bands can be pushed away from Γ'. $\Diamond$

We next translate this proposition into the context of relative handlebodies and n-relator 3-manifolds, in order to get the desired generalization of the original handle-addition lemma for one-relator 3-manifolds.

Given a relative handlebody (M, k), choose a simple arc, ℓ, in the boundary ∂M with $\ell \cap k = \partial \ell$ and meeting two different curves from $\partial U(k)$. Then the end-points of ℓ lie in curves from k, say k_1 and k_2, which may or may not be equal. In any case, consider

$$c := \partial U(k_1 \cup \ell \cup k_2).$$

It is easily checked that the 1-manifold c consists either of one or three curves according whether $\partial \ell$ lies in one or two curves from k. Let k_0 denote the component of c which is not parallel to k_1 or k_2. Define

$$k' := (k - k_1) \cup k_0.$$

If ℓ joins two different components of k, then, intuitively, we obtain k' from k by "sliding" k_1 along the sliding arc ℓ and across a 2-cell attached to k_2. Therefore we say that k' is obtained from k by a *relator-slide of k_1 along ℓ*. Two curve-systems, k, k', in ∂M are *slide-equivalent* if there is a finite sequence $k = k_1, k_2, ..., k_n = k'$ such that k_{i+1} is obtained from k_i by a relator-slide. Notice that slide-equivalent curve-systems need not have the same number of curves.

The following statement is a straightforward consequence of the definitions.

23.17. Proposition. *Let* Γ, Γ' *be two Heegaard graphs in a 3-manifold* M. *Then* Γ *and* Γ' *are slide-equivalent if and only if* $M_\Gamma := M_{\Gamma'}$ *and the associated curve-systems* $k_\Gamma, k_{\Gamma'} \subset \partial M_\Gamma$ *are slide-equivalent.* $\Diamond$

Now, using the previous notation, we get the following proposition as an immediate consequence of prop. 23.16. It generalizes the original handle-addition lemma (prop. 15.16).

23.18. Proposition. *Let* $M^+(k)$ *be an n-relator 3-manifold. If* $M^+(k)$ *is reducible or ∂-reducible, then there is a curve-system,* k', *in* ∂M, *slide-equivalent to* k, *so that* $\partial M - k'$ *is compressible in* M. $\Diamond$

23.19. *Reducibility of Heegaard Graphs.*

We are mainly interested in irreducible Heegaard graphs. These are studied in the next chapter. Here we collect a number of useful reducibility criteria for Heegaard graphs. We prove them using the the handle-addition lemma for n-relator 3-manifolds.

23.20. Definition. *Let Γ be a Heegaard graph in a 3-manifold M. We say Γ is reducible if there is a Heegaard graph, Γ', slide-equivalent to Γ, and a 2-sphere, S, in M such that $\#(\Gamma' \cap S) = 1$. If not, then Γ is irreducible.*

23.21. Proposition [Ha 3]. *If M is reducible, then Γ is reducible.*

Proof. This is a straightforward consequence of prop. 23.16. $\Diamond$

23.22. Proposition [Fro]. *Let Γ be a Heegaard graph in a 3-manifold M. Suppose there is a simple closed curve, $k \subset \Gamma$, and a 3-ball, $E \subset M$ such that $k \subset E$. Then either M is the 3-sphere or Γ is reducible.*

Proof. Suppose $M \neq S^3$, for otherwise we are done. Define $M' := (M - U(k))^-$ and $\Gamma' := \Gamma \cap M' = (\Gamma - U(k))^-$. Of course, Γ' is a Heegaard graph in M'. Notice further that $S := \partial E$ is an essential 2-sphere in M' since $M \neq S^3$. Thus, by prop. 23.16, there is a Heegaard graph $\Gamma'' \subset M'$, slide-equivalent to Γ', and an essential 2-sphere $S' \subset M'$ such that $\#(S' \cap \Gamma'') = 1$. Now, set $\Gamma^* := k \cup \Gamma''$. Notice that Γ^* is slide-equivalent to Γ since all slides of Γ' in M' can be realized by slides of Γ in M. Hence Γ is reducible since $\#(\Gamma^* \cap S') = 1$. $\Diamond$

23.23. Theorem [Wa 3]. *A Heegaard graph Γ in the 3-sphere is either a curve or reducible.*

Remark. For other proofs of Waldhausen's theorem see [Wa 3] and [Ot]. Recall further that a Heegaard graph in S^3 which is a closed curve must be the trivial knot.

Proof. The following proof is by induction on the Euler characteristic, $\chi\Gamma$, of Heegaard graphs Γ. If $\chi\Gamma = 0$, then Γ is a curve and the theorem is obviously true. This establishes the induction beginning.

For the induction step suppose the theorem holds for all Heegaard graphs, Γ, in S^3 with $-\chi\Gamma < n$. *Under this induction hypothesis the following reducibility-criterion is true:*

23.24. Lemma. *Let M be a closed 3-manifold and let Γ be a Heegaard graph in M with $-\chi\Gamma = n$. Suppose there is a subgraph, $\Gamma' \subset \Gamma$ with $\Gamma' \neq \Gamma$ and $(M - U(\Gamma'))^-$ ∂-reducible. Then M contains an incompressible surface or Γ is reducible.*

Proof of lemma. Suppose Γ' is chosen so that, in addition, $\chi(\Gamma')$ is as large as possible. Set $N := U(\Gamma')$.

Let $\mathcal{D}$ be a maximal system of essential, pairwise non-parallel discs in $(M - N)^-$. This is non-empty since $(M - N)^-$ is supposed to be ∂-reducible. Denote by N^+ the n-relator 3-manifold obtained by attaching $\mathcal{D}$ to N. Set $\Gamma^* := \Gamma \cap (M - N)^-$. Then Γ^* is a Heegaard graph in $(M - N)^-$.

Suppose ∂N^+ consists of 2-spheres. Set $S := \partial U(\partial N^+) \cap (M - N^+)^-$. If one 2-sphere from S is essential in $(M - N^+)^-$, then, by prop. 23.16, Γ^*, and so Γ, is reducible. Thus we may suppose every component of S is inessential in $(M - N^+)^-$. Then $\mathcal{D}$ splits $(M - N)^-$ into a system of 3-balls. Moreover, by prop. 23.14, we may suppose that Γ^* is slid so that afterwards $\Gamma^* \subset ((M - N)^- - U(\mathcal{D}))^-$. But $\chi\Gamma^* > n = \chi\Gamma$. So, by induction hypothesis, it follows that Γ^*, and so Γ, must be reducible.

Thus w.l.o.g. we may suppose ∂N^+ has a component, say S, which is not a 2-sphere. If S is incompressible, we are done.

Assume S is compressible, then S has a compression disc, say A. This compression disc must lie in N^+ since, by our maximal choice of $\mathcal{D}$, it cannot lie in $(M - N^+)^-$. But N^+ is an n-relator 3-manifold and so, by prop. 23.18, we may suppose the discs from $\mathcal{D}$ are slid so that $A \cap \partial \mathcal{D} = \emptyset$. Then $A \subset N$ and so $(N - U(A))^-$ is a handlebody. But the spine of N is a subgraph of Γ whose relative Euler characteristic is strictly larger than that of Γ'. So we get a contradiction to our maximal choice of $\chi\Gamma'$.

This proves the lemma. $\Diamond$

In order to continue, we first generalize Gabai's concept of "thin positions" from knots to Heegaard graphs (I am grateful to C. Frohmann and M. Scharlemann for pointing out this possibility; see also [SchT2]).

Let M be a 3-manifold and let G be a Heegaard surface in M. Then G splits M into two handlebodies. Let M' be the complement of the union of the spines of these two handlebodies. Then M' may be identified with $G \times \mathbf{R}$. The projection $h : G \times \mathbf{R} \to \mathbf{R}$ onto the second factor is called the *height function*. For every $t \in \mathbf{R}$, we call the pre-image $Q_t = h^{-1}t$ a *level surface (of height t)*.

Every Heegaard graph in M can be pushed into M', using a small general position isotopy. Let Γ be a Heegaard graph in M, contained in M'. A point $z \in \Gamma$ is a *critical point* if either z is a vertex of Γ, or the height function does not restrict to an embedding, for any neighborhood of z in Γ. W.l.o.g. suppose Γ has only finitely many critical points and no two critical points lie in the same level surface.

Associate a complexity to the position of Γ as follows. First, let $\mathcal{Q}$ be a *good level system*, i.e., a maximal system of level surfaces with the property that (1) none of them contains a critical point and that (2) between any two of them there lies *exactly one* critical point of Γ. Note that the number $w_G(\Gamma) := \#(\mathcal{Q} \cap \Gamma)$ is independent of the choice of $\mathcal{Q}$. Let $v_G(\Gamma) := \min \#(\mathcal{Q}' \cap \Gamma)$, where the minimum is taken over all *good sub-collections*, i.e., over all maximal collections, Q', of level surfaces from Q with the property that between any two surfaces from Q' there lies exactly one *vertex*. Define

$$c_G(\Gamma) := (\ v_G(\Gamma),\ w_G(\Gamma)\ ).$$

Now fix a 2-sphere, G, in S^3 once and for all. This 2-sphere will be our reference Heegaard surface in S^3. Let Γ be any Heegaard graph in S^3, different from a closed curve. Suppose $c_G(\Gamma)$ cannot be reduced, using slides or isotopies ("thin position"). Let Q be a good level system and let Q' be a good sub-collection of Q with $c_G(\Gamma) = (\#(Q' \cap \Gamma), \#(Q \cap \Gamma))$.

Let $(\mathcal{D}, \underline{d})$ be a meridian-system of essential discs in the handlebody $(M_\Gamma, \underline{m}_\Gamma)$. Then every face of $(\mathcal{D}, \underline{d})$ is mapped, under φ_Γ (see 23.5), into exactly one edge of Γ. But edges of Γ may contain the images of more than one face of $(\mathcal{D}, \underline{d})$.

23.25. Lemma. *There is a level plane, Q, from Q with the property that at least one arc from $Q \cap \mathcal{D}$ is outermost and neither bad nor inessential in $(\mathcal{D}, \underline{d})$.*

Proof. Assume the converse. Then for every plane, R, from Q all outermost arcs from $R \cap \mathcal{D}$ are bad or inessential in $(\mathcal{D}, \underline{d})$. Every outermost arc from $R \cap \mathcal{D}$ separates a disc from $\mathcal{D}$ which meets R in the given outermost arc alone. This disc lies either above or below R. It is a *high* disc if it lies above and a *low disc* if it lies below R. As in [Ga, lemma 4.4] it can now be shown that there must be at least one plane, Q, which has a high disc, say D_1, as well as a low disc, say D_2. Set $\ell_i := D_i \cap \partial \mathcal{D}$, $i = 1, 2$. The restrictions $\varphi_\Gamma|\ell_1$, $\varphi_\Gamma|\ell_2$ are, possibly singular, paths in Γ.

Case 1. $\varphi_\Gamma \partial \ell_1 = \varphi_\Gamma \partial \ell_2$.

Define $\Gamma' := \varphi_\Gamma(\ell_1) \cup \varphi_\Gamma(\ell_2)$. Note that the arcs $D_1 \cap Q$, $D_2 \cap Q$ are isotopic in $Q - \varphi_\Gamma \partial \ell_1$. It follows that $(S^3 - U(\Gamma'))^-$ is ∂-reducible. If $\Gamma' \neq \Gamma$, then we are done by lemma 23.24. If $\Gamma' = \Gamma$, then $\Gamma \cap Q = \varphi_\Gamma \partial \ell_1 = \varphi \partial \ell_2$. $\varphi_\Gamma \partial \ell_1$ (or $\varphi_\Gamma \partial \ell_2$) contains at least one vertex of Γ, for otherwise Γ is a curve and we are done. Thus, sliding an edge of Γ intersecting Q across the disc D_1, we get a Heegaard graph which intersects Q in exactly one point, and we are done.

Case 2. $\varphi_\Gamma \partial \ell_1 \neq \varphi_\Gamma \partial \ell_2$.

In this Case, $\varphi_\Gamma \partial \ell_1$ and $\varphi_\Gamma \partial \ell_2$ have at most one point in common, say z.

If both $\varphi_\Gamma \ell_1$ and $\varphi_\Gamma \ell_2$ contain at least one vertex of Γ, then Γ cannot be in a thin position, i.e., $c_G(\Gamma)$ can be reduced. Figure 23.2 illustrates how this can actually be achieved with appropriate slides.

23.26. Example.

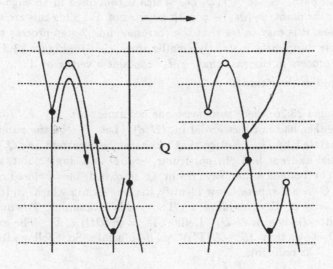

(Figure 23.2.)

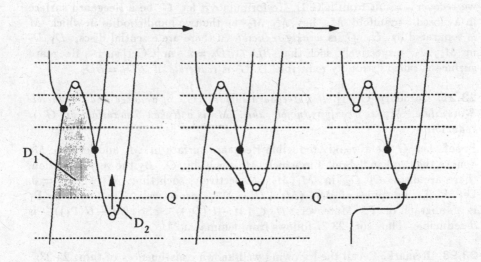

(Figure 23.3.)

If $\varphi_\Gamma \ell_2$ contains no vertex, then we reduce the complexity of Γ in a 2-step process as follows. First note that $\varphi_\Gamma | \ell_2$ must be an embedding. Thus $\varphi_\Gamma \ell_2$ can be pushed upwards across D_2 and across Q. This is our first step. The second step depends on whether $\varphi_\Gamma \ell_1$ contains a vertex of Γ. If not, then in the same way $\varphi_\Gamma \ell_1$ can be pushed downwards across D_1. If $\varphi_\Gamma \ell_1$ contains a vertex, then $\varphi_\Gamma \ell_1$ *cannot* be pushed as before. But then there is a unique non-singular path, α, in $\Gamma \cap \varphi_\Gamma D_1$ which is contained in an edge of Γ and which joins the point $\varphi_\Gamma(\partial \ell_1) - z$ with a vertex of Γ. Slide this arc across D_1. In both cases, it is easy to see that the corresponding 2-step process results in a graph whose complexity is strictly smaller than $c_G(\Gamma)$. Figure 23.3 illustrates the 2-step process in the case that $\varphi_\Gamma \ell_1$ contains a vertex of Γ.

This proves the lemma.

By lemma 23.25, there is at least one outermost arc, say k, from $\mathcal{D} \cap Q$ which is neither bad nor inessential in $(\mathcal{D}, \underline{\bar{d}})$. Let D_1 be the outermost disc in $\mathcal{D}$ separated by k. Notice that k must be recurrent in $Q - \Gamma$ since it is essential and not bad. In particular, $\varphi_\Gamma \partial k$ is a single point, say x, of $Q \cap \Gamma$. $\varphi_\Gamma(k)$ bounds a disc, say B, in Q since it defines a closed curve in Q and since Q is a 2-sphere. Now identify the two arcs in which $\varphi_\Gamma(\partial D_1 \cap \partial \mathcal{D})$ meets that edge of Γ containing x. This yields an annulus. The union of this annulus with B is a disc D. Define $\Gamma' := \varphi_\Gamma(\partial D) \subset \Gamma$. The existence of the disc D shows that $(S^3 - U(\Gamma'))^-$ is ∂-reducible. So it follows from lemma 23.24 that Γ is reducible.

This finishes the proof of the theorem. $\Diamond$

In particular, lemma 23.24 is now true without restriction. As a corollary we recover a result from [CG 1]. To formulate it let G be a Heegaard surface in a closed 3-manifold M. Let M_1, M_2 be the two handlebodies in which M is separated by G. G is *weakly reducible* if there are essential discs, D_1, D_2 in M_1, M_2, respectively, such that $\partial D_1 \cap \partial D_2 = \emptyset$ (in [CG 1] such a Heegaard surface is called "strongly reducible"). G is *reducible* if $\partial D_1 = \partial D_2$.

23.27. Corollary [CG 1]. *If a Heegaard surface, G, of a closed and irreducible 3-manifold, M, is weakly reducible, then M is a Haken 3-manifold or G is reducible.*

Proof. Let G be a weakly reducible Heegaard surface in M and let M_1, M_2 denote the two handlebodies in which M is split by G. By the very definition, there are discs D_1, D_2 in M_1, M_2, respectively, such that $\partial D_1 \cap \partial D_2 = \emptyset$. Let Γ, Γ_1 be spines of M_1, $(M_1 - U(D_1))^-$, respectively. Then w.l.o.g. Γ_1 is a subgraph of Γ. Moreover, $D_2 \subset (M - U(\mathcal{D}_1))^-$. So $(M - U(\Gamma_1))^-$ is ∂-reducible. Thus cor. 23.27 follows from lemma 23.24. $\Diamond$

23.28. Remark. Recall the following well-known consequences of thm. 23.23:
1. Every Heegaard graph in the 3-sphere, S^3, is slide-equivalent to some Heegaard graph contained in the standard 2-sphere of S^3.
2. It can be decided if a given 3-manifold is S^3, provided it can be decided whether Heegaard graphs are reducible.
3. Every Heegaard graph in a handlebody is slide equivalent to a system of trivial arcs (by the handle-addition lemma it is slide-equivalent to a Heegaard graph in the complement of a meridian-system).

23.29. *Good Pencils.*

Recall from section 14 the definition of (very) good pencils. Recall further from section 15 that strongly normal surfaces in one-relator 3-manifolds not only have good pencils, but that they were crucial for our study of surfaces in (generalized) one-relator 3-manifolds. Now, for higher-relator 3-manifolds, the situation is slightly more involved, but at the same time surprisingly similar to the one-relator case. Again good pencils do exist and play a crucial role. Indeed, our proof of our finiteness thm. 23.33 relies heavily on their properties. In fact, good pencils behave rather similar in the more general situation; the essential difference being that they may be a bit more difficult to find.

We begin with the following observation.

23.30. Lemma. *Let M be a 3-manifold. Let $\Gamma \subset M$ be a Heegaard graph and let $S^+ \subset M$ be a 2-manifold. Suppose $S := S^+ \cap M_\Gamma$ is strictly normal w.r.t. to some meridian-system $\mathcal{D} \subset M_\Gamma$. Then each recurrent arc from $\mathcal{D} \cap S$ is essential in S^+.*

Remark. Recall from 14.5 that an arc from $\mathcal{D} \cap S$ is *recurrent* if both its end-points lie in a single component of $S^+ \cap U(\Gamma)$. Such an arc is *essential* in S^+ if if its union with some arc in $S^+ \cap U(\Gamma)$ defines a non-contractible curve in S^+.

Proof of lemma. Assume the converse. Then there is at least one recurrent arc, say ℓ, from $\mathcal{D} \cap S$ which is inessential in S^+. Let B be that component from $S^+ \cap U(\Gamma)$ which contains $\partial \ell$. Then, by the very definition, ℓ separates a disc, say A_0, from $(S^+ - B)^-$. But A_0 has to contain at least one component of $S^+ \cap U(\Gamma)$, for ℓ is essential in S since S^+ is in strictly normal position. On the other hand, by prop. 23.3, there is a system of arcs from $S \cap \mathcal{D}$, essential *and* recurrent in S, which meets each component of $S^+ \cap U(\Gamma)$, especially each such component contained in A_0. It is easily verified that this is impossible since A_0 is a disc. $\Diamond$

We are now ready to establish the existence of very good pencils. To this end recall that every Heegaard graph, Γ, in a 3-manifold M fixes, dually, an n-relator structure for $(M - U(\Gamma^{(0)})^-$. Indeed, $(M - U(\Gamma^{(0)}))^- = M_\Gamma^+(k_\Gamma)$ (see 23.5). In particular, a surface S^+ in M lies in $M_\Gamma^+(k_\Gamma)$ if and only if it is in general position to Γ. In this case, *very good pencils* for S^+ are the very good pencil for $S^+ \subset M_\Gamma^+(k_\Gamma)$ as defined in in 14.2.

23.31. Proposition. *Let M be a 3-manifold. Let $\Gamma \subset M$ be a Heegaard graph and let $S^+ \subset M$ be a 2-manifold. Suppose S^+ is in strictly normal position w.r.t. Γ. Then S^+ admits a very good pencil, for every component of k_Γ.*

Remark. It will be apparent from the proof below that there is an essential meridian-system, $(\mathcal{D}, \underline{d})$, in $(M_\Gamma, \underline{m}_\Gamma)$ such that the union of all the very good pencils for components of k_Γ forms a sub-system of $S^+ \cap \mathcal{D}$.

Proof of proposition. Set $S := M \cap S^+$. Let $\mathcal{D}$ be a meridian-system of essential discs in the handlebody $(M_\Gamma, \underline{m}_\Gamma)$ so that S is strictly normal w.r.t. $\mathcal{D}$. Let k_i, $1 \leq i \leq n$, be a component of the curve-system k_Γ. We then have to construct a very good pencil for k_i, i.e., we have to find a component, I_0, from $I_i := U(k_i) \cap \partial \mathcal{D}$ satisfying the conditions of def. 14.2. Here it is to be understood that $\bigcup U(k_i) = A_\Gamma \subset \partial M_\Gamma$ (notation as in 23.5).

If $S \cap U(k_i) = \emptyset$, we are done.

Suppose $S \cap U(k_i) \neq \emptyset$. Recall that no arc from $\mathcal{D} \cap S$ has both its end-points in one component of the interval-system I_i since S^+ is strictly normal. Furthermore, by lemma 23.2, there is at least one non-recurrent arc, l, from $\mathcal{D} \cap S$ which has at least one end-point in I_i. Let D be the disc from $\mathcal{D}$ containing l. Moreover, let D' be one of the two discs in which D is separated by l. W.l.o.g. we may suppose that l and D' are both chosen so that, in addition, D' contains no non-recurrent arc from $D \cap S$ with an end-point in I_i. Define I_i' to be the union of all those component from I_i which are entirely contained in the arc $D' \cap \partial D$. Recalling our choice of l, it is easily checked that I_i' has to be non-empty (S^+ is strictly normal). Given I_i', let $\mathcal{L}_i$ denote the union of all those arcs from $\dot{D} \cap S$ which have an end-point in I_i'. Then $\mathcal{L}_i$ is non-empty since I_i' is non-empty and since $U(k_i)$ contains at least one component of ∂S. Furthermore, by our choice of l, it follows that $\mathcal{L}_i$ consists of recurrent arcs in S. In particular, $\mathcal{L}_i$ consists of arcs whose end-points lie in I_i.

We claim the required component, I_0, of I_i can be found in I_i'. If this were not the case, there were at least one arc, say l_0, from $\mathcal{L}_i$ joining two non-adjacent components of I_i. In this case, let l_0 be chosen to be minimal, i.e. in such a way that l_0 separates a disc, D_0, from D' not meeting l and with the property that every arc from $D_0 \cap S$, different from l_0, joins adjacent components of I_i. Then of course $D_0 \cap \partial D$ has to contain the required component I_0. This proves the claim. $\Diamond$

23.32. *Estimates.*

Following section 21, we now use the properties of very good pencils to estimate the minimal intersection of essential surfaces with Heegaard graphs. As a result we obtain a direct generalization of prop. 21.8 from one-relator 3-manifolds to general n-relator 3-manifolds (see cor. 23.34). In fact, technically, we obtain this generalization by reducing the n-relator case to the one-relator case.

Recall from section 21 that, given a surface S, the complexity $c(S)$ denotes the maximal number of pairwise disjoint, pairwise non-parallel, non-contractible and simple closed curves in S, i.e.,

$$c(S) = \begin{cases} 1, & \text{if } \chi S = 0 \\ -\tfrac{3}{2}\chi(S), & \text{otherwise} \end{cases}$$

23.33. Theorem. *Let M be a Haken 3-manifold and let $\Gamma \subset M$ be a Heegaard graph. Suppose S^+ is a surface in M which is strictly normal w.r.t. Γ. Then there is a product, h, of Dehn twists along essential tori in M and an ambient isotopy, α_t, $t \in I$, in M such that*

$$\#(hS^+ \cap \alpha_1 \Gamma)^- \leq 32 \cdot e(\Gamma) \cdot c(S^+),$$

where $e(\Gamma)$ denotes the total number of edges of Γ.

Proof. S is strictly normal. Hence, in the notation from 23.5, there is a meridian system $(\mathcal{D}, \underline{d})$ of essential discs in the handlebody $(M_\Gamma, \underline{m}_\Gamma)$ such that $S := S^+ \cap M_\Gamma$ is strictly normal w.r.t. $\mathcal{D}$.

Let k_i, $1 \leq i \leq n$, be a component of k_Γ. Consider the *one-relator* 3-manifold $M^+(k_i) \subset M$. Let E_i, denote the 2-handle of E_Γ attached to the curve k_i. Set $I_i := \mathcal{D} \cap U(k_i)$, set $S_i^+ := S^+ \cap M^+(k_i)$ and set $S_i := S_i^+ \cap M = S^+ \cap M$. Since $\#k_\Gamma = e(\Gamma)$, it remains to show that $\#(S^+ \cap E_i) \leq 4 \cdot c(S^+)$, modulo isotopy in $M^+(k_i)$ fixing boundary.

If $M^+(k_i)$ is irreducible and ∂-irreducible and if S_i^+ is strongly normal in $M^+(k_i)$, the claim follows from the proof of prop. 21.8.

Assume the converse. Indeed, suppose S_i^+ is not strongly normal in $M^+(k_i)$. Then recall S^+ is strictly normal in $M^+(k)$. Thus it follows the existence of an (outermost) arc, l, from $S_i^+ \cap \mathcal{D} = S^+ \cap \mathcal{D}$ which has both its end-points in one component of $(\partial \mathcal{D} - I_i)^-$. In particular, l separates a disc, D_0, from $\mathcal{D}$ with $D_0 \cap \partial \mathcal{D} \subset (\partial \mathcal{D} - I_i)^-$. Splitting S_i^+ along D_0, we obtain a surface which intersects E_i in the same number of discs as S_i^+ and whose Euler characteristic is not smaller than χS^+. Eventually, we find a strongly normal surface, G_i, in $M^+(k_i)$ with $\#(G_i \cap E_i) = \#(S^+ \cap E_i)$. Theorem 23.6 clearly follows when we show $\#(G_i \cap E_i) \leq 32 \cdot c(G_i)$, mod isotopy in $M^+(k_i)$ fixing boundary.

If $M^+(k_i)$ is reducible or ∂-reducible, then, by prop. 15.16, (M, k_i) is ∂-reducible. W.lo.g. one of the discs from $\mathcal{D}$ is a ∂-compression disc in (M, k_i). This disc cannot meet G_i since G_i is strongly normal. Thus, splitting along ∂-compression discs if necessary, we eventually get an irreducible and ∂-irreducible one-relator 3-manifold containing G_i as a strongly normal surface. Thus, by the proof of prop. 21.8, $\#(G_i \cap E_i) \leq 32 \cdot c(G_i)$ mod isotopy in $M^+(k_i)$ fixing boundary. $\Diamond$

Combining the previous theorem with prop. 23.11, we get the following estimate for the intersection of Heegaard graphs with essential surfaces.

23.34. Corollary. *Let M be a Haken 3-manifold. Let $\Gamma \subset M$ be a Heegaard graph and let $S^+ \subset M$ be an essential surface. Then there is a product, h, of Dehn twists along essential tori in M, a Heegaard graph, Γ', slide-equivalent to Γ, and an ambient isotopy, α_t, $t \in I$, such that*

$$\#(hS^+ \cap \alpha_1 \Gamma')^- \leq 32 \cdot e(\Gamma') \cdot c(S^+). \quad \Diamond$$

Remark. 1. Note that $e(\Gamma') \leq \# \partial\Gamma - \chi(\Gamma, \partial\Gamma)$ and that $e(\Gamma') = e(\Gamma)$ if $\partial M = \emptyset$.

2. The corollary gives an *upper bound* for the intersection of Heegaard graphs with essential surfaces. In contrast, reasonable *lower bounds* are harder to come by. The reason is the following phenomenon. Let F be a closed, orientable surface with non-trivial genus and let $\ell \subset F$ be a non-separating, simple closed curve. Consider the product $M := F \times S^1$. Note that F is a horizontal and that $T := \ell \times S^1$ is a vertical surface in M. Let F_n, $n \geq 1$, be the surface obtained from T and n copies of F by performing cut-and-paste along intersection curves. Now, let $\Gamma_F \subset F$ be a spine, i.e., a graph such that $F - \Gamma_F$ is an open disc. Let $x \in \Gamma_F$ and set $\Gamma := \Gamma_F \cup x \times S^1$. Then Γ is a Heegaard graph in M and it is easy to verify that, by sliding the edge corresponding to $x \times S^1$, we obtain a Heegaard graph Γ_n, slide-equivalent to Γ, such that $\#(\Gamma_n \cap F_n) = 1$.

23.35. *Heegard-Cylinders.*

We show that the intersection of strictly normal surfaces with Heegaard graphs is governed by the intersection of this surface with Heegaard cylinders. This observation is very useful for theoretical purposes since it reduces the study of intersections of essential surfaces with Heegaard graphs to the better known situation of intersections of codim 1 objects.

The following definition modifies 21.2 in order to accommodate Heegaard graphs.

23.36. Definition. *Let Γ be a Heegaard graph in a 3-manifold M. Let l be an edge from Γ. A Heegaard cylinder (for l) is a non-contractible annulus A in M, $A \cap \partial M = \emptyset$, such that $A \cap \Gamma = A \cap l$ is a non-trivial arc which is essential in $A(l)$. A Heegaard cylinder-system (for Γ) is any system of disjoint Heegaard cylinders.*

The following result concerning Heegaard cylinders generalizes proposition 21.4.

23.37. Proposition. *Let M be a Haken 3-manifold. Let $\Gamma \subset M$ be a Heegaard graph and let $S \subset M$ be a surface in M which are in strictly normal position. Then there is a Heegaard cylinder-system, $\mathcal{A}(\Gamma, S)$, associated to Γ, S such that the following holds:*

(1) Every edge of Γ meets exactly two cylinders from $\mathcal{A}(\Gamma)$.

(2) $S \cap \Gamma \subset S \cap \mathcal{A}(\Gamma)$ and $S \cap \mathcal{A}(\Gamma)$ consists of closed curves which are essential in both S and $\mathcal{A}(\Gamma)$.

Proof. The proof is similar to the proof of prop. 21.4.

Here is a quick construction of $\mathcal{A}(\Gamma, S)$. Γ and S are in strictly normal position. So, by def. 23.7, there is a meridian-system, $(\mathcal{D}, \underline{d})$, of essential discs in the handlebody $(M_\Gamma, \underline{m}_\Gamma)$ such that $S \cap M_\Gamma$ is strictly normal w.r.t. $\mathcal{D}$. Let $A_1, A_2, ..., A_n$ be the annuli from $\mathcal{A}_\Gamma$ (see 23.5). Set $K_i := (\partial\mathcal{D} - A_i)^-$, $1 \leq i \leq n$. Define $\mathcal{B}_i := U(\partial U(K_i) - \partial\mathcal{D})^-$, where the regular neighborhoods are taken in $\mathcal{D}$. Then $\mathcal{B}_i$ is a system of squares. These squares may or may not fit across an edge of Γ to form an annulus. Let $\mathcal{A}_i$ be the collection of all annuli obtained in this way from squares of $\mathcal{B}_i$. Note that all components of $\mathcal{A}_i$ are Heegaard cylinders. Moreover, applying a small general position isotopy if necessary, we may suppose that $\mathcal{A}_i \cap \mathcal{A}_j = \emptyset$ if $i \neq j$. Then define $\mathcal{A}(\Gamma, S) := \bigcup \mathcal{A}_i$.

$\mathcal{A}(\Gamma, S)$ is a Heegaard cylinder-system. It remains to show that it has the desired properties. But according to prop. 23.31, $S \cap \mathcal{D}$ contains a very good pencil. By lemma 23.30, this pencil corresponds to a system of essential closed curves in S ($\mathcal{D} \cap S$ has no bad arcs). Thus it has to lie in $\mathcal{A}(\Gamma, S)$. Properties (1) and (2) of prop. 23.37 then follow immediately. $\Diamond$

The proof for the next two results illustrates the relevance of the concept of Heegaard cylinder-systems.

23.38. Corollary. *Let M be a Haken 3-Manifold. Let $\Gamma \subset M$ be a Heegaard graph and let V be the characteristic submanifold in M. Then there is a Heegaard graph Γ', slide-equivalent to Γ, so that*

$$\#(\Gamma' \cap (\partial V - \partial M)^-) \leq 6 \cdot e(\Gamma').$$

Remark. See [Ko 2,3] for better estimates.

Proof. $(\partial V - \partial M)^-$ is an essential 2-manifold in M. Thus, by prop. 23.11, there is a Heegaard graph, Γ', which is slide-equivalent to Γ and strictly normal w.r.t. $(\partial V - \partial M)^-$. Then, by prop. 23.37, there is a Heegaard cylinder-system, $\mathcal{A} = \mathcal{A}(\Gamma', S)$, with properties (1) and (2) of 23.37.

$\mathcal{A}$ has at most $2 \cdot e(\Gamma')$ annuli. Let A be an annulus from $\mathcal{A}$. Then $A \cap \Gamma \subset A \cap S$ and $A \cap S$ is a system of essential curves in A. In particular, $A \cap S$ separates A into a collection of annuli. Let $A_i, 1 \leq i \leq n$, be all those annuli from them with $\partial A_i \subset A \cap S$. Then $A_i \subset V$ or $A_i \subset (M - V)^-$, and $\partial A_i \subset (\partial V - \partial M)^-$. Moreover, we may suppose A_i is essential in V resp. $(M - V)^-$, for otherwise $\#(\Gamma \cap S)$ can be reduced, by some isotopy of Γ (and the new graph has an obvious Heegaard cylinder-system). Therefore and since characteristic submanifolds are full (see [Joh 1, def. 8.2]), it follows that $\#(A \cap \Gamma) \leq 3$. Hence $\#(\mathcal{A} \cap \Gamma) \leq 3 \cdot \#\mathcal{A} \leq 6 \cdot e(\Gamma')$. $\Diamond$

23.39. Recall that the *Heegaard genus*, $g(M)$, of a closed 3-manifold, M, is defined to be the minimum taken over the genera of all Heegaard surfaces in M. According to Haken [IIa 3], the Heegaard genus for closed 3-manifolds is additive under taking connected sums. This fact is not necessarily true for manifolds glued (or split) along surfaces different from discs or 2-spheres. Using prop. 23.37, however, we can now give estimates for more general situations as well. To do this note first that the Heegaard genus, $g(M)$, is also defined for 3-manifolds, M, with non-empty boundary since, according to our convention (see def. 23.6), Heegaard surfaces are always closed.

23.40. Proposition. *Let S be an incompressible surface in a closed 3-manifold M. Set $M' := (M - U(S))^-$. Suppose M' is a simple 3-manifold. Then*

$$g(M') \leq 5 \cdot g(M) + 2 \cdot \text{genus } S \leq \begin{cases} 5 \cdot g(M') + 2 \cdot \text{genus } S, & \text{if } S \text{ is separating} \\ 5 \cdot (g(M) + 1) + 2 \cdot \text{genus } S, & \text{otherwise}. \end{cases}$$

Proof. Let Γ' be a Heegaard graph of M' with $g(M') = \text{genus } \partial U(\Gamma' \cup \partial M')$. Then $g(M') = \text{genus } \partial(U(\Gamma') \cup U(S))$. Now $U(S)$ is a product I-bundle, i.e., $U(S) \cong S \times I$. Let ℓ be an I-fiber of this product structure with $\partial \ell \cap \partial \Gamma' = \emptyset$. Set $U := U(\Gamma') \cup (U(S) - U(\ell))^-$. Then the genus of ∂U equals $g(M')$ or $g(M') + 1$ according whether S is separating or not. Now, $(U(S) - U(\ell))^-$ is a handlebody since it is the regular neighborhood of a surface with non-empty boundary. It follows that U as well as $(M - U)^-$ is a handlebody. So ∂U is a Heegaard surface for M. Hence

$$g(M) \leq \text{genus } \partial U = \begin{cases} g(M'), & \text{if } S \text{ is separating} \\ g(M') + 1, & \text{otherwise}. \end{cases}$$

For the other inequality of prop. 23.40, let Γ be a Heegaard graph of M with $g(M) = \text{genus } \partial U(\Gamma)$. Then, by prop. 23.11, we may suppose Γ has been slid so that it is strictly normal w.r.t. S. In particular, $(S \cap (M - U(\Gamma))^-$ is incompressible in $(M - U(\Gamma))^-$. So $(M' - U(\Gamma))^-$ is a handlebody since incompressible 2-manifolds split handlebodies into handlebodies. It follows that $\partial U(\Gamma \cup S)$ consists of Heegaard surface in M'. In particular, $g(M') \leq$ genus $\partial U(\Gamma \cup S)$. Now, M' is supposed to be simple and so has no essential annuli whatsoever. Thus it follows from prop. 23.37 that $\#(\Gamma \cap S) \leq 2 \cdot e(\Gamma)$, where $e(\Gamma)$ denotes the total number of edges of Γ (recall that, for closed 3-manifolds, $e(\Gamma)$ is an invariant of the slide-equivalence class of Γ). Hence

$$
\begin{aligned}
g(M') &\leq \text{genus } \partial U(\Gamma \cup S) = 1 - \tfrac{1}{2}\chi \partial U(\Gamma \cup S) = 1 - \chi U(\Gamma \cup S) \\
&= 1 - (\chi\Gamma + \chi S - \chi(\Gamma \cap S)) \\
&\leq 1 - \chi\Gamma - \chi S + 2 \cdot e(\Gamma), \quad \text{since } \chi(\Gamma \cap S) = \#(\Gamma \cap S), \\
&= 1 - 2 \cdot (1 - g(M)) - 2 \cdot (1 - \text{genus } S) + 3 \cdot (g(M) - 1), \\
&\qquad \text{since } 3 \cdot (\text{genus } \partial U(\Gamma) - 1) = e(\Gamma), \\
&= 5 \cdot g(M) + 2 \cdot \text{genus } S - 6 \\
&\leq 5 \cdot g(M) + 2 \cdot \text{genus } S. \ \Diamond
\end{aligned}
$$

In a similar way one can work out estimates for non-simple 3-manifolds and manifolds with boundaries.

23.41. *Deformations of Heegaard Graphs.*

In section 17 we have studied homotopies of generalized Heegaard strings. Homotopies of Heegaard graphs in general are much harder to understand. Here we consider only deformations of Heegaard graphs into the boundary of 3-manifolds. It turns out that this special problem is actually rather simple and can be handled by standard techniques. We give the relevant argument for completeness.

We begin with the following observation which we prove using an argument originally due to Stallings [Sta 2].

23.42. Lemma. *Let N be an irreducible 3-manifold with or without boundary and let $\Gamma \subset N$ be a Heegaard graph. Let $S \subset N$ be a closed, incompressible surface different from a 2-sphere. Then Γ cannot be deformed rel boundary into the complement of S.*

Proof. Assume the converse. Then there is homotopy, h_t, $t \in I$, of id_N which pushes Γ into $N - S$. In particular, $h_1(\partial M) \subset N - S$, where $M := (N - U(\Gamma))^-$. But M is a handlebody and every surface in a handlebody is compressible. Thus, by using the usual surgery-technique on the pre-image $h_1^{-1}(S)$ if necessary, we may suppose that $h_1^{-1}S = \emptyset$.

Let Γ' be the spine of M, i.e., a connected graph with $(M - U(\Gamma'))^- = \partial U(\Gamma') \times I$. Then, by what we have just seen, $h_1(\Gamma') \cap S = \emptyset$. So, in particular every curve in N can be deformed out of of S since every curve can be deformed into Γ'. It therefore follows that S must be separating. Let N_1, N_2 be the two sub-manifolds into which N is split by S. Let the indices be chosen so that $h_1(\Gamma') \subset N_2$. Now, let z be a vertex of Γ' and let $g = h_1|\Gamma'$ be deformed in N_1 so that $g(z) \in S$. Recall $g_* : \pi_1(\Gamma', z) \to \pi_1(N, gz)$ is a surjection. So, in particular, every loop in N_1, based in hz, can be deformed rel base-point into N_2. Since S is incompressible, it follows that every such loop can be deformed rel base-point into S. N_1 is aspherical since N is and since S is incompressible [Wa 2]. Hence there is no obstruction to construct a deformation retraction $p : N_1 \to S$, say. Given a simple closed curve $k \subset S$, p can be deformed, by the usual surgery argument, so that afterwards $p^{-1}k$ is an essential annulus in N_1. Given a system $\mathcal{L}$ of arcs in $S' := (S - U(k))^-$ which splits S' into a disc, p can be deformed rel k so that afterwards $p^{-1}\mathcal{L}$ is a system of discs (squares). Let U be the regular neighborhood of $p^{-1}k \cup p^{-1}\mathcal{L}$ in N_1. Then $(\partial U - \partial N_1)^-$ is an annulus. One boundary curve of this annulus bounds a disc in S and so the other bounds a disc in $S \cup \partial N$ since N_1 is ∂-irreducible and since S is incompressible. It follows that $(N_1 - U)^-$ is a 3-ball since N_1 is irreducible. Altogether, it follows that N_1 is the I-bundle over a surface. More precisely, it must be a product I-bundle since every curve in N_1 can be deformed into the boundary. Thus one component of ∂N_1 must be a component of ∂N. But Γ meets every boundary component of N, and so it cannot be deformed (rel boundary) into $N_2 = (N - N_1)^-$. Contradiction.
$\diamond$

23.43. Theorem. *Let N be an irreducible 3-manifold with boundary and let $\Gamma \subset N$ be a Heegaard graph. Then N is a handlebody if and only if Γ can be deformed into ∂N rel boundary.*

Remark. For a related result concerning *generalized* one-relator 3-manifolds see cor. 17.8 (see also cor. 17.9).

Proof. One direction follows immediately. For the other direction let $\mathcal{D} \subset N$ be a maximal system of essential discs. Since Γ can be deformed into ∂N, it follows from the previous lemma that no component from $\partial U(\mathcal{D} \cup \partial N) - \partial N$ can be an incompressible surface in N different from a 2-sphere. So every one of those components is a 2-sphere, and it follows that N is a handlebody since N is irreducible. $\Diamond$

24. The Unknotting Theorem.

In this section we are going to continue our study of 2-spheres in 3-manifolds. In section 23 we have seen that the existence of an essential 2-sphere gives rise to the existence of an essential 2-sphere which intersects a given Heegaard graph in one point only (modulo slides). This is Haken's classical 2-sphere result. We have seen how to extend this result to arbitrary *essential* surfaces. Here in turn we are interested in possible generalizations of it to *inessential* surfaces. In the proof of Waldhausen's theorem (thm. 23.23), we have already seen that one can make use of certain aspects of our theory of strictly normal surfaces for certain inessential 2-spheres. In this section we study yet another aspect in which inessential 2-spheres become relevant, namely the problem of local knots for Heegaard graphs. Specifically, we first establish the existence of local knots for Heegaard graphs - a phenomenon which does not occur for Heegaard strings. Then, as the main result of this section, we show that basically all local knots for Heegaard graphs can be resolved by slidings (see thm. 24.5). This is a result which has obvious relevance for any study of Heegaard graphs at large. It will indeed become crucial in the next section, where we give a refined version of the Reidemeister-Singer theorem, and in chapter V, where we will develop a deformation theory for Heegaard graphs.

24.1. *Existence of Local Knots.*

We are going to show how to create local knots for Heegaard graphs. But first some terminology.

Recall from section 14 that a closed surface, S^+, in M is *normal* (with respect to a Heegaard-graph $\Gamma \subset M$) if

(1) S^+ is in general position with respect to Γ, and
(2) the surface $S := (S^+ - U(\Gamma))^-$ is either a disc or incompressible in $M_\Gamma := (M - U(\Gamma))^-$.

Moreover, let us call S^+ *reasonably normal* if, in addition,

(3) S is not ∂-parallel in M_Γ.

Here we say S^+ is in *general position* with respect to Γ if S^+ does not meet any vertex of Γ and if it intersects edges of Γ transversally.

Every incompressible surface in an irreducible 3-manifold, M, can be isotoped into a reasonably normal surface (with respect to any Heegaard graph in M). But note that the very concept of (reasonably) normal surfaces does not require the surface itself to be incompressible. Thus *à priori* the notion of normal surfaces is much more general than that of incompressible surfaces. Indeed, (reasonably) normal surfaces are much more frequent than incompressible surfaces. They do often exist even when incompressible surfaces are absent. They are also important. Figure 24.1 illustrates a typical aspect of the phenomenon. It establishes the existence of local knots for Heegaard graphs, and so the existence of reasonably normal but inessential 2-spheres.

24.2. Example. (Existence of local knots)

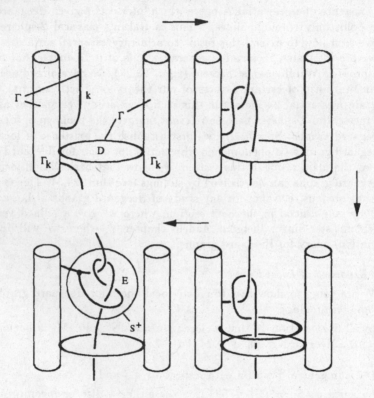

(Figure 24.1.)

To explain the construction, let Γ be an irreducible Heegaard graph in a closed 3-manifold, M, different from the 3-sphere. Set $\Gamma_k := (\Gamma - k)^-$, for every edge k of Γ. For our example let us make the (weak) assumption that there is an edge, k, of Γ such that $(M - U(\Gamma_k))^-$ contains a non-separating disc, say D. It is not difficult to construct examples of *irreducible* Heegaard graphs satisfying this hypothesis. Let us denote the arc $(k - U(\Gamma_k))^-$ by k again. Since D is non-separating, k can be isotoped in $(M - U(\Gamma_k))^-$ so that afterwards the end-points of k lie in different components of $U(D) - D$, where the regular neighborhood $U(D)$ is taken in $(M - U(\Gamma_k))^-$ (in particular, the Heegaard surface associated to Γ is irreducible, but weakly reducible).

The sequence of pictures in fig. 24.1 results in a Heegaard graph, Γ', and a 3-ball, E, as shown. We claim $S^+ := \partial E$ is reasonably normal with respect to Γ'. To see this observe that $S := (S^+ - U(\Gamma))^-$ is a thrice punctured 2-sphere.

In particular, every simple closed curve in S is parallel to some curve from ∂S. It follows that S cannot be compressible in $(M - U(\Gamma))^-$, for Γ, and so Γ', is irreducible. Moreover, observe that $\Gamma' \cap E$ contains a knotted arc. Indeed, the projection above shows an alternating arc in $\Gamma' \cap E$ and so this arc must be knotted in E (see e.g. [Cr]). It follows that $(E - U(\Gamma'))^-$ cannot be a parallelity region for S in $(M - U(\Gamma'))^-$. But $(M - (E \cup U(\Gamma')))^-$ cannot be a parallelity region either, for M is supposed to be different from the 3-sphere. This proves the claim. $\diamondsuit$

The previous example indicates that Haken's 2-sphere result ([Ha 3] or prop. 23.21) has no direct generalization to inessential 2-spheres. Specifically, irreducible Heegaard graphs may admit reasonably normal 2-spheres. On the other hand, we know from cor. 23.4 (see also [Joh 5]) that there are no 2-spheres at all which are strictly normal with respect to irreducible Heegaard graphs, i.e., the above phenomenon does not occur for those surfaces. This is our starting point.

24.3. *Resolving Local Knots.*

Having established the existence of local knots for Heegaard graphs, we next show how to untie them. For the convenience of the reader we begin by giving a self-contained formulation of def. 23.7.

24.4. Definition. *Let* Γ *be a Heegaard graph in a closed 3-manifold* M. *A reasonably normal 2-manifold* S^+ *in* M *is called* strictly normal *(with respect to* Γ) *if there is a meridian-system,* $\mathcal{D}$, *in* $(M - U(\Gamma))^-$ *such that, for every component,* D, *from* $\mathcal{D}$, *the following holds:*

(1) D intersects $S := (S^+ - U(\Gamma))^-$ in arcs alone and every arc from $D \cap S$ is essential in S (in the sense that it does not separate a disc from S),

(2) if b is an arc from $D \cap S$ which joins two different curves from $S \cap \partial U(\Gamma)$, then each (open) component of $\partial D - \partial b$ intersects both the latter curves.

We are now in the position to give our generalization of Haken's 2-sphere result to reasonably normal 2-spheres in the form it is needed later. For this let us call a (possibly disconnected) finite graph Γ in a 3-manifold N, $\Gamma \cap \partial N = \partial \Gamma$, *unknotted* (in N), if there is a system, $\mathcal{D}$, of discs in N, $\mathcal{D} \cap \partial N = \partial \mathcal{D}$, such that $\mathcal{D} \cap \Gamma = \emptyset$ and that the pair $((N - U(\mathcal{D}))^-, \Gamma)$ is homeomorphic to the pair $(U(\Gamma), \Gamma)$, where the regular neighborhoods are taken in N.

24.5. Unknotting Theorem. *Let* M *be an irreducible 3-manifold different from the 3-ball or the 3-sphere. Let* Γ *be an irreducible Heegaard graph in* M. *Suppose* N *is a system of (pairwise disjoint) 3-balls in* M *such that* ∂N *is in general position with respect to* Γ *and that* $(\Gamma - N)^-$ *consists of arcs. Then there is a Heegaard graph,* Γ', *in* M *such that*

(1) Γ' *is slide-equivalent to* Γ,

(2) $\Gamma' \cap (M - N)^- \subset \Gamma \cap (M - N)^-$ *and*

(3) $\Gamma' \cap N$ *is unknotted in* N.

Proof. In the course of the proof we will have to alter Γ and N. To do this in a controlled way we first introduce a complexity for "admissible pairs". Here a pair (Γ, N) is called an *admisssible pair* if the following holds:

(1) Γ is a Heegaard graph in M,

(2) N is a system of pairwise disjoint 3-balls in M each of which meets Γ,

(3) ∂N is in general position with respect to Γ, and

(4) $(\Gamma - N)^-$ consists of arcs.

A 3-ball, N_0, from N will be called a *special 3-ball* (for (Γ, N)) if N_0 contains a vertex, x, of Γ such that $(N_0, N_0 \cap \Gamma) \cong (U(x), U(x) \cap \Gamma)$. At this point recall that all Heegaard graphs are supposed to be tri-valent. So the boundary of a special 3-ball intersects Γ in exactly three points. Given an admissible pair (Γ, N), denote by N_u the union of all components of N which are *not* special 3-balls. Set

$$c(\Gamma, N) := \chi(\partial N_u - U(\Gamma))^-.$$

Note that $c(\Gamma, N) \leq 0$ since Γ is supposed to be irreducible. The proof is by induction on the complexity $c(\Gamma, N)$.

Let N and Γ be given as in the theorem. Then, by hypothesis, (Γ, N) is an admissible pair.

Suppose $c(\Gamma, N) = 0$. Let N_0 be a component of N_u. Then $\chi(\partial N_0 - U(\Gamma))^- = 0$ since no component of $(\partial N - U(\Gamma))^-$ has a strictly positive Euler characteristic (Γ is supposed to be irreducible). So $\#(\Gamma \cap \partial N_0) = 2$. Thus, by prop. 23.22, $\Gamma \cap N_0$ must be an arc since Γ is supposed to be irreducible. Moreover, $(\partial N_0 - U(\Gamma))^-$ is an incompressible annulus in $M_\Gamma = (M - U(\Gamma))^-$ since Γ is irreducible. Thus $(N_0 - U(\Gamma))^-$ is a solid torus since M_Γ is a handlebody. Therefore $\Gamma \cap N_0$ is an unknotted arc in N_0. It follows that $\Gamma \cap N$ is unknotted since $N - N_u$ consists of special 3-balls. The induction beginning is therefore established.

For the induction step, suppose that $c(\Gamma, N) < 0$ and that the theorem holds for all admissible pairs $c(\Gamma', N')$ with $c(\Gamma', N') > c(\Gamma, N)$.

Now, since $\Gamma \subset M$ is a Heegaard graph, we may choose a *meridian-collection* for Γ, i.e., a collection, $\mathcal{D}$, of discs in M with $\mathcal{D} \cap \Gamma = \partial \mathcal{D}$ and such that $\mathcal{D} \cap (M - U(\Gamma))^-$ is a meridian-system in the handlebody $(M - U(\Gamma))^-$. In particular, we may suppose that all self-singularities of $\mathcal{D}$ are in $\partial \mathcal{D}$. We consider $\partial \mathcal{D}$ as a collection of edge-paths in Γ. The edges of Γ induce a boundary-pattern for $\mathcal{D}$ (= collection of arcs in $\partial \mathcal{D}$). Thus $\mathcal{D}$ may be considered as a (possibly disconnected) polygon. The faces of this polygon lie in edges of Γ and neighboring faces lie in different edges. An arc, b, from $\mathcal{D} \cap \partial N'$ is called a *bad arc* for (Γ, N) if (1) it joins two different points, x, y, from $\Gamma \cap \partial N$ and if (2) it separates a disc, D_b, from $\mathcal{D}$ such that $D_b \cap \partial \mathcal{D}$ is an edge-path meeting x in its end-points only. A bad arc, b, from $\mathcal{D} \cap \partial N$ is a *special bad arc* if it separates a disc, D_b, from $\mathcal{D}$ with $D_b \cap \partial N = b$ and containing exactly one corner-point of the polygon $\mathcal{D}$.

Having established the above notation we are now ready to continue the proof. First we treat the subject under the following hypothesis.

Hypothesis A. Every component of ∂N_u is reasonably normal with respect to Γ.

We first claim that w.l.o.g. there is no special bad arc b from $\mathcal{D} \cap \partial N$ which is a component of $\mathcal{D} \cap \partial N_u$. Assume the converse and let b be such a special bad arc. Then b separates a disc $D_b \subset \mathcal{D}$ with $D_b \cap \partial N = b$ and containing exactly one corner point of $\mathcal{D}$. Define $N' := (N - U(U(D_b)))^- \cup U(D_b)$. Then (Γ, N') is an admissible pair with $c(\Gamma, N') > c(\Gamma, N)$. By induction, the theorem holds for (Γ, N'). It is easily verified that it then must hold for (Γ, N) as well since $U(D_b) \cap (N - U(D_b))^-$ is a disc which intersects Γ in exactly one point.

N_u is not empty since $c(\Gamma, N) < 0$. Moreover, by hypothesis A, every component from ∂N_u is reasonably normal. It follows from cor. 23.4 that ∂N_u is not strictly normal (Γ is supposed to be irreducible). Thus there is at least one bad arc, b, from $\mathcal{D} \cap \partial N_u$ which (by what we have just seen) is not a special bad arc. By definition, b connects two different points x, y in Γ and separates a disc, D_b, from $\mathcal{D}$ such that the path $D_b \cap \partial \mathcal{D}$ meets x, say, in its end-points only. Without loss of generality we may suppose that b is chosen to be outermost in the sense that b is the only bad arc from $\mathcal{D} \cap \partial N$, contained in D_b, which is not a special bad arc. Note that the arc $D_b \cap \partial \mathcal{D}$ contains more than one corner-point since b is not a special bad arc and since ∂N_u is reasonably normal. Thus $D_b \cap \partial k \neq \emptyset$, where k denotes that edge of Γ containing x.

In order to proceed, let $U(b)$ denote the regular neighborhood in $\mathcal{D}$, and let b', b'' be the two components of $(\partial U(b) - \partial \mathcal{D})^-$ with $b' \subset D_b$. Define

$$k' := (k - (D_b - U(b)) \cup b' \quad \text{and} \quad k'' := (k - (D_b \cup U(b))) \cup b''$$

and set

$$\Gamma' := (\Gamma - k)^- \cup k' \quad \text{and} \quad \Gamma'' := (\Gamma - k)^- \cup k''.$$

Since the arc b is bad, it follows that Γ is slide-equivalent to Γ' as well as Γ'' (simply slide $k \cap D_b$ across D_b and into b' resp. b'').

The intersection $D_b \cap \partial N$ is a system of arcs, and so splits D_b into a system of discs. Let A be that one of those discs containing b. Observe that A is either contained in N or in $(M - N)^-$.

Case 1. $A \subset N$.

Consider the Heegaard graph Γ'. Note that $B := D_b \cap U(b)$ is a disc with $B \cap (\Gamma' \cup \partial N) = \partial B$. More precisely, B is a 3-faced disc with two faces in neighboring edges of Γ' and the third face in ∂N. Moreover, $B \subset N$ since we are in Case 1. Thus $N \cap U(B)$ is a special 3-ball contained in N, where $U(B)$ denotes the regular neighborhood in M. Define

$$N' := (N - U(N \cap U(B)))^- \cup U(B).$$

Then of course (Γ', N') is an admissible pair with $c(\Gamma', N') > c(\Gamma, N)$. Thus, by induction, the theorem holds for (Γ', N'). In particular, there is a Heegaard graph, Γ^*, in M such that

(1) Γ^* is slide-equivalent to Γ',
(2) $\Gamma^* \cap (M - N')^- \subset \Gamma' \cap (M - N')^-$, and
(3) $\Gamma^* \cap N'$ is unknotted.

By (1), Γ^* is slide-equivalent to Γ since Γ' is slide-equivalent to Γ. Moreover,

$$\Gamma^* \cap (M - N)^- \subset \Gamma^* \cap (M - N')^- \subset \Gamma' \cap (M - N')^- \subset \Gamma \cap (M - N)^-.$$

Finally, $\Gamma^* \cap N$ is unknotted since $\Gamma^* \cap N'$ is unknotted and since N is the union of N' with the regular neighborhood of some arc from $\Gamma^* \cap (M - N')^-$. The existence of Γ^* therefore shows that (Γ, N) satisfies the conclusion of thm. 24.5. This completes the proof in Case 1.

Case 2. $A \subset (M - N)^-$

In this Case consider the Heegaard graph Γ''. Observe that (Γ'', N) is an admissible pair with $c(\Gamma'', N) > c(\Gamma, N)$. Thus, by induction, the theorem holds for (Γ'', N). As before it is easily verified that it then holds for (Γ, N) as well.

It remains to discuss the subject under the following hypothesis.

Hypothesis B. *Not every component from ∂N_u is reasonably normal with respect to Γ.*

By hypothesis, there is a component, say N_0, from N_u such that ∂N_0 is not reasonably normal w.r.t. Γ. Then, by definition of reasonably normal surfaces, $(\partial N_0 - U(\Gamma))^-$ is either ∂-parallel or compressible in $M_\Gamma := (M - U(\Gamma))^-$.

Case 1. $(\partial N_0 - U(\Gamma))^-$ *is ∂-parallel in M_Γ.*

The associated parallelity region is either contained in N_0 or in $(M - N_0)^-$. But it cannot lie in $(M - N_0)^-$. Otherwise ∂M would be a 2-spheres if M is bounded or $(M - N_0)^-$ would be a handlebody if M is closed. Thus M

would be either a 3-ball (M is irreducible) or a 3-sphere (N_0 is a 3-ball); contradicting the hypothesis of thm. 24.5. Thus the parallelity region lies in N_0. Thus N_0 is the regular neighborhood of the graph $N_0 \cap \Gamma$. Since N_0 is not a special 3-ball, there is at least one edge of Γ entirely contained in N_0. Let D be a disc in N_0, $D \cap \partial N_0 = \partial D$, which intersects this particular edge in exactly one point. Set $N' := (N - U(D))^-$, where the regular neighborhood is taken in M. Suppose D is chosen so that at least one component of N' is a special 3-ball (this choice is of course always possible). Then $c(\Gamma, N') > c(\Gamma, N)$. Thus, by induction, the theorem holds for (Γ, N'). It is easily verified that it then holds for (Γ, N) as well.

Case 2. $(\partial N_0 - U(\Gamma))^-$ *is compressible in* M_Γ

Let $B \subset (M - U(\Gamma))^-$ be a compression disc for $(\partial N_0 - U(\Gamma))^-$. Then B is contained either in N_0 or in $(M - N_0)^-$.

If $B \subset N_0$, define $N' := (N - U(B))^-$, where the regular neighborhood $U(B)$ is taken in M. Then of course (Γ, N') is an admissible pair with $c(\Gamma, N') > c(\Gamma, N)$. Thus, by induction, the theorem holds for (Γ, N'). It is easily verified that it then holds for (Γ, N) as well.

If $B \subset (M - N_0)^-$, set $N_0' := N_0 \cup U(B)$, where the regular neighborhood is taken in M. Then $\partial N_0'$ consists of two 2-spheres since ∂N_0 is a 2-sphere (N is a system of 3-balls). These 2-spheres bound 3-balls, say P, Q, since M is irreducible. Moreover, P cannot be disjoint to Q, for otherwise $(M - N_0)^- = P \cup U(B) \cup Q$ would be a 3-ball and M would be a 3-sphere. Thus we may suppose that $P \subset Q$ since $\partial P \cap \partial Q = \emptyset$. But $P \cap U(B)$ is a disc. Thus $\Gamma \cap P$ can be pushed out of P, using an ambient isotopy which is constant outside of Q. This isotopy deforms Γ into a Heegaard graph, Γ^*, which intersects $(M - N)^-$ in strictly fewer arcs, for $P \subset (M - N_0)^-$. Set $N^* := N - (N \cap int(P))$. It follows that (Γ^*, N^*) is an admissible pair with $c(\Gamma^*, N^*) > c(\Gamma, N)$. Thus, by induction, the theorem holds for (Γ^*, N^*). It is easily verified that it then holds for (Γ, N) as well.

This finishes the proof of thm. 24.5. $\Diamond$

25. Distances and Winding Numbers for Heegaard Graphs.

This section is concerned with the classical theorem of Reidemeister and Singer [Rei 1][Si] concerning the stable equivalence of Heegaard graphs. Actually, we are going to use the result from the last section in order to give a new proof of this theorem (thm. 25.4). The advantage of our approach is that it allows estimates for the minimal number of stabilizations needed to pass from one Heegaard graph to another. In this section we give this estimate in terms of the winding number of Heegaard graphs as introduced in 25.1.

25.1. *Definition of Distances and Winding Numbers.*

Let Γ be a Heegaard graph in a 3-manifold M. Let $D \subset int(M)$ be a disc such that $(\Gamma \cap int(D))^-$ is a proper arc in D. Then

$$\Gamma' := (\Gamma - D)^- \cup \partial D$$

is of course again a Heegaard graph in M. We say Γ' is obtained from Γ by an *edge-addition*. Two Heegaard graphs $\Gamma, \Gamma' \subset M$ are called *stably equivalent* if there is a finite sequence $\Gamma_1, \Gamma_2, ..., \Gamma_n$ such that

(1) Γ, Γ' is slide-equivalent to Γ_1, Γ_n, respectively, and
(2) Γ_{i+1}, $1 \leq i < n$, is obtained from Γ_i, or Γ_i is obtained from Γ_{i+1}, by an edge-addition.

A finite collection, $\mathcal{F}$, of Heegaard graphs with (1) and (2) above is a *stabilization sequence* for Γ, Γ'. The cardinality $\#\mathcal{F}$ is the *length* of the stabilization sequence. $\mathcal{F}(\Gamma, \Gamma')$ denotes the set of all stabilization sequences for Γ, Γ'. According to the Reidemeister-Singer theorem (see also thm. 25.4 below), any two Heegaard graphs in a given 3-manifold are stably equivalent, i.e., $\mathcal{F}(\Gamma, \Gamma')$ is always non-empty. This fact allows us to associate a "distance" to any pair of Heegaard graphs in a 3-manifold as follows.

25.2. Definition. *The <u>distance</u> between two Heegaard graphs, Γ, Γ', in a 3-manifold M is defined to be* $d(\Gamma, \Gamma') := \min\{ \#\mathcal{F} \mid \mathcal{F} \in \mathcal{F}(\Gamma, \Gamma') \}$.

Intuitively, the distance $d(\Gamma_1, \Gamma_2)$ gives a (crude) measure of how much the *embedding types* of the Heegaard graphs Γ_1, Γ_2 (or rather their associated Heegaard surfaces) differ from each other. Therefore it is interesting to calculate it.

Let Γ_1, Γ_2 be two Heegaard graphs in a closed 3-manifold M. Then $(M - U(\Gamma_i))^-$ is a handlebody and so contains a *meridian-system*, i.e., a system of proper discs which splits $(M - U(\Gamma_1))^-$ into a 3-ball. Let us define

$$d_{\Gamma_1}(\Gamma_2) := \min \#(\Gamma_2' \cap \mathcal{D}),$$

where the minimum is taken over all meridian-systems, $\mathcal{D}$, in $(M - U(\Gamma_1))^-$ and all Heegaard graphs, Γ_2', in M, slide-equivalent to Γ_2 and in general position to D.

25.3. Definition. *The <u>winding number</u> for two Heegaard graphs* Γ_1, Γ_2 *in a 3-manifold* M *is defined to be* $w(\Gamma_1, \Gamma_2) := \min \{ d_{\Gamma_1}(\Gamma_2), d_{\Gamma_2}(\Gamma_1) \}$.

The winding number for Heegaard graphs generalizes the winding number for curves. Intuitively, it measures how Γ_1 and Γ_2 "wind" around each other.

The distance and the winding number for Heegaard graphs are related. Indeed, it is the object of this section to show:

25.4. Theorem. *For every closed 3-manifold* M, *there is a polynomial function* $p_M(\mathbf{x}) \in \mathbf{Z}[x, y, z]$ *such that* $d(\Gamma_1, \Gamma_2) \leq p_M(\beta_1(\Gamma_1), \beta_1(\Gamma_2), w(\Gamma_1, \Gamma_2))$, *for all Heegaard graphs* $\Gamma_1, \Gamma_2 \subset M$.

As indicated before, we will prove this theorem without any reference to the Reidemeister-Singer theorem (and therefore giving an independent proof of that theorem). Our strategy is the following. First we show that a pair, Γ_1, Γ_2, of Heegaard graphs in M is stably equivalent to a pair, Γ'_1, Γ'_2, of Heegaard graphs with $\Gamma'_2 \subset \partial U(\Gamma'_1)$. Then we show that Γ'_2 is stably equivalent to Γ'_1. After that the above estimate is then a straightforward consequence of the construction.

We begin with the first step.

25.5. *Pushing Heegaard Graphs.*

Given two Heegaard graphs Γ_1, Γ_2 in a 3-manifold M, we want to push Γ_2 into $\partial U(\Gamma_1)$. It seems reasonable to expect that this is *not* always possible, using ambient isotopies alone. But, applying thm. 24.5, we are going to show that it can always be done, if we allow ambient isotopies *and* stabilizations.

25.6. Proposition. *Let* Γ_1, Γ_2 *be two irreducible Heegaard graphs in a closed, irreducible 3-manifold* M, *different from the 3-sphere. Then there is a Heegaard graph,* $\Gamma'_1 \subset M$, *stably equivalent to* Γ_1, *such that* Γ_2 *can be ambient isotoped into* $\partial U(\Gamma'_1)$.

In addition, $d(\Gamma_1, \Gamma'_1) \leq 2 \cdot w(\Gamma_1, \Gamma_2)$.

Proof. By definition of winding numbers, we may suppose Γ_1, Γ_2 are chosen, within their respective slide-equivalence classes, so that $\Gamma_2 \subset N := (M - U(\Gamma_1))^-$ and that
$$w(\Gamma_1, \Gamma_2) = \#(\Gamma_2 \cap \mathcal{D}),$$
w.r.t. some appropriate meridian-system, $\mathcal{D}$, in the handlebody N. Let $U = U(\mathcal{D})$ be the regular neighborhood of $\mathcal{D}$ in N. Then $U \cap \Gamma_2$ is a system of arcs, joining opposite discs from $(\partial U - \partial N)^-$. W.l.o.g. we may suppose that Γ_2 is isotoped so that the above holds and that, in addition, $\Gamma_2 \cap U \subset U \cap \partial N$. Set $S := \partial(N - U(\mathcal{D}))^-$ and $G := S \cap U(\mathcal{D})$. It follows from our choice of $\mathcal{D}$ that S is a 2-sphere and G is a system of discs in S with $(S - G)^- \subset \partial N = \partial U(\Gamma_1)$. The 2-sphere S is the boundary of the 3-ball $E := (N - U(\mathcal{D}))^-$. Set $\hat{\Gamma}_2 := \Gamma_2 \cap E$. Then $\hat{\Gamma}_2$ is a graph in E

which may or may not be connected. In any case thm. 24.5 tells us that this graph is not knotted in E. In particular, it can be pushed into S, using an ambient isotopy which fixes $\partial \hat{\Gamma}_2$. However, the resulting graph may run across G since $\hat{\Gamma}_2$ may still be braided in some complicated way (all isotopies of $\hat{\Gamma}_2$ have to be constant on $\partial \hat{\Gamma}_2$). But there is a system, $\mathcal{L}$, of arcs in G with $\mathcal{L} \cap \partial G = \partial \mathcal{L} \subset \partial G - \Gamma_2$ and such that Γ_2 can be ambient isotoped into the surface $\partial(U(\Gamma_1) \cup U(\mathcal{L}))$. Let $\mathcal{L} \subset G$ be chosen so that the above holds and that, in addition, the total number of arcs is as small as possible. Then $\#\mathcal{L} \leq 2 \cdot \#(\Gamma_2 \cap \mathcal{D}) = 2 \cdot w(\Gamma_1, \Gamma_2)$. Since $\mathcal{L}$ lies in a system of discs, namely G, it follows that $\partial(U(\Gamma_1) \cup U(\mathcal{L}))$ is a Heegaard surface. Adding the arcs from $\mathcal{L}$ to Γ_1, we get a Heegaard graph, say Γ_1', so that Γ_2 can be ambient isotoped into $\partial U(\Gamma_1')$. Moreover, $d(\Gamma_1, \Gamma_1') = \#\mathcal{L} = 2 \cdot w(\Gamma_1, \Gamma_2)$. So Γ_1' is the desired Heegaard graph. $\Diamond$

The second step of our strategy requires a little preparation.

25.7. Embeddings of N-Relator 3-Manifolds.

We begin with the following observation.

25.8. Lemma. *Let Γ be a Heegaard graph in a closed 3-manifold M different from the 3-sphere. Let N be a handlebody in M with $\Gamma \subset N$, and set $W := U(\Gamma)$. Suppose there is an essential disc, E, in $(N - W)^-$ with $E \cap W = \partial E$. Then there is a 2-sphere, S, in N such that $S \cap W$ is an essential disc in W.*

In particular, Γ is reducible.

Proof. Let $\mathcal{E}$ be a minimal system of proper discs in $(N - W)^-$ with the property that $\partial(W \cup U(\mathcal{E}))$ has no compression disc in $(N - (W \cup U(\mathcal{E})))^-$. Then $W^+ := W \cup U(\mathcal{E})$ is an n-relator 3-manifold, $n \geq 1$, since, by hypothesis, $\mathcal{E}$ is non-empty. Moreover, we claim ∂W^+ is not a system of 2-spheres. Assume the converse. Then recall that ∂W^+ lies in the handlebody $(M - U(\Gamma))^-$. So every component of ∂W^+ bounds a 3-ball in $(M - U(\Gamma))^-$. Since $\partial N \neq \emptyset$, it follows that there must be at least one component of ∂W^+ which does not bound a 3-ball in $(N - U(\Gamma))^-$. But this boundary-component must still bound a 3-ball in N since it lies in N and since N is a handlebody. Thus M is the union of two 3-balls. This, however, contradicts our hypothesis that M is not a 3-sphere.

Now, let $\mathcal{A}$ be a minimal system of proper discs in W^+ with the property that $(\partial W^+ - U(\mathcal{A}))^-$ has no compression disc in W^+. Note that $\mathcal{A}$ must be non-empty. Indeed, by what we have seen before, at least one component of ∂W^+ is not a 2-sphere. This component is a compressible surface in the handlebody N, and so, by our choice of $\mathcal{E}$, it must have a compression disc in W^+.

By the handle-addition lemma 23.18, we may suppose that $\mathcal{E}$ has been slid and $\mathcal{A}$ has been chosen so that $\mathcal{A} \subset W = U(\Gamma)$. Define $V^+ := (W^+ - U(\mathcal{A}))^-$. Then, by our choices of $\mathcal{E}$ and $\mathcal{A}$, no component of ∂V^+ is compressible in

N. But N is a handlebody. So ∂V^+ must consist of 2-spheres. Now these 2-spheres bound 3-balls in N since N is a handlebody. It follows that V^+ is a system of 3-balls since M is not a 3-sphere. $\partial \mathcal{E}$ is non-empty and contained in ∂V^+ (see above). Thus w.l.o.g. $\mathcal{A}$ is a meridian-system in $W = U(\Gamma)$. Let $\mathcal{D}$ be the meridian-system in $U(\Gamma)$ dual to Γ. Then, by cor. 1.6, $\mathcal{D}$ and $\mathcal{A}$ are slide-equivalent. Thus w.l.o.g. we may suppose Γ is slid so that $\mathcal{A} = \mathcal{D}$. Now, let Q be a component of V^+ which contains the boundary of a component, E, of $\mathcal{E}$. Then ∂E bounds a disc in Q which intersects Γ in exactly one point. The union of this disc with E is the desired 2-sphere.

The previous lemma is crucial in the proof of the following result.

25.9. Proposition. *Let* Γ_1, Γ_2 *be Heegaard graphs in a closed, irreducible 3-manifold* M *different from the 3-sphere. Suppose* $\Gamma_2 \subset \partial U(\Gamma_1)$. *Then there are Heegaard graphs* $\Gamma_1', \Gamma_2' \subset M$, *stably equivalent to* Γ_1, Γ_2, *respectively, such that the following holds:*

(1) $\Gamma_2' \subset \partial U(\Gamma_1')$ *and* $(\partial U(\Gamma_1') - U(\Gamma_2'))^-$ *consists of discs and incompressible surfaces in* $(M - U(\Gamma_2'))^-$.

(2) $d(\Gamma_1, \Gamma_1') \leq \beta_1(\Gamma_1)$ *and* $d(\Gamma_2, \Gamma_2') = 0$.

Remark. $\beta_1(\Gamma)$ denotes the first Betti number, i.e., $\beta_1(\Gamma) = \operatorname{rank} \pi_1 \Gamma$.

Proof. Let $F := (\partial U(\Gamma_1) - U(\Gamma_2))^-$. If F consists of discs and incompressible surfaces in $(M - U(\Gamma_1))^-$, we are done. So suppose the converse. Then there is a compression disc for F. This compression disc is contained either in $U(\Gamma_1)$ or $(M - U(\Gamma_1))^-$. W.l.o.g. it is contained in $(M - U(\Gamma_1))^-$. Set

$$U := U(\partial U(\Gamma_1)) \quad \text{and} \quad W := (U(\Gamma_1) - U)^-,$$

where the regular neighborhood U is taken to be a small neighborhood in $U(\Gamma_1)$. Define $N := (M - U(\Gamma_2))^-$. Then there is a compression disc in $(N - W)^-$ with boundary in W. So, by lemma 25.8, there is a 2-sphere, $S \subset N$, such that $S \cap W$ is an essential disc.

Set $G := S \cap U$. Then G is 2-manifold in U. Compressing G along discs in U if necessary, we may suppose that G is incompressible in U. But it may or may not be essential in U. We are going to treat these two cases separately.

Case 1. G is inessential in U.

Since G is incompressible in U, there is a ∂-compression disc for G in U. By definition, this is a disc, D, in U such that $G \cap D = G \cap \partial D$ is an arc in ∂D and $\ell_D := (\partial D - G)^-$ is an arc in ∂U. Since G is incompressible and since U is the product I-bundle over some orientable surface, it is not hard to see that there must be such a ∂-compression disc with $\ell_D \subset \partial U - W$ (G is basically the band-connected sum of a collection of vertical annuli and ∂-parallel discs).

Moreover, we claim D may be chosen so that ℓ_D joins two *different* components of ∂G. Indeed, if $\partial \ell_D$ is contained in one component of ∂G, then D splits off a component, G_0, from G with $\partial G_0 \subset F \subset \partial U - W$ and different from a disc (G consists of planar surfaces). But G_0 is incompressible in U since G is. So G_0 must be ∂-parallel (it is basically a band-connected sum of ∂-parallel discs in U). In particular, G_0 has a ∂-compression disc, D, such that ℓ_D joins different components of ∂G_0. But this is also a ∂-compression disc for G. So the claim is established.

S splits $\partial U(D)$ into two discs. Let D_0 be that one of them which meets ∂U. Set

$$S' := (S - U(D))^- \cup D_0.$$

Then notice, in particular, that the 2-manifold $S' \cap U$ is obtained from $G = S \cap U$ by the ∂-compression along D.

If $\ell_D \cap \Gamma_2 \neq \emptyset$, then Γ_2 splits ℓ_D in a collection of arcs. Let ℓ_1 be that one of them which joins Γ_2 with a component, say r_1, of ∂G. Let $U(\ell_1 \cup r_1)$ denote the regular neighborhood in $\partial U(\Gamma_1)$. Then $\ell_D \cap (\partial U(\ell_1 \cup r_1) - r_1)$ consists of exactly one point since ℓ_D is supposed to join different components of ∂G. In particular, there is an arc $\ell_1' \subset \partial U(\ell_1 \cup r_1)$, joining the two points of $\Gamma_2 \cap (\partial U(\ell_1 \cup r_1) - r_1)$ without meeting ℓ_D. Set $\Gamma_2' := (\Gamma_2 - U(\ell_1))^- \cup \ell_1'$. Then Γ_2' is a Heegaard graph. In fact, it is ambient isotopic to Γ_2 since r_1 bounds a disc (in S). Moreover, $\#(\Gamma_2' \cap \ell_D) < \#(\Gamma_2 \cap \ell_D)$, by construction. Thus, repeating the above procedure if necessary, we may suppose Γ_2 is ambient isotoped so that afterwards $\Gamma_2 \subset \partial U(\Gamma_1)$ and $\Gamma_2 \cap \ell_D = \emptyset$. So $\Gamma_2 \cap S' = \emptyset$.

Now, ∂-compressions as above turn G into an essential 2-manifold. Thus, repeating the above ∂-compressions process if necessary, we eventually arrive at the situation from Case 2.

Case 2. G is essential in U.

In this case, G is the disjoint union of one vertical annulus and a collection of ∂-parallel discs in U (see e.g. [Joh 1, prop. 5.6]). Notice that Γ_2 cannot lie in a 3-ball, for otherwise the boundary of that 3-ball bounds a 3-ball in the handlebody $(M - U(\Gamma_2))^-$ which in turn is impossible since M is not a 3-sphere. In particular, Γ_2 cannot be contained in any of the parallelity regions of the discs from G. So we may suppose S is isotoped so that $S \cap U$ is a vertical annulus. Then $S \cap U(\Gamma_1)$ is an essential disc. Moreover, S bounds a 3-ball $E \subset M$ since M is supposed to be irreducible. Set $F' := \partial(U(\Gamma_1) \cup E)$. Then F' is a Heegaard surface containing Γ_2 (recall that Γ_2 cannot lie in E). Let Γ_1' be the spine of $U(\Gamma_1) \cup E$. Then Γ_1' is a Heegaard graph for M. By thm. 23.23, Γ_1' is stably equivalent to Γ_1 (note that, in contrast to [Wa 3], our proof of thm. 23.23 does not use the Reidemeister-Singer theorem). So $\Gamma_2 \subset \partial U(\Gamma_1')$ and $d(\Gamma_1, \Gamma_1') \leq \beta_1(E \cap U(\Gamma_1))$.

By what we have just seen, the assumption that F is compressible yields Heegaard graphs Γ_1', Γ_2', stably equivalent to Γ_1, Γ_2, respectively, with $\beta_1(\Gamma_1') < \beta_1(\Gamma_1)$ and $d(\Gamma_1, \Gamma_1') \le \beta_1(\Gamma_1 \cap E)$, $d(\Gamma_2', \Gamma_2) = 0$ and $\Gamma_2' \subset \partial U(\Gamma_1)$. Thus the proposition follows by induction. $\Diamond$

25.10. *The Reidemeister-Singer Theorem.*

The following is a rather flexible tool for stable equivalence of Heegaard graphs.

25.11. Proposition. *Let* F *be a closed surface in a closed 3-manifold* M. *Let* Γ_1, Γ_2 *be two graphs in* F *which are Heegaard graphs for* M. *Suppose* $(F - U(\Gamma_i))^-$ *consists of discs and incompressible surfaces in* $(M - U(\Gamma_i))^-$, *for* $i = 1$ *and* 2. *Then* Γ_1 *and* Γ_2 *are stably equivalent.*

In addition, $d(\Gamma_1, \Gamma_2) \le \beta_1(\Gamma_1) + \beta_1(\Gamma_2) + 2 \cdot (2 - \chi F)$.

Proof. Consider $F_1 := (F - U(\Gamma_1))^- \subset F$. Fix a meridian-system, $\mathcal{D}_1$, for the handlebody $(M - U(\Gamma_1))^-$. Since F_1 consists of discs and incompressible surfaces, we may suppose that $\mathcal{D}_1$ is chosen so that $F_1 \cap \mathcal{D}_1$ is a system of arcs which splits F_1 into a system of discs. All these arcs are contained in $\mathcal{D}_1$. So $U(\Gamma_1) \cup U(F_1 \cap \mathcal{D}_1)$ is the regular neighborhood of some graph in F, stably equivalent to Γ_1. In particular, the complement of this graph in F consists of open discs, i.e., it contains a standard graph in the sense of 1.7. Choosing the number of stabilizations to be minimal, we find in this way a Heegaard graph $\Gamma_1' \subset F$ with $d(\Gamma_1, \Gamma_1') \le 2 - \chi F$ and such that $F - \Gamma_1'$ consists of open discs. In the same way, Γ_2 is stably equivalent to a Heegaard graph $\Gamma_2' \subset F$ such that $F - \Gamma_2'$ consists of open discs. W.l.o.g. we suppose $\beta_1(\Gamma_1') \le \beta_1(\Gamma_2')$. Adding $\beta_1(\Gamma_2') - \beta_1(\Gamma_1')$ trivial edges to Γ_1', we get a Heegaard graph $\Gamma_2'' \subset F$ with $\beta_1(\Gamma_2'') = \beta_1(\Gamma_2')$. Then, by lemma 1.8, Γ_2'' and Γ_2' are slide-equivalent (and so also stably equivalent). In particular, $d(\Gamma_2'', \Gamma_2') = 0$. So, altogether, we have $d(\Gamma_1, \Gamma_2) \le d(\Gamma_1, \Gamma_1') + d(\Gamma_1', \Gamma_2'') + d(\Gamma_2'', \Gamma_2') + d(\Gamma_2', \Gamma_2) \le (2 - \chi F) + (\beta_1 \Gamma_1' + \beta_1 \Gamma_2') + 0 + (2 - \chi F) \le \beta_1 \Gamma_1 + \beta_1 \Gamma_2 + 2 \cdot (2 - \chi F)$. $\Diamond$

25.12. Corollary. *Let* Γ_1, Γ_2 *be two Heegaard graphs in a closed, irreducible 3-manifold different from the 3-sphere. Suppose* $\Gamma_2 \subset \partial U(\Gamma_1)$. *Then* Γ_1 *and* Γ_2 *are stably equivalent.*

In addition, $d(\Gamma_1, \Gamma_2) \le 15 \cdot \beta_1(\Gamma_1) + \beta_1(\Gamma_2)$.

Proof. By prop. 25.9, there are Heegaard graphs Γ_1', Γ_2', stably equivalent to Γ_1, Γ_2, respectively, such that $\Gamma_2' \subset \partial U(\Gamma_1')$ and that $(\partial U(\Gamma_1') - U(\Gamma_2'))^-$ consists of discs and incompressible surfaces in $(M - U(\Gamma_2'))^-$. Now push Γ_1' into a graph $\Gamma_1'' \subset \partial U(\Gamma_1')$. Let $\mathcal{D}$ be a meridian-system in $U(\Gamma_1')$. W.l.o.g. every component of $\mathcal{D}$ intersects Γ_1'' in exactly one point. Set $\Gamma_1^* := \Gamma_1'' \cup \partial \mathcal{D}$. Then of course Γ_1^* is a Heegaard graph stably equivalent to Γ_1 and $(\partial U(\Gamma_1 - U(\Gamma_1^*)))^-$ is a disc. Thus, by prop. 25.11, Γ_1^* and Γ_2' are stably equivalent. So Γ_1 and Γ_2 are stably equivalent. It remains to put together the estimates from 25.9 and 25.11.

By 25.9, $d(\Gamma_1, \Gamma_1'') = d(\Gamma_1, \Gamma_1') \le \beta_1 \Gamma_1$ and $d(\Gamma_2, \Gamma_2') = 0$. Moreover, $d(\Gamma_1'', \Gamma_1^*) = \#\mathcal{D} = \beta_1 \Gamma_1' \le 2 \cdot \beta_1 \Gamma_1$. Finally, by 25.11, $d(\Gamma_1^*, \Gamma_2') \le \beta_1 \Gamma_1^* + \beta_1 \Gamma_2' + 2 \cdot (2 - \chi \partial U(\Gamma_1')) \le 2 \cdot \beta_1 \Gamma_1' + \beta_1 \Gamma_2' + 4 \cdot \beta_1 \Gamma_1' \le 4 \cdot \beta_1 \Gamma_1 + \beta_1 \Gamma_2 + 8 \cdot \beta_1 \Gamma_1$. Thus, altogether,

$$
\begin{aligned}
d(\Gamma_1, \Gamma_2) &\le d(\Gamma_1, \Gamma_1'') + d(\Gamma_1'', \Gamma_1^*) + d(\Gamma_1^*, \Gamma_2') + d(\Gamma_2', \Gamma_2) \\
&\le \beta_1 \Gamma_1 + 2 \cdot \beta_1 \Gamma_1 + (12 \cdot \beta_1 \Gamma_1 + \beta_1 \Gamma_2) + 0 \\
&= 15 \cdot \beta_1 \Gamma_1 + \beta_1 \Gamma_2. \quad \diamond
\end{aligned}
$$

The combination of prop. 25.6 and cor. 25.12 yields the following:

25.13. Corollary. *For every closed, irreducible 3-manifold M which is not the 3-sphere, there is a polynomial function $p_M(\mathbf{x}) \in \mathbf{Z}[x, y, z]$ such that $d(\Gamma_1, \Gamma_2) \le p_M(\beta_1(\Gamma_1), \beta_1(\Gamma_2), w(\Gamma_1, \Gamma_2))$, for all irreducible Heegaard graphs $\Gamma_1, \Gamma_2 \subset M$.* $\diamond$

25.14. Remark. Using 23.21 and 23.23, it is easily seen that this corollary holds for reducible 3-manifolds and reducible Heegaard graphs as well (and so the proof of thm. 25.4 is complete). We leave it as an exercise for the reader to make the (slight) modifications in the argument for showing that cor. 25.13 also holds for all 3-manifolds with non-empty boundary.

V. THE SPACE OF HEEGAARD GRAPHS

In order to understand a group G, it often helps to compare it with some known group H. This has been done either by means of injections $G \hookrightarrow H$ or by means of surjections $H \twoheadrightarrow G$. Moreover, for a deeper understanding of the group one needs to understand not only the individual homomorphisms of this type but also the associated moduli spaces

$$\mathcal{P}(G;H) := \mathrm{Aut}(H) \setminus \{H \twoheadrightarrow G\} \quad \text{and} \quad \mathcal{R}(G;H) := \{G \hookrightarrow H\}/I(H),$$

of all such homomorphisms. Here $I(H) \subset Aut(H)$ denotes the subgroup generated by all inner automorphisms, and note that the automorphism group, $\mathrm{Aut}(H)$, acts, by composition, on $\mathcal{P}(G;H)$ (from the left) and on $\mathcal{R}(G;H)$ (from the right). In the case of a 3-manifold group, $G = \pi_1 M$, the discrete injections $\pi_1 M \hookrightarrow \mathrm{PSL}_2\mathbb{C}$ (besides others) and the surjections $\mathcal{F} \twoheadrightarrow \pi_1 M$, $\mathcal{F} = $ free group, have been used for the purpose. $\mathcal{R}(\pi_1 M; \mathrm{PSL}_2\mathbb{C})$ and $\mathcal{P}(\pi_1 M; \mathcal{F})$ are the *algebraic representation* and the *algebraic presentation* space of $\pi_1 M$, respectively. Both these spaces have geometrically defined subspaces which are more accessible. Namely, the space of all geometric (i.e., faithful, discrete, geometric finite) representations and the space of all geometric presentations (i.e., Heegaard graphs). Beginning with the work of Thurston, the above representation spaces for 3-manifold groups have been studied extensively. In contrast, the presentation spaces for 3-manifold groups have attracted much less attention. It is the purpose of the present chapter to demonstrate that geometric presentation spaces for 3-manifold groups have similar (if not stronger) rigidity features than representation spaces.

As far as the representation theory is concerned, recall from [Th 2] that, for every torus-free Haken 3-manifold, M, the space of geometric representations of $\pi_1 M$ is non-empty. Moreover, we know that this space is a point [Mos] or the Teichmüller space over the non-torus components of ∂M [AhBe], according whether or not $\chi \partial M$ vanishes. Moreover, the algebraic representation space of $\pi_1 M$ is *compact* whenever M is annulus- and torus-free [Th 4]. In particular, it is then a compactification of the geometric representation space.

The space of geometric presentations of $\pi_1 M$ is the same as the space of all homotopy classes of Heegaard graphs in M. Instead of this space we better study the space of Heegaard graphs in M modulo isotopy. It is somewhat larger but still a good approximation and much easier to work with. More precisely, we work with the space of Heegaard graphs modulo isotopy and slides. Recall from 23.5 that the slide-equivalence class of every Heegaard graph is represented by its Heegaard surface. So, equivalently, we are interested in the space of all Heegaard surfaces. Note that this space, in contrast to the geometric representation space, is never empty since, by [Moi 1], every 3-manifold admits a triangulation. In fact, we will discuss a number of constructions which indicate that the space of Heegaard surfaces is usually quite rich (see section 26).

Observe that Heegaard surfaces and incompressible surfaces constitute two extreme types of surfaces in a 3-manifold. Indeed, surfaces of the first type allow no compression whatsoever while surfaces of the second type are totally compressible from both sides. Nevertheless, it turns out that both types have some interesting, common rigidity features.

In order to explain this statement a little more, let M be a Haken 3-manifold and let $S(M)$, resp. $\mathcal{H}(M)$ denote the set of isotopy classes of all incompressible surfaces resp. all Heegaard surfaces in M. The set $S(M)$ of all (isotopy classes of) incompressible surfaces has a natural decomposition by "layers" $S_n(M) := \{ S \in S(M) \mid -\chi S = n \}$. This decomposition is rigid. Indeed, by [Ha 1], every layer $S_n M$ is finite, modulo Dehn twists along annuli and tori. Recall that the underlying reason for this rigidity result is the fact that "normal" surfaces in 3-manifolds are generated from a finite list of building blocks (fundamental surfaces), using some simple operation (cut-and-paste) [Ha 1] (see also section 13). Now this process of passing from one layer to the next by means of cut-and-paste is numerically controlled by the growth rate of the *Haken spectrum* $\Theta_M : \mathbf{N} \to \mathbf{N}$ given by $\Theta_M(n) := \#(S_n(M)/T(M))$. Here $T(M)$ is the subgroup of the mapping class group generated by all Dehn twists along essential annuli or tori in M. The following is a consequence of Haken's results:

Theorem. *The Haken spectrum exists and has polynomial growth.*

In chapter III we have given some explicit upper bounds for the Haken spectrum of one-relator 3-manifolds.

We now change our point of view. Instead of studying incompressible surfaces with the help of Heegaard graphs, we now study Heegaard graphs with the help of incompressible surfaces. In particular, we will focus on the set $\mathcal{H}(M)$ of all isotopy classes of Heegaard surfaces in a given Haken 3-manifold, M (recall from chapter IV the definition of Heegaard surfaces in 3-manifolds with boundary).

For technical reason we here exclude all Haken 3-manifolds M which contain a non-trivial, essential Stallings fibration, i.e., a mapping torus over an orientable surface with incompressible boundary which cannot be isotoped into the regular neighborhood of ∂M. The special appearance of Stallings fibrations is by now a familiar feature in the study of Haken 3-manifolds (due to the nature of hierarchies). Recall that they also come up in the solution of the homeomorphism problem [He]Ha 1], in the classification of exotic homotopy equivalences [Joh 1] and in the existence of hyperbolic structures [Th 1] for Haken 3-manifolds. In all these cases Stallings fibrations demand special attention and methods. Now, it is true that every 3-manifold contains a closed curve whose complement is a Stallings fibration (open book decompositions for 3-manifolds) but notice that all those Stallings fibrations are inessential. On the other hand, Haken 3-manifolds without *non-trivial, essential* Stallings fibrations form a large and interesting class. E.g., all Haken 3-manifolds in this class are atoroidal (i.e., they contain no essential torus which is not ∂-parallel) and so, according to Thurston, they are hyperbolic. Vice versa, every hyperbolic 3-manifold whose fundamental group is a Kleinian group of the second kind (i.e., a Kleinian group whose limit set is not the entire sphere of infinity) and not a non-trivial free-product, is a Haken 3-manifold without non-trivial, essential Stallings fibration. In other words, all hyperbolic Haken 3-manifolds with infinite volume lie in our class.

Now, as for $\mathcal{S}(M)$, the set $\mathcal{H}(M)$, M a 3-manifold, is not only infinite but decomposes into "layers" $\mathcal{H}_n(M) := \{F \in \mathcal{H}(M) \mid -\chi F = n\}$, $n \geq -2$. Due to this this layer-structure, it is possible to assign various counting functions to the set $\mathcal{H}(M)$ as described next.

Recall that every Haken 3-manifold M has a hierarchy of surfaces. Now, any such hierarchy defines a 2-dimensional complex in M, which we call a *Haken 2-complex* (see section 27). Given a fixed Haken 2-complex $\Psi \subset M$, we define the *combinatorial length* of $F \in \mathcal{H}(M)$ by setting $length_\Psi(F) := \min \{ \#(\Gamma \cap \Psi) \}$, where the minimum is taken over all Heegaard graphs $\Gamma \subset M$ with $\partial U(\Gamma \cup \partial M) - \partial M$ isotopic to F. Moreover, the counting function

$$\mathcal{L}_{(M,\Psi)}(n) := \max \{ \, length_\Psi(F) \mid F \in \mathcal{H}_n(M) \, \}$$

is called the *length-spectrum* for $\mathcal{H}(M)$ w.r.t. Ψ.

Finiteness Theorem. (thm. 31.6) *Let M be a Haken 3-manifold (closed or not) without non-trivial, essential Stallings fibration and let $\Psi \subset M$ be a great and useful Haken 2-complex (see section 27). Then the length-spectrum of $\mathcal{H}(M)$ w.r.t. Ψ exists and has polynomial growth.*

Combining this theorem with our version of the Reidemeister-Singer theorem, we can give estimates for the Tietze spectrum. To define the Tietze spectrum, recall from the Reidemeister-Singer theorem (see cor. 25.13), that any two Heegaard surfaces in M differ by some finite number of handle-additions and -subtractions. Clearly, the minimal number of these stabilizations defines a metric on $\mathcal{H}_n(M)$ (and $\mathcal{H}(M)$). We call this metric the *Reidemeister-Singer metric*. Finally, we denote by $\operatorname{diam} \mathcal{H}_n(M)$ the largest distance in the Reidemeister-Singer metric. The arithmetic counting function defined by setting

$$\Psi_M(n) := \operatorname{diam} \mathcal{H}_n(M)$$

is called the *Tietze spectrum* for $\mathcal{H}(M)$.

Theorem. (thm. 31.9) *Let M be a Haken 3-manifold (closed or not) without non-trivial, essential Stallings fibration. Then the Tietze spectrum for $\mathcal{H}(M)$ exists and has polynomial growth.*

Finally, we count Heegaard surfaces. Given a 3-manifold, we define an arithmetic counting function by setting

$$\Phi_M(n) := \#\mathcal{H}_n(M)$$

and we call it the *Heegaard spectrum* for $\mathcal{H}(M)$. The Rigidity Theorem tells us that this counting function exists for Haken 3-manifolds without non-trivial, essential Stallings fibration.

Rigidity Theorem. (thm. 32.17) *Let M be a Haken 3-manifold (closed or not) without non-trivial, essential Stallings fibration. Then $\mathcal{H}_n(M)$ is finite, for every $n \geq 1$. In particular, the Heegaard spectrum for $\mathcal{H}(M)$ exists.*

In particular, the space of all geometric presentations $\mathcal{F}_n \twoheadrightarrow \pi_1 M$ modulo the action of $\mathrm{Aut}(\mathcal{F}_n)$ is always finite ($\mathcal{F}_n$ denotes the free group of rank n). As indicated before, this result complements the compactness result for representation-spaces of these groups. It also fits very well with the fact that the fundamental groups of simple 3-manifolds have finite outer automorphism groups [Joh 1]. Obviously, it would be interesting to know more about the above spectra and their relationships (if any). It would be especially interesting to learn more about the growth rate of the Heegaard spectrum. It would also be interesting to see good lower bounds. However, in contrast to incompressible surfaces, there is so far no systematic process known for creating higher layers of $\mathcal{H}(M)$ from lower ones. There is of course the trivial operation of handle-addition but, according to the Reidemeister-Singer theorem, this process is rather annihilating than creating Heegaard surfaces. The process which forces growth for the Heegaard spectrum still remains largely a mystery.

We finish this introduction with a few words regarding the proof of the Rigidity Theorem. First, we find it easier to work with Heegaard graphs rather than Heegaard surfaces because Heegaard graphs have more combinatorial structure. Recall that, by definition, a *Heegaard graph* in a Haken 3-manifold, M, is a finite (not necessarily connected) graph $\Gamma \subset M$ with $\Gamma \cap \partial M = \partial \Gamma$ and such that $(M - U(\Gamma))^-$ is a handlebody. The close relationship between Heegaard graphs, Heegaard diagrams and n-relator 3-manifolds has been been pointed out before (notably in section 23) and some relevant results concerning the latter have already been translated into the context of Heegaard graphs.

In the coming outline we only consider 3-manifolds, M, with *non-empty boundary* (and no non-trivial, essential Stallings fibration). Then, in particular, every Heegaard surface in M is represented by some *Heegaard string-system*, i.e., by some Heegaard graph which consists of arcs alone. These Heegaard string-systems will have to be changed frequently by edge-slides, and in the actual proof we will have to face the annoying problem that in general these edge-slides change Heegaard string-systems into Heegaard graphs which are not necessarily Heegaard string-systems anymore. But for the benefit of this introduction we better avoid this technical point by simply pretending that all edge-slides result in Heegaard string-systems.

Let $\mathcal{H}(M, m)$ be the set of all Heegaard string-systems in M with m strings, modulo isotopies and edge-slides. The goal is to show that $\mathcal{H}(M, m)$ is a finite and constructable set (see remark 21). Recall that it can be decided whether a finite graph in M is actually a Heegaard graph. Thus, for our purpose, it suffices to construct a finite set $\mathcal{G}(M, m)$ of graphs in M with $\mathcal{H}(M, m) \subset \mathcal{G}(M, m)$. But in order to construct the set $\mathcal{G}(M, m)$ the embedding types of certain linked and knotted arcs (and graphs) in M have to be controlled. In order to control these embedding types an appropriate reference structure for M is needed. For the 3-sphere and for lens spaces, certain foliations with singularities have been used for this purpose [Ot][Bo]. For Haken 3-manifolds, however, it suggests to take hierarchies. Recall that a hierarchy is a finite sequence $S_1, ..., S_n$, $n \geq 1$, of incompressible surfaces in M which eventually split M into a collection of 3-balls, say M_n. The union of all these surfaces defines a certain 2-dimensional CW-complex in M which we call a *Haken-complex*. This complex has various useful features, but for this outline it is sufficient to remember that Haken-complexes are certain 2-dimensional complexes in Haken 3-manifolds whose complement consists of 3-balls.

We continue by fixing a Haken 2-complex for every Haken 3-manifold. Having done so, we define $\mathcal{H}_i(M, m) \subset \mathcal{H}(M, m)$, $i \geq 0$, to be the subset of all Heegaard string-system from $\mathcal{H}(M, m)$ which intersect the Haken 2-complex in no more than i points, modulo isotopies and edge-slides. Given this setting, the proof now consists of two major steps. In the first step it will be shown that, for every Haken 3-manifold, M, there is a constant $\alpha := \alpha(M, m)$ with $\mathcal{H}(M, m) \subset \mathcal{H}_\alpha(M, m)$ and in the second step the finiteness of the set $\mathcal{H}_\alpha(M, m)$ will be demonstrated (by constructing $\mathcal{G}(M, m)$).

To carry out the first step, we use induction on the lengths of hierarchies. Corollary 23.34 may be interpreted as the induction beginning. Indeed, if $\Gamma \in \mathcal{H}(M, m)$ is any Heegaard string-system in M, then this corollary tells us that the intersection of Γ with the first incompressible surface S_1 from the specified hierarchy is limited, modulo isotopy and edge-slides. Next, we split M along the surface S_1, and we obtain a 3-manifold M_1. Moreover, the intersection $\Gamma_1 := M_1 \cap \Gamma$ defines a Heegaard graph in M_1 since S_1 is incompressible. So the above situation is repeated. Thus another application of cor. 23.34 tells us that Γ_1 intersects the second surface S_2 from the hierarchy in a limited number of points, modulo isotopy and edge-slides. At this point, however, we have to be very careful since we are dealing with isotopies and slides *in the split manifold* M_1 and these operations are not necessarily operations in the original manifold M as requested. To overcome this obstacle, it is in fact not enough to consider the intersection of Γ with the Haken-complex. At this point we are rather forced to consider the intersection of an appropriate meridian-system in $(M - U(\Gamma))^-$ with that complex. It turns out that the latter intersections would develop certain limiting patterns, if the intersection of Γ with the Haken-complex became large. A subsequent, careful analysis of these patterns shows that they are actually inconsistent with the topology of M. Having established this phenomenon, the induction is then easily completed. In particular, $\Gamma \in \mathcal{H}_\alpha(M, m)$, for some appropriate $\alpha \geq 0$ which is independent of the embedding type of Γ, and so $\mathcal{H}(M, m) \subset \mathcal{H}_\alpha(M, m)$.

Having completed the first step, it remains to show the finiteness of the set $\mathcal{H}_\alpha(M, m)$, for every integer $\alpha \geq 0$. For simplicity, we only consider the special case that $\alpha = 0$. Modulo isotopies and edge-slides, all Heegaard string-systems from $\mathcal{H}_0(M, m)$ are entirely contained in the 3-balls M_n (= complement of the specified Haken-complex). We finish the proof of the theorem by constructing a finite set, $\mathcal{G}_0(M, m)$, of arc-systems in M and an embedding $\mathcal{H}_0(M, m) \hookrightarrow \mathcal{G}_0(M, m)$. For this we use induction on the number of strings from Γ.

For the induction beginning, recall from [Joh 1] that the 3-ball M_n carries the boundary-pattern induced by the hierarchy, and that isotopies in M_n are called admissible if they respect this boundary-pattern. Note further that the admissible isotopy classes of unknotted arcs in M_n are in one-to-one correspondence to pairs of faces of M_n. It therefore follows that the set, $\mathcal{T}_1(M_n)$, of admissible isotopy classes of all unknotted arcs in M_n is finite. But every Heegaard string in M_n is also an unknotted arc, for its complement is a handlebody whose boundary is a torus. So $\mathcal{H}_0(M, 1) \subset \mathcal{T}_1(M_n)$. Thus we may define $\mathcal{G}_0(M, 1) := \mathcal{T}_1(M_n)$ to finish the induction beginning.

For the induction step assume that the set $\mathcal{H}_0(M', m-1)$ is finite, for *every* Haken 3-manifold M'. Let $\Gamma \in \mathcal{H}_0(M, m)$. Then Γ is a Heegaard string-system in the 3-ball M_n. Actually, according to the Unknotting Theorem, it is an *unknotted* Heegaard string-system. However, it may still be *braided* in some complicated fashion (recall that we consider string-systems modulo *admissible* isotopies and not just isotopies). We overcome this potential obstacle to finiteness by the following trick. Fix a string k from Γ, and define $\Gamma' := \Gamma - k$. Then, modulo admissible isotopy, k is contained in the finite set $\mathcal{T}_1(M_n)$ (see above). Moreover, Γ' is a Heegaard string-system in the Haken 3-manifold $M' := (M - U(k))^-$, and so contained in the set $\mathcal{H}_0(M', m-1)$, modulo isotopy and slides. By induction, the latter set is finite. Also every isotopy and slide of Γ' in M' can be realized by a slide of Γ in M. Altogether, it follows the existence of an embedding $\mathcal{H}_0(M, m) \hookrightarrow \mathcal{G}_0(M, m) := \mathcal{T}_1(M) \cup \bigcup \mathcal{H}_0((M - U(\ell))^-, m-1)$, where the union is taken over all arcs $\ell \in \mathcal{T}_1(M_n)$. The induction is now complete since $\mathcal{G}_0(M, m)$ is finite.

This finishes the outline for the Rigidity Theorem under our special assumptions. The general case requires considerable extra work but the basic strategy remains the same.

26. Constructing Heegaard Graphs.

Every 3-manifold (as always: compact and orientable) has at least one Heegaard graph, namely the 1-skeleton of a triangulation. So it has at least one irreducible Heegaard graph and at least one irreducible Heegaard surface. Here we take up the challenge of finding others.

Examples for 3-manifolds with non-unique Heegaard surfaces have proved hard to come by. We have already seen Waldhausen's theorem (thm. 23.23) which says that the 3-sphere has only one irreducible Heegaard surface. In [Bo] it has been shown that any other lens space has also only one irreducible Heegaard surface. Finally, by [SchT] and [Schul], it is known that the product of any orientable surface with S^1 has this property. The problem of finding 3-manifolds with non-unique Heegaard surfaces appears to be two-fold. First one needs a way for generating new candidates for Heegaard surfaces and second one needs a method for distinguishing between these candidates in order to find out which one of them are really new. The first partial solution for these problems appeared in [Eng] where it has been shown that the connected sum of two lens spaces has non-unique Heegaard decompositions. In [BGM] the first example of an irreducible 3-manifold with non-unique Heegaard decomposition is given. It is a Seifert fiber space. In [CG 2] examples for closed 3-manifolds have been constructed which admit irreducible Heegaard surfaces of arbitrary high genus. In [Ko 6] more examples of this type can be found. See also [BRZ],[Mor] and [Fu] for other examples.

In this section we discuss three explicit constructions for Heegaard graphs each of which illustrates a significant aspect. Indeed, the first construction produces 3-manifolds with *infinitely* many non-isotopic but homeomorphic Heegaard surfaces. The second construction produces an *abundance* of one-relator 3-manifolds which have non-homeomorphic Heegaard strings. Finally, the torsion-type invariant from [LM] is discussed for showing that the set of isotopy classes of Heegaard surfaces in certain Seifert fiber spaces is not only non-trivial, but grows *exponentially* with the number of exceptional fibers.

All these constructions (and all examples mentioned above) are ad hoc and provide only limited insight into the way Heegaard graphs come up in general. But they have the advantage of producing concrete examples. In fact, these constructions provide evidence for the expectation that uniqueness for Heegaard decompositions is most likely a rather uncommon feature for 3-manifolds. In section 32 we will show how to construct, in a systematic way, *all* Heegaard graphs of fixed relative Euler characteristic in a given Haken 3-manifold (not containing an essential Stallings fibration).

We begin this section by fixing some basic notation.

26.1. *Heegaard Graphs and Presentations.*

There is a close relationship between Heegaard graphs of 3-manifolds, M, on the one hand and presentations for the fundamental group, $\pi_1 M$, on the other. Indeed, Heegaard graphs may be considered as special, *geometric* presentations for $\pi_1 M$.   We discuss some first aspects of this relationship. In principal, this relationship opens the way for us to use known results about group presentations for the problem of distinguishing Heegaard graphs (unfortunately, problems concerning group presentations are known to be notoriously hard - and often unsolvable). Vice versa, the present theory of Heegaard graphs may hint at general properties of group presentations for wider classes of groups.

26.2. *Equivalences for Heegaard Graphs.*

Recall from 23.5 that a finite, tri-valent, possibly disconnected graph Γ in a 3-manifold M, $\Gamma \cap \partial M = \partial \Gamma$, is a *Heegaard graph* (for M) if $(M - U(\Gamma))^-$ is a handlebody. Two Heegaard graphs $\Gamma, \Gamma' \subset M$ are *slide equivalent* if $U(\Gamma \cup \partial M)$ is isotopic to $U(\Gamma' \cup \partial M)$. $G_\Gamma := \partial U(\Gamma \cup \partial M) - \partial M$ is the *Heegaard surface* associated to Γ. Recall further that the relative Euler characteristic $\chi_0(\Gamma) := \chi(\Gamma, \partial \Gamma) = \chi \Gamma - \chi \partial \Gamma$ is an invariant of the slide-equivalence class of Γ. When it creates no confusion, we denote the slide-equivalence class of Γ by Γ again.

26.3. Definition. *Let M be a 3-manifold. Then $H(M)$ denotes the set of all Heegaard graphs in M modulo slide-equivalence.*

In addition, for $m \in \mathbf{N}$:

$$H(M, m) := \{\, \Gamma \in H(M) \mid \chi(\Gamma, \partial \Gamma) = -m \,\}$$
$$\mathcal{H}^*(M, m) := \{\, \Gamma \in H(M, m) \mid \Gamma \text{ is irreducible} \,\}.$$

Note that the mapping class group $\pi_0 \text{Diff}(M)$ acts on $H(M, m)$ as well as $\mathcal{H}^*(M, m)$. Let $\mathcal{A}(M) \subset \pi_0 \text{Diff}(M)$ denote the subgroup generated by all Dehn twists along tori (i.e., all diffeomorphisms with support in the regular neighborhood of some essential torus). Define

$$\mathcal{H}(M, m) := \mathcal{H}^*(M, m)/\mathcal{A}(M).$$

Two Heegaard graphs $\Gamma, \Gamma' \subset M$ are *homeomorphic* if there is a homeomorphism $h : M \to M$ such that Γ and $h\Gamma$ are slide-equivalent. $\Gamma, \Gamma' \subset M$ are *homotopic* if there is a Heegaard graph $\Gamma'' \subset M$ which is homotopic to Γ' and slide-equivalent to Γ.

26.4. *The Map $\theta : H(M) \to H(M)$.*

Let $\Gamma \in H(M)$. Then, by definition, $W := (M - U(\Gamma))^-$ is a handlebody. Let Γ' be a spine of W. Then Γ' is a graph in W such that $(W - \Gamma')^- \cong \partial W \times [0, 1)$. If $\partial M \neq \emptyset$, let ℓ be a minimal system of fibers in the product structure $\partial W \times I$ with the property that ℓ joins every component of $(\partial M - U(\Gamma))^-$ with Γ'. Now set

$$\theta \Gamma := \begin{cases} \Gamma', & \text{if } \partial M = \emptyset \\ \Gamma' \cup \ell, & \text{otherwise.} \end{cases}$$

26.5. Lemma. *The assignment $\Gamma \mapsto \theta \Gamma$ defines a map $\theta : H(M) \to H(M)$.*

Remark. It is not known when this map is trivial or not. This question appears to be related to the Poincaré conjecture.

Proof. We first show that $\theta\Gamma$ is a Heegaard graph for M. This is clear if M is closed. So suppose the converse. Then w.l.o.g. we may suppose Γ is a system of arcs since it certainly is slide-equivalent to an arc-system. We have to show that $(M - U(\theta\Gamma))^- = (M - U(\Gamma' \cup \ell))^-$ is a handlebody. To see this note first that $\Gamma' \cup \ell \subset W = (M - U(\Gamma))^-$ and that, by construction, $N := (W - U(\Gamma' \cup \ell))^-$ is the product I-bundle, $N = F \times I$, over a bounded surface F (namely $F := (\partial W - U(\ell))^-$). It follows that N is a handlebody. Thus $(M - U(\theta\Gamma))^- = (W - U(\theta\Gamma))^- \cup U(\Gamma) = N \cup U(\Gamma)$ is an n-relator 3-manifold. Moreover, $U(\Gamma) \cap \partial W'$ is a system of annuli since Γ is supposed to be a system of arcs. Let k be the union of all cores of this particular collection of annuli (w.l.o.g. $k \subset F \times 0$). Then, in particular, $(M - U(\theta\Gamma))^- = N^+(k)$. To see that $N^+(k)$ is a handlebody, note that, by construction, every component of $(F \times 0) - k$ contains an end-point of ℓ. So there is a system, r, of pairwise non-parallel arcs in $F \times 0$, with $\partial r \subset U(\ell)$ and disjoint to k, splitting $F \times 0$ into a system of annuli. Then $r \times I$ is a system of discs in N, not meeting k, which splits $N = F \times I$ into a system of solid tori. Hence $(N - U(r \times I))^- \cup U(\Gamma))^-$ is a system of 3-balls. Thus $N^+(k) = (M - U(\theta\Gamma))^-$ is a handlebody. This proves that $\theta\Gamma$ is a Heegaard graph.

It remains to show that θ is well-defined. For this let Γ_1, Γ_2 be two slide-equivalent Heegaard graphs in M. Then, by definition, $U(\Gamma_1 \cup \partial M)^- = U(\Gamma_2 \cup \partial M)^-$ mod isotopy. Thus Γ_1', Γ_2' are spines of the same handlebody, namely $(M - U(\Gamma_1))^-$. Hence they are the same modulo edge-slides (apply cor. 1.6 to the dual disc-systems). It then follows that $\theta\Gamma_1$ and $\theta\Gamma_2$ are slide-equivalent, since any two systems of equal numbers of fibers (such as ℓ_1 and ℓ_2) in a product I-bundle over an orientable surface are isotopic. $\diamond$

Because of the previous result, we may call $\theta\Gamma$ the *dual Heegaard graph* for Γ. The map θ is an involution if $\partial M = \emptyset$. But not necessarily if $\partial M \neq \emptyset$. In any case, however, $\theta^3 = \theta$.

26.6. *Presentations Associated to Heegaard Graphs.*

We start again with a definition to fix notation.

26.7. Definition. *Let G be a group. A* presentation *for G is an epimorphism $P : \pi_1\Gamma \twoheadrightarrow G$, where Γ is a graph. Two presentations $P_1, P_2 : \pi_1\Gamma \twoheadrightarrow G$ are* Nielsen-equivalent *if there is an automorphism $\psi : \pi_1\Gamma \to \pi_1\Gamma$ with $P_1 \circ \psi = P_2$.*

Let $\mathcal{P}(G)$ denote the set of all presentations for G modulo Nielsen-equivalence.

Two presentations, P_1, P_2, for G are said to be *isomorphic* if there is an automorphism $\psi : G \to G$ such that $\psi \circ P_1$ is Nielsen-equivalent to P_2. P_1, P_2 are *conjugate* if they are isomorphic via an inner automorphism of G.

Observe that conjugate presentations are also Nielsen equivalent (but of course not vice versa).

Note that in order to turn P into a presentation in the usual sense (see e.g. [LS]) we would have to fix generating sets for both $\pi_1\Gamma$ and $\text{kern}(P)$. We prefer the above definition though since we do not especially care for any particular choice of generators for $\text{kern}(P)$. Now, P is a *finite presentation* for G if Γ is a finite graph and if there is a finite graph Γ' and a homomorphism $R : \pi_1\Gamma' \to \pi_1\Gamma$ with $\text{im}(R) = \text{kern}(P)$. We only consider finite presentations, i.e., the word "presentation" always means "finite presentation" for us. Let $\Gamma \subset \mathbf{R}^3$ and let $U(\Gamma)$ denote the regular neighborhood of Γ in $\mathbf{R}^3$. Then we say that P is a *geometric presentation* if there is an embedding $\Gamma' \hookrightarrow \partial U(\Gamma)$ which induces a homomorphism $R : \pi_1\Gamma' \to \pi_1 U(\Gamma) = \pi_1\Gamma$ with $\text{im}(R) = \text{kern}(P)$.

26.8. Proposition. *Let M be a simple 3-manifold. Then every geometric presentation $P : \pi_1\Gamma \twoheadrightarrow \pi_1 M$ is induced by an embedding $f : \Gamma \to M$ for which $f(\Gamma)$ is a Heegaard graph modulo edge-slides.*

Proof. Since P is a geometric presentation, there is an embedding $g : \Gamma' \hookrightarrow \partial U(\Gamma)$ which induces a homomorphism $R : \pi_1\Gamma' \to \pi_1\Gamma$ with $\text{im}(R) = \text{kern}(P)$. Let N^+ denote the 3-manifold obtained from $U(\Gamma)$ by first attaching 2-cells along $\partial U(g\Gamma') \subset \partial U(\Gamma)$ and then 3-cells along all sphere-components of the boundary. Let $f : \Gamma \to M$ be a map which induces the presentation $P : \pi_1\Gamma \twoheadrightarrow \pi_1 M$. Composing f with the contraction $U(\Gamma) \to \Gamma$, we get a map $U(\Gamma) \to M$. This map in turn can be extended to a map $f^+ : N^+ \to M$ since $\text{im}(R) = \text{kern}(P)$. By construction, f^+ induces a homomorphism $\pi_1 N^+ \to \pi_1 M$ which is a surjection as well as an injection. So it is an isomorphism. By the loop- and sphere-theorem it follows that N^+ is irreducible and ∂-irreducible since M is. Moreover, N^+ is a Haken 3-manifold since M is (see e.g. [Ja 0, thm. III.10]). In fact, N^+ must be a simple 3-manifold. Otherwise N^+ contains an essential surface, T, which is a torus or an annulus. But $f' : M \to N^+$ can be deformed so that afterwards $(f')^{-1}T$ is an essential annulus or torus, where f' is some homotopy inverse for f^+. This in turn is impossible since M is simple. Thus f^+ is a homotopy equivalence between simple 3-manifolds. Hence, by [Joh 1, thm. 24.2], it can be deformed into a homeomorphism $h : N^+ \to M$. Now, by construction, the co-cores of the attached 2-cells form a Heegaard string-system $\Gamma^* \subset N^+$ with $(N^+ - U(\Gamma^*))^- = U(\Gamma)$. It follows that $\Gamma \subset \theta\Gamma^* = \Gamma \cup \ell$ (see 26.4 for the definition of ℓ). By lemma 26.5, $\theta\Gamma^*$ is a Heegaard graph for N^+. So $h\theta\Gamma^*$ is a Heegaard graph for M and the claim follows. $\Diamond$

To any Heegaard graph Γ in M there is associated a (geometric) presentation $P_\Gamma : \pi_1\Gamma \twoheadrightarrow \pi_1 M$ as follows. First, recall $\theta\Gamma$ is a graph. More precisely, there is a graph Γ_0 and an inclusion $f : \Gamma_0 \hookrightarrow M$ with $f\Gamma_0 = \theta\Gamma$. Fix a vertex, v, for Γ_0. Now f is homotopic to a map $g : (\Gamma_0, v) \to (M, z)$, $z \in M$. Let $P_\Gamma := \pi_1(g) : \pi_1(\Gamma_0, v) \to \pi_1(M, z)$.

26.9. Lemma. *The assignment* $\Gamma \mapsto \wp(\Gamma) := P_\Gamma$ *defines a map* $\wp : H(M) \to \mathcal{P}\pi_1(M)$.

Proof. Notice that, modulo conjugacy, the presentation P_Γ is independent of the above map g (homotopic maps yield conjugate presentations). If Γ_1, Γ_2 are two slide-equivalent Heegaard graphs, then there is a sequence of edge-slides and homotopies joining one with the other. In particular, $P_{\Gamma_1}, P_{\Gamma_2}$ differ by Nielsen equivalences alone. Altogether, this shows that $\wp$ is well-defined. $\Diamond$

Note that the map $\wp : H(M) \to \mathcal{P}\pi_1 M$ need not be injective since there may well be Heegaard graphs which are homotopic but not isotopic. But in any case it follows that two Heegaard graphs Γ_1, Γ_2 are *not* slide-equivalent if the presentations $P_{\Gamma_1}, P_{\Gamma_2}$ are not Nielsen-equivalent. By this reason the problem of deciding whether two presentations are Nielsen equivalent becomes relevant for us. We will return to this question in 26.28.

26.10. *Turbulization for Heegaard Graphs.*

Given a Heegaard graph, Γ, in a 3-manifold M and a homeomorphism, $h : M \to M$, the image $h(\Gamma)$ is certainly a Heegaard graph again. Since the mapping class group of a 3-manifold may contain elements of infinite order, it suggests to use some of these infinite order homeomorphisms to generate infinitely many irreducible Heegaard graphs. It turns out that, borrowing an idea from [Ko 6], this program can indeed be made to work. This is our first construction.

26.11. *Construction:* Let B be a surface with genus$(B) > 1$ and two boundary-components. Let $N := B \times S^1$. Let Q be the link space, $Q = (S^3 - U(k))^-$, of a 2-bridge link, $k \subset S^3$. Let $g : \partial N \to \partial Q$ be a homeomorphism which maps $\partial(B \times 0)$ to the boundary of the meridian-disc in $\partial U(k)$. Let

$$M := N \cup_g Q.$$

We next construct a Heegaard surface $F \subset M$ as follows. First take $G := (\partial U(B \times 0) - \partial N)^-$, where $B \times 0 \subset B \times S^1 = N$. Since k is a 2-bridge link, there is a 2-sphere $S \subset S^3$ which separates k into two pairs of bridges (see example 26.12). In particular, $S' := (S - U(k))^-$ is a 4-punctured 2-sphere in $(S^3 - U(k))^-$. So $\partial S'$ has exactly four curves. Every component of $\partial U(k)$ contains exactly two of them. Thus w.l.o.g. $g(\partial G) = \partial S'$ since every component of $\partial U(k)$ contains exactly two components of ∂G as well. Hence the union $F := S' \cup_g G$ forms a closed surface in M. Now let ℓ be a non-separating curve in B. Let $T_\ell := \ell \times S^1$ be the vertical torus above ℓ. Let $h_\ell : M \to M$ be a fiber-preserving homeomorphism with support in $U(T_\ell)$ and so that $h_\ell | U(T_\ell)$ is not isotopic to the identity, by an isotopy fixing the boundary $\partial U(T_\ell)$. Define

$$F_n := h_\ell^n(F), \ n \in \mathbf{Z}.$$

26.12. Example. *Whitehead Link.*

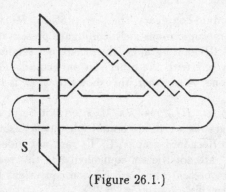

(Figure 26.1.)

26.13. Proposition. *For every integer $n \in \mathbf{Z}$, F_n is a Heegaard surface for M. Moreover, F_m is isotopic to F_n if and only if $m = n$.*

In particular, the total number of non-isotopic Heegaard decompositions of M is infinite.

Proof. We first verify that F is a Heegaard surface for M. To see this note that $U(B \times 0) \subset N$ is the product I-bundle over a surface with non-empty boundary and so a handlebody. By the same reason, $(N - U(B \times 0))^-$ is a handlebody. Thus G separates N into two handlebodies, say W_1, W_2. Moreover, the planar surface $S \subset Q$ splits Q into two handlebodies, say V_1, V_2. The intersection $A_i = U(k) \cap V_i$ is a pair of disjoint annuli in ∂V_i. But note that there is a meridian-system, $\mathcal{D}_i$, in V_i such that every annulus from A_i intersects exactly one disc from $\mathcal{D}_i$ in exactly one arc. It follows that, attaching V_i to another handlebody along A_i, yields a handlebody again. In particular, $V_i \cup_{A_i} W_i$, $i = 1, 2$, is a handlebody. So F is a Heegaard surface for M.

It remains to show that $F_m := h_\ell^m F$ is isotopic in N to F_n iff $m = n$. To see this claim recall $\ell \times 0 \subset B \times 0$ is non-separating. Thus there is a simple closed curve $k \subset B \times 0$ which intersects $\ell \times 0$ in exactly one point. Now, there is an $\alpha \in \mathbf{Z}$, such that the image $h_\ell^n(k)$, $n \in \mathbf{Z}$, is a curve which has intersection number $n \cdot \alpha$ with $B \times 0$. So the claim follows from the fact that the intersection number is a homotopy invariant. ◊

By what we have just seen, Dehn-twists along essential tori may generate a large set of Heegaard surfaces. In this context it is important to note that Dehn twists along essential *annuli* behave completely different. Indeed, our next result tells us that Dehn twists along annuli do not effect Heegaard surfaces at all. This is somewhat surprising since otherwise Dehn twists along annuli and tori share many similar properties. Specifically, for Haken 3-manifolds, compositions of Dehn twists are the only homeomorphisms which can possibly have infinite order [Joh 1, cor. 27.6].

26.14. Proposition. *Let* Γ *be an irreducible Heegaard graph in a 3-manifold* M. *Let* $h : M \to M$ *be a homeomorphism with support in the regular neighborhood of an essential annulus in* M. *Then* $h\Gamma$ *is slide-equivalent to* Γ.

Proof. Let A be an essential annulus in M and let $h_A : M \to M$ be a Dehn twist along A. Given a Heegaard graph Γ in M, we have to show that $h_A(\Gamma)$ is a Heegaard graph which is slide-equivalent to Γ.

By prop. 23.11, we may suppose that Γ and S are in strictly normal position. Let τ be an edge from Γ. Set $\Gamma' := (\gamma - \tau)^-$. Then $M' := (M - U(\Gamma'))^-$ is a one-relator 3-manifold and $S' := S \cap M'$ is a strongly normal surface in M' (S is strictly normal w.r.t. Γ). An inductive application of prop. 15.16 shows that M' must be irreducible since Γ is irreducible. Thus there is (see remark to prop. 15.20 and 15.21) a system of at most two annuli, C_1, C_2, in $M(\Gamma') := (N - U(\Gamma'))^-$ with the following properties, for $i = 1, 2$:

(1) $\partial C_i \subset A \cup \partial M'$ and $C_i \cap \partial M' \neq \emptyset$.

(2) $C_i \cap \tau$ is an arc in C_i joining both components of ∂C_i.

(3) $A \cap C_i$ consists of simple closed curves which are essential in both C_i and A.

(4) $A \cap \tau \subset C_i \cap \tau$.

Now, it is an immediate, but nevertheless remarkable, fact that the Dehn twist, h_A, along the annulus A maps the annulus-system $C := C_1 \cup C_2$ to itself (here we use the hypothesis that A is an annulus, for Dehn twists along tori need not have this property). In particular, $h_A(\tau \cap C) \subset C$ and $h_A(\tau - C)^- = (\tau - C)^-$. It follows from properties (1), (2) and (3) that $h_A(\tau)$ can be deformed into $\tau \cap C$, using an isotopy in C which fixes $\partial C - \partial M(\Gamma')$. But any such isotopy is easily seen to be a product of edge slides of Γ. Thus the proposition follows by an induction on the edges of Γ. $\diamond$

26.15. Remark. Not all Dehn twists along essential *tori* have an effect on Heegaard surfaces. Here is an example. Let M be a Seifert fiber space over the 2-sphere. Then in general the mapping class group of M is an infinite group generated by Dehn twists along essential and vertical tori in M (see [Joh 1,cor.27.6]). Given the collection, $s_1, s_2, ..., s_{n_i}$ of all exceptional fibers of M a Heegaard surface, F, of N can easily be constructed. For this let $A_2, ..., A_{n-1}$ be a system of vertical, pairwise disjoint annuli in $(M - \bigcup U(s_i))^-$ which join $U(s_i)$, $1 \le i \le n-1$, with $U(s_{i+1})$ and let t_j, $2 \le j \le n-1$, be an essential arc in A_j. Then $F := \partial(\bigcup U(s_i) \cup \bigcup U(t_j))$ is a Heegaard surface for M, and the following figure indicates that F is invariant (modulo isotopy) under Dehn-twists along essential, vertical tori in M.

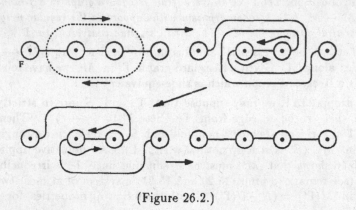

(Figure 26.2.)

26.16. *Exotic Heegaard Strings.*

We next construct examples of one-relator 3-manifolds which have non-unique Heegaard strings. The idea is to use "resolvents" for this purpose.

26.17. Definition. *Let* (M, k) *be a relative handlebody. An arc* ℓ *in* M, $\ell \cap \partial M = \partial \ell \subset \partial M - k$ *is called a __resolvent__ (for* (M, k)*), provided*

(1) $N_\ell := (M^+(k) - U(\ell))^-$ *is a handlebody, and*

(2) $M_\ell := (M - U(\ell))^-$ *is a handlebody.*

Of course, every resolvent for (M, k) is a Heegaard string for $M^+(k)$ but not necessarily vice versa. We say a resolvent is *exotic* if it is not homeomorphic to the Heegaard string in $M^+(k)$ dual to k. In the following we are looking for exotic resolvents.

To explain the choice of name we need the notion of "insignificant" curves. Here we say a curve $k \subset \partial M$ in the boundary of a handlebody, M, is *insignificant* (for M) if there is a meridian-system $\mathcal{D} \subset M$ such that $\#(k \cap \mathcal{D}) = 1$. Note that $M^+(k)$ is a handlebody if k is insignificant (and so $M^+(k)$ is not a real one-relator 3-manifold - hence the name "insignificant"). The next result tells us that a resolvent for (M, k) "resolves" the entanglement of the curve $k \subset \partial M$ (at the expense of an additional handle).

26.18. Proposition. *Let* (M, k) *be a relative handlebody. Then an arc* $\ell \subset M$, $\ell \cap \partial M = \partial \ell \subset \partial M - k$, *is a resolvent for* (M, k) *if and only if* $M_\ell := (M - U(\ell))^-$ *is a handlebody and the curve* k *is insignificant for* M_ℓ.

Proof. One direction is obvious. For the other direction suppose ℓ is a resolvent for (M, k). Then, by definition, M_ℓ and $N_\ell := (M^+(k) - U(\ell))^-$ are both handlebodies. Moreover, $genus(M_\ell) = genus(M) + 1$ and $genus(M^+(k) - U(\ell))^- = genus(M)$. Thus there is a system, $\mathcal{D}$, of n proper discs in $N(\ell)$

which splits $N(\ell)$ into a 3-ball. Now, the handlebody N_ℓ equals the one-relator 3-manifold $M_\ell^+(k)$. Thus a repeated application of the handle addition lemma (see prop. 15.16) shows that $\mathcal{D}$ can be chosen to be a system of discs in the handlebody M_ℓ. Then the complement $W := (M_\ell - U(\mathcal{D}))^-$ is a solid torus whose boundary contains k. Since $M_\ell^+(k)$ is a handlebody, it follows that the winding number of k w.r.t. W must be one. So, in particular, there is a meridian-disc $A \subset W$ with $\#(A \cap k) = 1$. Thus, altogether, the union $\mathcal{D}' := \mathcal{D} \cup A$ is a system of meridian-discs in M_ℓ with $\#(\mathcal{D}' \cap k) = 1$. Hence, by definition, k is insignificant for M_ℓ. $\Diamond$

At first it may appear doubtful that resolvents exists at all. But experience shows that they are actually quite frequent. In fact, we will see that there is already an abundance of interesting resolvents for relative handlebodies of genus two and many more can be expected for higher genus handlebodies.

In order to search for exotic resolvents for (M, k) we look at a special type of resolvents, namely those which lie in a fixed meridian-system of M. So let $\mathcal{D} \subset M$ be a meridian-system for M which we have fixed in advance. Let ℓ be an arc in $\mathcal{D}$. Then note that $M_\ell := (M - U(\ell))^-$ is automatically a handlebody since $\mathcal{D}_\ell := (\mathcal{D} - U(\ell))^-$ is a meridian-system for M_ℓ. This fact makes the arcs in $\mathcal{D}$ so attractive for our purpose. According to the previous proposition, it now remains to check whether the curve k is insignificant for M_ℓ. At this point recall from cor. 1.6 that any two meridian-systems in a handlebody are slide-equivalent. Thus it remains to check whether $\mathcal{D}_\ell$ is slide-equivalent to a meridian-system which intersects k in exactly one point. To do this it is convenient to use Heegaard diagrams. We therefore translate the whole situation into the context of Heegaard diagrams.

We first translate the operation of disc-slides into the context of Heegaard diagrams. For this let $\underline{D} = (D, h, \mathcal{K})$ be a Heegaard diagram. Let F denote the system of 2-spheres underlying $\underline{D}$. A disc $A \subset F$ is called a *sliding disc* if it has the following properties:

(1) $\partial A \cap D = \emptyset$ and $A \cap D \neq \emptyset$.

(2) there is at least one disc, D_0, from D with $D_0 \subset A$ and $h(D_0) \subset F - A$.

A sliding disc $A \subset F$ gives rise to a new Heegaard diagram as follows. Let $D_0 \subset A$ be a disc from the collection D with $h(D_0) \subset F - A$. Let $g : (A - D_0)^- \hookrightarrow h(D_0)$ be an embedding with $g|\partial D_0 = h|\partial D_0$. Define

$$D_A := (D - A) \cup A \cup (h(D_0) - g(A - D)),$$
$$\mathcal{K}_A := (\mathcal{K} - A)^- \cup g(\mathcal{K} \cap A).$$
$$h'_A(x) := \begin{cases} h(x), & x \in \partial(D - (A \cup h(A \cap D))), \\ g(x), & x \in \partial A \\ gh^{-1}(x), & x \in \partial(h(A \cap D) - h(D_0)) \\ hg^{-1}(x), & x \in \partial((h(D_0) \cap D_A) - g(A)) \\ g^{-1}(x), & x \in \partial g(A), \end{cases}$$

Let $h_A : D_A \to D_A$ be the extension of $h'_A : \partial D'_A \to \partial D'_A$. Then $\underline{D}_A :=$ $(D_A, h_A, \mathcal{K}_A)$ is a Heegaard diagram. We say $\underline{D}_A$ is obtained from $\underline{\underline{D}}$ by *sliding the meridian-disc, D_0, across A.* Notice that such a slide may well increase the complexity $c(\underline{D}) := \frac{1}{2}\#\partial\mathcal{K}$ of a Heegaard diagram.

26.19. Example. Figure 26.3 shows the effect of sliding the meridian-disc, B, across the sliding-disc, A, (the attaching map maps B, C to B', C' and dots to dots). We see that, given the sliding disc, A, the sliding process amounts to (1) expanding the disc B' to a large disc, (2) moving the content of the disc A over to the disc B' (using the extension of the attaching homeomorphism $\partial B \to \partial B'$), forgetting $\partial B'$ and the content of disc A and finally (4) shrinking A, A' to ordinary meridian-discs and pushing all of them to the standard positions.

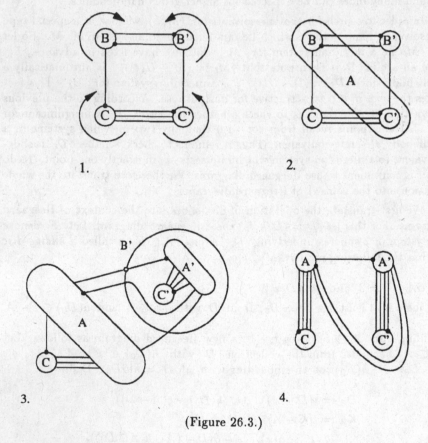

(Figure 26.3.)

In the following we will only need slides associated to recurrent arcs. Recall that, by definition, a *recurrent arc* for the diagram $\underline{D} = (D, h, \mathcal{K})$ is an arc, r, in $(F - D)^-$, disjoint to $\mathcal{K}$, such that ∂r is contained in one disc, say D_0, from D and that r joins two different components of $\partial D_0 - \partial \mathcal{K}$. Now r separates $(F - D_0)^-$ into two discs. Let B_r be that one of them not containing $h(D_0)$. Define $\underline{D}_r := \underline{D}_{A_r}$, where $A_r := U(D_0 \cup B_r)$. Then we say that $\underline{D}_r$ is the Heegaard diagram obtained *by sliding* $\underline{D}$ *over the recurrent arc* r. Notice that a slide over a recurrent arc reduces the complexity $c(\underline{D}) := \frac{1}{2} \cdot \#\partial\mathcal{K}$. We say a one-relator Heegaard diagram $\underline{D}$ is *insignificant* if there is a finite sequence $\underline{D} = \underline{D}_1, \underline{D}_2, ..., \underline{D}_n = (D_n, h_n, \overline{\mathcal{K}}_n)$ of Heegaard diagrams such that

(1) $\underline{D}_{i+1}$ is obtained by sliding $\underline{D}_i$ over some recurrent arc.

(2) $\#\mathcal{K}_n \le \frac{1}{2} \cdot \#D_n$ and $\mathcal{K}_n$ meets every disc from D_n in at most one point.

It is easy to decide whether a given Heegaard diagram is insignificant since recurrent arcs in Heegaard diagrams can be located by inspection. Thus, by the next proposition, it can be decided whether a given curve on a handlebody is insignificant.

26.20. Proposition. *Let* $\underline{D}$ *be a Heegaard diagram and let* $(M, k) = M(\underline{D})$ *the associated relative handlebody. Then the curve* k *is insignificant for* M *if and only if the Heegaard diagram* $\underline{D}$ *is insignificant.*

Proof. Suppose $\underline{D}$ is insignificant. Then, by definition, there is a sequence $\underline{D} = \underline{D}_1, ..., \underline{D}_n$ with (1) and (2) above. If $\#\mathcal{K}_n \ge 2$, let A be a component of $U(D_n \cup \mathcal{K}_n)$ containing an arc from $\mathcal{K}_n$. Then sliding $\underline{D}_n$ over A, results in a Heegaard diagram with smaller complexity. So w.l.o.g. $\#\mathcal{K}_n = 1$. More precisely, $\mathcal{K}_n$ is an arc which joins two discs from D_n whose union is invariant under the attaching homeomorphism h_n. Thus the disc-system D_n corresponds to a meridian-system, $\mathcal{D}$, in $M(\underline{D}_n)$ which intersects the relator curve of $\underline{D}_n$ in exactly one point. But, by (1), $M(\underline{D}_n)$ is homeomorphic to (M, k). Thus $\#(\mathcal{D} \cap k) = 1$ and so, by definition, k is insignificant for M.

For the other direction let $\mathcal{D}$ be the meridian-system for $M(\underline{D})$ corresponding to the meridian-discs from $\underline{D}$. Suppose k is insignificant for M. Then there is a meridian-system, $\mathcal{D}'$, in $M(\underline{D})$ with $\#(\mathcal{D}' \cap k) = 1$. Let $\mathcal{D}'$ be chosen so that $\mathcal{D}' \cap \mathcal{D}$ is a system of arcs whose total number is as small as possible. If $\mathcal{D}' \cap \mathcal{D} \ne \emptyset$, then $\mathcal{D}$ separates an outermost disc, D_0, from $\mathcal{D}'$ with $D_0 \cap k = \emptyset$. It follows that $D_0 \cap \partial\mathcal{D}'$ is a recurrent arc in $\underline{D}$. Sliding over recurrent arcs reduces the complexity of Heegaard diagrams. Thus, repeating this process if necessary, we may suppose that $\mathcal{D}' \cap \mathcal{D} = \emptyset$, and we are done. $\Diamond$

Let $\underline{D} = (D, h, \mathcal{K})$ be any Heegaard diagram. To any arc $\ell \subset D$ with $\ell \cap \partial D = \partial\ell$ and $\ell \cap \partial\mathcal{K} = \emptyset$ we associate a new Heegaard diagram $\underline{D}_\ell = (D_\ell, h_\ell, \mathcal{K}_\ell)$ by setting

$$D_\ell := (D - U(\ell \cup h\ell))^-, \quad h_\ell := h|D_\ell, \quad \mathcal{K}_\ell := \mathcal{K}.$$

We say $\underline{D}_\ell$ is obtained from $\underline{D}$ by *splitting* along the arc ℓ. An arc $\ell \subset D$ with $\ell \cap \partial D = \partial \ell$ and $\ell \cap \partial \mathcal{K} = \emptyset$ is a *resolvent* for $\underline{D}$ if the diagram $\underline{D}_\ell$ obtained by splitting along ℓ is insignificant. Of course, if ℓ is a resolvent for a Heegaard diagram, $\underline{D}$, then the arc, associated to ℓ, in the relative handlebody $M(\underline{D})$, associated to $\underline{D}$, is a resolvent for $M(\underline{D})$. Thus we can formulate the following search-procedure.

26.21. *Search-Procedure for Resolvents.* To search for resolvents apply the following steps:

1. Enumerate simple closed curves on a handlebody of a given genus (mod handlebody homeomorphism) by enumerating Heegaard diagrams.
2. Given a Heegaard diagram $\underline{D} = (D, h, \mathcal{K})$, enumerate all all arcs in D.
3. Given an arc in D, split $\underline{D}$ along this arcs and check whether or not the splitted diagram is insignificant.

To carry out step 3, minimize the complexity of $\underline{D}$, by using slides over recurrent arcs from $\mathcal{K}$. Then $\underline{D}$ is insignificant if and only if the resulting Heegaard diagram has a connected arc-system. This procedure produces resolvents and so candidates for exotic resolvents. But to complete the algorithm we still have to find out which one of them are exotic. In principal, this can always be done and in many cases this can be done quite effectively (see below).

To illustrate the method, we carry out the above procedure for genus two handlebodies. Curves on handlebodies of genus two (with incompressible complement) are easy to enumerate since they are all carried by one of only a small number of diagrams. In fact, every such curve is carried (mod handlebody homeomorphism) by one of the following two diagrams (this statement is easy to verify; e.g., note that we may suppose that these diagrams have no recurrent arcs).

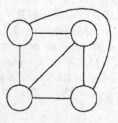 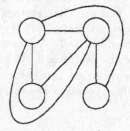

(Figure 26.4.)

To be more explicit consider one of the previous diagrams. Then, given a 6-tuple $x = (x_0, x_1, x_2, x_3, r_1, r_2) \in \mathbf{N}^6$, we construct a simple closed curve $k(x) \subset \partial M$ as follows (see fig. 26.5). Let D_x denote the system of four discs in the 2-sphere

F. Taking x_i copies for every arc labeled x_i (in the diagram on the left of fig. 26.5), we fix an arc-system $\mathcal{K}_x \subset (F - D_x)^-$ (see diagram on the right). This arc-system is the arc-system of a Heegaard diagram iff the labels, x_i, satisfy the equations

$$x_0 + x_1 + x_5 = x_0 + x_2 + x_4 \quad \text{and} \quad x_1 + x_2 + x_3 = x_3 + x_4 + x_5.$$

It follows that $x_1 = x_4$ and $x_2 = x_5$. In order to specify a Heegaard diagram, $\underline{D}_x = (D_x, h_x, \mathcal{K}_x)$, it remains to specify the attaching homeomorphism, h_x. This is done by the convention that the attaching homeomorphism maps dots of meridian-discs to dots. Then h_x is uniquely given by a pair of "rotation numbers", r_1, r_2. Here r_1, r_2 denote the differences of the distances of the above dots (see diagram on the right) to the standard dots (see diagram on the left), counted in the direction of the indicated orientations. Attaching the meridian-discs from D_x under the attaching homeomorphism, h_x, we get the relative handlebody $M(\underline{D}_x)$.

26.22. Example. *The Heegaard diagram $\underline{D}_x$, $x = (2, 3, 2, 3, 4, 3)$.*

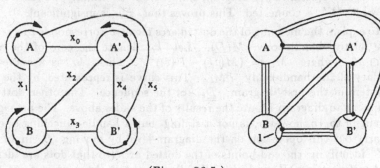

(Figure 26.5.)

Doing the same with the other diagram from fig. 26.4, we get all curves on handlebodies of genus two. In this way, curves on handlebodies of genus two (mod handlebody homeomorphisms) are parameterized by 7-tuples of non-negative integers (where the first entry is either 0 or 1 and indicates which one of the diagrams from fig. 26.4 has been selected as carriers). So, in particular, relative handlebodies can be enumerated by enumerating 7-tuples; a fact which we use for our search procedures.

Using the previous enumeration scheme, we can now carry out the search procedure for exotic resolvents. In fact, it is not hard to program a computer to do this. The following proposition gives a result of such a systematic computer search (the same program finds a lot more examples). To verify the output, we carry out by hand the relevant sliding-operations for the specific example given below.

26.23.   Proposition. *Resolvents for Heegaard diagrams, and so for relative handlebodies, do exist. In particular, the arc $\ell \subset B$ in the Heegaard diagram $\underline{D} := (0, 2, 3, 2, 3, 4, 3)$ (see figure 26.5) is a resolvent.*

Proof. By definition, we have to show that the Heegaard diagram $\underline{D}_\ell$, obtained from $\underline{D}$ by splitting along the arc ℓ from fig. 26.5, is insignificant. For this it suffices to verify that $\underline{D}_\ell$ can be turned, by successively sliding over recurrent arcs, into a Heegaard diagram $\underline{D}' = (D', h', \mathcal{K}')$ such that $\mathcal{K}'$ is connected. Consider the sequence of diagrams from fig. 26.6, disregarding the dotted arcs for a moment. The first diagram in this sequence is the diagram $\underline{D}_\ell$, and the sequence itself describes a sequence of slides beginning with $\underline{D}_\ell$. More precisely, every passage from an odd numbered diagram to the next (even numbered) diagram describes a slide. Consider for example the passage from diagram 1 to diagram 2. In this passage the big circle D' in diagram 1 has been contracted to the small circle D' in diagram 2. At the same time the circle A in diagram 1 has been enlarged to the big circle A in diagram 2 and the content of D' from diagram 1 has been moved over (using the attaching homeomorphism) to circle A from diagram 2. Recall from 26.19 that this defines a sliding-operation. The same rule describes the passage from diagram 3 (which equals diagram 2) to diagram 4 and so on. Eventually, we reach diagram 14 in which $\mathcal{K}'$ is connected. This proves that $\underline{D}_\ell$ is insignificant.

To explain the meaning of the dotted arcs recall ℓ corresponds to a resolvent in the relative handlebody $M(\underline{D})$. Let k_ℓ denote the core of the annulus $U(\ell) \cap M_\ell$, where $M_\ell := (M(\underline{D}) - U(\ell))^-$. Then k_ℓ is a curve in the boundary of the handlebody ∂M_ℓ. This curve is represented by the dotted arc-system in the first diagram, $\underline{D}_\ell$, of the sequence. The other dotted arc-systems, up to diagram 14, are the results of the slides above. The passage from diagram 14 to 15, however, is not a sliding- but a handle-cancelling operation. We carry out this operation on the diagram-level by moving B in 14 on top of B', identifying the end-points of the dotted arcs (so that dots are identified) and then forgetting B and B'. Finally, we obtain diagram 16 from 15 by a reflection.

Let $\underline{D}'$ denote diagram 16. Then $\underline{D}' := (0, 1, 1, 1, 4, 0, 4)$ and, by what we have shown above, the two one-relator 3-manifolds $M^+(\underline{D})$ and $M^+(\underline{D}')$ are homeomorphic.

26.24. Example.

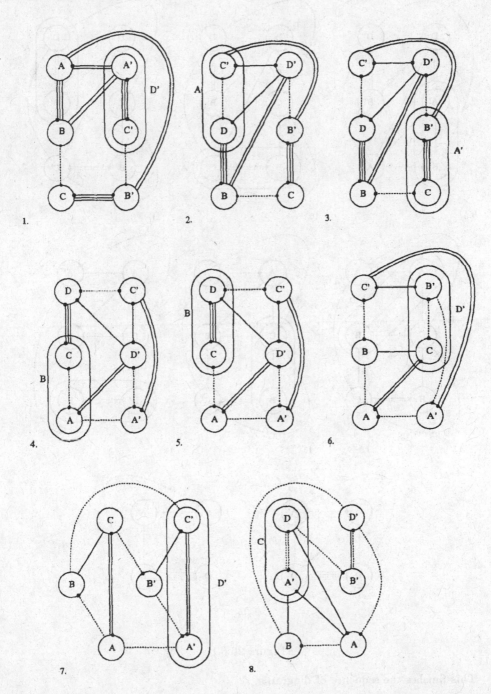

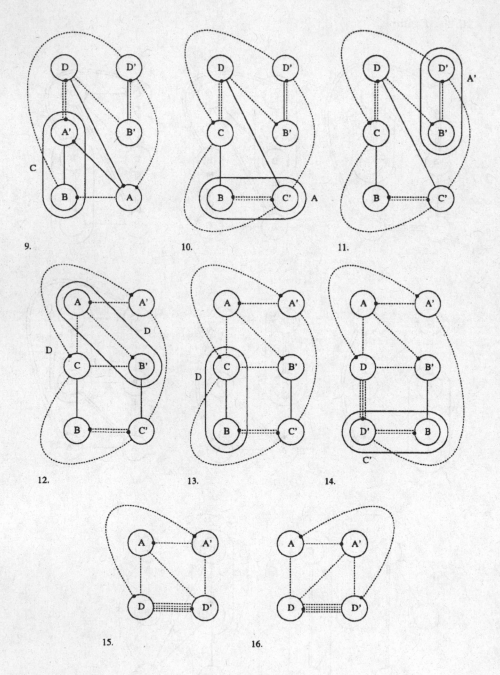

(Figure 26.6.)

This finishes the sequence of diagrams. ◊

The next proposition tells us that the resolvent given in the previous proposition is actually exotic (see 26.17 for definition), i.e., $M^+(0, 2, 3, 2, 3, 4, 3)$ and $M^+(0, 1, 1, 1, 4, 0, 4)$ are homeomorphic but $M(0, 2, 3, 2, 3, 4, 3)$ and $M(0, 1, 1, 1, 4, 0, 4)$ are not. So we have found a new example of non-homeomorphic Heegaard strings in one-relator 3-manifolds (we remark in passing that the examples from [BRZ] fit into this scheme as well). Note that this is also another counterexample to the Magnus conjecture [LySch] since the corresponding one-relator 3-manifold is simple and so every automorphism of its fundamental group is induced by a homeomorphism.

26.25. Proposition. *Let* $M(\underline{D})$ *be the relative handlebody associated to the Heegaard diagram* $\underline{D} = (D, h, \overline{\mathcal{K}}) = (0, 2, 3, 2, 3, 4, 3)$ *(see figure 26.5). Let* ℓ *be the arc in* $M(\underline{D})$, *associated to the resolvent* l *for* $\underline{D}$. *Then* ℓ *is an exotic resolvent for* $\underline{D}$.

Proof. We know that ℓ is a resolvent for $(M, k) = M(\underline{D})$. It remains to show that (M, k) is not homeomorphic to $(M_\ell^+(k_\ell), k_\ell)$, where k_ℓ is the core of the annulus $M_\ell \cap U(\ell)$. But we know from the proof of prop. 26.23 that $M_\ell^+(k_\ell) = M(\underline{D}')$, where $\underline{D}'$ is the last diagram from fig. 26.6.

Now, notice that $c(\underline{D}) > c(\underline{D}')$. Thus there cannot be any admissible homeomorphism $M(\underline{D}) \to M(\underline{D}')$. For otherwise, by the algorithm from 7.9, there must be a simple closed curve, s, in $(S - D)^-$ with $\#(\mathcal{K} \cap s) < \#(\mathcal{K} \cap \partial D_i)$, for some disc D_i from D. But inspection shows that this is not the case. $\diamond$

In the above proof we have appealed to Whitehead's theorem in the form of 7.9 to find out that a certain pair of Heegaard strings is non-homeomorphic. At this point we mention that the same problem can be translated into an isomorphism problem for fundamental groups.

26.26. Proposition. *Let* (M, k), (M, l) *and* (N, t) *be simple relative handlebodies with a single relator curve. Then* (M, k) *and* (M, l) *are homeomorphic if and only if the fundamental groups of* $P := (M, k) \sqcup (N, T)$ *and* $Q := (M, l) \sqcup (N, t)$ *are isomorphic, where* $(M, k) \sqcup (N, t)$ *denotes the 3-manifold obtained by attaching* k *and* t.

Proof. Note that the characteristic submanifolds of P is the regular neighborhood of a single annulus. The same for Q. Thus the claim follows from the classification of homotopy equivalences of Haken 3-manifolds [Joh 1]. $\diamond$

In this context recall that it has proved useful to consider representations into known finite groups in order to determine that two given groups are non-isomorphic. But this is usually a tedious process. In contrast, the Whitehead algorithm often gives fast results (as seen above). But the same algorithm is completely useless for checking the finer, but also important question whether Heegaard strings are *isotopic*. Here the following observation may prove to be helpful.

26.27. Theorem. *It can be decided whether two given Heegaard strings in a simple 3-manifold M are ambient isotopic, provided no component of ∂M is a torus.*

Proof. Let s_1, s_2 be two given Heegaard strings in a simple 3-manifold M. According to cor. 17.3, s_1, s_2 are ambient isotopic if and only if they are homotopic. Let DM denote the double of M and denote by $Ds_1, Ds_2 \subset DM$ the doubles of the arcs s_1, s_2, respectively. Then s_1, s_2 are homotopic in M if and only if Ds_1, Ds_2 are homotopic in DM. But the problem of checking whether closed curves in a 3-manifold are homotopic is the conjugacy problem for the fundamental group of that 3-manifold. Now doubling M yields a simple 3-manifold again. Hence, by Thurston [Th 2], it has a hyperbolic structure. But, according to [Can], the conjugacy problem is solvable for every 3-manifold which is known to have a hyperbolic structure with finite volume. $\Diamond$

26.28. *Exponential Growth.*

The results of the remainder of this section are due to Lustig and Moriah [LM]. They are included here because they show an important new phenomenon, namely the occurrence of exponential growth for Heegaard graphs. The result is that the number of slide-inequivalent Heegaard graphs can grow exponentially when the underlying 3-manifold is varied. This phenomenon occurs already for Seifert fiber spaces, in fact, for Seifert fiber spaces over the disc. Seifert fiber spaces may or may not have exceptional fibers and this alternative seems to be important for Heegaard decompositions. Indeed, while the product of *any* orientable surfaces with S^1 has unique Heegaard decompositions [Wa 3][SchT] [Schul], this is no longer true for Heegaard decompositions of Seifert fiber spaces over the disc with two exceptional fibers [BRZ].

It is very easy to construct a variety of Heegaard string systems in Seifert fiber spaces over the disc (the hard part is to show that they are slide-inequivalent). Here is a straightforward construction. Let X be the Seifert fiber space over the disc with $n+1$ exceptional fibers. Let $\mathcal{A} \subset X$, $\mathcal{A} \cap \partial X = \partial \mathcal{A}$, be a system of vertical annuli, $A_0, A_1, ..., A_n$, with the following properties, for all $0 \le i \le n$:

(1) A_i separates a solid torus, V_i, from X with $V_i \cap \mathcal{A} = A_i$ and containing exactly one exceptional fiber.

(2) The solid tori V_i are pairwise disjoint and every exceptional fiber of X is contained in one of them.

Taking the pre-image (under the fiber-projection map) of an appropriate arc-system in the orbit surface of X, the above annulus system is easy to construct. Now, for every i, $0 \le i \le n$, let s_i denote the spanning arc in the annulus A_i. By r_i denote the union of the exceptional fiber in V_i with some straight horizontal arc in V_i joining this exceptional fiber with the boundary ∂X.

We say s_i is a *regular string* and r_i is an *exceptional wand*. Define $\mathcal{T}_n := \{0,1,2\} \times \{0,1\}^{n-1}$ and, for every $(i_1, i_2, ..., i_n) \in \mathcal{T}_n$, set

$$\Gamma(i_1, i_2, ..., i_n) := \bigcup_{1 \le j \le n} p_j,$$

where $p_1 := \begin{cases} r_0, & \text{if } i_1 = 0 \\ s_0, & \text{if } i_1 = 1 \\ s_1, & \text{if } i_1 = 2 \end{cases}$ and where $p_j := \begin{cases} r_j, & \text{if } i_j = 0 \\ s_j, & \text{if } i_j = 1 \end{cases}, \ 2 \le j \le n.$

Note that, by construction, $\#\mathcal{T}_n = 3 \cdot 2^{n-1} = \frac{3}{2} \cdot 2^{\#\mathcal{A}}$. Finally, it is easy to see that $T(i_1, i_2, ..., i_n)$ is a Heegaard string-system for X. Denote by $\mathcal{T}_n(X)$ the subset of all slide-equivalence classes of Heegaard graphs $T(i_1, ..., i_n)$ with $(i_1 ..., i_n) \in \mathcal{T}_n$.

We wish to show that $\#\mathcal{T}_n(X)$ grows exponentially. Here is where the work of Lustig and Moriah comes in. To describe it let $\mathcal{F}_n = \mathcal{F}(x_1, ..., x_n)$ be the free group in n letters $x_1, ..., x_n$. Let G be the Fuchsian group given by

$$< q_1, ..., q_n \mid q_1^{\nu_1}, ..., q_n^{\nu_n} >, \quad \nu_i \text{ odd and pairwise coprime.} \tag{$*$}$$

Note that there is a ring homomorphism ρ from the integer group ring $\mathbf{Z}G$ to the ring $\mathcal{M}_2(\mathbf{C})$ of 2×2-matrices given as follows. Start with a faithful representation $G \to \mathrm{PSL}_2\mathbf{C}$ (see e.g. [Ka, thm. 4.3.2] for existence). Recall from [Cu] that this representation lifts to $G \to \mathrm{SL}_2\mathbf{C}$ since all α_i are odd. Let ρ be the linear extension of this lift to $\mathbf{Z}G$.

Now let $\frac{\partial}{\partial x_i} : \mathbf{Z}\mathcal{F}_n \to \mathbf{Z}\mathcal{F}_n$ be the i-the Fox derivative of the integer group ring of the free group $\mathcal{F}_n$. By definition, this is the unique **Z**-linear function satisfying

$$\frac{\partial}{\partial x_i} x_j = \delta_{i,j} \quad \text{and} \quad \frac{\partial}{\partial x_i}(ab) = \frac{\partial a}{\partial x_i} + a \frac{\partial b}{\partial x_i}, \text{ for } a, b \in \mathcal{F}_n.$$

Given $\alpha = (\alpha_1, ..., \alpha_n)$, define $P_\alpha : \mathcal{F}_n \to G$ by setting

$$P_\alpha(x_i) := q_i^{\alpha_i}.$$

Notice that P_α is a presentation if $(\nu_i, \alpha_i) = 1$, for all $1 \le i \le n$. It is crucial to observe [LM, lemma 1.5] that

$$\rho(P_\alpha \frac{\partial r_j}{\partial x_i}) = 0, \ r_j \in \mathrm{kern}(P_\alpha).$$

Using this observation, the following necessary criterion for Nielsen equivalence can be deduced.

26.29. Lemma. *Let* $P_\alpha, P_\beta : \mathcal{F}_n \to G$ *be presentations as above. Suppose* P_α *is Nielsen-equivalent to* P_β. *Let* $y_j \in \mathcal{F}_n$ *with* $P_\alpha(y_j) = P_\beta(x_j)$. *Then*

$$\det[\rho P_\alpha \frac{\partial y_j}{\partial x_i}]_{i,j} = 1.$$

Proof. P_α is Nielsen equivalent to P_β. Hence, by definition, there is an isomorphism $\psi : \mathcal{F}_n \to \mathcal{F}_n$ with $P_\beta = P_\alpha \circ \psi$. Set $z_j := \psi(x_j)$. Then $P_\alpha(z_j) = P_\alpha(\psi(x_j)) = P_\beta(x_j)$. Since both $z_1, ..., z_n$ and $x_1, ..., x_n$ are basis for $\mathcal{F}_n$, we get one from the other by some finite sequence of Nielsen transformations. In particular, the matrix $[P_\beta \frac{\partial z_j}{\partial x_i}]_{i,j}$ is a finite product of elementary matrices in $\mathbf{Z}G$ [CF]. Hence $[\rho P_\beta \frac{\partial z_j}{\partial x_i}]_{i,j}$ is such a product in $\mathcal{M}_2 \mathbf{C}$. But it is not hard to see that the determinant of any block matrix with entries in $\mathbf{SL}_2\mathbf{C}$ must be one. So $\det[\rho P_\beta \frac{\partial z_j}{\partial x_i}]_{i,j} = 1$.

It remains to show that $\det[\rho P_\beta \frac{\partial z_j}{\partial x_i}]_{i,j} = \det[\rho P_\beta \frac{\partial y_j}{\partial x_i}]_{i,j}$. To see this let $\mathrm{kern}(P_\beta)$ be normally generated by $r_1, ..., r_m \in \mathcal{F}_n$. Then $P_\beta(z_j) = P_\beta(y_j) \Rightarrow z_j y_j^{-1} \in \mathrm{kern}(P_\beta) \Rightarrow z_j = (\prod w_k s_k w_k^{-1}) y_j$, where $w_k \in \mathcal{F}_n$ and $s_k \in \{r_1^{\pm 1}, ..., r_m^{\pm 1}\}$. So

$$P_\beta(\frac{\partial z_j}{\partial x_i}) = P_\beta(\frac{\partial \prod w_k s_k w_k^{-1}}{\partial x_i} + P_\beta(\prod w_k s_k w_k^{-1}) P_\beta(\frac{\partial y_j}{\partial x_i})$$

$$= \sum [P_\beta(\frac{\partial w_k}{\partial x_i}) + P_\beta(w_k) P_\beta(\frac{\partial s_k}{\partial x_i}) + P_\beta(w_k) P_\beta(\frac{\partial w_k^{-1}}{\partial x_i})] + P_\beta(\frac{\partial y_j}{\partial x_i})$$

$$= \sum P_\beta(w_k) P_\beta(\frac{\partial s_k}{\partial x_i}) + P_\beta(\frac{\partial y_j}{\partial x_i}). \diamond$$

Now let $P_\alpha, P_\beta : \mathcal{F}_n \to G$ be presentations as above. Then note that there are number $\mu_i \geq 2$, $1 \leq i \leq n$ with $P_\alpha(x_i^{\mu_i}) = P_\beta(x_i)$. Thus, by the previous lemma, we have $\det[\rho P_\alpha \frac{\partial x_i^{\mu_i}}{\partial x_i}] = 1$, if P_α and P_β are Nielsen equivalent. Using the fact that the ν_i are pairwise coprime, it is calculated in [LM], that this is only possible if $\alpha = \beta$. Hence

26.30. Theorem [LM]. *If the presentations P_α, P_β are Nielsen-equivalent, then $\alpha = \beta$.* $\diamond$

The following theorem now shows the existence of exponential growth for Heegaard graphs:

26.31. Theorem. [LM] *Let X_n be a Seifert fiber space over the disc with n exceptional fibers. Suppose the orders of these fibers are odd and pairwise coprime. Then*

$$\#H(X_n, n) \geq e^n.$$

Proof. In order to point to the relationship between Heegaard graphs in X_n and presentations of Fuchsian groups, let us take another look at the Heegaard graphs $\Gamma(i_1, ..., i_n)$ introduced above. Recall the system $\mathcal{A}$ of vertical annuli in X_n used in the definition of $\Gamma(i_1, ..., i_n)$. Recall $\mathcal{A}$ separates solid tori $V_1, ..., V_n$ from X_n containing $\Gamma(i_1, ..., i_n)$. $A_i := U(\mathcal{A} \cap \partial V_i$ is an essential annulus in ∂V_i. $(\partial V_i - A_i)^- = V_i \cap \partial X_n$. $t_i := T \cap V_i$ is either a horizontal arc next to A_i or the union of the exceptional fiber contained in V_i with

§26 Constructing Heegaard Graphs 309

some horizontal arc joining it with ∂X_n. Let $(D_i, \underline{d_i})$ be an essential disc in (V_i, A_i), where $\underline{d_i}$ is the boundary-pattern induced by A_i. Note that (V_i, A_i) is the mapping torus of an appropriate finite rotation $r_i : (D_i, \underline{d_i}) \to (D_i, \underline{d_i})$. Let z_i be a point in $\partial D_i - \underline{d_i}$ and let t_i' be a simple arc in ∂D_i joining z_i with $r_i(z_i)$. It is not hard to see that there is an edge-slide which pushes t_i into t_i' and whose sliding arc lies in $t_i \cup \partial X_n - A$.

Now let F be the orbit surface of X_n and x the collection of all exceptional points in F. The fiber projection maps the arcs t_i' to, possibly singular, arcs $s_i' \subset F' := F - \bigcup x_i$. Let $z \in \partial F$ be any point. Push all end-points of the arcs s_i' to one point, by contracting $\partial F - z$ to one point. In this way we get a collection of arcs with base-point.

Construct a Fuchsian surface complex F^+ by attaching the boundary of discs to $\partial U(x)$, using covering maps whose number of sheets equals the order of the corresponding exceptional point. Set $P(i_1, ..., i_n)(x_i) = s_i'$. Then $P(i_1, ..., i_n)$ is a presentation for $\pi_1 F^+$. But $\pi_1 F^+$ is a Fuchsian group G as in (*). Moreover, $\Gamma(i_1, ..., i_n)$ and $\Gamma(j_1, ..., j_n)$ are slide-inequivalent if $P(i_1, ..., i_n)$ and $P(j_1, ..., j_n)$ are Nielsen inequivalent. So thm. 26.31 follows from an application of thm. 26.30 to these presentations. ◇

27. Haken Complexes and Heegaard Complexes.

Let us turn our attention to the proof of the Rigidity Theorem. As indicated in the introduction to this chapter the proof for this theorem falls into two steps, and the next few sections are concerned with the first step.

Recall that in chapter IV we studied the interaction of essential surfaces with Heegaard-graphs. Basically, the result obtained there says that the intersection of an essential surface with a Heegaard graph is limited, modulo edge-slides (and isotopies). Up to now this has been considered as a statement about essential surfaces in 3-manifolds. But, changing viewpoint, we find that it may also be viewed as a statement about Heegaard graphs. Indeed, the same result may be read as saying that the intersection of *any* Heegaard graph with a given essential surface is limited (mod slides) by some number which depends on the genus of the graph (and the Euler characteristic of the surface), but not on the *embedding type* of the graphs involved. Thus it is even more than just a statement about individual graphs - it is a statement about the *set of all Heegaard graphs* (of given genus) in a 3-manifold. In the next few sections we generalize this statement from surfaces to certain surface-complexes, the so called Haken 2-complexes. In the present section we introduce the notion of Haken 2-complexes and establish their elementary properties. These Haken 2-complexes are closely related to hierarchies of surfaces in 3-manifolds and in fact their very existence characterizes Haken 3-manifolds. Moreover, to every Heegaard graph, we associate a Heegaard 2-complex by attaching to it a complete system of meridian-discs. Finally, we introduce the notion of a useful position for Heegaard 2-complexes w.r.t. a given Haken 2-complex. The main result of this section is the theorem (thm. 27.26) that every Heegaard 2-complex in a Haken 3-manifold can be isotoped into a useful position. In later sections we will begin with useful Heegaard 2-complexes and improve their positions more and more until they are in some sort of normal form.

27.1. *Haken 2-Complexes.*

Recall that a hierarchy is a device for splitting a Haken 3-manifold into 3-balls, by splitting it along a sequence of incompressible surfaces. Given the existence of incompressible surfaces for Haken 3-manifolds it has been observed by Haken that the very existence of a hierarchy is a consequence of the Kneser-Haken Finiteness Theorem (see e.g. [Hem]). The existence of hierarchies in turn is the underlying reason for various other finiteness results for Haken 3-manifolds. However, using hierarchies for splitting 3-manifolds is not good enough for our purpose. Here we wish to use them for constructing Haken complexes.

27.2. Definition. *A <u>Haken n-complex</u> $\Psi \subset N$ in an $(n+1)$-manifold N is a CW-complex together with a manifold-filtration.*

Here a *manifold-filtration*, $\{\Psi_i\}_{1 \leq i \leq m}$, for $\Psi \subset N$ (of *depth* m) is a sequence

$$\Psi_0 \subset \Psi_1 \subset ... \subset \Psi_m = \Psi$$

recursively defined by setting $\Psi_0 = \emptyset$ and $\Psi_i := \Psi_{i-1} \cup_{f_i} \sigma_i$, where σ_i is an n-manifold (possibly disconnected) and where $f_i : \sigma_i \to N$ is a map with $f_i^{-1}(\Psi_{i-1} \cup \partial N) = \partial \sigma_i$ which is in general position to Ψ_i and which is an embedding on the interior of σ_i.

In the following we will only be concerned with low-dimensional Haken complexes, i.e., with Haken 1- or 2-complexes. Haken 1-complexes are also called *Haken graphs* and the manifold filtration of a Haken graph is a *curve-filtration*.

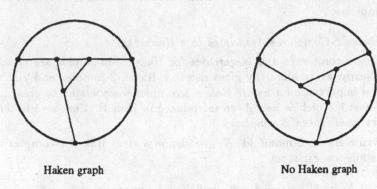

Haken graph   No Haken graph

(Figure 27.1.)

Globally, a Haken 1- or 2-complex can be quite complicated but locally it is always rather simple. This is clear for Haken graphs. For Haken 2-complexes, the following picture gives the complete list of all four possible local models.

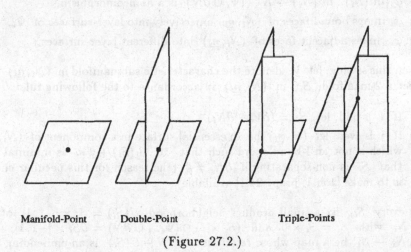

Manifold-Point   Double-Point   Triple-Points

(Figure 27.2.)

A Haken complex, Ψ, comes equipped with a fixed manifold filtration $\{\Psi_i\}_{1 \le i \le m}$. This filtration is part of the definition of Ψ. The manifold $S_i := (\Psi_i - \Psi_{i-1})^-$, $1 \le i \le m$, is called the *layer* (*layer-curve*, *layer-surface* etc.) of *index* i. A *Haken sub-complex* $\Psi' \subset \Psi$ is not just a Haken complex per se but a complex for which $\{ \Psi' \cap \Psi_i \}_{1 \leq i \leq m}$ is the manifold filtration of Ψ'. Notice that a Haken 2-complex Ψ has a canonical manifold stratification as well: the 0-manifolds (vertices) are given by the triple-points of Ψ, the 1-manifolds (edges) are given by double-arcs joining triple-points (and double-arcs joining triple-points with ∂N) and the complementary parts are surfaces. We are particularly interested in Haken 2-complexes in irreducible 3-manifolds for which the latter surfaces are discs. Those particular Haken 2-complexes are also CW-complexes.

27.3. *Haken 2-Complexes Associated to a Hierarchy.*

Haken 2-complexes and hierarchies for Haken 3-manifolds are closely related. Clearly, every hierarchy gives rise to a Haken 2-complex and vice versa. Of special importance for us are Haken 2-complexes associated to great hierarchies. Great hierarchies have been introduced in [Joh 1]. Here we use them to construct *great Haken 2-complexes*.

Given a Haken 3-manifold N, we define a great Haken 2-complex recursively as follows. First, set

$$N_0 := N, \quad \underline{n}_0 := \emptyset, \quad \Psi_0 := \emptyset \quad \text{and} \quad \varphi_0 := id : N_0 \to N.$$

Next, let $(N_i, \underline{n}_i)$, $i \geq 0$, be a 3-manifold with useful boundary-pattern, let $\Psi_i \subset N$ be a Haken 2-complex and let $\varphi_i : N_i \to N$ be an immersion such that

(1) $\varphi_i | int(N_i) : int(N_i) \to N - (\Psi_i \cup \partial N)$ is a homeomorphism,

(2) φ_i maps bound faces of $(N_i, \underline{n}_i)$ injectively into layer-surfaces of Ψ_i, and

(3) φ_i maps adjacent faces of $(N_i, \underline{n}_i)$ into different layer-surfaces.

Given this setting, let V_i denote the characteristic submanifold in $(N_i, \underline{n}_i)$ and select a 2-manifold, S_i, in $(N_i, \bar{\underline{n}}_i)$ in accordance to the following rule:

(1) If i is odd, let $S_i := (\partial V_i - \partial N_i)^-$.

(2) If i is even, let $(S_i, \underline{s}_i)$ be an essential surface in a component of $(N_i, \bar{\underline{n}}_i)$ which is not an I-bundle and such that $10 \cdot \beta_1(S_i) + \#\underline{s}_i$ is minimal and that S_i is non-separating if $\partial N_i \neq \emptyset$ (the reason for this peculiar choice is to make [Joh 1, prop. 24.1] available).

For every S_i, fix a small product neighborhood $U(S_i) = S_i \times [-1,1]$ of S_i in N_i with $S_i = S_i \times 0$ and $(S_i \times I) \cap (\Psi_{i-1} \cup \partial N) = \partial S_i \times [-1,1]$. Let $r_i : N_i \to N_i$ be a map whose restriction to $N - U(S_i)$ is an embedding and whose restriction to $S \times [-1,1]$ maps every $x \times [-1,1]$, $x \in S_i$, to x. Then define

$$\Psi_{i+1} := \Psi_i \cup \varphi_i(S_i), \quad \varphi_{i+1} := \varphi_i \circ (r_i | (N_i - U(S_i)))^-$$

and let $(N_{i+1}, \underline{n}_{i+1})$ be the manifold obtained from $(N_i, \underline{n}_i)$ by splitting along S_i.

Let m be the first index such that the closure of the components of $N - \Psi_m$ are 3-balls. Define $\Psi := \Psi_m$. Note that Ψ depends on the choices of both S_i and φ_i.

27.4. Definition. *A Haken 2-complex* $\Psi \subset N$ *in a Haken 3-manifold* N *is* <u>complete</u> *P if* $(N - U(\Psi))^-$ *consists of 3-balls. A complete Haken 2-complex* $\Psi \subset N$ *as constructed above is called a* <u>great Haken 2-complex</u>.

27.5. Useful Haken 2-Complexes.

For every Haken 2-complex $\Psi \subset N$ with manifold-filtration $\{\Psi_i\}_{1 \leq i \leq m}$ and every layer-surface $S := (\Psi_i - \Psi_{i-1})^-$, $1 \leq i \leq m$, from Ψ, set

$$\theta_S := (\, S \cap (\Psi \cap (S \times [0,1])) - \partial S \,)^- \subset S,$$
$$\theta'_S := (\, S \cap (\Psi \cap (S \times [-1,0])) - \partial S \,)^- \subset S.$$

We say θ_S, θ'_S are the two *Haken graphs in* S *induced by* Ψ.

Given two Haken graphs θ, θ' in a surface S (and in general position) we say a disc $D \subset S$ is an *i-faced*, $i \geq 1$, *compression-cell for* $\theta \cup \theta'$ if ∂D is an edge-path of exactly i edges (faces) with the following properties:

(1) ∂D is not contained in ∂S.

(1) Every edge lies either in ∂S or in a layer-arc from θ or θ'.

(2) No two adjacent faces of D lie in ∂S or in a single layer-curve from θ or θ'.

Note that ∂D may well be contained in θ (or θ') alone.

27.6. Definition. *We say two Haken graphs* θ, θ' *in a surface* S *are in a* <u>useful position</u> *if they are in a general position and if the following holds:*

(1) There is no 1- or 2-faced compression-cell for $\theta \cup \theta' \subset S$, *and no 3-faced compression-cell for* θ *or* θ'.

(2) There is no 3-faced compression-cell for $\theta \cup \theta' \subset S$ *which is disjoint to* ∂S.

(3) If D *is a 4-faced compression-cell for* θ *(resp.* θ'*), disjoint to* ∂S, *then exactly two faces of* D *are layer-arcs of* θ *(resp.* θ'*).*

A Haken 2-complex Ψ *is called a* <u>useful Haken 2-complex</u> *if, for every layer-surface* S *of* Ψ, *the two induced Haken graphs* $\theta_S, \theta'_S \subset S$ *are in a useful position.*

27.7. Example. Fig. 27.3 shows all four possible types of 3-faced compression-cells for $\theta_S \cup \theta'_S$, and fig. 27.4 shows all three possible types of 4-faced compression-cells for θ_S (resp. θ'_S). A useful Haken 2-complex can have only the first two 3-faced compression-cells and only 4-faced compression-cells of type I (the latter compression-cells, however, can in general *not* be avoided).

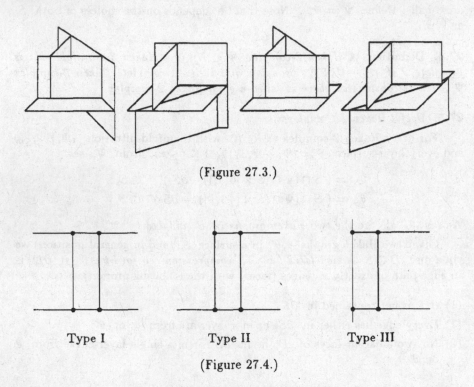

(Figure 27.3.)

Type I   Type II   Type III

(Figure 27.4.)

Useful Haken 2-complexes are very important for us. The next result establishes their existence.

27.8. Proposition. *Every Haken 3-manifold has a great and useful Haken 2-complex.*

Proof. Let N be a Haken 3-manifold and let $\Psi \subset N$ be a great Haken 2-complex. By definition, Ψ is constructed from a great hierarchy, but its actual construction rests on various choices. This in turn leaves room for improvement, and we will see that Ψ can always be "modified" so that it becomes useful. To be more specific, we need the notion of "local slides" for Haken 2-complexes. Here is the rigorous definition.

Let $\{\Psi_i\}_{1 \leq i \leq m}$ be the surface-filtration of Ψ. For every i, $1 \leq i \leq m$, denote by $\Psi_i \subset \Psi$, the *complementary Haken 2-complex* of Ψ_i, i.e., the union of all layer-surfaces of Ψ whose indices are strictly larger than i. Recall from 27.3 the map $\varphi_i : N_i \to N$ with $\varphi_i(\partial N_i) = \Psi_i \cup \partial N$ and such that $\varphi_i|int(N_i) : int(N_i) \to N - \Psi_i$ is a homeomorphism. Let E be a 3-ball in N_i such that $E \cap \partial N_i$ is a non-empty disc and that $\varphi_i|E : E \to N$ is an embedding (it is important to note that the disc $\varphi_i(E \cap \partial N_i)$ need not be contained in a single layer-surface of Ψ). Now let β_t, $t \in [0,1]$, be an ambient isotopy of E which is the identity on $(\partial E - \partial N_i)^-$, and let α_t be the extension of the isotopy $\varphi_i \circ \beta_t$ by the identity. Then α_t is an ambient isotopy of N, constant outside of $\varphi_i(E)$, which we call a *local slide* of $\bar{\Psi}_i$ (or simply: a *local slide of* Ψ) along $\Psi_i \cup \partial N$. Intuitively, a local slide "slides" $\bar{\Psi}_i$ around while keeping $\partial\bar{\Psi}_i$ in $\Psi_i \cup \partial N$. Note further that a local slide of $\bar{\Psi}_i$ keeps Ψ_i unchanged.

We claim that Ψ can be turned into a useful Haken 2-complex, using local slides. We prove this statement by an induction on the surface-filtration $\{\Psi_i\}_{1 \le i \le m}$ of Ψ. The induction beginning is trivial. For the induction step fix an index i, $i \ge 1$. Suppose Ψ is modified, using local slides alone, so that afterwards all layer-surfaces of Ψ_{i-1} satisfy properties (1)-(3) of def. 27.6. Let S be the layer-surface of Ψ of index i, and let $\theta := \theta_S$ and $\theta' := \theta'_S$ be the induced Haken graphs. W.l.o.g. $\theta \cup \theta'$ has at least one j-faced, $1 \le j \le 4$, compression-cell, say D, as excluded in def. 27.6 (for otherwise we are done). We next have to discuss the various possibilities for D.

Suppose D is either a 2- or 3-faced compression-cell for θ, or a 4-faced compression-cell for θ as excluded in (3) of def. 27.6 (1-faced compression-cells for θ are somewhat peculiar in the present context and will be left for the end of this proof). Observe that exactly one face, say l, of D is a layer-arc from θ. Let S_l be the layer-surface of Ψ with $l \subset \partial S_l$. Note that each face of D, different from l, lies in a layer-surface of Ψ whose index is strictly larger than that of S (recall $D \cap \partial S = \emptyset$) and strictly smaller than that of S_l. It follows that the total number of layer-curves of θ can be reduced by sliding $\bar{\Psi}_i$ along Ψ_i in such a way that l is pushed across D and out of S. Of course, this local sliding operation does not affect Ψ_{i-1} at all. So w.l.o.g. all 2-, 3- and 4-faced compression-cells of θ and θ' satisfy properties (1) and (2) of def. 27.6.

We next deal with compression-cells whose boundaries lie neither in θ nor θ'. For this we introduce a complexity for the pair (θ, θ'). To do this let $\{\theta_i\}_{1 \le i \le m}$ and $\{\theta'_i\}_{1 \le i \le m}$ be the arc filtrations of θ and θ', respectively. Denote

$$\gamma(\theta_i, \theta'_j) := \#(\ (\theta_i - \theta_{i-1})^- \cap (\theta'_j - \theta'_{j-1})^-\ ),$$

$$\gamma(\theta_i; \theta') := (\ \gamma(\theta_i, \theta'_1), ..., \gamma(\theta_i, \theta'_m)\ ) \quad \text{and} \quad \gamma(\theta; \theta') := (\ \gamma(\theta_1; \theta'), ..., \gamma(\theta_m; \theta')\ ).$$

Given this notation, define

$$c(\theta, \theta') := (\ \gamma(\theta; \theta'),\ \gamma(\theta'; \theta)\ ).$$

All complexities are taken with respect to the lexicographical order.

We now show that $c(\theta, \theta')$ can be reduced whenever there is a 2- or 3-faced compression-cell for $\theta \cup \theta'$ which does not satisfy (1) or (2) of def. 27.6.

(1) Let D be 2-faced compression-cell for $\theta \cup \theta'$ (see fig 27.5). W.l.o.g. suppose D is chosen to be innermost in the sense that it contains no proper

2-faced compression-cell for $\theta \cup \theta'$. By construction, the layer surfaces of Ψ are images of essential surfaces. It follows that ∂D cannot be contained in $\theta \cup \partial S$ or $\theta' \cup \partial S$. Hence, by definition, one face of D lies in a layer-arc, say k, of θ of index p, say, while the other lies in a layer-arc, say l, of θ' of index q, say. Note that both $k \cap U(D)$ and $l \cap U(D)$ are connected since D is innermost. Let α_t, $t \in [0,1]$, be an ambient isotopy of S, constant outside of $U(D)$, with $(\alpha_1(k) \cap l) \cap U(D) = \emptyset$. Define $\theta^* := \alpha_1(\theta)$. Then θ^* is a Haken graph with arc-filtration $\{ \theta_i^* := \alpha_1(\theta_i) \}_{1 \leq i \leq m}$. Now, $\gamma(\theta_j^*; \theta') = \gamma(\theta_j, \theta')$, for $j < p$, and $\gamma(\theta_j'; \theta^*) = \gamma(\theta_j'; \theta)$, for $j < q$. Moreover, $\gamma(\theta_p^*; \theta') < \gamma(\theta_p; \theta'$ and $\gamma(\theta_q'; \theta^*) < \gamma(\theta_q'; \theta)$. Thus $\gamma(\theta^*; \theta') < \gamma(\theta; \theta')$ and $\gamma(\theta'; \theta^*) < \gamma(\theta'; \theta)$. So, $c(\theta^*; \theta') < c(\theta, \theta')$. Note finally that α_t can be realized by a local slide of Ψ which leaves Ψ_{i-1} unaffected.

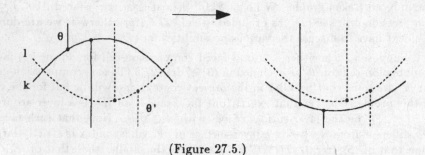

(Figure 27.5.)

(2) Let D be a 3-faced compression-cell for $\theta \cup \theta'$ disjoint to ∂S. Then one face of D lies in a layer-arc, say l, of θ' and the two other faces lie in layer-arcs, say k_1, k_2, of θ. Let D be chosen to be innermost and let the indices be chosen so that the index of k_1 is higher than that of k_2. Let p be the index of the layer-surface of Ψ, different from S, which contains k_2. Then, using a local slide of the complementary Haken 2-complex $\bar\Psi_p$, push $k_1 \cap U(D)$ across D. The next figure indicates the effect of this local slide on $\theta \cup \theta'$.

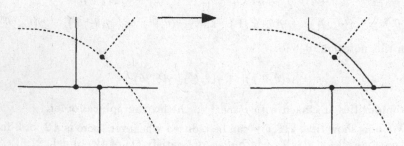

(Figure 27.6.)

A moments reflection as above shows again that this local slide reduces $c(\theta, \theta')$. (For curiosity we also remark that in general this operation does not only remove the 3-faced compression-cell D but, at the same time, creates many others. However, all the newly emerging 3-faced compression-cells lie in a layer-surface whose index is larger than that of S - and meet the boundary of that layer-surface).

By what we have seen so far, the existence of a 2- or 3-faced compression cell for $\theta \cup \theta'$ as excluded in def. 27.6, gives rise to a local slide for Ψ which reduces $c(\theta, \theta')$ without affecting Ψ_{i-1}. Thus we may suppose Ψ is slid so that all layer-surfaces of Ψ_i satisfy the requirements of def. 27.6, except possibly the one concerning 1-faced compression cells.

It therefore remains to deal with 1-faced compression-cells for θ (or θ'). So let D be such a 1-faced cell in S. Then ∂D bounds a layer-surface, say T, of Ψ which is a disc since all layer-surfaces are incompressible. Moreover, T is parallel to D since N is irreducible. But we started with a great Haken 2-complex Ψ and modified it, using local slides. Thus it follows from the choice of layer-surfaces for great hierarchies that T must originally come from a component of the frontier of a characteristic submanifold. But, by the definition of the above sliding processes, T is only being affected if one of its faces has originally been a face of a 2-, 3- or 4-faced compression-cell as excluded by 27.6. But, by definition of characteristic submanifolds in Haken 3-manifolds with boundary-patterns, this is excluded for the components of the frontier of characteristic submanifolds. So D does not exist after all.

This completes the induction step. Inductively, it therefore follows that the great Haken 2-complex Ψ can be turned into a useful Haken 2-complex, by using local slides. Of course, it is then great and useful. ◊

27.9. *Graphs and Haken Graphs.*

By definition, the induced Haken graphs in every layer-surface of a useful Haken 2-complex are in useful position. This makes the useful position of Haken graphs so important for us. We now establish a very general property of Haken graphs in useful position. The statement of this property may perhaps look somewhat technical but its special significance for our study of Heegaard graphs will soon become obvious.

27.10. Definition. *Let S be a surface. Let $\theta \subset S$ be a Haken graph and let $\Gamma \subset S$ be any finite graph in general position to θ. Then we say Γ is* <u>*essential*</u> *(or: <u>in essential position</u>) <u>w.r.t.</u> θ if there is no 2-faced disc in S with one face in an edge of Γ and the other face in a layer-curve of θ.*

The next theorem is the one of the main results of this section. It tells us that, given two Haken graphs in useful position, every finite graph can be pushed into a position which is *simultaneously essential* w.r.t. to both of them.

27.11. Theorem. *Let S be an orientable surface and let $\theta, \theta' \subset S$ be two Haken graphs in useful position. Let $\Gamma \subset S$ be an arbitrary, finite graph. Then there is an ambient isotopy α_t, $t \in I$, in S, fixing ∂S, such that $\alpha_1 \Gamma$ is in essential position with respect to both θ and θ'.*

Proof. It suffices to prove this theorem for irreducible graphs. Here we say a graph $\Gamma' \subset S$ is *irreducible* if there is no disc $D \subset S$ such that $\Gamma' \cap int(D) \neq \emptyset$ and that $\Gamma' \cap \partial D$ is a single vertex of Γ'. Clearly, every finite graph is either contained in a disc or contains a unique irreducible graph (which does not lie in a disc). In the following we suppose w.l.o.g. that Γ itself is irreducible.

It is not hard to isotope Γ so that it is in essential position w.r.t. *one* Haken graph, but it is a challenge to show that this can be done in such a way that the position of Γ is essential w.r.t. to *both* of them. The idea for addressing this challenge is to minimize an appropriate invariant which measures the complexity of the position of Γ w.r.t. $\theta \cup \theta'$. We next introduce just such an invariant.

Given any finite graph $\Gamma \subset S$ in general position w.r.t. θ (resp. θ'), we say a disc $D \subset S$ is a *2- or 3-faced disc for* Γ w.r.t. θ if one face of D lies in Γ and the other (one or two) faces lie in layer-curves of θ, but no two different faces in one layer-curve. Notice though that we do not require the interior of D to be disjoint to either Γ or θ. For later reference we call the face in Γ the *free face* and the faces in θ the *bound faces* of D. The free face is a non-singular edge-path in Γ and the bound faces lie in different layer-curves. Define the *index*, $i_\theta(D)$, of D w.r.t. θ to be:

$$i_\theta(D) := \text{the largest index of all layer-curves of } \theta \text{ containing a face of } D.$$

It turns out that this complexity is not quite good enough for our purpose. To enhance it, we must also record whether D is "good" or "bad". Here a 2- or 3-faced disc, D, for Γ is said to be *good* if no arc from $(\Gamma \cap (U(D) - D))^-$ meets the (interior of the) free face of D, otherwise it is called *bad*.

27.12. Example. *Good 2-faced disc and bad 3-faced disc for Γ w.r.t. θ.*

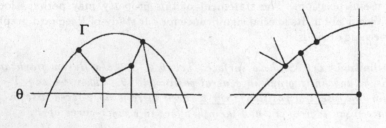

(Figure 27.7)

Taking the above notion into account, define

$$l_\theta(D) := \begin{cases} 2 \cdot i_\theta(D), & \text{if } D \text{ is a good disc,} \\ 1 + 2 \cdot i_\theta(D), & \text{if } D \text{ is a bad disc} \end{cases}$$

This definition makes sure that (1) good and bad discs never have the same complexity and that (2) a bad disc is always worth more than any good disc of the same index.

27.13. Definition. We say $D \subset S$ is a 2-*faced sliding-disc for* Γ *w.r.t.* θ if it is a good 2-faced disc for Γ w.r.t. θ. We say D is a 3-*faced sliding-disc for* Γ *w.r.t.* θ if it is a 3-faced disc for Γ w.r.t. θ and if the interior of D meets no layer-curve from θ whose index is strictly smaller than the index $i_\theta(D)$.

Thus a 2-faced sliding-disc is always good while a 3-faced sliding-disc may be good or bad. It is important to keep in mind the asymmetry in this definition: a 2-faced sliding-disc satisfies a certain condition for its free face while a 3-faced sliding-disc satisfies a condition for its interior.

For every j, $1 \le j \le m$, set

$$\gamma_j(\Gamma, \theta) := \begin{cases} \text{the total number of 2- or 3-faced sliding-discs for } \Gamma \text{ w.r.t. } \theta \\ \text{with } \ell_\theta(D) = j. \end{cases}$$

Given this notation, define

$$c(\Gamma, \theta) := (\ \gamma_1(\Gamma, \theta), \ ..., \ \gamma_m(\Gamma, \theta)\ )$$

with respect to the lexicographical order. Similarly, define $c(\Gamma, \theta')$. Note that by design this complexity goes down whenever a sliding-disc is replaced by any number of sliding-discs of higher index and whenever a good 3-faced sliding-disc is replaced by any number of bad 3-faced sliding-discs of the same index. The next lemma expresses a crucial property of the above complexity,

27.14. Lemma. Let D be a 2-faced sliding-disc for Γ w.r.t. θ. Then there is an ambient isotopy α_t, $t \in I$, in S, fixing ∂S, such that $c(\alpha_1 \Gamma, \theta) < c(\Gamma, \theta)$.

Proof of lemma. Let k be the proper arc in $U(D)$ with $k \subset \theta$ and containing the bound face of D. Let α_t, $t \in [0,1]$, be an ambient isotopy of $U(D)$, constant on $\partial U(D)$, with $\alpha_1(\Gamma) \cap k = \emptyset$ (see fig. 27.9). To calculate $c(\alpha_1 \Gamma, \theta)$ note that every 2- or 3-faced sliding-disc for $\alpha_1 \Gamma$, disjoint to $U(D)$, is also a sliding-disc for Γ. Moreover, every 2- or 3-faced sliding-disc for $\alpha_1 \Gamma$ which meets $U(D)$ has at least one face in a layer-arc meeting k. Hence it has a face in a layer-arc whose index is strictly higher than that of k. Thus $\ell_\theta(D') > \ell_\theta(D)$, for every sliding-disc, D', for $\alpha_1 \Gamma$, meeting $U(D)$. This means that the sliding-disc D is replaced by sliding-discs with larger complexity. It therefore follows that $c(\alpha_1 \Gamma, \theta) < c(\Gamma, \theta)$. This establishes the lemma.

Now, let Γ *be isotoped so that* $c(\Gamma, \theta)$ *is as small as possible.* Then, according to the previous lemma, Γ has no 2-faced sliding-disc w.r.t. θ whatsoever. Next, we wish to remove the 2-faced sliding-discs of Γ w.r.t. θ' by means of an appropriate isotopy. More precisely, we wish to do this in such a way that $c(\Gamma, \theta)$ will not be increased, for this would make sure that no new 2-faced sliding-discs will be created for Γ w.r.t. θ. We will achieve this goal by a delicate analysis of the situation at hand. This analysis is the content of the proof of prop. 27.19. We prepare for this proof by first establishing a few elementary properties concerning induced Haken graphs in 2-faced sliding discs.

To fix ideas let D be a 2-faced sliding-disc of Γ w.r.t. θ'. Then, by definition, D is a 2-faced disc with one face in Γ and the other one in a layer-curve from θ'. Suppose D is chosen to be *innermost*, in the sense that it contains no 2-faced sliding-disc of Γ w.r.t. θ' different from D. Given D, we introduce the following notations:

$$w \;\; = w(D) \;\; := \text{the free face of } D, \text{ i.e., the face contained in } \Gamma,$$
$$w' = w'(D) \;\; := (\partial D - w)^- \subset \theta',$$
$$\tau \;\; = \tau(D) \;\; := \text{the Haken graph } \theta \cap D \text{ in } D,$$
$$\mathcal{K} \;\; = \mathcal{K}(D) \;\; := \text{the collection of all layer-arcs from } \tau \text{ meeting } w'(D).$$

Note that $\mathcal{K}$ consists of pairwise disjoint arcs since $\theta \cup \theta'$ has no 3-faced compression-cell disjoint to ∂S (since θ, θ' are in a useful position). Furthermore, it is to be understood that the arc-filtration $\{\tau_i\}_i$ of τ is the arc-filtration induced by θ, i.e., $\tau_i = \theta_i \cap D$. In other words, every layer-arc from τ has the same index as the corresponding layer-arc from θ containing it.

The Haken graph $\tau = \tau(D)$ has an interesting sub-graph which we call the "forest" of τ. By definition, the *forest of* τ is the unique Haken sub-graph $T \subset \tau$ whose arc-filtration $\{T_i\}_{1 \leq i \leq p}$ is recursively given as follows: T_0 is the system of all layer-arcs from τ joining $w(D)$ with $w'(D)$, and $T_{i+1} = T_i \cup \mathcal{L}_i$, where $\mathcal{L}_i$ denotes the system of all layer-arcs from τ, not in T_i, joining T_i with $w(D)$.

27.15. Example. *A forest of trees.*

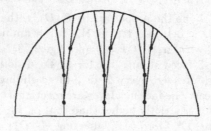

(Figure 27.8.)

Note that every component of the forest $T = T(D)$ is a tree (hence the name "forest"). The layer-arcs from T_0 are the *stems* and all other layer-arcs from T are the *branches* of T. Every layer-arc from τ which is proper in D is a stem of T, for D is innermost and $\theta \cup \theta'$ has no 2-faced compression discs disjoint to ∂S (θ, θ' are in a useful position).

To say more we need the notion of "descending chains". Let k, l be two layer-arcs in a given Haken graph. Then a *descending chain from k to l* is a sequence, $k = k_1, ..., k_p = l$ of pairwise different layer-arcs (in that Haken graph) such that k_i has an end-point in k_{i+1}, for all $1 \leq i \leq p - 1$. Note that the index of k is strictly larger than that of l, provided $k \neq l$ and there is a descending chain from k to l.

27.16. Lemma. *Let k be a layer-arc from $K = K(D)$. Then one of the following holds:*

(1) k is a stem from the forest T, or

(2) there is a descending chain from k to a branch from T, or

(3) k has an end-point in a layer-arc from τ which joins two different stems from T.

Proof of lemma. Suppose k is not a stem. Then we have to show that k has property (2) or (3). To do this recall T_0 is the system of all layer-arcs from τ which are proper in D. This system does not contain k since k is not a stem. But T_0 splits D into a collection of discs, and one of them, say C_0, contains k. Let $\mathcal{L}_0 \subset C_0$ be the system of all layer-arcs from τ which are proper arcs in C_0. Since $\theta \cup \theta'$ has no 2- or 3-faced compression-cell disjoint to ∂S (θ, θ' are in useful position), it follows that every arc from $\mathcal{L}_0$ is either a branch, or joins two different stems from T. $\mathcal{L}_0$ splits C_0 into discs and let C_1 be that one of them which contains k.

Suppose at least one arc from $\mathcal{L}_0$ is not a branch. Then this arc joins two different stems. In particular, C_1 is a square. Moreover, by our maximal condition on $\mathcal{L}_0$, there is no proper layer-arc from τ which is proper in C_1 and joins two stems from T. It follows that every layer-arc from τ in C_1 must join w with the opposite face face of C_1 (to see this recall $\theta \cup \theta'$ has no 2- or 3-faced compression-cells and θ has no 4-faced compression-cell of type II or III disjoint to ∂S since θ, θ' are in useful position). In particular, k has property (3).

Suppose $\mathcal{L}_0$ consists of branches. Let $\mathcal{L}_1$ be the collection of all layer-arcs from τ which are proper in C_1. It follows that every layer-arc from $\mathcal{L}_1$ meets $\mathcal{L}_0$, since $\mathcal{L}_0$ is maximal and since θ, θ' are in useful position. If $\mathcal{L}_1$ contains k, then stop. If not, then note that $\mathcal{L}_1$ splits C_1 into a collection of discs and that one of them, say C_2, contains k. Denote by $\mathcal{L}_2$ the collection of all layer-arcs from τ which are proper in C_2. As before we conclude that every layer-arc from $\mathcal{L}_2$ meets $\mathcal{L}_0 \cup \mathcal{L}_1$. Thus we can repeat the previous construction, and so on. Of course, the process has to stop after finitely many steps. The result is a finite sequence $\mathcal{L}_0, \mathcal{L}_1, ..., \mathcal{L}_n$ of collections of layer-arcs from τ with the property that k is a layer-arc from $\mathcal{L}_n$. By construction, every layer-arc from $\mathcal{L}_n$ has a descending chain to a branch from T. Thus, in particular, k has property (2).

This proves the lemma.

27.17. Lemma. *For every branch of the forest $T = T(D)$, there is at least one 3-faced sliding-disc $D^* \subset D$ for Γ w.r.t. θ whose index, $i_\theta(D^*)$, is not larger than the index of the layer-curve of θ containing that branch.*

Proof of lemma. Let b be a branch of T. Then clearly b is the face of a 3-faced disc $D_0 \subset D$ whose other faces lie in w and in a stem or branch of T. If this 3-faced disc happens to be a sliding-disc for Γ, we are done. If not, then, by definition, its interior meets a layer-curve, say k_0, of θ whose index is strictly smaller than $i_\theta(D_0)$. Let k_0 be chosen so that, in addition, its index is as small as possible. Then k_0 intersects D_0 in proper arcs. Every one of them must meet w since $\theta \cup \theta'$ has no 2- or 3-faced compression-cell. But none of these arcs can have both end-points in w since D contains no 2-faced sliding-disc of Γ w.r.t. θ. Finally, none of these arcs can meet b, for the index of the layer-curve of θ, containing b, equals $i_\theta(D_0)$. Thus it follows that $k_0 \cap D_0$ consists of branches of T. Replacing b by an outermost of those branches and repeating the previous construction, we eventually find the desired 3-faced sliding-disc. This proves the lemma.

27.18. Lemma. *Let l be a layer-arc from $\tau = \tau(D)$ which joins two different stems from $T = T(D)$. Suppose that $l \cap \Gamma \neq \emptyset$ and that Γ has no 2-faced sliding-disc w.r.t. θ. Then there is a 3-faced sliding-disc $D^* \subset D$ for Γ w.r.t. θ whose index $i_\theta(D^*)$ is not larger than the index of l.*

Proof of lemma. Let s_1, s_2 be the two stems from T joined by l. Let C be the square in D which has l as a face and whose other faces lie in s_1, s_2 and w, respectively. Since $\Gamma \cap l \neq \emptyset$, there is a non-empty component, say Γ_C, of $\Gamma \cap C$ with $\Gamma_C \cap l \neq \emptyset$. Note that $\partial \Gamma_C$ cannot be contained in l since Γ has no 2-faced sliding-disc w.r.t. θ (here we use the fact that Γ is supposed to be irreducible). Since the face of C opposite l lies in w and so in Γ, it therefore follows the existence of an arc, c, in Γ_C which joins l with s_1 (or s_2). This arc separates a 3-faced disc, E, from C and the index of this disc equals the index of l. Hence we are done, if E is a 3-faced sliding-disc for Γ w.r.t. θ.

Thus suppose E is not a 3-faced sliding-disc for Γ. Then, by definition, the interior of E meets a layer-arc from τ whose index is strictly smaller than the index of l. Consider the Haken graph $\tau_C = (\tau \cap int(C))^-$ (with the arc-filtration given by τ). Let l_1 be a layer-arc from τ_C whose index is as small as possible. Then l_1 must be a proper in C. But it cannot meet l since it has an index which is strictly smaller than the index of l. If l_1 is a branch, then the lemma follows from lemma 27.17. If not, then l_1 joins the two stems s_1 and s_2. Let C_1 be the square separated from C by l_1 and containing *l.* If $l_1 \cap \Gamma = \emptyset$, then replace C by C_1 and repeat the above argument. If $l_1 \cap \Gamma \neq \emptyset$, then replace l by l_1 and repeat the above argument. This process must stop after finitely many steps and the lemma follows by induction.

Having established the above preliminaries, we are now ready to proceed with the proof of thm. 27.11. In order to prove this theorem it suffices to show the following proposition.

27.19. Proposition. *Let* Γ *be isotoped so that* $c(\Gamma, \theta)$ *is as small as possible. Suppose there is a 2-faced sliding-disc for* Γ *w.r.t.* θ'. *Then there is an ambient isotopy* α_t, $t \in [0,1]$, *of* S *such that* $c(\alpha_1 \Gamma, \theta') < c(\Gamma, \theta')$ *and* $c(\alpha_1 \Gamma, \theta) \leq c(\Gamma, \theta)$.

Proof of proposition. Let D be a 2-faced sliding-disc for Γ w.r.t. θ', and let D be chosen to be innermost. Then, in particular, lemmas 27.16 - 27.18 are available to us. For convenience, we consider the regular neighborhood $U(D) \subset S$ rather than D. For simplicity, we denote the Haken graph $\theta \cap U(D)$ by τ again. We also denote the forest of this Haken graph by T, and we denote by $\mathcal{K}$ the collection of all layer-arcs of τ meeting w'. Let k' be the proper arc in $U(D)$ with $w' \subset k' \subset \theta'$, and set

$$A = A(D) := \text{the disc in } U(D) \text{ separated by } k' \text{ and not containing } D.$$

Of course, $\tau \cap A$ is a system of pairwise disjoint arcs which join $A(D) \cap \partial U(D)$ with $(\partial A(D) - \partial U(D))^-$.

Let α_t, $t \in [0,1]$, be an isotopy in S which pushes $\Gamma \cap U(D)$ into $A(D)$ and which is constant on $(S - U(D))^-$. W.l.o.g. suppose α_t is chosen so that it preserves the stems of T, and that, more generally, it keeps the arcs from $\mathcal{K}$ in their respective layer-arcs from θ. Set

$$\Gamma' := \alpha_1(\Gamma).$$

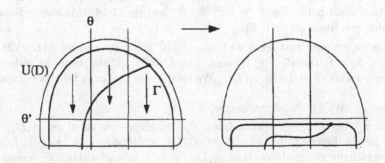

(Figure 27.9.)

We know from lemma 27.14 that $c(\Gamma', \theta') < c(\Gamma, \theta')$. It therefore remains to show that $c(\Gamma', \theta) \leq c(\Gamma, \theta)$. For this it suffices to show that, for every 2- or

3-faced sliding-disc, D', for Γ' w.r.t. θ there is a 2- or 3-faced sliding disc, D^*, for Γ w.r.t. θ such that one of the following holds:

(1) $\alpha_1(D^*) = D'$, or

(2) $D^* \subset D$ and $l_\theta(D^*) < l_\theta(D')$.

It is easily verified that then $c(\Gamma', \theta) \leq c(\Gamma, \theta)$. To see this, note that any disc, D^*, as in (2) will be removed from the list of 2- or 3-faced sliding-discs for Γ since $\alpha_1 \Gamma \cap D = \emptyset$.

To start the actual proof of prop. 27.20, let D' be an arbitrary 2- or 3-faced sliding disc for Γ' w.r.t. θ (keep in mind that this sliding-disc is taken w.r.t. θ and not θ'). Given D', set

$$v = v(D') := \text{the free face of } D', \text{ i.e., the face contained in } \Gamma',$$
$$v' = v'(D') := (\partial D' - v)^- \subset \theta,$$

If $D' \cap U(D) = \emptyset$, then we are done since α_t is constant outside of $U(D)$. So suppose $D' \cap U(D) \neq \emptyset$.

Case 1. D' is a 2-faced sliding-disc for Γ'.

Note first that every component of $v' \cap U(D)$ which does not contain an end-point of v' is a proper arc in $U(D)$. So it is a stem of the forest $\mathcal{T}$ since v' lies in a layer-arc of θ (we are in Case 1).

We claim $U(D) \cap \partial v \neq \emptyset$. Assume the converse. Then $v' \cap U(D) \subset \theta$ is a collection of stems of $\mathcal{T}$. Recall α_t is an ambient isotopy which is constant outside of $U(D)$ and which preserves the stems. Set $\gamma_t := \alpha_t^{-1}|v$. Then γ_t is an isotopy of the arc v with $\gamma_t(v) \cap v' = \partial v'$ and $\gamma_t|\partial v = \text{id}|\partial v$ (recall that, by assumption, $\partial v \subset (S - U(D))^-$ and α_t is constant on $(S - U(D))^-$). Moreover, $\gamma_1(v) \subset \Gamma$ since $v \subset \Gamma' = \alpha_1\Gamma$. It follows that $\gamma_1(v)$ is a free face of a 2-faced sliding-disc for Γ w.r.t. θ. By lemma 27.14, this contradicts our minimality condition on $c(\Gamma, \theta)$.

By what we have just seen, we have $U(D) \cap \partial v \neq \emptyset$, and so $A(D) \cap \partial v \neq \emptyset$ since $D \cap \partial v = \emptyset$ (recall $v \subset \Gamma'$ and $\Gamma' \cap D = \emptyset$). More precisely, either one or two end-points of v lie in $A(D)$. We treat these two possibilities separately.

Sub-Case 1. $A(D) \cap \partial v$ is connected.

Let x_1 be the (only) point from ∂v which is contained in $A(D)$, and let k_1 be the layer-arc from $\tau = \theta \cap U(D)$ which contains x_1. It follows, by the argument given above, that $\gamma_1(x_1) \notin k_1$. In particular, k_1 cannot be a stem. If there is a descending chain from k_1 to a branch from $\mathcal{T}$, then, by lemma 27.17, there is 3-faced sliding-disc for Γ w.r.t. θ which is contained in D and whose index is strictly smaller than that of k_1 and so of D'. Thus we have found the desired sliding disc whose complexity is strictly smaller than

$l_\theta(D')$, and we are done. Thus w.l.o.g. we may suppose there is no descending chain from k_1 to a branch. Then, by lemma 27.16, k_1 has an end-point in a layer-arc, say b, of τ which joins two different stems, say s_1, s_2, of T.

Now consider the component, v_1, of $v \cap A(D)$ which has x_1 as an end-point. Let $B \subset D$ be the square which has b as a face and the other faces in s_1, s_2 and w, respectively. Then $\gamma_1(x_1) \in B$ since α_t preserves stems and since $\gamma_1(x_1) \notin k_1$ (see above). Moreover, note that v cannot be entirely contained in $A(D)$, for $\theta \cup \theta'$ has no 2-faced compression-cell disjoint to ∂S. It follows that v_1 joins x_1 with a point in $A(D) \cap \partial U(D)$, and so $v_1^* := \gamma_1(v_1)$ is an arc in Γ which joins a point in B with a point in $A(D) \cap \partial U(D)$. (recall α_t is constant outside of $U(D)$). In particular, v_1^* must meet b or $s_1 \cup s_2$.

If $v_1^* \cap b \neq \emptyset$, then it follows from lemma 27.18 the existence of a 3-faced sliding-disc for Γ w.r.t. θ whose index is strictly smaller than the index of k_1 and so of D', and we are done again.

Thus suppose $v_1^* \cap b = \emptyset$. Then v_1^* meets $s_1 \cap B$ (or $s_2 \cap B$). Moreover, $v_1^* \cap s_1$ has to be a single point. In fact, $\gamma_1(v) \cap s_1$ has to be a single point since Γ has no 2-faced sliding-disc w.r.t. θ (see lemma 27.14 and our minimality condition on $c(\Gamma, \theta)$). Let $y_1 := \gamma_1(v) \cap s_1$ be that point. Let z_1 be the end-point of v different from x_1. Let k be the layer-arc from τ which contains k_1. Let $l_1 \subset k \cup b \cup s_1$ and $l_2 \subset \gamma_1(v)$ be the arcs joining y_1 with z_1. Then $l_1 \cap l_2 = \partial l_1 = \partial l_2 = y_1 \cup z_1$. To see this note that all components of $v' \cap U(D)$ different from k_1 are proper in $U(D)$ and so stems of T and recall α_t preserves stems. Hence the union $l_1 \cup l_2$ is a non-singular curve. Moreover, it bounds a disc, say C, in S since D' is a disc. According to this analysis, we have the situation depicted on the left of fig. 27.10.

We claim this particular situation is impossible. To see this consider the intersection $\theta_C := \theta \cap C$. Consider the layer-arc r_1 from θ_C which contains $s_1 \cap C$. Note that r_1 cannot be a proper arc in C since Γ has no 2-faced sliding-disc w.r.t. θ and since θ is a Haken graph. Thus r_1 ends in another layer-arc, r_2, of θ_C. Suppose r_2 is proper. Then, by the argument above, it has to have an end-point in k. Following the arc l_1, we see that the index of r_2 is strictly larger than that of r_1. But r_2 contains an end-point of r_1. So we have a contradiction. If r_2 is not proper, then it ends in a layer-arc r_3 from θ_C with even smaller index, and we get a contradiction as before, and so on. Thus we get an infinite sequence of layer-arcs and this is certainly impossible.

This finishes the proof of prop. 27.20 in Sub-Case 1.

Sub-Case 2. $A(D) \cap \partial v$ *is disconnected.*

Let x_1, x_2 be the two end-points from ∂v. Then $x_1, x_2 \in A(D)$ (we are in Sub-Case 2). Let k_i, $i = 1, 2$, be the layer-arc from τ which contains x_i. By the argument used right before Sub-Case 1, we can exclude the case that both k_1 and k_2 are stems from T. If k_1 (or k_2) is a stem, then, following the analysis from Sub-Case 1, we find that we are in the situation on the left of fig. 27.10 which in turn has been shown to be impossible.

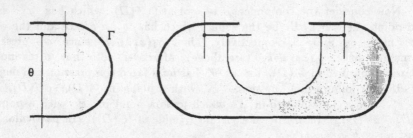

(Figure 27.10.)

Thus we may suppose that in fact neither k_1 nor k_2 is a stem. Then, following the analysis from Sub-Case 1 once more, we find that we are in the situation on the right of fig. 27.10. But this situation can be shown to be impossible in much the same way as in Sub-Case 1. We only have to recall that θ has no 4-faced compression discs of type II or III (see fig. 27.4) since θ, θ' are supposed to be in useful position.

Case 2. D' *is a 3-faced sliding-disc.*

We distinguish between the following two sub-cases.

Sub-Case 1. $U(D) \cap \partial v = \emptyset$.

Consider $v^* := \gamma_1(v) = \alpha_1^{-1}(v)$. Note that $\partial v^* = \partial v = \partial v'$ and that all components of $v^* \cap U(D) = \gamma_1(v \cap A(D))$ are proper arcs in $U(D)$ (we are in Sub-Case 1).

Suppose $v^* \cap v' = \partial v'$. Then $v^* \cup v'$ is the boundary of a 3-faced disc, D^*, for Γ w.r.t. θ and $\alpha_1(D^*) = D'$. Thus we are done, if D^* is a sliding-disc for Γ w.r.t. θ. If not, then, by definition, the interior of D^* meets a layer-arc from θ whose index is strictly smaller than $i_\theta(D^*)$. This layer-arc does not meet D' since D' is a 3-faced sliding-disc and $i_\theta(D^*) = i_\theta(D')$. So it meets the disc $D^* \cap D$ and so D (recall α_t is constant outside of $U(D)$). Thus at least one of the components of the intersection of D with the layer-arc at hand is a layer-arc, k, from τ which meets $int(D^*)$ and whose index is strictly smaller than $i_\theta(D')$. It follows, that k cannot be an arc from $\mathcal{K} = \mathcal{K}(D)$ (recall α_t maps $\mathcal{K}$ to itself and D' is a 3-faced sliding-disc). In particular, it is not a stem from $\mathcal{T}$. But the stems from $\mathcal{T}$ split D into discs and one of those discs, say C_1, must contain k. Let l be a layer-arc from τ which is contained in C_1 and whose index is as small as possible. Then l is a layer-arc which is proper in C_1 and whose index is not larger than that of k. Now, ∂l cannot lie in w, for D is innermost, and l cannot join a stem with $w' \subset \theta'$, for $\theta \cup \theta'$ has no 3-faced compression-cell disjoint to ∂S. If l is a branch from $\mathcal{T}$, then we get the desired sliding-disc from lemma 27.17. Thus suppose the converse. Then l must join two different stems from $\mathcal{T}$, say s_1, s_2. If $l \cap \Gamma \neq \emptyset$, then we get the desired sliding-disc from lemma 27.18. Thus suppose $l \cap \Gamma = \emptyset$. Then note that $l \neq k$, for otherwise $k \subset D^*$ (since $k \cap D^* \neq \emptyset$) and hence s_1 would meet D^* and so D' which in turn is impossible since the index of s_1 is strictly smaller than that of k and since D' is a 3-faced sliding-disc. It follows that l separates a disc, say C_2, from C_1 which contains k. Now, C_2 cannot meet $w \subset \Gamma$. To see this note that otherwise $v^* \cap (s_1 \cup s_2) \neq \emptyset$ (since $k \cap D^* \neq \emptyset$ and $v^* \cap l \subset \Gamma \cap l = \emptyset$) which in turn yields again a contradiction to the fact that D' is a 3-faced sliding disc (the indices of s_1, s_2 are strictly smaller than $i_\theta(D')$ since there is a descending chain from k to s_1 and s_2). It follows that C_2 must meet $A(D)$. Let $\mathcal{L}_2$ be the system of all layer-arcs from τ which lie in C_2 and which join s_1 with s_2. If k is a layer-arc from $\mathcal{L}_2$, then notice $k \cap v^* \neq \emptyset$ (k meets D^*, $v^* \cap (s_1 \cup l \cup s_2) = \emptyset$ and $\alpha_1 D^* = D'$) and so the desired 3-faced sliding-disc for Γ follows from lemma 27.18 again. If k is not contained in $\mathcal{L}_2$, then let $C_3 \subset C_2$ be any square with $C_3 \cap (s_1 \cup s_2 \cup \mathcal{L}_2 \cup w) = \partial C_3$. A moments reflection shows that every layer-arc in C_3 must join w with $\mathcal{L}_2$ (use that $\theta \cup \theta'$ has no 2- or 3-faced compression-cell and that θ has no 4-faced compression-cell of type II or III). But this is impossible since k is not an arc from $\mathcal{K}(D)$ (see above).

Suppose $v^* \cap v' \neq \partial v$. Since we are in Sub-Case 1, it follows that that there is a component, say v_1^*, of $v^* \cap U(D)$ and a component, v_1', of $v' \cap U(D) \subset \theta$ such that (1) v_1^* and v_1' are proper arcs in $U(D)$ and that (2) $v_1^* \cap v_1' \neq \emptyset$ (recall α_t is constant outside of $U(D)$). Now v_1' lies in the union of at most two layer-arcs from τ, say k_1, k_2. W.l.o.g. the index of k_1 is strictly smaller than the index of k_2. Consider k_1. Note that k_1 may or may not meet $A(D)$. We treat these two cases separately.

Suppose $k_1 \cap A(D) = \emptyset$. Then, in particular, k_1 is not a stem from $\mathcal{T}$. But the stems from $\mathcal{T}$ split D into discs and at least one of those discs, say C, contains k_1. Note that there is a descending chain from k_1 to a layer-arc l (possibly equal to k_1) from τ which is a proper arc in C. It follows that l is either a branch from $\mathcal{T}$, or it joins two stems from $\mathcal{T}$ (recall D contains no 2-faced sliding-disc for Γ and θ, θ' are in useful position). Assume for a moment l joins two stems. Then $l \neq k_1, k_2$ since v_1' lies in the union of at most two layer-arcs from τ. By the same reason, both k_1 and k_2 must meet w (note that $\partial v_1' \not\subset w'$ since θ, θ' are in useful position). It follows that v_1' separates a 3-faced disc from D with one face in w and the other two faces in k_1, k_2, respectively. Now, let $C_1 \subset C$ be the disc separated from C by l and meeting w. Then, more precisely, the above 3-faced disc lies in C_1. Moreover, by an argument from lemma 27.17, this disc contains a 3-faced sliding-disc for Γ. Thus there is at least one 3-faced sliding-disc for Γ w.r.t. θ which is contained in $C_1 - U$, where U denotes the regular neighborhood of $(\partial C_1 - w)^-$ in C_1. We conclude that there is an ambient isotopy β_t, $t \in I$, with $\beta_t | U(C_1) \cap (S - D)^- = id | U(C_1) \cap (S - D)^-$ and $\beta_1(C_1) = U$, such that $c(\beta_1 \Gamma, \theta) < c(\Gamma, \theta)$ (use the result from lemma 27.17 and the fact that θ has no 4-faced compression-cell of type II and III). But this is a contradiction to our minimality condition on $c(\Gamma, \theta)$. Hence our assumption is wrong, and l does not join two stems. Thus it is a branch (possibly equal to k_1). Then it follows from lemma 27.17 the existence of the 3-faced sliding-disc whose index is not larger than the index of k_1 and so strictly smaller than $i_\theta(D')$. This is the desired 3-faced sliding-disc whose complexity is strictly smaller than that of D'.

Suppose $k_1 \cap A(D) \neq \emptyset$. Note that k_1 cannot end in a layer-arc from τ which joins two different stems, for v_1' lies in the union of at most two layer-arcs from τ. Moreover, we may suppose there is no descending chain from k_1 to a branch, for otherwise the desired sliding-disc, D^*, for Γ follows from lemma 27.17 (recall the index of k_1 is strictly smaller than $i_\theta(D')$). Hence it follows from lemma 27.16 that k_1 must be a stem. Assume for a moment that $v^* \cap k_1 = \emptyset$. Then $k_1 \cap int(D') \neq \emptyset$ since α_t preserves stems. Hence the index of k_1 cannot be strictly smaller than $i_\theta(D')$ since D' is a 3-faced sliding-disc for Γ. But $i_\theta(D')$ equals the index of k_2 and k_2 has an end-point in k_1. It follows that k_2 must be empty. Hence $v_1' = k_1$, and so $v \cap v' \neq \partial v$ (recall α_t preserves stems). This, however, is a contradiction (see our definition of v and v'). Thus our assumption is wrong, and so $v^* \cap k_1 = \emptyset$. It then follows that k_2 is non-empty since $v_1^* \cap v_1' \neq \emptyset$ and $v_1' \subset k_1 \cup k_2$. Now, k_2 cannot join two stems since v_1' lies in the union of only two layer-arcs from τ. Moreover, we may suppose that k_2 has no descending chain to a branch unless it is a branch itself, for otherwise the desired sliding, D^*, for Γ follows from lemma 27.17 again. Therefore k_2 itself is a branch. This branch is a face of a 3-faced disc $D^* \subset D$ for Γ w.r.t. θ and this disc has the same index as D'. Note further that D' cannot be a good 3-faced sliding, for otherwise we find a 2-faced sliding disc for Γ w.r.t. θ in D^* (recall $v_1^* \cap (k_1 \cup k_2) = v_1^* \cap v_1' \neq \emptyset$ and $v^* \cap k_1 = \emptyset$) which is impossible since D is innermost. But, of course, D^* is good since D is a 2-faced sliding-disc and so good. Thus, altogether, we have $i_\theta(D^*) = i_\theta(D')$, D' is bad and D^* is good. Hence $l_\theta(D^*) < l_\theta(D')$. Therefore D^* is the desired 3-faced sliding-disc for Γ in D, provided it is a sliding-disc at all. If not, then such a sliding-disc is provided by lemma 27.17. Thus, in any case, we find a sliding-disc for Γ in D whose complexity is strictly smaller than that of D'.

This finishes the proof in Sub-Case 1.

Sub-Case 2. $U(D) \cap \partial v \neq \emptyset$.

In this Sub-Case, either one or two end-points of ∂v lies in $U(D)$. Let x_1, x_2 (possible $x_2 = \emptyset$) be the end-points contained in $U(D)$. Then, more precisely, x_i, $i = 1, 2$, lies in $A(D)$, for, by definition, $v \subset \Gamma'$ and $\Gamma' \cap D = \emptyset$. In particular, there is a layer-arc k_i, $i = 1, 2$, from $\mathcal{K}$ which contains x_i. Note that the indices of k_1 and k_2 are not strictly larger than $i_\theta(D')$. Now, if $\alpha_1^{-1} x_i \in k_i$, for $i = 1$ *and* $i =$, then there is a 3-faced sliding-disc, D^*, for Γ w.r.t. θ such that $\alpha_1(D^*) = D'$ (recall α_t maps $\mathcal{K}$ into itself). So w.l.o.g.

$\alpha_1^{-1} x_1 \notin k_1$. In particular, k_1 is not a stem (recall α_t preserves stems). If there is a descending chain from k_1 to a branch from T, then the existence of the desired 3-faced sliding-disc (whose index is strictly smaller than $i_\theta(D')$) follows from lemma 27.17, and we are done. Thus suppose the converse. Then, by lemma 27.16, k_1 has an end-point in a layer-arc, l, from τ which joins two different stems, say s_1, s_2, from T. Note that the indices of these stems are both strictly smaller than $i_\theta(D')$. Consider the component, v_1, of $v \cap A(D)$ which has x_1 as an end-point. Note that v cannot be entirely contained in $A(D)$, for $\theta \cup \theta'$ has no 3-faced compression-cell disjoint from ∂S. It follows that v_1 must join x_1 with $A(D) \cap \partial U(D)$. But it cannot meet $s_1 \cup s_2$, for $v_1 \subset D'$ and D' is a 3-faced sliding-disc (and so $int(D')$ cannot meet a layer-arc from θ whose index is strictly smaller than $i_\theta(D')$). It follows that $\alpha^{-1} v_1$ meets l, since $\alpha_1^{-1} x_1 \notin k_1$ and since α_t is constant on $\partial U(D)$. But $\alpha_1^{-1} v_1 \subset \Gamma$, and so the existence of the desired 3-faced sliding-disc follows from lemma 27.18.

This finishes the proof in Case 2. Hence, altogether, prop. 27.19 is established, and so the proof of thm. 27.11 is finally finished. $\Diamond$

27.20. *Heegaard 2-Complexes.*

Let us now turn our attention to Heegaard 2-complexes.

27.21. Definition. *A 2-dimensional CW-complex, Ω, embedded in a 3-manifold, N, is a <u>Heegaard 2-complex</u> if $(N - U(\Omega))^-$ is a single 3-ball.*

Intuitively, a Heegaard 2-complex is the union of a Heegaard graph with one of its meridian-systems. The close relationship between Heegaard graphs and Heegaard 2-complexes is crucial for us. Here is a more rigorous description of it and its converse. Let $\Omega \subset N$ be a given Heegaard 2-complex. Set

$$\Gamma_\Omega = \Omega^{(1)} \quad \text{and} \quad \mathcal{D}_\Omega := (\Omega - U(\Gamma_\Omega))^-,$$

where as usual $\Omega^{(1)}$ denotes the union of all 1-cells of Ω. Notice that $\Gamma_\Omega \cup \partial N$ must be connected. Moreover, $\mathcal{D}_\Omega$ is a system of discs in $(N - U(\Gamma_\Omega))^-$ which splits $(N - U(\Gamma_\Omega))^-$ into a single 3-ball. Thus Γ_Ω is a Heegaard graph for N and $\mathcal{D}_\Omega$ is a meridian-system for $(N - U(\Gamma_\Omega))^-$. Vice versa, given a Heegaard graph $\Gamma \subset N$, recall from 23.5 the natural contraction $\varphi_\Gamma : N \to N$. This map is characterized by the requirement that $\varphi_\Gamma | N - U(\Gamma)$ is an embedding with $\varphi_\Gamma | \Gamma = id_\Gamma$ and that $\varphi_\Gamma(U(\Gamma^{(0)})) = \Gamma \cap U(\Gamma^{(0)})$ and $\varphi_\Gamma(U(\Gamma^{(1)}) - U(\Gamma^{(0)}))^- = (\Gamma^{(1)} - U(\Gamma^{(0)}))^-$. Define an equivalence-relation on $U(\Gamma)$ by setting $x \sim y \Leftrightarrow \varphi_\Gamma(x) = \varphi_\Gamma(y)$, for all $x, y \in U(\Gamma)$. Then every selection of a meridian-system $\mathcal{D} \subset (N - U(\Gamma))^-$ gives rise to a CW-complex by setting $\Omega_{(\Gamma, \mathcal{D})} := (\Gamma \cup \mathcal{D})/ \sim$. More precisely, $\Omega_{(\Gamma, \mathcal{D})}$ is a Heegaard 2-complex canonically embedded in N. Of course $\Omega = \Omega_{(\Gamma_\Omega, \mathcal{D}_\Omega)}$, for every Heegaard 2-complex $\Omega \subset N$.

W.l.o.g. we will always suppose that $\varphi_\Gamma^{-1}(\Gamma_\Omega^{(0)}) \cap \partial \mathcal{D}_\Omega$ is a finite collection of points and that the restriction $\varphi_\Gamma | \partial \mathcal{D}_\Omega : \partial \mathcal{D}_\Omega \to \Gamma_\Omega$ is an embedding on all components of $\partial \mathcal{D}_\Omega - \varphi_\Gamma^{-1}(\Gamma_\Omega^{(0)})$. Moreover, we denote by $\underline{d}_\Omega$ the boundary-pattern for $\mathcal{D} := \mathcal{D}_\Omega$ to be the set given by the union of the *sets* of all components of $\varphi_\Gamma^{-1}(e)$, where the union is taken over all edges, e, of the Heegaard graph Γ_Ω. For notational convenience, we often do not distinguish between $\mathcal{D}_\Omega$ and its image under φ_Γ. This will cause no confusion as long as we realize that then every component of $\bigcup \underline{d}_\Omega$ is an edge-path in the Heegaard graph Γ_Ω whose edges coincide with those of Γ_Ω. More precisely, every bound face of $(\mathcal{D}_\Omega, \underline{d}_\Omega)$ is an edge of Γ_Ω and every free face lies in ∂N. But keep in mind that adjacent bound faces of $(\mathcal{D}_\Omega, \underline{d}_\Omega)$ may lie in the same edge and that the boundary of every component of $\mathcal{D}_\Omega$ may run several times across any given edge of Γ_Ω (and does so for every edge if Γ_Ω is irreducible).

27.22. *Useful Heegaard 2-Complexes.*

Let $\Psi \subset N$ be a Haken 2-complex with manifold-filtration $\{\Psi_i\}_{1 \le i \le m}$ and let $\Omega \subset N$ be a Heegaard 2-complex in general position to Ψ. Define

$$\Delta_\Omega := \Psi \cap \mathcal{D}_\Omega \subset \mathcal{D}_\Omega, \text{ and}$$
$$\Gamma_S := S \cap \Omega \subset S,$$

for every layer-surface, S, of Ψ. We say that Δ_Ω is the *Haken graph in* $\mathcal{D}_\Omega$ *induced by* Ψ. Moreover, we say that Γ_S is the *(Heegaard) graph in* S *induced by* Ω. It is a Heegaard graph (or better: standard graph) in a surface as considered in section 1 but, of course, it is *not* a Haken graph and it is *not* a Heegaard graph in N. Note further that θ_S, θ_S' and Γ_S are in general position.

Theorem 27.11 is the key to our investigation of Heegaard 2-complexes. Indeed, it allows us to bring Heegaard 2-complexes into an essential as well as into a useful position.

27.23. Definition. *Let* N *be a Haken 3-manifold with Haken 2-complex* $\Psi \subset N$. *Then a Heegaard 2-complex* $\Omega \subset N$ *is* <u>essential</u> *(or:* <u>in essential position</u>*)* <u>w.r.t.</u> Ψ, *if, for every layer-surface* S *of* Ψ, *the induced graph* $\Gamma_S \subset S$ *is in essential position w.r.t. both Haken graphs* θ_S, θ_S' *induced by* Ψ *(see 27.5 and def. 27.10).*

27.24. Proposition. *Let* N *be a Haken 3-manifold and let* $\Psi \subset N$ *be a useful Haken 2-complex. Then every Heegaard 2-complex* $\Omega \subset N$ *is isotopic to a Heegaard 2-complex which is in essential position w.r.t.* Ψ.

Proof. The proof is by induction on the surface-filtration, $\{\Psi_i\}_{1 \le i \le m}$, of Ψ. The induction beginning is trivial. For the induction step suppose Ω is isotoped so that it is in essential position w.r.t. Ψ_{i-1}. Let S be the layer-surface from Ψ of index i and let θ_S and θ_S' be the induced Haken graphs in S (see 27.5). By definition, θ_S, θ_S' are in useful position since Ψ is useful. Thus, by thm. 27.11, the induced Heegaard graph $\Gamma_S := \Omega \cap S \subset S$ can be deformed, using an ambient isotopy of S which is constant on ∂S, so that afterwards Γ_S is in essential position w.r.t. both θ_S and θ'_S. But such an isotopy can be extended to an ambient isotopy of N which does not affect $\Omega \cap \Psi_{i-1}$ at all. Thus Ω can be isotoped so that afterwards it is in essential position w.r.t. Ψ_i. This finishes the induction step. $\Diamond$

By prop. 27.24, we can push every Heegaard 2-complex $\Omega \subset N$ into an essential position. Even better are useful positions:

27.25. Definition. *Let* N *be a Haken 3-manifold with Haken 2-complex* $\Psi \subset N$. *Then a Heegaard 2-complex* $\Omega \subset N$ *is* useful *(or:* in useful position*) w.r.t.* Ψ *if the following holds:*

(1) Ω *is essential w.r.t.* Ψ,

(2) *every layer-curve of* Δ_Ω *is an arc, and*

(3) *no layer-arc of* Δ_Ω *and no (bound or free) face of* $(\mathcal{D}_\Omega, \underline{d}_\Omega)$ *contains both end-points of another layer-arc of* Δ_Ω.

27.26. Theorem. *Let* Ψ *be a useful Haken 2-complex in a Haken 3-manifold* N. *Then every Heegaard 2-complex* $\Omega \subset N$ *can be isotoped so that afterwards it is in useful position w.r.t.* Ψ.

Proof. The proof is again by induction on the surface-filtration $\{\Psi_i\}_{0 \leq i \leq m}$ of Ψ. The induction beginning is trivial. For the induction step suppose Ω is isotoped so it is useful w.r.t. Ψ_{i-1} (instead of Ψ). Denote by $\{\Delta_i\}_{0 \leq i \leq m}$ the curve-filtration of Δ_Ω induced by the surface-filtration of Ψ. Then, by induction, the Haken 2-complex Δ_{i-1} has properties (1) - (3) of def. 27.25. Let S be the layer-surface of Ψ of index i.

Suppose there is a layer-curve of Δ_i which is a closed curve. Then this curve bounds a disc, D_0, in $\mathcal{D}_\Omega$. Since $\partial D_0 \subset S$ and since S is incompressible, it follows that ∂D_0 bounds a disc in S. Since N is irreducible, D_0 is parallel to that disc. Pushing D_0 across this parallelity region, using an ambient isotopy which is constant outside of this parallelity region, clearly reduces the number of closed curves of Δ_i without effecting $\Omega \cap \Psi_{i-1}$ or $\Gamma_\Omega \cap \Psi_i$ (again this process may increase the intersections with layer-surfaces of indices larger than i). Thus, after finitely many of such steps, we may suppose that the above holds and that, in addition, Ω has property (2) of def. 27.25 w.r.t. Ψ_i.

Suppose there is a layer-arc of Δ_i which has both its end-points in a single face of $\mathcal{D}_\Omega$. Then this layer-arc has index i since (3) of def. 27.25 holds for Δ_{i-1}. It also separates a disc, D_0, from $\mathcal{D}_\Omega$ such that $D_0 \cap \partial \mathcal{D}_\Omega$ lies in a free or bound face of $(\mathcal{D}_\Omega, \underline{d}_\Omega)$, and so either in ∂N or in a single edge of the Heegaard graph Γ_Ω. If $D_0 \cap \partial \mathcal{D}_\Omega$ lies in ∂N, then, using an innermost-disc-argument as before, we find an isotopy for Ω which decreases its intersection with Ψ_i without increasing its intersection with Ψ_{i-1}. If $D_0 \cap \partial \mathcal{D}_\Omega$ lies in an edge of Γ_Ω, then, pushing this edge across D_0, we decrease $\#(\Gamma_\Omega \cap \Psi_i)$

without changing the intersection $\Gamma_\Omega \cap \Psi_{i-1}$ (although we may, and in general will, in the process increase the intersection of Γ_Ω with layer-surfaces of Ψ with indices higher than i). Thus, after finitely many of such steps, no layer-arc of Δ_i has its boundary in a single face of $\mathcal{D}_\Omega$.

The induced Haken graphs $\theta_S, \theta'_S \subset S$ are in useful position since Ψ is supposed to be a useful Haken 2-complex. Thus, by thm. 27.11, we can isotope Ω, using an ambient isotopy which preserves S and which is constant outside of the regular neighborhood of $(S - U(\partial S))^-$, so that afterwards the graph $\Gamma_S \subset S$ is in essential position w.r.t. θ_S and θ'_S. This isotopy has no effect on $\Omega \cap \Psi_{i-1}$ and it does not change $\Gamma_\Omega \cap \Psi_i$ or the number of closed curves of Δ_i. Thus we may suppose that the above holds and that, in addition, Ω is essential w.r.t. Ψ_i.

To finish the proof of thm. 27.26, assume for a moment there is a layer-arc, say k, of Δ_i which has both end-points in another layer-arc, say l, of Δ_i. Then k, together with some arc in l, bounds a disc, D_0, in $\mathcal{D}_\Omega$. W.l.o.g. we may suppose D_0 has been chosen to be innermost. Then D_0 is a 2-faced disc with $D_0 \cap \Psi_i = \partial D_0$. Since Ψ is a great Haken 2-complex, it follows that all layer-surfaces of Ψ are essential (see 27.3). Hence, in particular, ∂D_0 bounds a disc $D'_0 \subset \Psi_i$ which lies in the union of two layer-surfaces of Ψ_i. One of them is S. It follows that k is the face of a 2-faced sliding-disc for the Heegaard graph $\Gamma_S := S \cap \Omega$ w.r.t. $\theta_S \cup \theta'_S$. This, however, contradicts the fact that Ω is essential w.r.t. Ψ_i.

Thus, altogether, Ω is isotoped so that it satisfies the theorem w.r.t. Ψ_i. This completes the induction step. $\Diamond$

27.27. Remark. Note that, in the terminology of 27.5, thm. 27.26 says that the Haken graph $\Delta_\Omega \subset \mathcal{D}_\Omega$ has no 2-faced compression cells. But note also that Δ_Ω may well have 3-faced compression-cells. In fact, it is very likely that 3-faced compression-cells cannot be avoided for Heegaard 2-complexes at all.

28. Short Companions for Haken Graphs.

Having established general properties for Haken 2-complexes and induced Haken graphs of Heegaard 2-complexes, we next examine abstract Haken graphs somewhat more closely. Haken graphs in general can be quite complicated and interesting in themselves. But here we are interested only in those Haken graphs which can occur as induced Haken graphs of useful Heegaard 2-complexes. This makes the task of exploring Haken graphs much easier. For instance, we can restrict our attention to Haken graphs in discs (with boundary-patterns) which have no 2-faced compression-cells. Given this setting, we are interested in finiteness results concerning sequences of very large Haken graphs. To make progress we first have to develop a fair amount of terminology. Using this terminology, however, we will be able to see that very large Haken graphs look and behave more and more like arc-systems - at least qualitatively. As a result we obtain a crucial finiteness result for Haken graphs, namely the existence of short companions for long faces. For convenience, this result is first explained for the special case of arc-systems (prop. 28.3) before it is deduced for more general Haken graphs (thm. 28.36).

28.1. *Short Companion-Arcs.*

Throughout this section let $(D, \underline{d})$ be a disc with boundary-pattern and let Δ be a Haken graph in $(D, \underline{d})$ with arc-filtration (i.e., no layer-curve from Δ is closed). A Haken graph Δ with arc-filtration in $(D, \underline{d})$ will be called *useful* if, in addition, no layer-arc of Δ and no bound face of $(D, \underline{d})$ contains both end-points of another layer-arc of Δ (free faces on the other hand may well contain the boundary of layer-arcs). Useful Haken graphs will be most important for us since they appear as induced Haken graphs of useful Heegaard 2-complexes (see def. 27.25).

To every given path $w : I \to D$ (in general position to Δ) we associate a *length* (or Δ-*length*) by setting

$$length_\Delta(w) := \# \, w^{-1}\Delta,$$

The length of a finite collection of paths is the sum of the lengths of all paths in that collection. As usual, we then associate a *distance* (or Δ-*distance*) to every pair of points $x, y \in D$ by setting

$$dist_\Delta(x, y) := \min \{ \, l_\Delta(w) \mid w \text{ is a path in } D \text{ joining } x \text{ with } y \, \},$$

and we define the Δ-*distance of two subsets* $A, B \subset D$ by

$$dist_\Delta(A, B) := \min \{ \, dist_\Delta(x, y) \mid x \in A \text{ and } y \in B \, \}.$$

Given an arc $k \subset D$, we write $dist_\Delta(\partial k)$ for the Δ-distance of the end-points of k.

Note that the distance-function, $dist_\Delta$, depends on the choice of the reference-system, Δ. Since we will have to work with various choices of Δ, it is important for us to always specify the reference-system currently in use.

28.2. Definition. *We say a non-empty collection* $\mathcal{F} \subset \underline{d}$ *of bound faces of* $(D, \underline{d})$ *has an* α-*companion,* $\alpha \in \mathbf{N}$, *if there is a face* $d \in \mathcal{F}$ *and an arc* $\ell(d) \subset \partial D$ *such that*

(1) *$dist_\Delta(\ \partial \ell(d)\ ) \leq \alpha$, and*

(2) *$\ell(d)$ contains d, but no other face from $\mathcal{F}$.*

An arc $c \subset D$ with $\partial c = \partial l(d)$ and $length_\Delta(c) \leq \alpha$ is an α-companion for d.

Theorem 28.36 will establish the existence of appropriate companions for a large class of Haken graphs. This class will be rich enough for our purpose. It will include, e.g., the induced Haken graphs for all Heegaard 2-complexes of interest to us. It also includes all Haken graphs which are arc-systems. Actually, the existence of companions is easy to establish for arc-systems (a fact which was implicitly used in chapter III). The reader may find it instructive to see here a quick proof for this fact before we take up the general situation.

28.3. Proposition. *Suppose Δ is an arc-system in $(D, \underline{d})$. Then every non-empty collection of faces of $(D, \underline{d})$ has a 0-companion.*

Proof. Let $\mathcal{F} \subset \underline{d}$ be a non-empty collection of faces. We are asked to construct a 0-companion for $\mathcal{F}$.

For every face $d \in \mathcal{F}$, denote by $K(d)$ the collection of all arcs from Δ with exactly one end-point in d. For every arc, k, from $K(d)$, denote by $D_1(k), D_2(k)$ the two discs into which D is split by k. We say k is *bad* if both $D_1(k)$ and $D_2(k)$ contain at least one face from $\mathcal{F}$.

We claim there is a face $d \in \mathcal{F}$ such that $K(d)$ has no bad arc. To find such a face, fix an end-point, z, of some face of $(D, \underline{d})$. Let d_1 be any face from $\mathcal{F}$. Suppose there is a bad arc, k_1, in $K(d_1)$ and suppose w.l.o.g. that the indices are chosen so that $z \in D_2(k_1)$. Then select a face $d_2 \in \mathcal{F}$ contained in $D_1(k_1)$. If $K(d_2)$ contains a bad arc, repeat the above process, and so on. In this way we get a nested sequence of discs $D_1(k_1) \supset D_1(k_2) \supset ...$ which contain fewer and fewer faces from $\mathcal{F}$. Thus the process has to stop after finitely many steps and the claim follows.

Let $d \in \mathcal{F}$ be any face for which $K(d)$ has no bad arc. Then we may suppose the indices are chosen so that, for every arc, k, from $K(d)$, the disc $D_1(k)$ contains no face from $\mathcal{F}$. Under this convention set $B := \bigcap D_2(k)$, where the intersection is taken over all arcs, k, from $K(d)$. Then $B' := U(d) \cup (D - B)^-$ is a disc which contains d but no other face from $\mathcal{F}$. Moreover, by definition of $K(d)$, it follows that the algebraic intersection number of the arc $\ell := (\partial B' - \partial D)^-$ with Δ is zero. Hence and since Δ is an arc-system, there must be at least one arc which is disjoint to Δ and which joins two points in ∂D next to the end-points of ℓ. This arc is the desired 0-companion. $\Diamond$

28.4. *Heaps.*

The highest index, m, of the arc-filtration, $\{\Delta_i\}_{1 \le i \le m}$, of a Haken graph Δ is called the *depth* of Δ (notation: $depth(\Delta) = m$). The depth is a first in a series of complexities we are going to assign to Haken graphs. We wish to study Haken graphs with limited depth (but arbitrarily large numbers of layer-arcs). It turns out that the concept of "heaps" is a convenient organizing principle for looking at such graphs. To introduce this concept we need some notation.

A *sub-division* of a Haken sub-graph $\Delta' \subset \Delta$ is a finite graph, θ, which equals Δ' as a subspace and whose vertices contain the vertices of Δ'. In particular, every graph is its own sub-division. The *standard sub-division* of Δ' is Δ' viewed as an ordinary graph. A *cell* of θ is a disc $C \subset D$ with $C \cap (\Delta \cup \partial D) = \partial C$. It is to be understood that a cell has the boundary-pattern given by the edges of θ contained in ∂C and the components of the intersections of ∂C with the faces of $(D, \underline{d})$. Cells of Δ' are the cells of the standard sub-division of Δ'. Note that every cell, C, of a Haken sub-graph $\Delta' \subset \Delta$ contains the induced Haken graph $\Delta_C := (\Delta \cap int(D))^-$. Cells of Δ should not be confused with the compression-cells for Δ introduced in 27.5 (the interior of a compression-cell for Δ may meet Δ and also its boundary-pattern may be different).

Every Haken sub-graph $\sigma \subset \Delta_C$ has at least one layer-arc, k, with the property that $(\sigma - k)^-$ is again a Haken graph. Such a layer-arc is called a *top-arc* for σ. Top-arcs are also characterized by the property that they contain no vertices of σ other than their end-points. We are particularly interested in Haken sub-graphs with only one top-arc.

28.5. Definition. *Let C be the cell of a Haken sub-graph of Δ. A Haken sub-graph $\sigma \subset \Delta_C$ is called a <u>heap</u> in C (or in Δ_C) if*

(1) σ has exactly one top-arc, and

(2) there is a disc $E \subset C$ with $E \cap (\sigma \cup \partial C) = \partial E$ which meets both the boundary ∂C and the top-arc of σ.

28.6. Example.

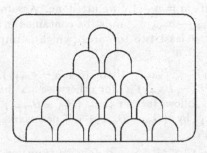

(Figure 28.1.)

If E_1, E_2 (E_2 possibly empty) denote the cells of σ which meet both ∂C and the top-arc of σ, then set

$$A(\sigma) := (C - E_1)^- \cap (C - E_2)^- \quad \text{and} \quad e(\sigma) := A(\sigma) \cap \partial C.$$

Note that $\sigma = A(\sigma)$ iff σ is a layer-arc. The components of $e(\sigma)$ are called the *ends* of σ. A heap has one or two ends. For later use note also that the set of all heaps is closed under intersections.

We next collect some elementary properties of heaps. We formulate these properties for heaps in the disc D, but of course they are also valid for heaps in all cells of Δ_C.

28.7. Lemma. *Let $\sigma \subset \Delta$ be a heap with arc-filtration $\{\sigma_i\}_{0 \le i \le n}$, and set $\sigma' := (\sigma - k)^-$, where k denotes the top-arc of σ. Then the following holds:*

(1) σ is connected.

(2) There is a nested sequence $B_0 \supset B_1 \supset B_2 \supset ... \supset B_n$ of discs such that $B_0 = D$ and that $(\sigma_{i+1} - \sigma_i)^- = (\partial B_{i+1} - \partial B_i)^-$, for all $0 \le i < n$.

(3) For every layer-arc of σ, there is exactly one Haken sub-graph of σ which is a heap and which has this layer-arc as its top-arc.

(4) $\sigma' \subset \sigma$ is a Haken sub-graph.

(5) Every top-arc of σ' contains an end-point of the top-arc k of σ.

(6) If σ' is disconnected, then it is the disjoint union of two heaps.

(7) If σ' is connected and has only one top-arc, then σ' is a heap.

(8) If σ' is connected and has two top-arcs, then $(\sigma' - \ell)^-$ is a heap, for every top-arc, ℓ, of σ'.

Proof. *ad (1)* Every Haken graph has at least one top-arc and every heap has only one top-arc.

ad (2) Let m, $0 \le m < n$, be the largest index with the property that there are discs $B_0 \supset ... \supset B_{m+1}$ with $B_0 = D$ and $(\sigma_{i+1} - \sigma_i)^- = (\partial B_{i+1} - \partial B_i)^-$, for all $0 \le i \le m$. If $m = n - 1$, we are done. Assume $m < n - 1$. Then observe first that $(\sigma_{m+2} - \sigma_{m+1})^-$ must be contained in a single cell of σ_{m+1}, for otherwise σ has at least two top-arcs (which is impossible since σ is a heap).

If $(\sigma_{m+2} - \sigma_{m+1})^- \not\subset B_{m+1}$, then $(\sigma_{m+2} - \sigma_{m+1})^-$ must be contained in a component of $(B_m - B_{m+1})^-$, for otherwise Δ has two top-arcs again. By the same reason, it follows further that $(B_m - B_{m+1})^-$ must be connected. Thus, replacing B_{m+1} by $(B_m - B_{m+1})^-$ if necessary, we may suppose that $(\sigma_{m+1} - \sigma_m)^- \subset B_{m+1}$.

If $(\sigma_{m+2} - \sigma_{m+1})^- \subset B_{m+1}$, it follows from our assumption that the arc-system $(\sigma_{m+2} - \sigma_{m+1})^-$ must contain a collection, $\mathcal{K}$, of at least three, pairwise parallel arcs (in the sense that there is a proper arc in B_{m+1} which intersects each one of those arcs in exactly one point). The union $\sigma_{m+1} \cup \mathcal{K}$ is a Haken graph with at least three top-arcs. Since all layer-arcs from σ, not contained in $\sigma_{m+1} \cup \mathcal{K}$, lie in cells of this graph, it follows that σ must have at least two top-arcs. But this is a contradiction since σ is a heap. Thus our assumption is wrong, and (2) is established.

ad (3) Let ℓ be a layer-arc from σ and let m, $1 \le m \le n$, be that index for which ℓ is a component of $(\sigma_{m+1} - \sigma_m)^-$. Recall from (2) that there is a sequence $B_0 \supset B_1 \supset B_2 \supset ... \supset B_{n+1}$ of discs with $B_0 = D$ and $(\sigma_{i+1} - \sigma_i)^- = (\partial B_{i+1} - \partial B_i)^-$, for all $0 \le i < n$. Thus ℓ is a component from $(\partial B_{m+1} - \partial B_m)^-$. In particular, $\ell \subset B_m$. Set $\mathcal{L}_{m+1} := \ell$. Given $\mathcal{L}_{i+1}$, $1 \le i \le m$, let $\mathcal{L}_i$ denote the union of all components from $(\partial B_i - \partial B_{i-1})^-$ meeting $\mathcal{L}_{i+1}$. Then $\mathcal{L} := \bigcup_{1 \le i \le m} \mathcal{L}_i$ is a Haken sub-graph of σ. It is the smallest Haken sub-graph containing ℓ. Note that ℓ is a top-arc for $\mathcal{L}$. We claim $\mathcal{L}$ is a heap. For this it remains to verify that one of the two cells of $\mathcal{L}$, which contain ℓ, meet ∂D. Now, by construction, $\ell \subset B_m$ and $(\mathcal{L} - \ell)^- \subset (D - B_m)^-$. Therefore it suffices to show that B_m meet ∂D. To see this note that B_m contains the top-arc of σ as well as both the cells of σ containing this top-arc. At least one of the latter cells has to meet ∂D since σ is supposed to be a heap. Thus it follows that B_m has to meet ∂D too.

ad (4) Since k contains no vertex of σ other than its end-points, it follows that $\{\sigma_i\}_{0 \le i \le n-1}$ is an arc-filtration for σ'. So σ' is a Haken graph in D.

ad (5) Every top-arc of σ' must contain an end-point of k for otherwise it is a top-arc for σ different from k.

ad (6) By (5), σ' cannot have more than two top-arcs. Moreover, by (4), σ' is a Haken graph in D. So also every component of σ' is a Haken graph in D. Thus every component of σ' has at least one top-arc. It follows that every component of σ' has exactly one top-arc if σ' is disconnected. At least one cell of each component of σ' contains its top-arc and meets ∂D, for this holds for the top-arc, k, of σ. Hence (6) follows.

ad (7) and (8) These facts follow by a similar argument as used in (6). $\diamond$

28.8. Lemma. *Let k be a layer-arc of Δ. Then k is the top-arc of a (unique) heap in Δ if and only if there is a Haken subgraph $\Delta' \subset \Delta$ and a cell C' of Δ' with $k \subset \partial C'$ and $C' \cap \partial D \ne \emptyset$.*

Proof. One direction is obvious. For the other direction let Δ' and C' be given as in the lemma. Then, in particular, $k \subset \Delta'$. Let $\sigma(k)$ be the Haken sub-graph of Δ which contains k and which has a minimal number of layer-arcs. Then $\sigma(k)$ has exactly one top-arc. Moreover, $k \subset \sigma(k) \subset \Delta'$. In particular, there is a cell C of $\sigma(k)$ which contains C'. So $k \subset C$ and

$C \cap \partial D \neq \emptyset$. Thus, by definition, $\sigma(k)$ is a heap. Uniqueness follows from the next lemma. $\Diamond$

28.9. Lemma. *Let* σ_1, σ_2 *be heaps in* Δ. *Then* $\sigma_2 \subset \sigma_1$ *if* σ_1 *contains the top-arc of* σ_2.

Proof. Assume $\sigma_2 \not\subset \sigma_1$. Then there is a cell C of σ_1 with $\sigma_2 \cap int(C) \neq \emptyset$. Now, $(\sigma_2 \cap int(C))^-$ is a Haken graph in C and so has a top-arc. This top-arc is also the top-arc for σ_2. It follows that the top-arc of σ_2 is not contained in σ_1. $\Diamond$

28.10. Lemma. *Every heap in* Δ *has at most* $2^d - 1$ *layer-arcs and at most* 2^d *end-points, where* $d := depth(\Delta)$.

Proof. The lemma is obvious for heaps of depth 1. Suppose Δ has depth n. Let k be the top-arc of Δ. Then, by lemma 28.7 (6)-(8), $(\Delta - k)^-$ is the union of at most two heaps of depth $n - 1$ (whose top-arcs contain the end-points of k). Thus, inductively, the total number of layer arcs of Δ is less than or equal to $1 + (2^{n-1} - 1) + (2^{n-1} - 1) = 2^n - 1$ and the total number of end-points is less than or equal to $2^{n-1} + 2^{n-1} = 2^n$. $\Diamond$

28.11. Lemma. *Let* Δ *be a useful Haken graph. Let* $\sigma \subset \Delta$ *be a Haken sub-graph and let* C *be a cell of* σ. *Suppose* σ *is the smallest Haken sub-graph in* Δ *which has* C *as a cell. Then* σ *has at most* $\varphi(depth(\sigma), \#\partial\sigma)$ *layer-arcs.*

Here φ *is some known arithmetic function, strictly increasing in both variables and independent of the choice of* Δ.

Proof. Let σ_1 be the union of all layer-arcs in σ which are proper arcs in $C_0 := D$, and let C_1 be the cell of σ_1 which contains C. Then $(\sigma - \sigma_1)^- \subset C_1$, by our minimality condition on σ. Let σ_2 be the union of σ_1 with all layer-arcs from σ which are proper arcs in C_1, and let C_2 be the cell of σ_2 which contains C, etc. In this way we get a finite sequence of Haken sub-graphs $\sigma_1, \sigma_2, ..., \sigma_m$ with $\sigma = \sigma_m$. Denote by $l(\sigma_i)$ the number of layer-arcs of σ_i, Then of course $l(\sigma_{i+1})$ does not exceed the sum of $l(\sigma_i)$ with the number of all components from $(\partial C_i - \partial C_{i-1})^-$ (σ is minimal). So $l(\sigma_1) \leq \#\partial\sigma$ and $l(\sigma_{i+1}) \leq 3 \cdot l(\sigma_i)$ (recall that no layer-arc from Δ contains the two end-points of another layer-arc). Hence the lemma follows, by induction, since $l(\sigma) = l(\sigma_m)$, $m \leq depth(\sigma)$. $\Diamond$

28.12. *Pseudo Arcs, Bundle-Paths and Combs.*

Using heaps, we next generalize the notion of layer-arcs to "pseudo-arcs". We then use these pseudo-arcs to introduce bundles, bundle-paths and combs.

28.13. *Pseudo arcs.* Every layer-arc is a heap. Vice versa, a heap with two ends has a top-arc which joins two disjoint sub-heaps (see lemma 28.7). If we think of the top-arc as very long and the sub-heaps as very small, then the original heap almost equals its top-arc, i.e., it is almost an arc itself. So we call it a "pseudo-arc". Intuitively, a pseudo-arc is an arc with somewhat fuzzy ends. Here is the formal definition.

28.14. Definition. *Let θ be a sub-division of a Haken sub-graph of Δ and let C be a cell of θ. A heap $\sigma \subset \Delta_C := (\Delta \cap int(C))^-$ is a* <u>*pseudo-arc*</u> *in C if*

(1) σ has two different ends, and

(2) every end of σ lies in a face (bound or free) of C, but no face of C contains $\partial\sigma$.

We also say that σ is a pseudo-arc for θ. Recall from 28.4 that $e(\sigma)$ denotes the union of the two ends of σ (since σ is a heap). We say a pseudo-arc *joins* its ends, or the faces of C containing the ends. An end is either an arc or a point, and a pseudo-arc is an ordinary arc if and only if its ends are points. Note that, for useful Haken graphs, an end of a pseudo-arc which is not a point must lie in a face of C which is contained in ∂D.

Pseudo-arcs cannot really intersect. Indeed, given a pseudo-arc $\sigma \subset C$, there is no other pseudo-arc which joins the two components of $\partial C - e(\sigma)$. But, as shown in the next picture, pseudo-arcs need not be disjoint either.

28.15. Example.

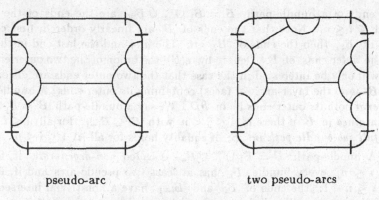

pseudo-arc   two pseudo-arcs

(Figure 28.2.)

28.16. *Bundle-paths.* Let θ be a sub-division of a Haken sub-graph of Δ and let $(C, \underline{c})$ be a cell of θ. A *bundle* (or better: a *bundle of pseudo-arcs*) in C is a union of pseudo-arcs in C which all join the same pair of different faces of $(C, \underline{c})$ *and* which is a system of ordinary arcs as long as it is disjoint to ∂D (note that the latter condition is automatic if Δ is a useful Haken graph). A *bundle for θ* is a bundle in some cell of θ. By definition, the *ends of a bundle*, $\mathcal{B} \subset C$, joining the two faces c_1, c_2 of C are the arcs $e_i = e_i(\mathcal{B})$, $i = 1, 2$, with $e_i \subset c_i$, $\partial e_i \subset \mathcal{B}$ and $\mathcal{B} \cap c_i \subset e_i$.

28.17. Definition. *Let* θ *be a sub-division of some Haken sub-graph of* Δ. *A* <u>*bundle-path*</u> <u>*for*</u> θ *is a union* $B = B_1 \cup ... \cup B_n$ *of bundles* B_i *for* θ *such that the following holds:*

(1) B_i *and* B_{i+1} *lie in different cells of* θ, *for* $1 \le i < n$,

(2) *any edge of* θ *which contains an end of* B_i *contains an end of* B_{i+1}, *for* $1 \le i < n$,

(3) *every edge of* θ *contains an end of at most two bundles* B_i *from* B.

Let $\Delta' \subset \Delta$ be a Haken sub-graph. Then we say B is a bundle-path for Δ' if there is a sub-division θ of Δ' such that B is a bundle-path for θ. Moreover, we say B is a bundle-path for Δ if there is a Haken sub-graph $\Delta' \subset \Delta$ such that B is a bundle-path for Δ'. Intuitively, a bundle-path is a collection of bundles stacked together in a linear fashion. In particular, every bundle is a bundle-path. Keep in mind that any bundle, and so any bundle from a bundle-path, consists of ordinary arcs, provided it does not meet ∂D. Given a bundle-path B for Δ, define

$$supp(B) := \left\{ \begin{array}{l} \text{the smallest Haken sub-graph } \Delta' \subset \Delta \text{ such that } B \text{ is a} \\ \text{bundle-path for } \Delta'. \end{array} \right.$$

$$\mathcal{K}(B) := \left\{ \begin{array}{l} \text{the collection of all layer-arcs from } supp(B) \text{ containing} \\ \text{end-points of layer-arcs from } B. \end{array} \right.$$

The ends of a bundle-path $B = B_1 \cup ... \cup B_n$ are the ends of the bundles B_i, $1 \le i \le n$. Note that the ends of B are linearly ordered; first come the ends of B_1, then the ends of B_2 etc. The first and the last end in that order are the *outer ends* of B. Denote by $e(B)$ the union of the two outer ends of B (we will only be interested in the case that the two outer ends are disjoint). We say B *joins* the layer-arcs (or faces) containing its outer ends. A bundle-path is *proper* if both its outer ends lie in ∂D. We say a bundle-path $B' = B'_1 \cup ... \cup B'_p$ is *contained in* B if there is $1 \le j < n$ with $B'_i \subset B_{i+j}$, for all $1 \le i \le p$. B' is a *full sub-bundle-path* in B, if equality holds for all i, $1 \le i \le p$.

A bundle-path $B = B_1 \cup ... \cup B_n$ is called *non-degenerate* if, for every $1 \le i \le n$, every bundle B_i has at least two pseudo-arcs and if, for every $1 \le i \le n-1$, the ends of B_i and B_{i+1} have a non-trivial intersection. In the following we will only be interested in non-degenerate bundle-paths. Given a non-degenerate bundle-path B, let $A(B) \subset D$ be the unique disc with

(1) $B \subset A(B)$ and $\partial A(B) \subset B \cup supp(B)$,

(2) $A(B)$ is contained in every disc with (1).

Given this notation, we define the *regular neighborhood of* B to be the regular neighborhood

$$U(B) := U(A(B))$$

in D. Note that $U(\mathcal{B}) \cap supp(\mathcal{B})$ is a system of arcs in $U(\mathcal{B})$. This arc-systems splits $U(\mathcal{B})$ into a collection of discs. Let

$$A'(\mathcal{B}) := \left\{ \begin{array}{l} \text{the union of the closure of all components of } (U(\mathcal{B}) - supp(\mathcal{B}))^- \\ \text{which contain a bundle from } \mathcal{B}. \end{array} \right.$$

Finally, define the *frontier of* $\mathcal{B}$ to be $fr(\mathcal{B}) := A'(\mathcal{B}) \cap \partial U(A(\mathcal{B}))$ and note that $\mathcal{K}(\mathcal{B}) \cap A'(\mathcal{B})$ is a system of arcs which join the two components of the frontier.

28.18. Complexities for bundle-paths. Given a bundle-path $\mathcal{B} = \mathcal{B}_1 \cup ... \cup \mathcal{B}_n$, denote

$$length(\mathcal{B}) := \text{the total number of full bundles in } \mathcal{B},$$

i.e., $length(\mathcal{B}) = n$. Apart from this length, a non-degenerate bundle-path (and so every bundle) has also a "width". To define it let e_1, e_2 be the outer ends of $\mathcal{B}$ and let σ be the Haken graph $\sigma := A'(\mathcal{B}) \cap (\mathcal{B} \cup supp(\mathcal{B}))$. Then define

$$width(\mathcal{B}) := \min\{ dist_\sigma(\partial e_1), \ dist_\sigma(\partial e_2) \}, \quad \text{and}$$
$$size(\mathcal{B}) := \min\{ length(\mathcal{B}), \ width(\mathcal{B}) \},$$

where the distances are taken in the disc $A'(\mathcal{B})$. It is important to remember that $width(\mathcal{B})$ and $size(\mathcal{B})$ are defined via σ-distances in $A'(\mathcal{B})$ and *not* via Δ-distances in D (in this way we do not have to worry about whether $\mathcal{B}$ is proper). Given, in addition, a layer arc r of Δ, set

$$b(\mathcal{B}, r) := \text{the total number of layer-arcs from } \mathcal{K}(\mathcal{B}) \text{ with an end-point in } r,$$

and define the *base-length of* $\mathcal{B}$ to be

$$base(\mathcal{B}) := max\{ b(\mathcal{B}, r) \mid r \text{ is layer-arc of } \Delta \ \text{or bound face of } (D, \underline{d}) \}.$$

28.19. Inessential bundle-paths. We say a bundle-path $\mathcal{B} = \mathcal{B}_1 \cup ... \cup \cup \mathcal{B}_n$ is *inessential for* Δ if there is a layer-arc, k, from $supp(\mathcal{B})$ with

(1) k contains both outer ends of $\mathcal{B}$,

(2) k contains no other end from $\mathcal{B}$, and

(3) $\mathcal{B}_1$ and $\mathcal{B}_n$ meet k from the same side.

28.20. Example. The following picture shows a bundle-path with large width and large base-length (left), and an inessential bundle-path with large width (right):

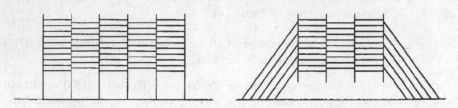

(Figure 28.3)

28.21. Lemma. *Let B be a bundle-path for Δ whose two outer ends lie in the same layer-arc from Δ. Then B contains a full sub-bundle-path which is inessential for Δ.*

Proof. By definition, there is a sub-division θ of $supp(B)$ such that $B = B_1 \cup \ldots \cup B_n$ is a bundle-path for θ. For every i, $1 \le i \le n$, there is a cell C_i of θ such that B_i is a bundle in C_i joining two different edges of θ. Let l_i be the straight arc in C_i which joins the mid-points of these two edges. Then, by assumption, the union $l := l_1 \cup \ldots \cup l_n$ is a simple closed curve in D which intersects $supp(B)$ transversely (except possibly for the one intersection point corresponding to the outer ends of B). Let $D' \subset D$ be the disc bounded by this curve. Consider the induced Haken graph $\sigma' := (supp(B) \cap int(D'))^-$ in D'. If σ' is empty, then it follows that B itself is inessential since at most one point from $l \cap supp(B)$ is non-transversal. If σ is non-empty, then there is at least one layer-arc, k', of σ' which is a proper arc in D'. Let k' be chosen to be outermost, i.e., in such a way that it separates a disc, say D'_0, from D' with $D'_0 \cap l' = k'$. Then the full sub-bundle-path in B', corresponding to the arc $D'_0 \cap \partial D'$, is clearly inessential for Δ. $\diamond$

28.22. Combs. Let Δ be a Haken graph in $(D,\underline{d})$. A Haken sub-graph $C \subset \Delta$ is a *comb* in D if there are two disjoint arcs, $d_1, d_2 \subset \partial D$ (not necessarily faces of $(D,\underline{d})$) such that (1) C is a bundle of pseudo-arcs in $(D, \{d_1, d_2\})$ joining d_1 with d_2 and that (2) no bound face of $(D,\underline{d})$ meets more than one pseudo-arc from C. Given a comb C in Δ define

$$width(C) := \text{the width of the bundle } C \text{ in } (D, \{d_1, d_2\}).$$

We say a comb is *top-free* if none of its top-arcs are top-arcs of Δ.

28.23. Proposition. *Let $\Delta \subset D$ be a useful Haken graph. Then, for every bundle-path B in Δ with $size(B) > \varphi(\alpha, depth(\Delta))$, one of the following holds:*

(1) B contains a full, inessential sub-bundle-path, or

(2) the base-length, $base(B)$, of B is larger than α, or

(3) $supp(B)$ contains a top-free comb whose width is larger than α.

Here φ *is some known arithmetic function, strictly increasing in both variables and independent of the choice of* Δ.

Proof. Given $\alpha \in \mathbf{N}$, we are asked to produce a constant $\varphi(\alpha, depth(\Delta))$ with the property that, for every bundle-path with size larger than $\varphi(\alpha, depth(\Delta))$, one of the conclusions (1)-(3) of the proposition holds. So suppose $depth(\Delta)$ is some fixed number. We view α and $depth(\mathcal{B})$ as small. In contrast, let $\mathcal{B}$ be a bundle-path whose size is "sufficiently" large and which does not satisfy (1) or (2).

Recall $\mathcal{K} = \mathcal{K}(\mathcal{B})$ denotes the collection of all layer-arcs of $supp(\mathcal{B})$ containing ends of $\mathcal{B}$. If $\#\mathcal{K}$ is small, then it follows that $\mathcal{B}$ contains a full sub-bundle-path $\mathcal{B}'$ with $supp(\mathcal{B}') \subset (supp(\mathcal{B}) - \mathcal{K})^-$ and which still has large length (recall our assumption on $\mathcal{B}$ and use lemma 28.21). But $depth(supp(\mathcal{B}')) < depth(supp(\mathcal{B}))$, and so we have a reduction since $depth(supp(\mathcal{B}))$ is small. Hence w.l.o.g. we suppose $\#\mathcal{K}$ is large. Applying lemma 28.21 once more, it follows that $supp(\mathcal{B})$ must have a large number of layer-arcs and so $\#\partial(supp(\mathcal{B}))$ must be large (recall Δ is useful and so, by definition, it has no 2-faced compression-cells disjoint to ∂D).

For every $k \in \mathcal{K}$, let $\sigma(k)$ denote the smallest Haken sub-graph of Δ containing k. Then $\#\partial\sigma(k)$ is small. But $\#\partial(supp(\mathcal{B}))$ is large and the base-length of $\mathcal{B}$ is supposed to be small. It therefore follows that there is a large subset $\mathcal{K}' \subset \mathcal{K}$ such that $\{ \sigma(k') \mid k' \in \mathcal{K}' \}$ forms a collection of *pairwise disjoint* Haken graphs in D. Now, select a sequence $k_1, k_2, ..., k_m \subset \mathcal{K}$ recursively as follows. First, give $\mathcal{K}$ the linear ordering coming from the linear ordering of the bundles of $\mathcal{B}$. Next, set $\Delta_1 := \bigcup \sigma(k')$, $k' \in \mathcal{K}'$, and let k_1 be the first arc from $\mathcal{K}$ contained in Δ_1. Let $\Delta(k_1)$ be the component of Δ_1 containing k_1. Set $\Delta_2 := \Delta_1 - \Delta(k_1)$ and let k_2 be the first arc from $\mathcal{K}$ contained in Δ_2, and so on. Note that $\sigma(k_i)$, $i \geq 2$, is a heap with top-arc k_i since $\sigma(k_i)$ is minimal and since one cell, say C_i, of $\sigma(k_i)$ contains k_i and $\sigma(k_{i-1})$ (and so meets ∂D). Set $D_i := (D - C_i)^-$. Then $\sigma(k_i) \subset D_i$ and $(\partial D_i - \partial D)^-$ is connected, i.e., an arc. But, by our assumption, $\mathcal{B}$ contains no inessential bundle-path. So it follows that $\mathcal{B}$ cannot contain a full bundle-path which lies in D_i and which has both its outer ends in $(\partial D_i - \partial D)^-$ (recall lemma 28.8 (3) and use induction). Thus we have a descending sequence $D_1 \supset D_2 \supset ... \supset D_m$ of discs with $\sigma(k_i) \subset D_i - D_{i+1}$. More precisely, $\sigma(k_i)$ meets both components of $(D_i - D_{i+1}) \cap \partial D$. It follows that $D_i - D_{i+1}$ contains a pseudo-arc τ_i which joins the two components of $(D_i - D_{i+1}) \cap \partial D$. Set $\tau := \bigcup_i \tau_i$. Then τ is a is a collection of pairwise disjoint pseudo-arcs. Moreover, every bound face of $(D, \underline{d})$ meets only a small number of pseudo-arcs from τ, for $base(\mathcal{B})$ is supposed to be small. Thus τ contains a comb with large width which is also top-free since $\tau_i \subset supp(\mathcal{B})$, for all i. Hence we have found the desired comb satisfying (3) of the proposition. $\Diamond$

28.24. *Complexities for Haken Graphs.*

Haken graphs in general can be very complicated and hard to control. Fortunately, we have to consider only Haken graphs of limited complexities. The complexities in question are designed to restrict certain aspects of combs and bundle-paths. Here are their precise definitions.

$$base(\Delta, \alpha) := \max \{ \ base(\mathcal{B}) \ | \ \mathcal{B} \ \text{bundle-path in} \ \Delta \ \text{with} \ width(\mathcal{B}) > \alpha \ \},$$
$$bundle(\Delta) := \max \{ \ width(\mathcal{B}) \ | \ \mathcal{B} \ \text{inessential bundle-path in} \ \Delta \ \},$$
$$comb(\Delta) := \max \{ \ width(\mathcal{C}) \ | \ \mathcal{C} \ \text{top-free comb in} \ \Delta \ \},$$

There is one more complexity for Δ. To define it let $\mathcal{B}$ be a bundle in Δ disjoint to ∂D. Given a number $\alpha \in \mathbf{N}$, define

$$square(\Delta, \mathcal{B}, \alpha) := \left\{ \begin{array}{l} \text{the total number of cells of } \mathcal{B} \cup supp(\mathcal{B}) \text{ in } A'(\mathcal{B}) \\ \text{which contain more than } \alpha \text{ layer-arcs of } \Delta. \end{array} \right.$$

Using this notation, define

$$square(\Delta, \alpha) := \max \{ \ square(\Delta, \mathcal{B}, \alpha) \ \}$$

where the maximum is taken over all bundles, $\mathcal{B}$, in Δ disjoint to ∂D.

28.25. Definition. *A Haken graph Δ in $(D, \underline{d})$ is called α-dominated, $\alpha \in \mathbf{N}$, if*

$$depth(\Delta), \ base(\Delta, \alpha), \ bundle(\Delta), \ comb(\Delta), \ square(\Delta, \alpha)$$

are all smaller than α.

The following lemma is an immediate consequence of this definition and prop. 28.23. It expresses a crucial property of α-dominated Haken graphs.

28.26. Lemma. *Let Δ be a useful and α-dominated Haken graph in $(D, \underline{d})$ and let $\mathcal{B}$ be any bundle-path in Δ. Then $size(\mathcal{B}) \leq \varphi(\alpha)$, where $\varphi(\alpha)$ is a constant depending on α alone.* $\diamondsuit$

28.27. *Short Companions for Haken Graphs.*

Given a constant $\alpha \in \mathbf{N}$, we single out the set of all useful and α-dominated Haken graphs in discs. This set is of course infinite. But it turns out that all its Haken graphs have short companions. In fact, we are now ready to construct those companions. The construction is in several steps. The first step is crucial. In this step we show (prop. 28.28) that from every long face there starts a proper bundle-path with large width (provided the boundary-pattern of the underlying disc has small size). Now, a bundle-path with large width in an α-dominated Haken graph is always short, and, intuitively, short bundle-paths with large widths cannot intersect. We will see that this is an important property. It will allow us to let those bundle-paths play the rôle of "arcs" and to construct short companions along the lines of the proof of prop. 28.3.

28.28. Proposition. *Let $\alpha, \beta, \gamma \in \mathbf{N}$ be constants. Suppose Δ is a useful and α-dominated Haken graph in $(D, \underline{d})$. Suppose further that $\#\underline{d} \leq \gamma$ and* *that d is a bound face of $(D, \underline{d})$ with $dist_\Delta(\partial d) > \varphi(\alpha, \beta, \gamma)$. Then there is a bundle-path $\mathcal{B}$ in Δ with*

$$width(\mathcal{B}) > \beta,$$

which joins d either with d or some other bound or free face of $(D, \underline{d})$.

Here φ is some known arithmetic function, strictly increasing in all variables and independent of the choice of $\underline{d}$ and Δ.

Proof. Estimates are crucial in this proof but we will avoid *explicit* estimates as much as possible (for they would only confuse the matter). We rather follow the convention that a number such as length, distance etc. is *small* if it is smaller than a known constant depending on α, β and γ alone. All other numbers are *large*. In particular, we can assume them to be as large as we want. In this sense $dist_\Delta(\partial d)$ is large while $bundle(\Delta)$, $comb(\Delta)$ etc. are small.

We are asked to construct a *proper* bundle-path with large width and starting at d, and we are going to construct such a bundle-path from a "start-bundle" by a process of successive "bundle-continuations". Before we explain these concepts, we introduce two more notations first. Let $\mathcal{B}$ be any bundle-path for Δ whose two outer faces lie neither in the same layer-arc of Δ nor in the same face of $(D, \underline{d})$, and suppose r is either a face or a layer-arc for which we know that it contains an outer end of $\mathcal{B}$. Then denote

$$e(\mathcal{B}; r) := \text{the outer end of } \mathcal{B} \text{ contained in } r.$$

Given a non-proper bundle-path $\mathcal{B}$ which starts in d, denote

$$breadth_\Delta(\mathcal{B}) := dist_\Delta(\partial e(\mathcal{B}; d)).$$

The start-bundle. A *start-bundle* is a bundle $\mathcal{B}$ in Δ such that one end of $\mathcal{B}$ lies in d. Our starting point is the following existence result.

28.29. Lemma. *If $dist_\Delta(\partial d)$ is large, then there exists at least one start-bundle $\mathcal{B}$ such that $width(\mathcal{B})$ as well as $breadth_\Delta(\mathcal{B})$ is large.*

Proof of lemma. Let $\mathcal{B}_0$ be the bundle of all pseudo-arcs in $(D, \underline{d})$ with one end in d. Recall from the definition of pseudo-arcs - or the fact that Δ is useful - that d cannot contain both ends of $\mathcal{B}_0$. If $width(\mathcal{B}_0)$ and $breadth_\Delta(\mathcal{B}_0)$ are both large, then $\mathcal{B}_0$ itself is the desired start-bundle, and we are done.

So suppose $width(\mathcal{B}_0)$ or $breadth_\Delta(\mathcal{B})$ is small. Of course, $width(\mathcal{B}_0)$ is small if $breadth_\Delta(\mathcal{B}_0)$ is. So in any cace $width(\mathcal{B}_0)$ is small. Then consider the collection $\mathcal{L}$ of all top-arcs from $\mathcal{B}_0$. Note that $\mathcal{L}$ consists of layer-arcs which are pairwise disjoint and which meet d. Set $\Delta(\mathcal{B}_0) := (\mathcal{B}_0 - \mathcal{L})^-$. Then $\Delta(\mathcal{B}_0)$ is a union of heaps. Moreover, $width(\mathcal{B}_0)$ is small. Hence it follows from lemma 28.10 that $\Delta(\mathcal{B}_0)$ contains only a small number of layer-arcs. Hence $\mathcal{L}$ decomposes into a small number of (maximal) bundles, $\mathcal{L}_1, \mathcal{L}_2, ..., \mathcal{L}_m$. If

$width(\mathcal{L}_i)$ and $breadth_\Delta(\mathcal{L}_i)$ are both large, for some index i, $1 \leq i \leq m$, then $\mathcal{L}_i$ is the desired start-bundle, and we are done.

So suppose that, for all $1 \leq i \leq m$, $width(\mathcal{L}_i)$ or $breadth_\Delta(\mathcal{L}_i)$ is small. Let Δ_0 be the smallest Haken graph which contains all $\mathcal{L}_i$ with small width. Then Δ_0 has a small number of layer-arcs, and we can replace D by the cells of Δ_0 meeting d, and so on. But $dist_\Delta(\partial d)$ is large and $depth(\Delta)$ is small. Hence the process has to stop after a small number of steps. But it only stops if it finds bundles with large width. Hence w.l.o.g. we may suppose for simplicity that $width(\mathcal{L}_i)$ is large, for all $1 \leq i \leq m$. Then $breadth_\Delta(\mathcal{L}_i)$ is small, for all $1 \leq i \leq m$. But $dist_\Delta(\partial d)$ is supposed to be large. Hence there is at least one component a of $(e - \bigcup_i e(\mathcal{L}_i; d))^-$ such that $dist_\Delta(\partial a)$ is large. For every such component a, let $D(a)$ be the cell of $\mathcal{B}_0$ containing a, and let $\Delta(a) = (\Delta \cap int(D(a)))^-$. Then we try the same procedure for all pairs $(D(a), \Delta(a))$ (instead of the pair (D, Δ)) in order to find a start-bundle in $D(a)$ with one end in a. Either we find such a start-bundle or we repeat the process. In each step we get an obvious reduction (of $depth(.)$), and we therefore get a start-bundle after a small number of steps (cf. also the proof of lemma 28.31).

This establishes the lemma.

The bundle-continuation. The bundle-continuation is a certain process which takes a non-proper bundle-path, starting in d, and extends it to some longer bundle-path. To describe this process, let

$$\mathcal{B} = \mathcal{B}_1 \cup ... \cup \mathcal{B}_n$$

be any non-proper bundle-path for Δ with one outer end in d. Then, more precisely, there is a sub-division, θ, of some Haken sub-graph of Δ such that $\mathcal{B}$ is a bundle-path for θ. In particular, $\mathcal{B}$ joins d with some edge, k, of θ. Let C be the cell of θ which contains the end $e(\mathcal{B}; k)$, but not the bundle, $\mathcal{B}_n$, of $\mathcal{B}$ which has $e(\mathcal{B}; k)$ as an end. We say C is the *head-cell* of $\mathcal{B}$ w.r.t. θ.

Assume there is a bundle $\mathcal{K} \subset C$ with one end, say $e(\mathcal{K}; k)$, contained in k (recall a bundle consists of pseudo-arcs and pseudo-arcs are ordinary arcs unless they meet ∂D). Set $e^* := e(\mathcal{B}; k) \cap e(\mathcal{K}; k)$ and, using e^*, define a bundle-path $\mathcal{B}_1^* \cup ... \cup \mathcal{B}_n^*$ contained in $\mathcal{B}$ as follows: Let $\mathcal{B}_{n+1}^*$ be the bundle of all pseudo-arcs from $\mathcal{K}$ which meet e^*. Let $\mathcal{B}_n^* \subset \mathcal{B}_n$ be the bundle of all layer-arcs which meet e^*. Let e_n^* be the end of $\mathcal{B}_n^*$ different from e^*. Let $\mathcal{B}_{n-1}^* \subset \mathcal{B}_{n-1}$ be the bundle of all layer-arcs which meet e_n^*, and so on. Finally, define

$$\mathcal{B} \sqcup \mathcal{K} := \mathcal{B}_1^* \cup ... \cup \mathcal{B}_n^* \cup \mathcal{B}_{n+1}^*.$$

Then $\mathcal{B} \sqcup \mathcal{K}$ is a bundle-path. By definition, it is the *bundle-continuation* of $\mathcal{B}$ by $\mathcal{K}$ (and into C).

28.30. Example.

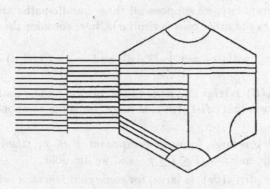

(Figure 28.4.)

We say a non-proper bundle-path $\mathcal{B}$, starting at d, has a *small head-cell* if there is a sub-division θ of some Haken sub-graph of Δ such that $\mathcal{B}$ is a bundle-path for θ and that the head-cell of $\mathcal{B}$ w.r.t. θ has a small number of faces (recall from 28.4 the definition of boundary-patterns for cells of θ). A *square* for $\mathcal{B}$ is a square $A \subset D$, $A \cap (\mathcal{B} \cup supp(\mathcal{B})) = \partial A$, which has two opposite faces in $\mathcal{B}$ and the remaining two faces in $supp(\mathcal{B})$. Thus, for all non-proper bundle-paths $\mathcal{B}$ which have an end in d, the squares are exactly the closures of the components of $A(\mathcal{B}) - (\mathcal{B} \cup supp(\mathcal{B}))$.

28.31. Lemma. *Let $\mathcal{B}$ be a bundle-path for Δ which has one outer end in d and which has a small head-cell, say C. Suppose $breadth_\Delta(\mathcal{B})$ is large, but every square for $\mathcal{B}$ contains only a small number of layer-arcs from Δ. Then there is a bundle $\mathcal{K} \subset C$ such that $\mathcal{B} \sqcup \mathcal{K}$ is a bundle-path with a small head-cell and such that $breadth_\Delta(\mathcal{B} \sqcup \mathcal{K})$ is large.*

Proof of lemma. The problem is that the desired bundle $\mathcal{K}$ may be hidden deep in the structure of $\Delta_C := (\Delta \cap int(C))^-$. In particular, there is no reason to expect $\mathcal{K}$ to be proper in C. To uncover a non-proper $\mathcal{K}$, we have to resort to an iteration-process (similar to the one used for lemma 28.29).

To begin with let $\mathcal{C}$ be the collection of all pseudo-arcs joining k with some other face of the cell C. Then $\mathcal{C}$ is the disjoint union $\mathcal{C} = \mathcal{C}_1 \cup ... \cup \mathcal{C}_p$ of (maximal) bundles in C (see our convention above concerning C). Since the number of faces of C is small, we have that $\mathcal{C} = \mathcal{C}_1 \cup ... \mathcal{C}_p$ is actually the union of only a small number of bundles. Set

$$\mathcal{P}_i := \mathcal{B} \sqcup \mathcal{C}_i, \ 1 \leq i \leq p.$$

Note that $\mathcal{C}_i$ cannot have an end in one of the layer-arcs from $\mathcal{B}$ (otherwise we find as in 28.21 that $\mathcal{B}$ contains an inessential sub-bundle-path with large width which in turn is impossible since $bundle(\Delta)$ is small). Thus $\mathcal{P}_i$, $1 \leq i \leq p$, is a bundle-path which joins d with some other layer-arc of Δ (or some face of

$(D, \underline{d})$). For simplicity, we suppose all these bundle-paths are non-degenerate (the argument in the other case is similar). Now, consider the decomposition

$$e(\mathcal{B}; d) = e_1 \cup e_2 \quad \text{with} \quad e_1 := \bigcup_i e(\mathcal{P}_i; d) \quad \text{and} \quad e_2 := (\ e(\mathcal{B}; d) - \bigcup_i e(\mathcal{P}_i; d)\ )^-.$$

Since $breadth_\Delta(\mathcal{B})$ is large and since the number of bundle-paths $\mathcal{P}_i$, $1 \leq i \leq p$, is small, it follows that $dist_\Delta(\partial e)$ is large, for some component e of e_1 or e_2.

If $dist_\Delta(\partial e)$ is large, for some component e of e_1, then $breadth_\Delta(\mathcal{P}_i; d)$ is large, for some index i, $1 \leq i \leq p$, and we are done.

So suppose $dist_\Delta(\partial e)$ is large, for some component e of e_2. Let B be the component of $(A(\mathcal{B}) - \bigcup_i A'(\mathcal{P}_i))^-$ which contains e. Define

$$\mathcal{B}' := \mathcal{B} \cap B.$$

Then $breadth_\Delta(\mathcal{B}')$ is large since $dist_\Delta(\partial e)$ is large (recall that every square for $\mathcal{B}$ contains only a small number of layer-arcs). To make further progress consider the component, E', of $(C - \bigcup_i A'(\mathcal{K}))^-$ which contains $e(\mathcal{B}'; k)$. Finally, define

$\sigma(k) :=$ the union of all heaps in E' which are disjoint to k and whose

boundary is not contained in a face of E'.

$C' :=$ the cell of $\sigma(k) \subset E'$ which contains the edge k.

Then $e(\mathcal{B}'; k) \subset C'$. Note that E' is a cell of $\mathcal{K} \cup supp(\mathcal{B})$ (or rather its subdivision). An application of lemma 28.10 shows that E' has a small number of faces. Hence an application of lemma 28.11 shows that C' has only a small number of faces.

The previous construction yields a bundle-path $\mathcal{B}' \subset \mathcal{B}$ with a small head-cell C' and such that $breadth_\Delta(\mathcal{B}')$ is still large. Hence, replacing $\mathcal{B}$, C by $\mathcal{B}'$, C', we may repeat the above construction, and so on. But notice that $depth(\Delta \cap int(C))^- < depth(\Delta \cap int(C'))^-$. So the process must stop after at most $depth(\Delta) \leq \alpha$ number of steps, i.e., after a small number of steps. But the process only stops if it finds a bundle-path $\mathcal{B}^* \subset \mathcal{B}$ and a bundle $\mathcal{K}^* \subset C$ such that $breadth_\Delta(\mathcal{B}^* \sqcup \mathcal{K}^*)$ is large.

To finish the proof of the lemma, it remains to show that $\mathcal{B}^* \sqcup \mathcal{K}^*$ has a small head-cell. This is trivial if $\mathcal{B}^*$ is a proper bundle-path. If not, then let C^* be the head-cell of $\mathcal{B}^*$ w.r.t. $supp(\mathcal{B}^*)$. We claim C^* has only a small number of faces. To see this let $\sigma \subset supp(\mathcal{B}^*)$ be the smallest Haken sub-graph which contain the outer end of $\mathcal{B}^*$ not contained in d. Then C^* is clearly a cell of σ. Hence, it follows from lemma 28.11 that C^* has a small number of faces. This establishes the claim.

Therefore the proof of the lemma is finished.

Construction of proper bundle-paths. We apply the above results as follows. First, we consider a start-bundle $\mathcal{B}_1$ with $width(\mathcal{B}_1)$ and $breadth_\Delta(\mathcal{B}_1)$ large as given by lemma 28.29. If this bundle is proper, then we are done. So suppose it is not. Then, in particular, $\mathcal{B}_1$ consists of ordinary arcs. We also *assume that every square for $\mathcal{B}_1$ contains only a small number of layer-arcs* (at the end of this proof we indicate the necessary modification for the other case).

By definition, $breadth_\Delta(\mathcal{B}_1)$ is large. By lemma 28.31, any bundle-path $\mathcal{B}_i$, starting at d, can be continued to a bundle-path $\mathcal{B}_{i+1} := \mathcal{B}_i \cup \mathcal{K}$ as long as $breadth_\Delta(\mathcal{B}_i)$ large and its squares contain only a small number of layer-arcs. Since $square(\Delta, \alpha) \leq \alpha$, it follows that $\mathcal{B}_{i+1}$ can have only a small number of squares which contain a large number of layer-arcs. So, passing to a sub-bundle-path, if necessary, we may suppose every square for $\mathcal{B}_{i+1}$ contains only a small number of layer-arcs. Thus a repeated application of lemma 28.31, produces a sequence of bundle-paths $\mathcal{B}_1, \mathcal{B}_2, \ldots$ with strictly increasing lengths and large breadth and such that every square for $\mathcal{B}_i$ contains only a small number of layer-arcs.

28.32. Lemma. *If $\mathcal{B}_i$ is a non-proper bundle-path. Then $width(\mathcal{B}_i)$ is large.*

Proof of lemma. Assume the converse. Let e_1, e_2 be the two outer ends of $\mathcal{B}_i$ and let a_1, a_2 be the two components of $(\partial A(\mathcal{B}_i) - e_1 \cup e_2)^-$. W.l.o.g. let the indices be chosen so that $e_1 = e(\mathcal{B}_i; d)$. The closure of every component of $A(\mathcal{B}_i) - (\mathcal{B}_i \cup supp(\mathcal{B}_i)$ is a square for $\mathcal{B}_i$ ($\mathcal{B}_i$ is non-proper) and every square for $\mathcal{B}_i$ contains only a small number of layer-arcs from Δ (see construction of $\mathcal{B}_i$). It follows that ∂a_1 (and ∂a_2), can be joined in $A(\mathcal{B}_i)$ by an arc, say α_1, whose Δ-length is small (see construction of bundle-continuation). By assumption, $width(\mathcal{B}_i)$ is small. So, by definition, there is an arc $w \subset A'(\mathcal{B}_i)$ which joins the end-points either of e_1, or of e_2, and which meets $\mathcal{B}_i \cup supp(\mathcal{B}_i)$ in a small number of points only. Since every square for $\mathcal{B}_i$ contains only a small number of layer-arcs from Δ, it follows that w can be isotoped, fixing the intersection $w \cap (\mathcal{B}_i \cup supp(\mathcal{B}_i))$, so that afterwards $\#(w \cap \Delta)$ is small. Composing w with α_1 and α_2 if necessary, we see that ∂e_1 can be joined by an arc with small Δ-length. This, however, contradicts the fact that $breadth_\Delta(\mathcal{B}_i)$ is large (see construction of $\mathcal{B}_i$). So our assumption was wrong, and the lemma is established.

Thus if $dist_\Delta(\partial d)$ is large enough, then, according to the previous lemma, we have $width(\mathcal{B}_i) > \varphi(\alpha)$, where $\varphi(\alpha)$ is the constant from lemma 28.26. Moreover, Δ is supposed to be useful and α-dominated and so, by lemma 28.26, we have $size(\mathcal{B}_i) \leq \varphi(\alpha)$. But $length(\mathcal{B}_i) = i$. It therefore follows that no bundle-path, $\mathcal{B}_i$, can be longer than $\varphi(\alpha)$. This means that the above process has to stop after at most $\varphi(\alpha)$ steps. Hence there is an index m, $1 \leq m \leq \varphi(\alpha)$, such that
$$\mathcal{B}' := \mathcal{B}_m$$
is a proper bundle-path in D.

This does not quite finish the proof of the proposition though because $width(\mathcal{B}')$ may not be large (the problem arises with the very last bundle added by the continuation-process). In this situation we will have to apply another cycle of iteration. So suppose $breadth_\Delta(\mathcal{B}')$ is large and $width(\mathcal{B}')$ is small. Let $\mathcal{B}' = \mathcal{B}'_1 \cup ... \cup \mathcal{B}_n$ be the decomposition of $\mathcal{B}'$ into its bundles. Then $\mathcal{B}'$ may be viewed as a bundle-continuation of $\mathcal{B}'_1 \cup ... \cup \mathcal{B}'_{n-1}$ by $\mathcal{B}'_n$. We are going to replace this particular bundle-continuation by a different one.

Since $width(\mathcal{B}')$ is small, there is an arc $w \subset A'(\mathcal{B}')$ which joins the end-points of the end $e(\mathcal{B}';d)$ and which intersects $\mathcal{B}' \cup supp(\mathcal{B}')$ in a small number of points. Of course, $\#(w \cap \Delta) \geq breadth_\Delta(\mathcal{B}')$, i.e., w must intersect Δ in a large number of points. W.l.o.g. we may suppose that $w \cap \mathcal{B}' = \emptyset$ and that $length_\Delta(w - C)^-$ is small, where C is the head-cell of $\mathcal{B}'_1 \cup ... \cup \mathcal{B}'_{n-1}$ (recall $square(\Delta)$ is small). Consider $w \cap C$. Note that w is the union $w := w_1 \cup ... \cup w_p$ of proper arcs in C. Moreover, $\partial w_i \subset k$ and $w_i \cap \mathcal{B}'_n = \emptyset$, for $1 \leq i \leq p$, where k is the layer-arc from Δ containing an outer end of $\mathcal{B}'_1 \cup ... \cup \mathcal{B}'_{n-1}$.

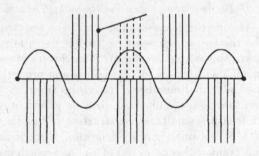

(Figure 28.5)

Let C_i be the cell of $C \cap \mathcal{B}'$ which contains w_i. Then $\mathcal{B}'_1 \cup ... \cup \mathcal{B}'_{n-1}$, C_i is a pair as in the definition of bundle-continuation. Hence, as defined there, C_i is the small head-cell of some bundle-path $\mathcal{B}^*_1 \cup ... \cup \mathcal{B}^*_{n-1}$ contained in $\mathcal{B}'_1 \cup ... \cup \mathcal{B}'_{n-1}$. Applying lemma 28.31 to this new situation, we find a bundle-continuation for $\mathcal{B}^*_1 \cup ... \cup \mathcal{B}^*_{n-1}$ into C_i (indicated by the dotted arcs in fig. 28.5). In fact, by the construction used before, $\mathcal{B}^*_1 \cup ... \cup \mathcal{B}^*_{m-1}$ can be extended to a proper bundle-path, say $\mathcal{C}_1$, in D. Applying this process over and over again, we get a sequence $\mathcal{C}_1, \mathcal{C}_2, ...$ of proper bundle-paths in D whose lengths, $length(\mathcal{C}_i)$, are strictly increasing. Moreover, by construction, the bundle-path $\mathcal{C}_i$ contains a non-proper full sub-bundle-path of length $length(\mathcal{C}_i) - 1$ and large width. But, by lemma 28.26, the sizes of all bundle-paths are bounded by $\varphi(\alpha)$. So $length(\mathcal{C}_i) \leq \varphi(\alpha)$. Thus the process has to stop after a small number of steps (equivalently, we could have deduced this from the fact that the Haken graph $(\Delta \cap int(C_i))^-$ has a strictly smaller depth than $(\Delta \cap int(C))^-$). But the process only stops if it results in a proper bundle-path C_p with $width(C_p)$ large. Thus, altogether, C_p is the desired bundle-path.

We have finished the proof under the overall assumption that every square of the start-bundle B_1 contains only a small number of layer-arcs. But this may not be true. In fact, B_1 may have a large number of squares for which this does not hold. In that case we do not get the bundle-continuation off the ground because there is nothing to start with. To overcome this obstacle, we begin the same process with start-bundles which do not start in d but at that end k of the start-bundle B_1 not contained in d. The squares of these new bundles contain only a small number of layer-arcs since $square(\Delta, \alpha) \leq \alpha$. So the above process constructs a bundle-path B which starts at k and ends in ∂D. It is then not hard to construct the desired bundle-path as a union of B with an appropriate bundle-path in $A(B_1)$ (use the fact that $base(\Delta)$ is small).

The proof of the proposition is therefore finished. $\Diamond$

A bundle-path, B, as given by the previous proposition, may still have a long frontier and in that case it is not good enough for our purpose. We are now going to fix this deficiency. To do this, however, we find it more convenient to look at the "band" $A(B)$ instead of the bundle-path B. The idea is to slim down this band to a band of large width and short frontier. Here a *band* in $(D, \underline{d})$ is a disc (square) $B \subset D$ with the property that $B \cap \partial D$ consists of exactly two components which in turn are entirely contained in faces of $(D, \underline{d})$. We say B *joins the faces* which contain the components of $B \cap \partial D$. The *width* (or: Δ-*width*) of B is given by

$$width_\Delta(B) := dist_\Delta(b_1, b_2),$$

where b_1, b_2 are the two components of $(\partial B - \partial D)^-$. It is important to keep in mind that, in contrast to bundle-paths, the width of bands is measured w.r.t. to the Haken graph Δ.

28.33. Proposition. *Let* $\alpha, \beta, \gamma \in \mathbf{N}$ *be constants. Suppose* Δ *is a useful and* α-*dominated Haken graph in* $(D, \underline{d})$. *Suppose further that* $\#\underline{d} \leq \gamma$ *and that* d *is a bound face of* $(D, \underline{d})$ *with* $dist_\Delta(\partial d) > \varphi(\alpha, \beta, \gamma)$. *Then there is band* B *in* D *with*

$$length_\Delta(\partial B - \partial D)^- \leq \beta \quad and \quad width_\Delta(B) > 2\beta,$$

which joins d *either with* d, *or some other bound or free face of* $(D, \underline{d})$.

Here φ *is some known arithmetic function, strictly increasing in all variables and independent of the choice of* $\underline{d}$ *and* Δ.

Proof. By prop. 28.28, there is a bundle-path, B_1, with large width which joins d with some face of $(D, \underline{d})$. Set

$$B_1 := A(B_1).$$

Then, in the terminology above, B_1 is a band.

We claim $width_\Delta(B_1)$ is large. Assume the converse. Then there is an arc $\ell \subset B_1$ which joins the two components of $(\partial B_1 - \partial D)^-$ and whose Δ-length is small. In particular, ℓ intersects $\mathcal{B}_1 \cup supp(\mathcal{B}_1)$ in a small number of points. But the length of $\mathcal{B}_1$ is small since the width of $\mathcal{B}_1$ is large and since w.l.o.g. $size(\mathcal{B}_1) \leq \alpha$. Thus there is a pair of arcs $k, k' \subset B_1$ which joins $\partial \ell$ with the end-points of $B_1 \cap d$ and which intersect $\mathcal{B}_1 \cup supp(\mathcal{B}_1)$ in a small number of points. The union $k \cup \ell \cup k'$ is a path which joins the end-points of $B_1 \cap d$ and which intersects $\mathcal{B}_1 \cup supp(\mathcal{B}_1)$ in a small number of points. Thus $width(\mathcal{B}_1)$ is small. This contradicts our choice of $\mathcal{B}_1$. The claim is therefore established.

Let b_1, b_1' be the two components of $(\partial B_1 - \partial D)^-$. If both $dist_\Delta(\partial b_1)$ and $dist_\Delta(\partial b_1')$ are small, then, by definition, there are arcs, l_1, l_1', with small Δ-length and $\partial l_1 = \partial b_1$ and $\partial l_1' = \partial b_1'$. Actually, using cut-and-paste, we may suppose that these two arcs are disjoint. The union $l_1 \cup (B_1 \cap \partial D) \cup l_1'$ is a simple closed curve. The disc bounded by this curve is the desired band, and we are done. Thus w.l.o.g. we suppose that $dist_\Delta(\partial b_1)$ is large.

Now, define a boundary-pattern for B_1, given by the components of $B_1 \cap \partial D$ and $(\partial B_1 - \partial D)^-$. Then B_1 is a 4-faced disc. Moreover, $dist_\Delta(\partial b_1)$ is large (and so also large when measured in B_1). Thus we are in the situation of prop. 28.28. According to that proposition there is at least one bundle-path, say $\mathcal{B}_2$, in B_1 which has large width and which starts at b_1 and ends at b_1 or some other face of B_1. Define

$$B_2 := A(\mathcal{B}_2).$$

Then B_2 is a band in B_1 which joins the face b_1 with b_1 or some other face of B_1. So B_2 joins b_1 either with b_1, or with d, or with $(\partial D - d)^-$, or with the other component of $(\partial B_1 - \partial D)^-$.

Suppose B_2 joins b_1 with b_1 or d. Then $\mathcal{B}_2$ consists of ordinary arcs (Δ is useful) and, in particular, $A(\mathcal{B}_2)$ is a union of 4-faced cells of $\mathcal{B} \cup supp(\mathcal{B})$. Using the hypothesis that $square(\Delta, \alpha)$, $base(\Delta, \alpha)$ and $bundle(\Delta)$ are all small, it is not hard to construct a band $B_2' \subset B_2$ with $B_2' \cap \partial B_2 = B_2 \cap \partial B_1$ such that $length_\Delta(\partial B_2' - \partial B_2)^-$ is small. Let B_1' be that component of $(B_1 - B_2')^-$ which meets both components of $B_1 \cap \partial D$. Let $\mathcal{B}_1'$ be the bundle-path $\mathcal{B}_1$ minus all layer-arcs from $\mathcal{B}_1$ which meet $(B_1 - B_1')^-$. Then a moments reflection shows that $\mathcal{B}_1'$ has still large width since $length_\Delta(\partial B_2' - \partial B_1)^-$ is small. Iterating this construction if necessary, we may suppose that $\mathcal{B}_1$ is chosen so that, in addition, there is no bundle-path $\mathcal{B}_2$ in B_1 with large width which joins a component of $(\partial B_1 - \partial D)^-$ with d.

We claim B_2 cannot join b_1 with the other component of $(\partial B_1 - \partial D)^-$. Assume the converse. Then $B_2 \cap \partial D = \emptyset$. This implies that $\mathcal{B}_2$ consists of ordinary arcs. As above we therefore find a band $B_2' \subset B_2$ such that $length_\Delta(\partial B_2' - B_1)^-$ is small. But this contradicts the fact that $width_\Delta(\mathcal{B}_1)$ is large (see above).

Thus, by what we have seen so far, we may suppose $\mathcal{B}_2$ joins b_1 with $(\partial D - d)^-$. Let C_2 be that component of $(C_1 - A'(\mathcal{B}_2))^-$ which contains d. Then C_2 has a large width, and we are done if the Δ-length of $(\partial C_2 - \partial D)^-$ is small. If not, then we can certainly iterate the construction by choosing the next band in C_2, etc. Of course, the process has to stop after finitely many steps since Δ has only finitely many layer-arcs. Every step of this process yields a band. So the whole process results in a band. The resulting band has large width. But if we have to perform a large number of steps, then this band will still have a long frontier, and nothing is gained. The next figure shows that this can really happen.

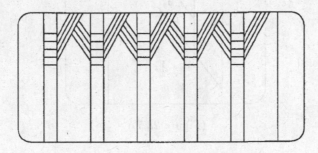

(Figure 28.6.)

In order to make progress nevertheless we have to select the new bands slightly more carefully. Here is the construction (see fig. 28.8 for the schematics of this construction). As before we start with the band

$$C_1 := B_1 := A(\mathcal{B}_1).$$

Given the band $B_2 \subset B_1$ with large width which joins the component, $c_1 := b_1$, of $(\partial B_1 - \partial D)^-$ with $(\partial D - d)^-$, we set B_2' to be the unique disc in B_1 such that $B_2' \neq B_1$, $B_2 \subset B_2'$, $B_2' \cap \partial B_1$ is connected and $\partial B_2' \subset \partial B_1 \cup \partial B_2$. Given this notation, define $C_2 := (C_1 - B_2')^-$, and let c_2 be the component of $(\partial C_2 - \partial D)^-$ which meets c_1. If $dist_\Delta(\partial c_2)$ is small, we stop.

Suppose $dist_\Delta(\partial c_2)$ is large. Then w.l.o.g. we may suppose B_2 is chosen so that $dist_\Delta(\partial(c_2 \cap c_1))$ is small and that $dist_\Delta(\partial(c_2 - c_1)^-)$ is large. Then, by a similar argument as used before, there is a band $B_3 \subset B_2$ which joins $C_2 \cap B_2$ with ∂D and which has large width. Define B_3' as above and define $C_3 := C_2 \cup B_3'$. Let c_3 be the component of $(\partial C_3 - \partial D)^-$ which meets c_1. If $dist_\Delta(\partial c_3)$ is small, we stop. If not, we repeat the process. Thus, in general, we are given C_i, a band $B_i \subset B_{i-1}$ and a band $B_{i+1} \subset B_i$ with large width which joins $B_i \cap (\partial C_i - \partial D)^-$ with ∂D. Then define B_{i+1}' as above, i.e.,

$$B_{i+1}' := \left\{ \begin{array}{l} \text{the disc in } B_i \text{ such that } B_{i+1}' \neq B_i, \ B_{i+1} \subset B_{i+1}', \\ B_{i+1}' \cap \partial B_i \text{ is connected and } \partial B_{i+1}' \subset \partial B_i \cup \partial B_{i+1}, \end{array} \right.$$

and define

$$C_{i+1} := \begin{cases} (C_i - B'_{i+1})^-, & \text{if } i \text{ is odd,} \\ C_i \cup B'_{i+1}, & \text{if } i \text{ is even.} \end{cases}$$

Then select a band $B_{i+1} \subset C_{i+1}$. Recursively, this defines a finite sequence of discs:

$$C_1 \subset C_2 \supset C_3 \subset C_4 \supset ... C_m,$$

as well as a sequence, $B_1 \subset B_2 \subset ... \subset B_m$, of bands.

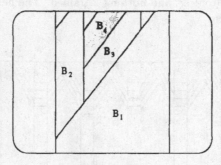

(Figure 28.7.)

It remains to show that this process stops after a small number of steps, i.e., that the index m is small (and not just finite). For then the end-points of the component of $(\partial C_m - \partial D)^-$, meeting b_1, have small Δ-distance and, applying the same process to the other component of $(\partial C_m - \partial D)^-$ (i.e. b'_1) if necessary, we may suppose that the end points of every component of $(\partial C_m - \partial D)^-$ have a small Δ-distance. Then we finish the proof as discussed before.

To show that the above process stops after a small number of steps consider the sequence $B_1, B_2, ..., B_m$ of bands. Recall that, for every band B_i, there is a bundle-path $\mathcal{B}_i$ with $B_i = A(\mathcal{B}_i)$. Every bundle-path is made out of bundles and every bundle is made out of pseudo-arcs (but recall that the pseudo-arcs of every bundle not meeting ∂D are layer-arcs). Let $ind(\mathcal{B}_i)$ denote the smallest index taken over the indices of all top-arcs of pseudo-arcs in $\mathcal{B}_i$ (recall that, by definition, the index of a layer-arc of Δ is the index j such that $(\Delta_{j+1} - \Delta_j)^-$ contains this layer-arc). Given this complexity it is now easy to finish the proof in the following special case.

Case 1. Δ has no 2-faced compression-cells.

In this case we finish the proof by showing that $ind(\mathcal{B}_1), ind(\mathcal{B}_2), ...,$ forms a sequence of strictly increasing numbers. To show this claim, fix an integer j, $1 \le j \le m$. For notational convenience, set $\mathcal{C} := \mathcal{B}_j$ and and $\mathcal{C}' := \mathcal{B}_{j+1}$. Let $\mathcal{C} = \mathcal{C}_1 \cup ... \cup \mathcal{C}_p$ be the decomposition of the bundle-path $\mathcal{C}$ into its bundles. Let the indices be chosen so that $\mathcal{C}_p$ meets ∂D. Now $square(\Delta, \alpha) \le \alpha$ and so w.l.o.g. each square-cell of the bundles $\mathcal{C}_2, ..., \mathcal{C}_{p-1}$ contains only a small number of layer-arcs. It follows that $\mathcal{C}'$ is entirely contained in the region

$A'(\mathcal{C}_p)$. Thus every layer-arc from $\mathcal{C}'$ has an end-point in a layer-arc from $\mathcal{C}_p$. Hence, for every layer-arc from $\mathcal{C}'$, there is a layer-arc from $\mathcal{C}$ whose index is strictly smaller. Thus $ind(\mathcal{B}_{j+1}) > ind(\mathcal{B}_j)$. This establishes the claim since j has been chosen arbitrarily. But $ind(\mathcal{B}_i) \le depth(\Delta) \le \alpha$, for all $1 \le i \le m$, and so prop. 28.33 follows in Case 1.

Case 2. Δ *has 2-faced compression-cells.*

In this Case we have to argue differently since the indices, $ind(\mathcal{B}_i)$, need no longer form an increasing sequence. A concrete example of this phenomenon can be constructed by iterating the following picture.

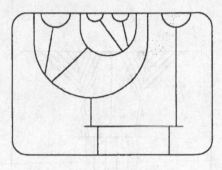

(Figure 28.8.)

To make progress we have to analyze the bundle-paths $\mathcal{B}_i$. Every bundle-path $\mathcal{B}_i$ has exactly one outer end in $\partial D - d$ and is itself the union of a linearly ordered collection of bundles. The bundle, $\mathcal{E}_i$, meeting $(\partial D - d)^-$ is called the *last bundle* of $\mathcal{B}_i$. Now $\mathcal{E}_i$ is the union of pseudo-arcs and these pseudo-arcs may not be layer-arcs (we are in Case 2). But every pseudo-arc from $\mathcal{E}_i$ is the union of its top-arc with a heap (see lemma 28.7) and the boundary of this heap lies in $(\partial D - d)^-$. Now, for every heap $\mathcal{H}$ with $\partial \mathcal{H} \subset (\partial D - d)^-$, there is a unique disc $D(\mathcal{H}) \subset D$ given by $\mathcal{H} \subset D(\mathcal{H})$ and $\partial D(\mathcal{H}) \subset \mathcal{H} \cup (\partial D - d)^-$. We call $D(\mathcal{H})$ the *region of the heap* $\mathcal{H}$.

Now recall $A(\mathcal{B}_{i+1}) = B_{i+1} \subset B_i = A(\mathcal{B}_i)$, for all $1 \le i \le m$. Moreover, by an argument from Case 1, we may suppose that $\mathcal{B}_{i+1} \subset A(\mathcal{E}_i)$. Note further that $\mathcal{B}_{i+1}$ lies in the region of a heap from $\mathcal{E}_i$ if this region contains at least one bundle from $\mathcal{B}_{i+1}$. To see this recall that $width(\mathcal{B}_i)$ is supposed to be large and $bundle(\Delta)$ is small.

In order to proceed note that $ind(\mathcal{B}_{i+1}) > ind(\mathcal{B}_i)$ if there is no heap from $\mathcal{E}_i$ whose region contains a bundle of $\mathcal{B}_{i+1}$. Since $ind(\mathcal{B}_i) \le depth(\Delta) \le \alpha$, it follows that there is a sub-sequence $\mathcal{B}_{i_1}, \mathcal{B}_{i_2}, ..., \mathcal{B}_{i_p}$ of $\mathcal{B}_1, ..., \mathcal{B}_m$ with $i_j - i_{j+1}$ small, for all $j \ge 1$, and such that every bundle-path $\mathcal{B}_{i_{j+1}}$ has at least one bundle in the region of some heap from $\mathcal{E}_{i_j}$. Thus, selecting appropriate heaps $\mathcal{H}_j$, from $\mathcal{E}_{i_j}$, we get a sequence of heaps whose regions form a nested and strictly decreasing sequence $D(\mathcal{H}_1) \supset D(\mathcal{H}_2) \supset ...$ of discs. But, by lemma

28.10, every heap has only a small number of layer-arcs. So we may suppose
w.l.o.g. that the sub-sequence of bundle-paths above has been chosen in such a
way that the heaps $\mathcal{H}_1, \mathcal{H}_2, \ldots$ are pairwise disjoint. But then this sequence
can have only a small length since $comb(\Delta) \leq \alpha$ (cf. the argument from
prop. 28.23). Hence the sequence $\mathcal{B}_{i_1}, \mathcal{B}_{i_2}, \ldots$ has only a small length. So,
finally, it follows that the sequence $\mathcal{B}_1, \ldots, \mathcal{B}_m$ has only a small length since
$i_j - i_{j+1}, \ j \geq 1$, is small. This proves the claim. Therefore prop. 28.33 is
established in Case 2.

Altogether, the proof of prop. 28.33 is finished. $\Diamond$

The following picture indicates how a band, as constructed in the previous
proof, might look like.

28.34. Example.

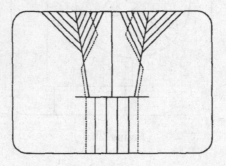

(Figure 28.9.)

28.35. Proposition. *Let* $\alpha, \beta, \gamma \in \mathbf{N}$ *be constants. Suppose* Δ *is a useful and*
α-*dominated Haken graph in* $(D, \underline{d})$. *Suppose further that* $\#\underline{d} \leq \gamma$ *and that*
d *is a bound face of* $(D, \underline{d})$ *with* $dist_\Delta(\partial d) > \varphi(\alpha, \beta, \gamma)$.

Then there is a system $\mathcal{P}(d)$ *in* D *of at most* γ *pairwise disjoint bands,*
joining d *with some other bound or free faces of* $(D, \underline{d})$, *such that*

(1) $dist_\Delta(\partial e) \leq \varphi(\alpha, \beta, \gamma)$, *for every component* e *of* $(d - \mathcal{P}(d))^-$,

(2) $width_\Delta(B) \geq 2\beta$, *for every band* B *from* $\mathcal{P}(d)$,

(3) $length_\Delta(\partial B - \partial D)^- \leq \beta$, *for every band* B *from* $\mathcal{P}(d)$.

Here φ *is a known arithmetic function, strictly increasing in all variables and*
independent of the choice of $\underline{d}$ *and* Δ.

Proof. We first show that there is at least one band $B \subset D$ with $length_\Delta(\partial B -$
$\partial D)^- \leq \beta$ and $width_\Delta(B) > 2\beta$, which joins d with a bound or free face
from $(D, \underline{d})$ different from d. To see this let $B_1, \ldots, B_n$ be a maximal
system of pairwise disjoint bands in D with $length_\Delta(\partial B_i - \partial D)^-$ small.
Let $A_1, \ldots, A_m \subset D$ be a minimal system of discs with $A_j \cap \partial D$ connected,

$\bigcup_i B_i \subset \bigcup_j A_j$ and $(\partial \bigcup_j A_j - \partial D)^- \subset (\partial \bigcup_i B_i - \partial D)^-$. Then it follows from prop. 28.33 that $dist_\Delta(\partial e)$ must be small, for all components e from $(d - A)^-$. Since $dist_\Delta(\partial d)$ is large and $length_\Delta(\partial B_i - \partial D)^-$ is small, it follows that the number m of discs A_j must be large. But $bundle(\Delta)$ is small and Δ is useful. So every Haken graph $\Delta_j := A_j \cap \Delta$ must contain at least one layer-arc a_j which joins $(\partial A_j - \partial D)^-$ with d. The large collection of arcs a_j lie in a bundle of pseudo-arcs in Δ which has large width. Starting with this bundle as start-bundle and using the bundle-continuation from the proof of prop. 28.28, we get a proper bundle-path of large width, and so also a band of large width and short frontier (see proof of prop. 28.33). If this band joins d with some other face of $(D, \underline{d})$, then we are done. If not, then we repeat the process. In each step we get an obvious reduction, and so in the end we will always find the desired band.

To continue let Q be the union of all bands, B, in $(D, \underline{d})$ with $width_\Delta(B) \geq 2\beta$ and $length_\Delta(\partial B - \partial D)^- \leq \beta$ and which join d with some other face of $(D, \underline{d})$. Let $D(Q)$ be the union of all components of $(D - Q)^-$ which contain a vertex of $(D, \underline{d})$. Define $\mathcal{P}(d) := (D - D(Q))^-$. We claim $\mathcal{P}(d)$ is the desired system of bands. Of course, $\mathcal{P}(d)$ is a system of at most $\#\underline{d} \leq \gamma$ bands joining d with other faces of $(D, \underline{d})$. Moreover, by construction, $\mathcal{P}(d)$ satisfies (2) and (3) of the proposition. Assume it does not satisfy (1). Then there is a component, d_0, of $(d - \mathcal{P}(d))^-$ such that $dist_\Delta(\partial d_0)$ is large. Let D_0 be the component of $D(Q)$ containing d_0. Let the boundary-pattern of D_0 be given by the components of $(\partial D_0 - \partial D)^-$ and the intersection of D_0 with the faces of $(D, \underline{d})$. Then D_0 is a disc with at most $\#\underline{d} + 2 \leq \alpha + 2$ faces. Thus we are in the situation of prop. 28.33. According to that proposition, there is a band, A_0, in D_0 with $width_\Delta(A_0) \geq 2\beta$ and $length_\Delta(\partial A_0 - \partial D) \leq \beta$ and joining d_0 with some other face of D_0. This other face cannot be a component of $(\partial D_0 - \partial D)^-$ since $width_\Delta(\partial A_0 - \partial D)^- \geq 2\beta$ and $length_\Delta(l) \leq \beta$, for every component l of $(\partial \mathcal{P}(d) - \partial D)^-$. Thus A_0 is a band which joins d with some other face of $(D, \underline{d})$. But A_0 is a band in $(D - \mathcal{P}(d))^-$, and so we have a contradicts to our maximal choice of $\mathcal{P}(d)$. So our assumption is wrong and $\mathcal{P}(d)$ does satisfy properties (1)-(3) of the proposition.

This finishes the proof. ◊

We now use the band-system, P, from the previous proposition to construct short companions for α-dominated Haken graphs.

28.36. Theorem. *Let $\alpha \in \mathbf{N}$ be a constant. Let Δ be a useful and α-dominated Haken graph in $(D, \underline{d})$. Then every non empty collection of bound faces of $(D, \underline{d})$ has a $\varphi(\alpha)$-companion.*

Here φ is a known, strictly increasing arithmetic function which is independent of the choice of $\underline{d}$ and Δ.

Proof. Let $\mathcal{F} \subset \underline{d}$ be a non-empty collection of faces. Suppose $dist_\Delta(d) > \varphi(\alpha)$, for all $d \in \mathcal{F}$, where $\varphi(\alpha)$ is some sufficiently large constant which depends on α alone. Then we have to construct an α-companion for $\mathcal{F}$. To do this we follow the strategy of prop. 28.3. In fact, all concepts introduced so far have been designed to make this strategy possible for Haken graphs.

First some more notation. Let us be given a face $d \in \mathcal{F}$, and let $d' \in \underline{d}$ be a face different from d. Then let $\underline{\underline{d}}(d, d')$ denote the boundary-pattern given by d, d'. Note that $(D, \underline{\underline{d}}(d, d'))$ is a 4-gon and that we are therefore in the situation of prop. 28.35. Let $\mathcal{P}(d; d') \subset D$ be a band-system starting at d and satisfying the conclusion of prop. 28.35. For every arc, k, from $(\partial\mathcal{P}(d; d') - \partial D)^-$ denote by $D_1(k)$, $D_2(k)$ the two discs into which D is split by k. We say k is *bad* if both $D_1(k)$ and $D_2(k)$ contain at least one face from $\mathcal{F}$. Moreover, we say $\mathcal{P}(d; d')$ is *bad* if $(\partial\mathcal{P}(d; d') - \partial D)^-$ has a bad component; otherwise it is *good*.

We claim there is at least one face $d \in \mathcal{F}$ with a good band-system $\mathcal{P}(d; d')$, for some $d' \in \underline{d} - \{d\}$. To construct such a face, fix an end-point z of an arbitrary face of $(D, \underline{d})$. Now, start with any face $d_1 \in \mathcal{F}$. Let $\mathcal{P}(d_1; d_1')$, $d_1' \in \underline{d}$, be a band-system satisfying prop. 28.35. If $\mathcal{P}(d_1; d_1')$ is good, we are done. If not, then $(\partial\mathcal{P}(d_1; d_1') - \partial D)^-$ has a bad component, say k_1. W.l.o.g. $z \in D_2(k_1)$. Select a face $d_2 \in \mathcal{F}$ contained in $D_1(k_1)$. Let the boundary-pattern of $D_1(k_1)$ be given by d_2 and the components of the regular neighborhood $U(\partial k_1)$ in $\partial D_1(k_1)$. Then $D_1(k_1)$ is a 6-gon and prop. 28.35 applies again. Let $\mathcal{Q}(d_2; k_1)$ be the band-system in $D_1(k_1)$ starting at d_2 and satisfying the conclusion of prop. 28.35. Then note that no band, B, from $\mathcal{Q}(d_2; k_1)$ ends in k_1 since $length_\Delta(k_1)$ is short (see (3) of prop. 28.35) and $width_\Delta(B)$ is large (see (2) of prop. 28.35). Thus $\mathcal{Q}(d_2; k_1)$ is actually a band-system $\mathcal{P}(d_2; d_2')$ in D (where $d_2' \in \underline{d}$ is some face contained in $D_2(k_1)$). If $\mathcal{P}(d_2; d_2')$ is bad, repeat the above process, and so on. The result is a sequence of discs $D_1(k_1) \supset D_1(k_2) \supset \dots$ which contain a strictly decreasing number of faces from $\mathcal{F}$. Thus the process has to stop after finitely many steps and the claim follows.

Let $d \in \mathcal{F}$ be any face for which there is a good band-system $\mathcal{P}(d; d') \subset D$, $d' \in \underline{d}$. Then there is a component, say C', of $(D - \mathcal{P}(d; d'))^-$ such that $C = U(D - C')^-$ contains d but no other face from $\mathcal{F}$. It then follows from (1) and (3) of prop. 28.35 that $c := (\partial C - \partial D)^-$ is the desired short companion. This finishes the proof of the theorem. $\Diamond$

29. Invariant Regions for Long Arc-Sequences.

Before the finiteness result of the previous section can be made available for the study of Heegaard 2-complexes, we have to show that induced Haken graphs are α-dominated. This will be done in the next section. As a preparation for that section, we here study global aspects of sequences of "long", but non-proper, arcs in surfaces. The actual result is technical and its relevance will become clear only in the next section. Basically, we will show (see thm. 29.14) that in a very concrete sense any infinite sequence of pairwise disjoint arcs "fills up" an invariant sub-surface in a very specific way. But to make this statement more precise we need some preparation.

29.1. *Train-Tracks.*

Recall that curves in surfaces are best studied by means of "train-tracks" as introduced in [Th 1]. We begin by adapting this concept to our needs.

Let S be an orientable surface with or without boundary and let $\theta \subset S$ be a finite graph with $\theta \cap \partial S = \partial \theta$ (later θ will be a Haken graph but all we need to know here is that θ is a graph). A finite graph $\tau \subset S$ is a *train-track (branched 1-manifold) in* (S, θ) if the following holds:

(1) τ intersects θ in a finite collection of points none of which is a vertex of θ.

(2) θ splits τ into a system of arcs (edges).

(3) τ intersects every component from $(\partial U(x) - \theta)^-$, for every $x \in (\tau \cap \theta) - \partial \tau$.

29.2. Example

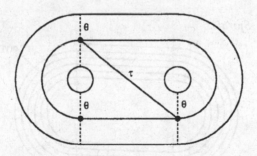

(Figure 29.1.)

A train-track τ is *essential* (w.r.t. θ) if every edge from τ is essential Here an arc $k \subset S$, $k \cap \theta = \partial k$, is *inessential* (w.r.t θ) if there is an arc l in some edge from θ such that $k \cap l = \partial k = \partial l$ and that $l \cup k$ bounds of a disc in S. If not, then k is *essential*. Finally, set

$$e(\tau) := \text{number of all components of } \tau - \theta.$$

In the following we are interested in long 1-manifolds carried by τ. Here we say that a 1-manifold $\gamma \subset U(\tau)$, $\partial\gamma \subset \theta$, is *carried by* τ if it is in general position to θ (but not necessarily proper) and if, for every component, k, from $\gamma - \theta$, there is a component, l, from $\tau - \theta$ and a disc (square) $A \subset U(\tau)$ with $A \cap (\theta \cup (k \cup l)) = \partial A$ and $(k \cup l) \subset \partial A$. The closures of the components of $\gamma - \theta$ are called the *edges* of γ. So γ is a collection of edge-paths. Set

$$\ell_\theta(\gamma) := \text{number of all edges of } \gamma.$$

We view $\ell_\theta(\gamma)$ as a measure for the (combinatorial) *length of* γ.

29.3. *Webs for 1-Manifolds.*

Let τ be a train-track in (S, θ) and let γ be a 1-manifold carried by τ. Then denote by $\mathcal{A}(\gamma)$ the set of all parallelity-regions for γ. Here a component, A, of $U(\tau) - (\gamma \cup \theta)$ is a *parallelity-region* for γ (w.r.t. θ) if ∂A is an edge-path of exactly four arcs which are alternately closures of components of $\gamma - \theta$ and $\theta - \gamma$ (∂A denotes the set of boundary points of A in S). Two edges from γ are *parallel* if they are opposite sides of a parallelity-region. Given $\mathcal{A} \subset \mathcal{A}(\gamma)$, set

$$\sigma(\gamma, \mathcal{A}) := \gamma \cup \bigcup_{A \in \mathcal{A}} A^- \quad \text{and} \quad \sigma^+(\gamma, \mathcal{A}) = \sigma(\gamma, \mathcal{A}) \cup \bigcup D^-,$$

where the latter union is taken over all components, D, of $U(\tau) - \sigma(\gamma, \mathcal{A})$ which are open discs. We say $\sigma(\gamma, \mathcal{A})$ is a *web* for γ (since it spans γ like a cobweb). γ may have various webs; all of them contained in the maximal web $\sigma(\gamma) := \sigma(\gamma, \mathcal{A}(\gamma))$. Note also that $\partial\gamma$ lies in $\partial\sigma(\gamma, \mathcal{A})$ but not necessarily in $\partial\sigma^+(\gamma, \mathcal{A})$.

29.4. Example. *Spirals.*

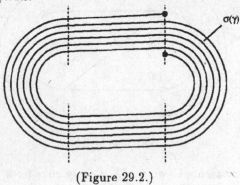

(Figure 29.2.)

The picture shows a spiral. A *spiral* is, by definition, a web $\sigma(\gamma)$ with the property that $\sigma(\gamma)$ is the closure of one component of $\sigma(\gamma) - \gamma$. Spirals are very simple webs. Other webs may be more complicated.

In any case, most of γ is always hidden in the interior $int(\sigma(\gamma))$. This is a crucial property of webs. It implies, among other things, that the number of components of $\gamma - \sigma(\gamma)$ is limited. Here is a more explicit estimate for this fact. The important point being that the estimates do not depend on the length of γ.

29.5. Lemma. *Let γ be a 1-manifold carried by τ. Then*

(1) $\ell_\theta(\gamma - int(\sigma(\gamma)) \le 8 \cdot e(\tau) \cdot \#\gamma$, and

(2) $\sigma(\gamma) - \gamma$ has at most $2 \cdot (e(\tau) + \#\gamma)$ components.

Proof. (1) Let $\gamma' := \gamma - (\sigma(\gamma))$. If $\ell_\theta(\gamma') > 4 \cdot e(\tau) \cdot \#\partial\gamma$, then there is at least one collection of edges from γ' which are all pairwise parallel and whose number is bigger than $3 \cdot \#\partial\gamma$. It follows that at least one of the edges of this collection is the face of two parallelity-regions which do not meet $\partial\gamma$. But such an edge cannot lie in $\gamma' = \gamma - int(\sigma(\gamma))$. So $\ell_\theta(\gamma') \le 4 \cdot e(\tau) \cdot \#\partial\gamma$.

(2) Consider the set $\mathcal{E}$ of all components of $\partial\sigma(\gamma) - \gamma$. Let $\mathcal{E}' \subset \mathcal{E}$ be the subset of all those components which do not meet $\partial\gamma$. Let $k \in \mathcal{E}'$. Then k is an arc (in θ). Let x, y be its end-points. Since x is not an end-point of γ, it must be the common point of a pair of adjacent edges from τ, say l_x and r_x. The same with y. Moreover, the edges l_x, l_y, say, are parallel while the edges r_x, r_y are not. Hence the edges l_x, r_x, l_y, r_y are parallel to some triple, Y_k, of edges from τ which meet in a single point. In fact, the assignment $k \mapsto Y_k$ defines an injection from $\mathcal{E}'$ into the set, T, of all such triples. The estimate (2) therefore follows since $\#(\sigma(\gamma) - \gamma) \le \#(\partial\sigma(\gamma) - \gamma) \le \#\mathcal{E}' + \#\partial\gamma \le \#T + \#\partial\gamma \le 2 \cdot e(\tau) + 2 \cdot \#\gamma$. ◊

29.6. Lemma. *Let γ, γ' be two 1-manifolds carried by τ with $\gamma \subset \gamma'$. Then $\sigma^+(\gamma) \subset \sigma^+(\gamma')$.*

Proof. This is an easy consequence of the definition. ◊

29.7. Lemma. *Let γ be a connected 1-manifold carried by τ. Suppose $\ell_\theta(\gamma) \ge 16 \cdot (e(\tau))^2$. Then $U(\sigma^+(\gamma))$ is an incompressible surface in $U(\tau)$ different from a disc.*

Proof. Since $\ell_\theta(\gamma) \ge 16 \cdot e(\tau) \cdot e(\tau)$, there must be a collection of at least $16 \cdot e(\tau)$ pairwise parallel edges of γ. Let A be the union of the corresponding parallelity-regions. Let A' be the union of all those parallelity-regions in A which do *not* meet $\partial\gamma$. Then A' consists of at most three components since A is connected and since γ is an arc. Moreover, $A' \subset \sigma(\gamma)$. Let A'_0 be a component of A' which contains not fewer parallelity-regions than the other two components of A'. Hence $A'_0 \cap (\sigma(\gamma) - \gamma) = A'_0 - \gamma$ contains more that $4 \cdot e(\tau)$ components. So, by lemma 29.5 (2), at least two of those components, say C_1, C_2, must lie in one component, say B, of $\sigma(\gamma) - \gamma$. Then there is an arc $l_1 \subset B$ which joins C_1 with C_2 and there is an arc $l_2 \subset A'_0$ with $l_1 \cap l_2 = \partial l_1 = \partial l_2$. The union $l_1 \cup l_2$ is a closed curve in $\sigma(\gamma)$ and this curve is not contractible in $U(\tau)$. In particular, $U(\sigma^+(\gamma))$ cannot be a disc. Moreover, $U(\sigma^+(\gamma))$ is not compressible in $U(\tau)$, for, by definition, no component of $(U(\tau) - \sigma^+(\gamma))^-$ is a disc. $\Diamond$

It follows from lemmas 29.6 and 7 that the regular neighborhood $U(\sigma^+(\gamma))$ $\subset U(\tau)$ is not a system of discs, whenever γ is a 1-manifold with a sufficiently long component.

29.8. *Spliced Graphs.*

In order to avoid certain auxiliary complications (see e.g. example 29.10), we find it convenient to splice the graph θ into a graph $\tilde{\theta}$, and to use this spliced graph instead. More specifically, given a graph $\theta \subset S$, we define

$$\tilde{\theta} := (\theta - U_\theta)^- \cup \partial U_\theta,$$

where $U_\theta := U(\theta - U(\theta^{(0)}))^-$ (recall $\theta^{(0)}$ denotes the collection of all vertices of θ). We say $\tilde{\theta}$ is a *spliced graph*. It is the spliced graph associated to θ. Intuitively, we get $\tilde{\theta}$ from θ by splicing all its edges in two. We use the following convention concerning spliced graphs $\tilde{\theta}$. We say $\tilde{\tau}$ is a train-track in $(S, \tilde{\theta})$ if it is a train-track in $(S, \tilde{\theta})$ (as in 29.1) *and* if, in addition, $\tilde{\tau} \cap U_\theta$ is a system of *proper* arcs in U_θ. A 1-manifold, $\tilde{\gamma}$, is carried by $\tilde{\tau}$ in $(S, \tilde{\theta})$ if it is carried by $\tilde{\tau}$ (as in 29.1) *and* if, in addition, $\tilde{\gamma} \cap U_\theta$ is a system of *proper* arcs in U_θ.

The next lemma establishes a crucial property for spliced graphs (which in general is not true for other graphs). To formulate it let $\tilde{\theta} \subset S$ be a spliced graph and let $\tilde{\tau}$ be a train-track in $(S, \tilde{\theta})$. Let $\tilde{\gamma}, \tilde{\gamma}'$ be two 1-manifolds carried by $\tilde{\tau}$. We say $\tilde{\gamma}'$ *lies in a neighborhood of* the web $\sigma(\tilde{\gamma}, A)$, $A \subset \mathcal{A}(\tilde{\gamma})$, if $\tilde{\gamma}'$ can be deformed into the regular neighborhood $U(\sigma(\tilde{\gamma}, A))$, using an isotopy in $U(\tilde{\tau})$ which preserves $\tilde{\theta} \cap U(\tilde{\tau})$.

29.9. Lemma. Let $\tilde{\theta} \subset S$ and $\tilde{\tau}$ be given as above. Let $r \geq 1$ and let $\tilde{\gamma}_1, \tilde{\gamma}_2, ..., \tilde{\gamma}_n$, $n \geq 2r \cdot e(\tilde{\tau})$, be a collection of pairwise disjoint and connected 1-manifolds which are carried by $\tilde{\tau}$. Then there is an index q, $1 \leq q \leq n - r$, such that $\tilde{\gamma}_{q+1}, ..., \tilde{\gamma}_{q+r}$ all lie in a neighborhood of $\sigma(\tilde{\gamma}_1 \cup ... \cup \tilde{\gamma}_q)$ in $U(\tilde{\tau})$.

Proof. Since $\tilde{\theta}$ is a spliced graph, we have the following criterion. Let $\tilde{\gamma}, \tilde{\gamma}'$ be two 1-manifolds carried by $\tilde{\tau}$ with $\tilde{\gamma} \subset \tilde{\gamma}'$. Then $\tilde{\gamma}' \subset U(\sigma(\tilde{\gamma}))$ if and only if every edge from $\tilde{\tau}$ which is parallel to an edge from $\tilde{\gamma}'$ is also parallel to an edge from $\tilde{\gamma}$. Given $\gamma_i^* := \tilde{\gamma}_1 \cup ... \cup \tilde{\gamma}_i$, $1 \leq i \leq n$, denote by $\mathcal{E}(\gamma_i^*)$ the set of all edges of $\tilde{\tau}$ which are parallel to edges of γ_i^*. Then, of course, $\mathcal{E}(\gamma_i^*) \subset \mathcal{E}(\gamma_{i+1}^*)$, i.e., every edge from $\tilde{\tau}$ which is parallel to an edge from γ_i^* is also parallel to an edge from γ_{i+1}^*. Therefore $\#\mathcal{E}(\gamma_i^*) \leq \#\mathcal{E}(\gamma_{i+1}^*)$. Moreover, $\#\mathcal{E}(\gamma_i^*) \leq e(\tilde{\tau})$ and $n \geq 2r \cdot e(\tilde{\tau})$. Thus, by a counting argument, there must be an index q, $1 \leq q \leq n-r$, such that $\#\mathcal{E}(\gamma_q^*) = \#\mathcal{E}(\gamma_{q+1}^*) = ... = \#\mathcal{E}(\gamma_{q+r}^*)$, and so $\mathcal{E}(\gamma_q^*) = \mathcal{E}(\gamma_{q+1}^*) = ... = \mathcal{E}(\gamma_{q+r}^*)$. Then $\tilde{\gamma}_{q+1}, \tilde{\gamma}_{q+2}, ..., \tilde{\gamma}_{q+r} \subset U(\sigma(\gamma_q^*))$, by the above criterion. $\Diamond$

29.10. Example. Fig. 29.3 shows (on the left) an arc-system, γ, with respect to a graph, θ, as well as (on the right) the associated arc-system, $\tilde{\gamma}$, w.r.t. to its associated spliced graph, $\tilde{\theta}$. Note that $\sigma^+(\gamma)$ is a disc and that $\sigma^+(\tilde{\gamma})$ is an annulus. Observe that the outer arc of γ does *not* lie in the neighborhood of the maximal web of the remaining arc-system. This is easily seen to be independent of the number of arcs of γ. It therefore follows that lemma 29.9 is in general false for non-spliced graphs. The use of spliced graphs avoids this and similar problems. In particular, notice that the outer arc of $\tilde{\gamma}$ does indeed always lie in the neighborhood of the maxim

splicing

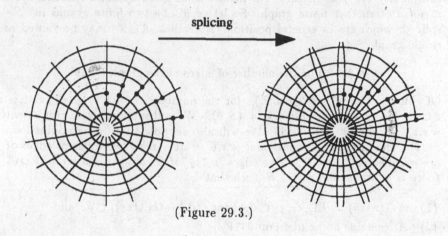

(Figure 29.3.)

29.11. Lemma. *Let $\tilde{\theta} \subset S$ and $\tilde{\tau}$ be given as in lemma 29.9. Let $\tilde{\gamma}$ be a 1-manifold carried by $\tilde{\tau}$. Let $\tilde{\gamma}_0$ be component of $\tilde{\gamma}$ and let k be a component of $(\tilde{\gamma}_0 - \sigma(\tilde{\gamma} - \tilde{\gamma}_0))^-$ with $k \cap \sigma(\tilde{\gamma} - \tilde{\gamma}_0) = \partial k$. Suppose k can be deformed (rel ∂k) into $\sigma(\tilde{\gamma} - \tilde{\gamma}_0)$. Then $\ell_\theta(k) \leq 2 \cdot e(\tilde{\tau})$.*

Proof. By our hypothesis on k, it follows that k lies in a neighborhood U of $\sigma := \sigma(\tilde{\gamma} - \tilde{\gamma}_0)$ in $U(\tilde{\tau})$. Let B be the closure of the component of $U - \sigma$ containing k. Then B is an annulus. W.l.o.g. we suppose B is isotoped rel $B \cap \sigma(\tilde{\gamma} - \tilde{\gamma}_0)$ so that no component of $\tilde{\theta} \cap B$ (resp. $\tilde{\theta} \cap (U(\tau) - B)^-$) is an arc which has both end-points in $\partial B - \sigma$ and separates a disc from B (resp. $U(\tau) - B)^-$). Then, in particular, $\tilde{\theta} \cap B$ is a system of arcs none of which has both end-points in $\partial B - \sigma$. Moreover, k is a proper arc in B with both end-points in one component of ∂B which intersects every arc from $\tilde{\theta} \cap B$ in at most one point since $\tilde{\gamma}_0$ is carried by τ and since ∂k is contained in one component of ∂B. In particular, k intersects at most two arcs from $\tilde{\theta} \cap B$ which are inessential in B. We claim the essential arcs from $\tilde{\theta} \cap B$ lie in components from $\tilde{\theta} \cap (U(\tau) - \sigma)$ which meet $\partial U(\tau)$. To see this let ℓ be a component from $\tilde{\theta} \cap (U(\tau) - \sigma)^-$ with $\partial \ell \subset \sigma$. Using that $\tilde{\theta}$ is a spliced graph, it is easily seen that ℓ must be inessential in the component from $U(\tau) - \sigma$ containing it. Hence ℓ separates a disc from $U(\tau) - \sigma$. Thus it follows from our choice of B that $\ell \cap B$ is either empty or an inessential arc in B. This proves the claim. But $\#(\tilde{\theta} \cap \partial U(\tau)) \leq 2 \cdot e(\bar{\tau})$, and so, altogether, the lemma follows. $\Diamond$

From now on we will use spliced graphs etc. without using special symbols (such as " $\sim$ ") for them.

29.12. *Mixing-Factors and Invariant Surfaces.*

In order to formulate the main result of this section we need to refine the complexity $\ell_\theta(\gamma)$ somewhat. For this let us be given a second reference-system in form of another finite graph. So let θ, θ' be two finite graphs in S, θ spliced, which are in general position. Note that $\theta \cup \theta'$ may be viewed as a single graph. Set

$$e(\theta \cup \theta') := \text{the number of edges of the graph } \theta \cup \theta'.$$

Of course, $v(\theta \cup \theta') \leq e(\theta \cup \theta')$, for the number $v(\theta \cup \theta')$ of all vertices of $\theta \cup \theta'$. Let τ be a train-track in (S, θ). We will suppose that τ is essential w.r.t. θ *and* θ' (see 29.1). We will also suppose that τ is minimal w.r.t. $\theta \cup \theta'$. Here we say τ is *minimal* w.r.t. $\theta \cup \theta'$. if no two different edges of τ are parallel w.r.t. $\theta \cup \theta'$. Two edges e_1, e_2 of τ are *parallel* w.r.t. $\theta \cup \theta'$ if there is a disc (square) $A \subset S$ such that

(1) $A \cap (\tau \cup \theta) = \partial A$, $e_1, e_2 \subset \partial A$ and $(\partial A - (e_1 \cup e_2)) \subset \theta$, and

(2) A contains no point from $\theta \cap \theta'$.

Note that θ and θ' intersect $U(\tau)$ in two disjoint arc-systems. Given the above setting, define the *mixing-factor*, $m(\gamma)$, of a 1-manifold γ, carried by τ, to be

$$m(\gamma) := \text{number of components of } \gamma - \theta \text{ meeting } \theta'.$$

Intuitively, the mixing-factor measures how thoroughly the points from $\gamma \cap \theta$ and $\gamma \cap \theta'$ are mixed. To continue let $\bar{\gamma}$ be a 1-manifold which is carried by τ (and recall from 29.8 our conventions regarding spliced graphs). Let $\gamma_1, ..., \gamma_{n+1}$, $n \geq 1$, be the components of $\bar{\gamma}$. Given $1 \leq q \leq n$, denote

$$\bar{\gamma}_q := \gamma_1 \cup \gamma_2 \cup ... \cup \gamma_q.$$

29.13. Definition. *A <u>web-extension for</u>* $\bar{\gamma}_q$, $1 \leq q \leq n$, *is an embedding* $g : \sigma(\bar{\gamma}_q) \hookrightarrow S$ *such that* $g(\theta \cap \sigma(\bar{\gamma}_q)) \subset \theta'$ *and that, for every* $1 \leq i \leq q$, γ_{i+1} *is an arc in* $g(\gamma_i)$ *with* $\theta \cap g(\gamma_i) \subset \gamma_{i+1}$ *and* $\partial \gamma_{i+1} \subset \theta \cap g(\gamma_i)$.

At this point we remark that it may happen that $m(gk) < m(k)$, for some sub-arc $k \subset \gamma_i \subset \bar{\gamma}$, and that, at the same time, $m(gl) > m(l)$, for some other sub-arc $l \subset \gamma_i$.

Now, suppose $\bar{\gamma}_n$ has a web-extension g (then also every $\bar{\gamma}_q$, $1 \leq q \leq n$, has a web-extension). Recall from lemma 29.6 that $\sigma^+(\bar{\gamma}_n)$ is an incompressible surface if $\bar{\gamma}_n$ is large enough. But note that this surface need not be invariant under g. In general, g does not map $\sigma(\bar{\gamma}_n)$ to itself. In fact, $g(\sigma(\bar{\gamma}_n))$ may not even lie in $U(\tau)$. Of course, $g(\bar{\gamma}_n)$ is always contained in $U(\tau)$ (since $\bar{\gamma}_{n+1}$ is carried by τ) but $g(\sigma(\bar{\gamma}_n) - \bar{\gamma}_n)$ need not lie in $U(\tau)$. Actually, we may not have any control over $g|\sigma(\bar{\gamma}_n)$ whatsoever. This however changes if size and mixing-factor of $\bar{\gamma}$ gets larger and indeed the situation looks much better if they are large enough. This is the content of our next theorem. It provides us with a set of reasonable criteria, involving size and mixing-factors, under which there is *some* index q and *some* web for $\bar{\gamma}_q$ which is mapped to itself under g, mod isotopy (it is reminiscent of the observation that an individual point of a set is not necessarily invariant under a bijection of this set but that its orbit always is).

29.14. Theorem. *Let* $\theta, \theta', \tau \subset S$ *be given as above. Suppose* τ *is essential w.r.t.* θ *and* θ', *and minimal w.r.t.* $\theta \cup \theta'$. *Then there are numbers* $a, b, c \in \mathbf{N}$, *depending on* $e(\tau)$ *and* $e(\theta \cup \theta')$ *alone, such that the following holds. Suppose* $\bar{\gamma} = \gamma_1, ..., \gamma_{n+1}$ *is a 1-manifold carried by* τ *and* $g : \sigma(\bar{\gamma}_n) \hookrightarrow S$ *is a web-extension. Suppose* $\#\bar{\gamma} \geq a$ *and* $m(\gamma_q) \geq b$, *for every* $1 \leq q \leq a$.

Then there is an index, q, $1 \leq q \leq a$, *and a collection* $\mathcal{A} \subset A(\bar{\gamma}_q)$ *(possibly empty) with* $\#(\sigma(\bar{\gamma}_q, \mathcal{A}) - \bar{\gamma}_q) \leq \#(\sigma(\bar{\gamma}_q) - \bar{\gamma}_q)$ *and the following two properties:*

(1) *There is a collection* $\mathcal{P} \subset \bar{\gamma}'_q := \bar{\gamma}_q - int(\sigma(\bar{\gamma}_q, \mathcal{A}))$ *of points with* $\partial\bar{\gamma}'_q \subset \mathcal{P}$, $\#\mathcal{P} \leq c$ *and, for every arc* $l \subset \bar{\gamma}_q$ *with* $l \cap \mathcal{P} = \partial l$, *there is an* i, $1 \leq i \leq b$, *with* $m(g^i l) \leq c$.

(2) *There is an isotopy* $g_t : \sigma(\bar{\gamma}_n) \to S$, $t \in I$, *of a web-extension* $g = g_0$ *for* $\bar{\gamma}_n$ *such that* $g_1 U(\sigma^+(\bar{\gamma}_q, \mathcal{A})) = U(\sigma^+(\bar{\gamma}_q, \mathcal{A}))$.

Proof. The main problem in this proof is to control "bands". By definition, a band of a web $\sigma(\gamma, \mathcal{A})$, $\mathcal{A} \subset A(\bar{\gamma}_q)$, is a component of $\sigma(\gamma, \mathcal{A}) - \gamma$. Note that a band is a union of elements from $\mathcal{A}$. Note further that any band B for $\sigma(\gamma, \mathcal{A})$ has a natural product-fibration $B = b \times int(I)$ (where b is an arc) with $(b \times I) \cap \gamma = b \times \partial I \subset \gamma$. We say that $b \times 0$, $b \times 1$ are the *sides* of the band B. Note that the sides of B are two arcs in γ and that these two arcs may intersect themselves. Denote by $\mathcal{B}(\gamma)$ the *set* of all components of $\sigma(\gamma) - \gamma$, i.e, $\mathcal{B}(\gamma)$ is the set of all bands for $\sigma(\gamma)$. Given $\mathcal{B} \subset \mathcal{B}(\gamma)$, we set

$$\sigma(\gamma, \mathcal{B}) := \gamma_q \cup \bigcup_{B \in \mathcal{B}} B.$$

Note that $\sigma(\gamma) = \sigma(\gamma, \mathcal{B}(\gamma))$ (recall from 29.3 the definition of $\sigma(\gamma)$).

Throughout this proof we require $\partial\bar{\gamma}_{q+1} \cap int(\sigma(\bar{\gamma}_q)) = \emptyset$, for all $1 \leq q \leq n$. This convention is made for convenience, because it ensures that $\bar{\gamma}_{q+1} \cap \sigma(\bar{\gamma}_q)$, $1 \leq q \leq n$, is a system of proper arcs in $\sigma(\bar{\gamma}_q)$ (recall θ is spliced and see 29.8 for our conventions regarding spliced graphs). Now, the requirement itself can be made without loss of generality, for we can always realize it by stretching the arc-system $\bar{\gamma}$ somewhat if necessary. Here is the construction. Suppose q is the first index for which $\partial\bar{\gamma}_{q+1} \cap int(\sigma(\bar{\gamma}_q)) \neq \emptyset$. Then, in particular, there is a band $B \in \mathcal{B}(\bar{\gamma}_q)$ with $\partial\bar{\gamma}_{q+1} \cap int(B) \neq \emptyset$. Set $z_B := \partial\bar{\gamma}_{q+1} \cap int(B)$ and let $\mathcal{L}_B$ be a collection of pairwise disjoint arcs in B with $\mathcal{L}_B \cap \bar{\gamma}_q = z_B$ which joins the points from z_B with the components from $\partial B_i - \bar{\gamma}_{q+1}$. Clearly, $\mathcal{L}_B$ can be chosen (see fig. 29.4) so that $\bar{\gamma}'_{q+1} := \bar{\gamma}_{q+1} \cup \bigcup_{B \in \mathcal{B}} \mathcal{L}_B$ is an arc-system carried by τ with $\partial\bar{\gamma}'_{q+1} \cap int(\sigma(\bar{\gamma}'_q)) = \emptyset$ (here $\bar{\gamma}'_q$ denotes the union of all arcs from $\bar{\gamma}'_{q+1}$ containing an arc from $\bar{\gamma}_q$). Notice further that every arc $l \in \mathcal{L}_B$, $B \in \mathcal{B}$, lies next to an arc l' from $\bar{\gamma}_q$. So, using $g|l'$, for all $l \in L_B$ and $B \in \mathcal{B}$, we can extend $g(\bar{\gamma}_{q+1}) = \bar{\gamma}_{q+2}$ to an arc-system γ_{q+2} carried by τ, and we can extend the homeomorphism $g|\bar{\gamma}_{q+1} : \bar{\gamma}_{q+1} \to \bar{\gamma}_{q+2}$ to a homeomorphism from $\bar{\gamma}'_{q+1}$ to $\bar{\gamma}'_{q+2}$. In the same way we extend $\bar{\gamma}'_{q+3}$ and $g|\bar{\gamma}'_2$ etc. This process eventually yields an arc-system $\bar{\gamma}^*$ which has the same property as $\bar{\gamma}$ but for which the first index q^* with $\partial\bar{\gamma}_{q^*+1} \cap int(\sigma(\bar{\gamma}_q^*)) \neq \emptyset$ is strictly larger than q. So, recursively, we get the desired arc-system.

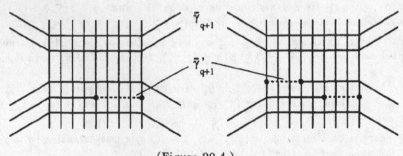

(Figure 29.4.)

We wish to find an index q for which there is a collection $\mathcal{B} \subset \mathcal{B}(\bar{\gamma}_q)$ of bands such that $\sigma(\bar{\gamma}_q, \mathcal{B})$ satisfies the conclusion of the theorem. Here is our strategy. First, recall from lemma 29.9 that there must be an index p with $g(\bar{\gamma}_p) \subset U(\sigma(\bar{\gamma}_p))$. But g may not map all bands for $\bar{\gamma}_p$ into $U(\sigma(\bar{\gamma}_p))$. We try to fix this by removing all those bands which are not mapped into $U(\sigma(\bar{\gamma}_p))$. But then we may be left with a subset $\mathcal{B} \subset \mathcal{B}(\bar{\gamma}_p)$ such that $g(\bar{\gamma}_p)$ is no longer contained in $U(\sigma(\bar{\gamma}_p, \mathcal{B}))$ (see fig. 29.6 for two examples). In that case, we apply the same procedure to the next larger 1-manifold $\bar{\gamma}_{p+1}$, and then to $\bar{\gamma}_{p+2}$, etc. It turns out that if we are careful in removing bands and if we carry out this process sufficiently often, then the process eventually stops at the desired index.

For every index q, $q \geq p$, the last arc $\gamma_{q+1} = g(\gamma_q)$ is critical in this strategy. For instance, the above process is halted whenever a collection $\mathcal{B} \subset \mathcal{B}(\bar{\gamma}_q)$ is reached for which γ_{q+1} does not lie in $U(\sigma(\bar{\gamma}_q, \mathcal{B}))$. To say more, we need to distinguish between proper and non-proper components of $(\gamma_{q+1} - \sigma(\bar{\gamma}_q, \mathcal{B}))^-$, where as usual a component of $(\gamma_{q+1} - \sigma(\bar{\gamma}_q, \mathcal{B}))^-$ is *proper* if and only if both its end-points lie in $\sigma(\bar{\gamma}_q, \mathcal{B})$. Note that every non-proper component contains an end-point of γ_{q+1}, i.e., all but at most two components are proper. We write

$$\gamma_{q+1} \sqsubset \sigma(\bar{\gamma}_q, \mathcal{B}) \quad \text{resp.} \quad \gamma_{q+1} \prec \sigma(\bar{\gamma}_q, \mathcal{B})$$

if every component, resp. every proper component, of $(\gamma_{q+1} - \sigma(\bar{\gamma}_q, \mathcal{B}))^-$ can be pushed in $U(\tau)$ into $\sigma(\bar{\gamma}_q, \mathcal{B})$, using a homotopy which fixes all end-points contained in $\sigma(\bar{\gamma}_q, \mathcal{B})$ and which preserves $U(\tau) \cap (\theta \cup \theta')$. Of course, $\gamma_{q+1} \prec \sigma(\bar{\gamma}_q)$ if $\gamma_{q+1} \sqsubset \sigma(\bar{\gamma}_q)$. Both notations are meant to indicate that γ_{q+1} is "almost" contained in the regular neighborhood $U(\sigma(\bar{\gamma}_q, \mathcal{B}))$. Indeed, if $\gamma_{q+1} \sqsubset \sigma(\bar{\gamma}_q, \mathcal{B})$ (resp. $\gamma_{q+1} \prec \sigma(\bar{\gamma}_q, \mathcal{B})$) then every component (resp. every proper component) of $(\gamma_{q+1} - \sigma(\bar{\gamma}_q, \mathcal{B}))^-$ can be pushed into $U(\sigma(\bar{\gamma}_q, \mathcal{B}))$, using an isotopy in $U(\tau)$ which fixes points in $\sigma(\bar{\gamma}_q, \mathcal{B})$ *and* which preserves the arcs from $(\theta \cup \theta') \cap U(\tau)$. This is easily verified since θ is supposed to be spliced. Also, every non-proper component, meeting $\sigma(\bar{\gamma}_q, \mathcal{B})$, can be contracted in itself into $U(\sigma(\bar{\gamma}_q, \mathcal{B}))$, but unfortunately we will not be allowed to use the latter operation.

To proceed set, for every $r \geq 1$,

$$p = p(r) := \text{the smallest index } p \text{ with } \gamma_{p+1}, \gamma_{p+2}, ..., \gamma_{p+r+1} \sqsubset \sigma(\bar{\gamma}_p).$$

Recall from lemma 29.9, that $p \leq 2(r + 2) \cdot e(\tau)$. By definition of p, we also have $\gamma_{p+i} \prec \sigma(\bar{\gamma}_p)$, for all $1 \leq i \leq r$. Hence, by lemma 29.11, the length of every proper component from $\gamma_{q+1} - int(\sigma(\bar{\gamma}_q))$ is limited, for every $p \leq q \leq p + r$. In contrast, we have no control over the lengths of non-proper components. Thus non-proper components may or may not be very long, and so may or may not have large mixing-factors. We will have to distinguish between these two possibilities. To allow some extra flexibility we introduce the following notation.

We say a 1-manifold $\gamma \subset \bar{\gamma}_n$ has *order* (or: an *arc-decomposition of order*) (c, d) if there is a collection $\mathcal{P}$ of points in γ with $\partial\gamma \subset \mathcal{P}$ and $\#\mathcal{P} \leq c$ such that, for every arc $l \subset \gamma$ with $l \cap \mathcal{P} = \partial l$, there is an integer i, $0 \leq i \leq d$, with $m(g^i l) \leq d$.

Set $r := 100 \cdot (e(\tau))^{100}$, and let $c_0 = d_0 = 16 \cdot (p(r) + r) \cdot e(\tau) \cdot e(\theta \cup \theta')$ (the actual values are irrelevant, we only need constants which are sufficiently large and depend on $e(\tau)$ and $e(\theta \cup \theta')$ alone). Now, according to lemma 29.5, every component from $\bar{\gamma}_q - int(\sigma(\bar{\gamma}_q))$ has order (c_0, d_0), but this need not be true for $\gamma_{q+1} - int(\sigma(\bar{\gamma}_q))$. We say an index q, $1 \leq q \leq n$, is *good* if every component from $\gamma_{q+1} - int(\sigma(\bar{\gamma}_q))$ has order (c_0, d_0). Otherwise it is *bad*. (In the following, we view (c_0, d_0) as a small order since it depends only on $e(\tau)$ and $e(\theta \cup \theta')$, but not on $\bar{\gamma}$).

Case 1. All indices q, $p(r) \leq q \leq p(r) + r$, are good.

In this Case we proceed in three steps:

Step 1. Given $\mathcal{B} \subset \mathcal{B}(\bar{\gamma}_q)$, we set

$$Good(\mathcal{B}) := \{ B \in \mathcal{B} \mid \text{there is a component } k \text{ from } \theta \cap B$$
$$\text{with } g(k) \subset \sigma(\bar{\gamma}_q, \mathcal{B}) \}.$$

29.15. Lemma. *Suppose* q, $p(r) \leq q < q(r) + r$, *is an index such that* $\bar{\gamma}_{q+1} - int(\sigma(\bar{\gamma}_q, \mathcal{B}))$ *has order* (c, d). *Then* $\bar{\gamma}_{q+1} - int(\sigma(\bar{\gamma}_q, Good(\mathcal{B})))$ *has order* (c', d'), *where* $c' := c + (8c + 8) \cdot e(\theta \cup \theta') \cdot \#\mathcal{B}$ *and* $d' := d + 1$.

Proof of lemma. Let $B \in \mathcal{B} - Good(\mathcal{B})$ and let s be one of the two sides of the band B (see above for definition). The lemma follows when we show that s (and so also the other side of B) has an arc-decomposition of order $((2c + 2) \cdot e(\theta \cup \theta'), d + 1)$. Now, $g(s) \subset \bar{\gamma}_{q+1}$ since $g(\bar{\gamma}_q) \subset \bar{\gamma}_{q+1}$. Moreover, g is an embedding. Thus, pulling back an arc-decomposition of $g(s)$ to s, we see that it suffices to show that $g(s)$ has an arc-decomposition of order $((4c + 4) \cdot e(\theta \cup \theta'), d)$.

To verify this claim consider the image $g(B)$ (see fig. 29.5). Set $\mathcal{G} := \theta \cap gB$ and $\mathcal{G}' := \theta' \cap gB$. Then $\mathcal{G}, \mathcal{G}'$ are two proper graphs in gB whose intersection may well be non-empty (recall gB need not lie in $U(\tau)$). Let $\mathcal{K}, \mathcal{K}'$ be the collection of all components of $\mathcal{G}, \mathcal{G}'$, resp., which are arcs. Then $\mathcal{K}, \mathcal{K}'$ are two arc-systems in gB which join $g(s)$ with $g(s_2)$ (recall τ is supposed to be essential w.r.t. θ and θ'). Let $\mathcal{L}' \subset \mathcal{K}'$ be the union of all those arcs from $\mathcal{K}'$ which do *not* meet $\mathcal{G}$. Then $\mathcal{L}'$ splits gB into a collection of squares. Let C denote the union of all those squares from this collection which do *not* contain a vertex of $\theta \cup \theta'$. Then C has at most $1 + v(\theta \cup \theta') \leq e(\theta \cup \theta')$ components since every component from $gB - C$ contains at least one vertex of $\theta \cup \theta'$. Moreover, $g(s) - C$ contains at most $4 \cdot e(\theta \cup \theta')$ end-points of $\mathcal{G}'$ since there are no two adjacent arcs from $\mathcal{K}'$ which are disjoint to C without meeting $\mathcal{G}$. Thus, in particular, the mixing-factor of $g(s) - C$ is at most $4 \cdot e(\theta \cup \theta')$. Hence it remains to show that every component of $C \cap g(s)$ has an arc-decomposition of order $(4c, d)$.

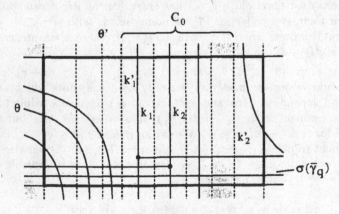

(Figure 29.5.)

Let C_0 be a component of C and set $s_0 := C_0 \cap g(s)$. Then C_0 is a square and s_0 is one of its faces. It remains to show that s_0 has order $(4c, d)$.

Suppose $\bar{\gamma}_q \cap int(C_0) = \emptyset$. Then $g(s) \cap C_0 \subset \bar{\gamma}_{q+1} - int(\sigma(\bar{\gamma}_q, \mathcal{B}))$ (recall $B \notin Good(\mathcal{B})$), and so the claim follows directly from the hypothesis that every component of $\bar{\gamma}_{q+1} - int(\sigma(\bar{\gamma}_q, \mathcal{B}))$ has an arc-decomposition of order (c, d).

Suppose $\bar{\gamma}_q \cap int(C_0) \neq \emptyset$. Then recall $\bar{\gamma}_q \cap B = \emptyset$ and $\gamma_{i+1} \subset g(\gamma_i)$, for all $1 \leq i \leq n - 1$ (see def. 29.13). Hence $\gamma_i \cap gB = \emptyset$, for all $2 \leq i \leq q + 1$. So $\bar{\gamma}_q \cap gB = \gamma_1 \cap gB$ and, in particular, $\bar{\gamma}_q \cap C_0 = \gamma_1 \cap C_0$.

If $C_0 \cap \partial \gamma_1 = \emptyset$, then every component from $C_0 \cap \gamma_1$ is an arc which joins the two components from $\mathcal{K}'$ contained in ∂C_0. Since τ is supposed to be minimal w.r.t. $\theta \cup \theta'$, it follows that the maximal number of pairwise parallel edges of τ in C_0 is two. Moreover, no arc from $g(\theta \cap B) \subset \theta' \cap gB$ lies in $\sigma(\bar{\gamma}_q, \mathcal{B})$ (recall $B \notin Good(\mathcal{B})$). Altogether, we may therefore conclude that at least one component of $\gamma_1 \cap C_0$ is an arc in C_0 which joins the two components of $\partial C - (s_1 \cup s_2)$ *and* which lies in $\bar{\sigma}_q - int(\sigma(\bar{\sigma}_q, \mathcal{B}))$. Thus, by hypothesis, this particular component has an arc-decomposition of order (c, d). But any component from $C_0 \cap \gamma_1$ is parallel in C_0 to s_0. It therefore follows that s_0 has an arc-decomposition of order (c, d).

If $C_0 \cap \partial \gamma_1 \neq \emptyset$, denote by K_0 the union of all those components from $C_0 \cap \mathcal{K}$ containing an end-point of $\partial \gamma_1$. Then K_0 consists of at most two arcs. Let k be one of them. Let k_1', k_2' be the two components of $\mathcal{K}'$ lying in ∂C_0. Note that, for every index $i = 1, 2$, there is an arc $l_i(k) \subset \gamma_1$ joining k with k_i'. Exchanging k for the other arc from K_0 if necessary, we may suppose w.l.o.g. that both arcs $l_1(k), l_2(k)$ lie in $\bar{\gamma}_q - int(\sigma(\bar{\gamma}_q, \mathcal{B}))$ and in the same component of $\sigma(\bar{\gamma}_q, \mathcal{B})$ containing s_0 (fig. 29.4 illustrates a typical situation). Thus, by hypothesis $l_i(k)$ has order (c, d), for $i = 1, 2$. Now, $l_i(k)$, $i = 1, 2$, is parallel to an arc, say t_i, in s_0 via some parallelity-regions in $\sigma(\bar{\gamma}_q, \mathcal{B})$. Hence $l_i(k)$ and t_i have arc-decompositions of the same order. Therefore and since $s_0 = t_1 \cup t_2$, it follows that s_0 has an arc-decomposition of order $(2c, d)$. The lemma is henceforth established.

The preceding lemma tells us that $Good(..)$ is a selection-process which still allows the control of arc-decompositions. The next step shows how to make further progress by iterating this selection-process.

Step 2. Given $\mathcal{B} \subset \mathcal{B}(\bar{\gamma}_q)$, define

$$Good^0(q) := \mathcal{B}(\bar{\gamma}_q) \quad \text{and} \quad Good^{i+1}(q) := Good(Good^i(q)), \ i \geq 0.$$

Using this construction, set

$$RB(q) := Good^{r(q)}(q),$$

where $r(q) := \min \{ i \in \mathbf{N} \mid Good^{i+1}(q) = Good^i(q) \}$.

We are looking for an index q, $p \leq q \leq p + r$, with $\gamma_{q+1} \prec \sigma(\bar{\gamma}_q, R\mathcal{B}(q))$. As mentioned before, this is *not* an automatic consequence of the fact that $\gamma_{q+1} \prec \sigma(\bar{\gamma}_q)$, $p \leq q \leq p + r$. Fig. 29.6 indicates two typical situations in which the property $\gamma_{q+1} \prec \sigma(\bar{\gamma}_q, \mathcal{B})$ will be lost after removing a single band:

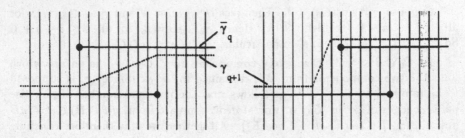

(Figure 29.6.)

The next lemma provides the necessary tool for finding the desired index.

29.16. Lemma. *Suppose* q, $p(r) \leq q < p(r) + r$, *is an index with* $\gamma_{q+1} \not\prec \sigma(\bar{\gamma}_q, R\mathcal{B}(q))$. *Then* $\chi\sigma^+(\bar{\gamma}_{q+1}, R\mathcal{B}(q+1)) < \chi\sigma^+(\bar{\gamma}_q, R\mathcal{B}(q))$.

Proof of lemma. Since $\gamma_{q+1} \not\prec \sigma(\bar{\gamma}_q, R\mathcal{B}(q))$, there is, by definition, at least one proper component of $(\gamma_{q+1} - \sigma(\bar{\gamma}_q, R\mathcal{B}(q)))^-$ which cannot be pushed into $\sigma(\bar{\gamma}_q, R(\mathcal{B}(q))$, using a homotopy in $U(\tau)$ fixing end-points. It therefore follows that $\chi(\gamma_{q+1} \cup \sigma^+(\bar{\gamma}_q, R\mathcal{B}(q))) < \chi\sigma^+(\bar{\gamma}_q, R\mathcal{B}(q))$. Thus the lemma follows when we show that $\gamma_{q+1} \cup \sigma^+(\bar{\gamma}_q, R\mathcal{B}(q)) \subset \sigma^+(\bar{\gamma}_{q+1}, R\mathcal{B}(q+1))$. But $\gamma_{q+1} \subset \sigma(\bar{\gamma}_{q+1}, R\mathcal{B}(q+1))$. Hence it remains to show that $B \in R\mathcal{B}(q) \Rightarrow B \subset \sigma^+(\bar{\gamma}_{q+1}, R\mathcal{B}(q+1))$. Of course, this follows when we show that, for all $1 \leq i \leq r(q)$:

$$B \in Good^i(q) \;\Rightarrow\; B \subset \sigma^+(\bar{\gamma}_{q+1}, Good^i(q+1)). \qquad (*)$$

We prove the latter statement by induction on the exponent i.

First note that $B \subset \sigma^+(\bar{\gamma}_{q+1})$, for every $B \in \mathcal{B}(\bar{\gamma}_q)$. This establishes the induction beginning $i = 0$. To carry out the induction step fix an index i. Suppose $(*)$ holds for $i \geq 1$, and let $B \in Good^{i+1}(q)$.

Now, $\gamma_{q+1} \cap B$ is either empty or a system of arcs joining the two components of $\partial B - \gamma_q$ (recall our convention on $\bar{\gamma}$ made at the beginning). In particular, γ_{q+1} splits B into a collection of bands for $\bar{\gamma}_{q+1}$. Let B_0 be any one of them. Then there is a unique $C_0 \in \mathcal{B}(\bar{\gamma}_{q+1})$ with $B_0 \subset C_0$. To show that $B \subset \sigma^+(\bar{\gamma}_{q+1}, Good^{i+1}(q+1))$, it now suffices to verify that $C_0 \in Good^{i+1}(q+1)$. But $B \in Good^{i+1}(q) = Good(Good^i(q))$. So there is a component, l, of $\theta \cap B$ with $l' := g(l) \subset \sigma(\bar{\gamma}_q, Good^i(q))$. More precisely, γ_q splits l' into arcs each of which lies in a band from $Good^i(q)$. By induction, every band from $Good^i(q)$ is contained in $\sigma^+(\bar{\gamma}_{q+1}, Good^i(q+1))$. Thus, setting $l_0 := l \cap C_0$, we have that $g(l_0)$ lies in $\sigma^+(\bar{\gamma}_{q+1}, Good^i(q+1))$. Hence, by definition, $C_0 \in Good(Good^i(q+1)) = Good^{i+1}(q+1)$. Thus the induction step is complete.

The lemma is therefore established.

29.17. Lemma. *Suppose* q, $p(r) \leq q < p(r) + r$, *is an index with* $\gamma_{q+1} \prec \sigma(\bar{\gamma}_q, R\mathcal{B}(q))$. *Then the web-extension* g *can be isotoped so that afterwards*

$$g(U\sigma(\bar{\gamma}_q, R\mathcal{B}(q))) = U\sigma(\bar{\gamma}_q, R\mathcal{B}(q)).$$

Proof of lemma. Let $B \in R\mathcal{B}(q)$. Then $B \in Good(R\mathcal{B}(q))$ since $R\mathcal{B}(q) = Good(R\mathcal{B}(q))$. Hence, by definition, there is a component, say k_B, of $\theta \cap B$ such that $g(k_B)$ is entirely contained in $\sigma(\bar{\gamma}_q, R\mathcal{B}(q))$. Set

$$K_q := \bigcup_B k_B,$$

where the union is taken over all $B \in R\mathcal{B}(q)$. Note that $g(K_q) \subset \sigma(\bar{\gamma}_q, R\mathcal{B}(q))$ (by construction) and that $\gamma_{q+1} \prec \sigma(\bar{\gamma}_q, R\mathcal{B}(q))$ (by hypothesis). It follows that $g|\bar{\gamma}_q \cup K_q$ can be deformed into $\sigma(\bar{\gamma}_q, R\mathcal{B}(q))$, using an isotopy which fixes K_q (see def. 29.13 for our convention regarding γ_{q+1} and $g(\gamma_q)$). Moreover, k_B joins the two sides of B, for every band $B \in R\mathcal{B}(q)$. So $\bar{\gamma}_q \cup K_q$ is a deformation retract of $\sigma(\bar{\gamma}_q, R\mathcal{B}(q))$. Hence the restriction $g|U\sigma(\bar{\gamma}_q, R\mathcal{B}(q))$ can be deformed into a map $h : U\sigma(\bar{\gamma}_q, R\mathcal{B}(q)) \rightarrow U\sigma^+(\bar{\gamma}_q, R\mathcal{B}(q))$. Moreover, h can be extended to a map $h^+ : U\sigma^+(\bar{\gamma}_q, R\mathcal{B}(q)) \rightarrow U\sigma^+(\bar{\gamma}_q, R\mathcal{B}(q))$ since $U\sigma^+(\bar{\gamma}_q, R\mathcal{B}(q))$ is incompressible in $U(\tau)$. But h^+ is homotopic to an embedding into $U(\tau)$. Using again the fact that $U\sigma^+(\bar{\gamma}_q, R\mathcal{B}(q))$ is incompressible in $U(\tau)$, it follows that h^+ can be deformed in $U\sigma^+(\bar{\gamma}_q, R\mathcal{B}(q))$ into an embedding. Thus w.l.o.g. h^+ is an embedding. So it is a homeomorphism onto its image. Comparing Euler characteristics, we find that the complement of the image of h^+ in $U\sigma^+(\bar{\gamma}_q, R\mathcal{B}(q))$ must be a collection of annuli. Hence h^+ can be deformed into an embedding onto $U\sigma^+(\bar{\gamma}_q, R\mathcal{B}(q))$. This proves the lemma.

We now finish the proof in Case 1 as follows.

Step 3. By lemma 29.16 and since $\chi U(\tau) \leq \chi\sigma^+(\bar{\gamma}_m, R\mathcal{B}(m))$, for all m, it follows that there is an index q, $p(r) \leq q \leq p(r) + r$, such that $\gamma_{q+1} \prec \sigma(\bar{\gamma}_q, R\mathcal{B}(q))$. Moreover, note that $R\mathcal{B}(q) \subset \mathcal{B}(\bar{\gamma}(q))$. Thus there is a subset $\mathcal{A} \subset \mathcal{A}(\bar{\gamma}_q)$ with

$$\sigma(\bar{\gamma}_q, \mathcal{A}) = \sigma(\bar{\gamma}_q, R\mathcal{B}(q)).$$

It remains to show that q and $\mathcal{A}$ satisfy the conclusion of theorem 29.14. To show this recall that $p(r) + r$ is a number which depends on $e(\tau)$ alone. Thus, by lemma 29.5, the number of components of $\sigma(\bar{\gamma}_q) - \bar{\gamma}_q$ is limited by some number depending on $e(\tau)$ and $q \leq p(r) + r$, and so on $e(\tau)$ alone. Moreover, it follows from lemma 29.5 again, that every component of $\bar{\gamma}_{q_1} - int(\sigma(\bar{\gamma}_{q_1}))$ has order which depends on $e(\tau)$ alone. Now, we are in Case 1. Hence an induction involving lemma 29.15 and 29.16 reveals that $\bar{\gamma}_q - int(\sigma(\bar{\gamma}_q, A))$ has an arc-decomposition of order (c, d), where c, d are numbers depending on $e(\tau)$ and $e(\theta \cup \theta')$ alone. Thus $\sigma(\bar{\gamma}_q, A)$ satisfies (1) of theorem 29.14. By lemma 29.17, it also satisfies (2). Hence, altogether, the theorem is established in Case 1.

Case 2. There is at least one index q, $p(r) \le q \le p(r) + r$, which is bad.

Again we proceed in three steps.

Step 1. Suppose q, $p(r) \le q \le p(r) + r$, is a bad index. Then, by definition, there is a component of $(\gamma_{q+1} - \sigma(\bar{\gamma}_q))^-$ which is *bad* in the sense that its order is not (c_0, d_0). It follows from lemmas 29.5 and 29.11 (see also our definition of $p(r)$), that this component is non-proper, and so it contains at least one end-point of $\partial\gamma_{q+1}$ (which in turn is not contained in $\sigma(\bar{\gamma}_q)$). Hence there are at most two such components. For convenience, we assume that there is only one (the proof in the other case is similar). Let

$$l = l(q) := \text{the component of } (\gamma_{q+1} - int(\sigma\bar{\gamma}_q))^- \text{ whose order is not } (c_0, d_0).$$

Roughly speaking, we may view $m(g^i l(q))$ as large, for a large number of exponents i (see definition of arc-decompositions). So $l(q)$ is a long arc in γ_{q+1} which lies outside of $\sigma(\bar{\gamma}_q)$ (and which remains long under a large number of iterations of the map g). But $p \le q \le p + r$, and so we may suppose $\gamma_{q+1} \subset U(\sigma(\bar{\gamma}_q))$, where $U(\sigma(\bar{\gamma}_q))$ denotes the regular neighbourhood of $\sigma(\bar{\gamma}_q)$ (see our choice of p). Hence $l \subset (U(\sigma(\bar{\gamma}_q)) - \sigma(\bar{\gamma}_q))^-$. Now, $(U(\sigma(\bar{\gamma}_q)) - \sigma(\bar{\gamma}_q))^-$ is a system of annuli intersecting $\theta \cup \theta'$ in arcs joining both boundary curves. It follows that $\sigma(l)$ must be a spiral.

Recall from 29.4 that an arc $l \subset \bar{\gamma}$ is a spiral if there is a band B in $\sigma(l)$ with $\sigma(l) = B^-$ (= closure of B). Given $1 \le j \le r$, set

$$l_j = l_j(q) := \text{the arc } l_j \text{ in } g^{j-1}(l) \text{ with } l_j \subset g^{j-1}(l) \cap \theta \text{ and } \partial l_j \subset \theta.$$

It is not hard to check that $l_1, ..., l_r$ are all spirals (since $l_1 = l$ is a long spiral). So, in particular, $\sigma(l_1), \sigma(l_1), ..., \sigma(l_r)$ is a collection of annuli (since $\ell_\theta(l_i) > e(\tau)$). All these annuli lie in $U(\sigma(\bar{\gamma}_q)) \subset U(\tau)$ (see our definition of $p(r)$). But other than that we have little control over them. In particular, we do not know anything about the way they are embedded in $U(\sigma(\bar{\gamma}_q))$. However, we can make the following observation.

29.18. Lemma. *If $i \ne j$, then the annuli $\sigma(l_i)$ and $\sigma(l_j)$ are disjoint modulo isotopy.*

Proof of lemma. Assume the converse. Then $\sigma(l_j)$ cannot be isotoped out of $\sigma(l_i)$. Let s_j be a component of $\partial\sigma(l_j)$. Then s_j cannot be isotoped out of $\sigma(l_i)$. In particular, there must be a component s'_j of $s_j \cap \sigma(l_i)$ which joins the two boundary-components of the annulus $\sigma(l_i)$. Now, s'_j is the union of a sub-arc l'_j of l_j with a sub-arc t_j of θ. Of course, $t_j \cap l'_j = \partial t_j = \partial l'_j$ and $l'_j \cap l_i = \emptyset$ (l_i and l_j lie in different components of $\bar{\gamma}_n$). Hence and since $\sigma(l_i), \sigma(l_j)$ are spirals, it easily follows that $t_j \cap l_i$ consists of at most one point, and so $\ell_\theta(l_i) \le \ell_\theta(l')$. But $\ell_\theta(l'_j) \le 2 \cdot e(\tau)$ since no more than two edges of l'_j can be simultaneously parallel (recall $l'_j \subset \partial\sigma(l_j)$). Thus, altogether, $\ell_\theta(l_i) \le 2 \cdot e(\tau)$. This, however, contradicts the fact (see above) that $\ell_\theta(g^i l) > d_0$, for all $0 \le i \le r - 1$. The lemma is henceforth established.

Recall that no system of pairwise non-parallel, simple closed curve in a surface S can have more than $\frac{3}{2} \cdot |\chi S|$ components. So, in particular, $|\chi U(\tau)|$ has at most $3 \cdot e(\tau)$ pairwise disjoint curves. It therefore follows from lemma 29.18 the existence of an index s such that $\sigma(l_s)$ is isotopic to an annulus from $\sigma(l_0), \sigma(l_1), ..., \sigma(l_s)$. Let

$$s = s(q) := \left\{ \begin{array}{l} \text{the smallest index } s \text{ for which the annulus } \sigma(l_{s+1}) \\ \text{is isotopic to an annulus from } \sigma(l_1), \sigma(l_2), ..., \sigma(l_s). \end{array} \right.$$

Intuitively, the collection $\sigma(l_0), \sigma(l_1), ..., \sigma(l_s)$ forms a cycle of pairwise disjoint annuli. This whole cycle is invariant under g mod isotopy, although the individual members are not (unless $s = 0$). We next add this cycle to $\sigma(\bar{\gamma}_q)$ in order to control the troublesome component from $\gamma_{q+1} - int(\sigma(\gamma_q))$, i.e., that one which does not have order (c_0, d_0).

More precisely, we use the above cycle of annuli to adjust the definitions from Case 1 as follows: For every index q, $p(r) \le q \le p(r) + r$, and every collection $\mathcal{B} \subset \mathcal{B}(\bar{\gamma}_q)$, we set

$$Good(\mathcal{B}) := \{ B \in \mathcal{B} \mid \text{there is a component } k \text{ from } \theta \cap B$$
$$\text{with } g(k) \subset \sigma(\bar{\gamma}_{q+s} \cup l_{s+1}, \mathcal{B}) \}.$$

Using this new definition, we set again:

$$Good^0(q) := \mathcal{B}(\bar{\gamma}_q) \text{ and } Good^{i+1}(q) := Good(Good^i(q)), \ i \ge 0, \text{ and}$$

$$RB(q) := Good^{r(q)}(q),$$

where $r(q) := \min \{ i \in \mathbb{N} \mid Good^{i+1}(q) = Good^i(q) \}$.

Step 2. In the second step we show that, with the above modifications, results analogous to those from Case 1 can be deduced. We begin with the following property concerning $RB(q + s)$ and the cycle $\sigma(l_1), ..., \sigma(l_s)$.

29.19. Lemma. *The cycle* $\sigma(l_1) \cup ... \cup \sigma(l_s)$ *is contained in* $\sigma(\bar{\gamma}_q, RB(q + s))$.

Proof of lemma. We have to verify that $\sigma(l_j) \subset \sigma(\bar{\gamma}_{q+s}, RB(q + s))$, for every $1 \le j \le s$. We know that $\sigma(l_j) \subset \sigma(\bar{\gamma}_{q+s})$, for every $1 \le j \le s$. Now, let $\mathcal{B} \subset \mathcal{B}(\gamma_{q+s})$ with $\sigma(l_j) \subset \sigma(\bar{\gamma}_{q+s}, \mathcal{B})$, for all $1 \le j \le s$. Then it remains to show that $\sigma(l_j) \subset \sigma(\bar{\gamma}_{q+s}, Good(\mathcal{B}))$, for all $1 \le j \le s$. To verify this claim fix an index $1 \le j \le s$. Recall that both $\sigma(l_j)$ and $\sigma(l_{j+1})$ are spirals. Thus, by definition, there are bands B_j, B_{j+1} from $\sigma(l_j), \sigma(l_{j+1})$ with $\sigma(l_j) = B_j^-, \sigma(l_{j+1}) = B_{j+1}^-$, respectively. Recall further that $l_{j+1} \subset g(l_j)$. So $g(B_j) \cap l_{j+1} = \emptyset$, for g is an embedding. Since l_{j+1} is a (long) spiral, it follows that $g(B_j) \cap B_{j+1} \ne \emptyset$. More precisely, there must be a component, d_j, of $\theta \cap B_j$ with $g(d_j) \subset B_{j+1}$. Now, $l_j, l_{j+1} \subset \bar{\gamma}_{q+s} \cup l_{s+1}$ and so $\bar{\gamma}_{q+s} \cup l_{s+1}$ splits B_j and B_{j+1} into bands (recall our convention that $\partial \bar{\gamma}_{m+1} \cap int(\sigma(\bar{\gamma}_m)) = \emptyset$, for all m). Hence, in particular, $\sigma(l_j), \sigma(l_{j+1}) \subset \sigma(\bar{\gamma}_{q+s} \cup l_{s+1})$. It follows that $\bar{\gamma}_q$ splits d_j into arcs which are mapped under g into bands from $\sigma(\bar{\gamma}_{q+s} \cup l_{s+1})$. Thus, by definition, every band from $\sigma(\bar{\gamma}_q)$ which meets l_j is also a band form $Good(\mathcal{B})$. So $\sigma(l_j) \subset \sigma(\bar{\gamma}_{q+s}, Good(\mathcal{B}))$. This proves the lemma.

With the help of the previous lemma we can now establish the analogon of lemma 29.17 from Case 1.

29.20. Lemma. *Suppose* $\gamma_{q+s+1} \prec \sigma(\bar{\gamma}_q, RB(q+s))$, *for* $p \le q+s \le p+r$. *Then the restriction* $h := g|U\sigma(\bar{\gamma}_{q+s}, RB(q+s))$ *can be isotoped, using an isotopy which fixes* $\bar{\gamma}_{q+s}$, *such that afterwards* $h(U\sigma(\bar{\gamma}_{q+s}, RB(q+s))) = U\sigma(\bar{\gamma}_{q+s}, RB(q+s))$.

Proof of lemma. Let $B \in RB(q+s)$. Then $B \in Good(RB)(q+s))$ since $RB(q) = Good(RB(q+s))$. Hence, by definition, there is a component, say k_B, from $\theta \cap B$ with $g(k) \subset \sigma(\bar{\gamma}_{q+s} \cup l_{s+1}), \mathcal{B})$. Let

$$K_{q+s} := \bigcup k_B,$$

where the union is taken over all $B \in RB(q+s)$.

By an argument from lemma 29.17, it remains to show that the restriction $g|\bar{\gamma}_q \cup K_{q+s}$ can be deformed into $U(\sigma(\bar{\gamma}_{q+s}, RB(q+s))$. Now, $g(K_{q+s}) \subset \sigma(\bar{\gamma}_{q+s} \cup l_{s+1}, RB(q+s))$ (by construction) and $g(\gamma_{q+s}) \prec \sigma(\bar{\gamma}_{q+s} \cup l_{s+1}, RB(q+s))$ (by hypothesis). Hence, g can be deformed so that afterwards $g(\sigma(\bar{\gamma}_{q+s}, RB(q+s))) \subset \sigma(\bar{\gamma}_{q+s} \cup l_{s+1}, RB(q+s))$. It remains to show that $\sigma(\bar{\gamma}_{q+s} \cup l_{s+1}, RB(q+s))$ can be deformed into $\sigma(\bar{\gamma}_{q+s}, RB(q+s))$. To do this recall $\sigma(l_{s+1})$ is isotopic to one of the annuli $\sigma(l_j)$ from the cycle $\sigma(l_1), ..., \sigma(l_s)$ (see our choice of s). More precisely, there is a component B from $(U(\tau) - (\sigma(l_j) \cup \sigma(l_{s+1})))^-$ which is either an annulus or a 2-faced disc and for which the union $\sigma(l_j) \cup B \cup \sigma(l_{s+1})$ is an annulus (see the argument from lemma 29.18). Moreover, by lemma 29.19, we have $\sigma(l_j) \subset \sigma(\bar{\gamma}_{q+s}, RB(q+s))$. Thus, altogether, $(\sigma(\bar{\gamma}_{q+s} \cup l_{s+1}, RB(q+s)) - \sigma(\bar{\gamma}_{q+s}, RB(q+s))^-$ consists of annuli. Hence $\sigma(\bar{\gamma}_{q+s} \cup l_{s+1}, RB(q+s))$ can be isotoped into $\sigma(\bar{\gamma}_{q+s}, RB(q+s))$ as required. The lemma is therefore established.

We are now ready to finish the proof.

Step 3. Consider the sequence $p(r), p(r)+1, ..., p(r)+r$. Some of these indices may be good and others may be bad. If q, $p(r) \le q \le p(r)+r$, is good, then form

$\sigma(\bar{\gamma}_q, RB(q))$ as in Case 1, and continue with index $q+1$. If it is bad, then form $\sigma(\bar{\gamma}_{q+s(q)}, RB(q+s(q)))$ as in Case 2, and continue with index $q+s(q)+1$. In this way we get a sequence $\sigma(\bar{\gamma}_{q_1}, RB(\bar{\gamma}_{q_1})),\ \sigma(\bar{\gamma}_{q_2}, RB(\bar{\gamma}_{q_2})),\ ...$ of webs. Copying the argument for lemma 29.16, we first verify that $\chi\sigma^+(\bar{\gamma}_{q_{i+1}}, RB(q_{i+1})) < \chi\sigma^+(\bar{\gamma}_{q_i}, RB(q_i))$, for every index i for which $\gamma_{q_{i+1}} \not\prec \sigma(\bar{\gamma}_{q_i}, RB(q_i))$. From this fact we deduce, inductively, the existence of an index q, $p \leq q \leq p+r$, with $\gamma_{q+1} \prec \sigma(\bar{\gamma}'_q, RB(q))$. Moreover, note that $RB(q) \subset B(\bar{\gamma}(q))$. Thus there is a subset $\mathcal{A} \subset \mathcal{A}(\gamma_q)$ with

$$\sigma(\bar{\gamma}_q, \mathcal{A}) = \sigma(\bar{\gamma}_q, RB(q)).$$

It remains to show that q and $\mathcal{A}$ satisfy the conclusion of theorem 29.14. To show this recall $p(r) + r$ is a number which depends on $e(\tau)$ alone. Thus, by lemma 29.5, the order of $\sigma(\bar{\gamma}_q) - \bar{\gamma}_q$ depends on $e(\tau)$ and $q \leq p(r)+r$, and so on $e(\tau)$ alone. Now, suppose every component from $\bar{\gamma}_q - int(\sigma(\bar{\gamma}_q, \mathcal{B}))$, $\mathcal{B} \subset \mathcal{B}(\bar{\gamma}_q)$, has an order depending on $e(\tau)$ and $e(\theta \cup \theta')$ alone. Then this is also true for $\bar{\sigma}_q - int(\sigma(\bar{\gamma}_q, Good(\mathcal{B})))$. Indeed, this follows directly from lemma 29.15 if q is a good index and if we use the old definition of $Good(..)$ (i.e., the one from Case 1). If q is a bad index, then this claim is easily seen to follow as in the proof of lemma 29.15 if we use the new definition of $Good(..)$ (i.e., the one from Case 2). Hence, inductively, it follows that every component of $\bar{\gamma}_q - int(\sigma(\bar{\gamma}_q, \mathcal{A}))$ has an order which depends on $e(\tau)$ and $e(\theta \cup \theta')$ alone. Moreover, by lemma 29.5, we have $\#\mathcal{B}(\bar{\gamma}_q) \leq 2 \cdot (e(\tau) + q)$. It therefore follows that $\bar{\gamma}_q - int(\sigma(\bar{\gamma}_q, \mathcal{A}))$ has an order which depends on $e(\tau)$ and $e(\theta \cup \theta')$ alone. Thus $\sigma(\bar{\gamma}_q, \mathcal{A})$ satisfies (1) of theorem 29.14. By lemma 29.17, it also satisfies (2). Hence, altogether, the theorem is established in Case 2 as well. $\Diamond$

30. The Universal Limit for Induced Haken Graphs.

We now return to our study of Heegaard 2-complexes. We already know from thm. 27.26 that every Heegaard 2-complex Ω in a Haken 3-manifold N with useful Haken 2-complex $\Psi \subset N$ can be isotoped into a useful position w.r.t. Ψ. We also know that the induced Haken graph $\Delta_\Omega \subset \mathcal{D}_\Omega$ of a useful Heegaard 2-complex Ω is a useful Haken graph (see section 28.1 for definition). But this is not good enough for our purpose. We want Δ_Ω to be α-dominated, for some appropriate constant α not depending on Ω, so that we can use the results from section 28. Of course, for every $\alpha \in \mathbb{N}$, there is always a Haken graph which is not α-dominated. But it turns out that for induced Haken graphs, as opposed to abstract Haken graphs, several crucial complexities are subject to severe limitations. For instance, the depth of an induced Haken graph Δ_Ω is never larger than the number of all layer-surfaces from Ψ, i.e., $depth(\Delta_\Omega)$ is dominated by some constant which does not depend on the embedding-type of $\Omega \subset N$. In this section we will show that similar upper bounds hold for various other complexities of Δ_Ω as well. To organize the relevant results of this section in a coherent way, we introduce the concept of an α-dominated Heegaard 2-complex (see def. 30.13). Using this notion, we show (see thm. 30.15) the existence of a universal constant $\alpha = \alpha(N, \Psi)$ with the property that every Heegaard 2-complex in N can be isotoped into a useful and α-dominated Heegaard 2-complex. In the next section we will see that for all practical purposes the induced Haken graphs of α-dominated Heegaard 2-complexes are α-dominated themselves.

Throughout this section N is a Haken 3-manifold and $\Psi \subset N$ is a useful Haken 2-complex (def. 27.6). We will make free use of the notations for Haken graphs as introduced in section 28. Specifically, recall from section 28 the definitions for pseudo-arcs, bundles and bundle-paths.

30.1. *Base-Length of Bundle-Paths.*

We are interested in bundle-paths of large size. To discuss them in their proper context, let $D \subset N$ be a disc such that $\Delta := D \cap \Psi$ is a Haken graph. Let

$$\mathcal{B} = \mathcal{B}_1 \cup \mathcal{B}_2 \cup ... \cup \mathcal{B}_n$$

be a bundle-path for Δ not meeting ∂D. Then recall from section 28.16 that each bundle $\mathcal{B}_i$ is a system of layer-arcs joining two other layer-arcs from Δ, say γ_i, γ_{i+1}. A layer-arc k from $\mathcal{B}_i$ is also a proper arc in some layer-surface S from Ψ. We say k is *inessential in* Ψ if it separates a 2- or 3-faced disc from S, i.e., a disc $B \subset S$ such that $B \cap \partial S$ lies in the union of at most two layer-surfaces from Ψ different from S. Otherwise k is essential in Ψ. Moreover, the bundle $\mathcal{B}_i$ is *essential in* Ψ if

(1) all layer-arcs from $\mathcal{B}_i$ are essential in Ψ, and

(2) γ_i, γ_{i+1} are in essential position in their respective layer-surfaces.

Following def. 27.10, we say that a curve γ in a surface S with Haken graph

$\theta \subset S$ is in *essential position w.r.t.* θ if there is no 2-faced disc in S with one face in γ and the other face in a layer-curve from θ. We say a curve γ in a layer-surface S is *essential in* S if it is in essential w.r.t. both Haken graphs $\theta_S \subset S$ and $\theta'_S \subset S$ (see 27.5 for definition of the latter). Finally, we say the bundle-path $\mathcal{B}$ is *essential in* Ψ if all its bundles are.

30.2. Example. Essential bundle-paths with large size (see def. 28.18) come up quite naturally in certain special 3-manifolds. Here are two constructions.

Firstly, let T be a torus with a useful Haken graph $\Delta \subset T$ which consists of a single closed curve and a bundle of layer-arcs joining this curve with itself. Then $N := T \times I$ is a Haken 3-manifold with Haken 2-complex $\Psi := \Delta \times I$. Now, let $\mathbf{R}^2 \hookrightarrow T \times I$ be an embedding such that its composition with the projection $T \times I \to T$ is a covering map, and set $\tilde{\Delta} := f^{-1}(\Psi)$. Then we find many discs in $\mathbf{R}^2$ which contain bundle-paths for $\tilde{\Delta}$ with arbitrarily large size.

Secondly, consider a mapping torus over an orientable surface, also known as *Stallings fibration.* To fix ideas, let S be a hyperbolic surface and let $h : S \to S$ be a pseudo-Anosov diffeomorphism. Then, by definition, the Stallings fibration N with monodromy h is homeomorphic to $(S \times I)/h$, where the quotient is taken with respect to the equivalence relation $(x, 0) \sim (h(x), 1)$. A Stallings fibration has a canonical hierarchy, given by the fiber-surface S and a collection of vertical annuli and squares in the complement of S. Let Ψ be the Haken 2-complex associated to this hierarchy. Now, recall that there is at least one bi-infinite geodesic, l, in S such that $h(l) = l$ (modulo isotopy of h) (see e.g. [CaB]). Let $\bigcup_i A_i \subset S \times I$ be an arbitrarily large collection of pairwise parallel copies of $l \times I$. Then, after a small isotopy which is constant outside of a regular neighborhood of S, all these copies fit together to form a large, non-proper band A. The induced Haken graph $\Delta_A := A \cap \Psi$ is then itself a bundle-path of arbitrarily large size.

The previous constructions are typical. Indeed, the next proposition tells us vice versa that essential bundle-paths of large size can only occur in essential Stallings fibrations. By definition, a *Stallings fibration* X in N is *essential* if no torus from ∂X is compressible in N. For later use we say that an essential Stallings fibration in n is *trivial* if it can be isotoped into the regular neighborhood $U(\partial N)$. Otherwise it is *non-trivial.*

30.3. Proposition. *Let* N *be a Haken 3-manifold and let* $\Psi \subset N$ *be a great and useful Haken 2-complex. Then there is a constant* $\alpha = \alpha(N, \Psi) \in \mathbf{N}$ *with the following property: If* $D \subset N$ *is a disc such that* $\Delta := D \cap \Psi$ *is a Haken graph, then every bundle-path for* Δ*, which has large size and which is essential in* Ψ*, is contained in an essential Stallings fibration in* N.

Proof. To get started let $\mathcal{B} = \mathcal{B}_1 \cup ... \cup \mathcal{B}_n \subset D$, $n \geq 1$, be any bundle-path, not meeting ∂D, which is essential in Ψ and for which $size(\mathcal{B})$ is large (i.e., as large as we want). Then, as mentioned before, every bundle $\mathcal{B}_i$ is a system of layer-arcs joining two different layer-arcs from Δ, say γ_i, γ_{i+1}. Denote by $A(\mathcal{B}_i) \subset D$ the unique disc such that $\mathcal{B}_i \subset A(\mathcal{B}_i)$ and that $\partial A(\mathcal{B}_i) \subset \gamma_i \cup \mathcal{B}_i \cup \gamma_{i+1}$. Then $A(\mathcal{B}_i)$ is a square, and $\mathcal{B}_i$ is a system of arcs joining opposite faces of this square. In particular, $\mathcal{B}_i$ splits $A(\mathcal{B}_i)$ into a collection of squares. We call them the *squares for* $\mathcal{B}_i$ (or $\mathcal{B}$).

To proceed observe that most squares for $\mathcal{B}$ are actually "admissibly parallel", i.e., not that many squares are really different. Here we say that two disjoint discs $C_1, C_2 \subset D$ with $\partial C_i \subset \Delta$, $i = 1, 2$, are *admissibly parallel in* Ψ if there is an annulus, A, embedded in Ψ such that $\partial A = \partial C_1 \cup \partial C_2$ and that A contains no vertex of Ψ. The unique 3-ball in the irreducible 3-manifold N bound by the 2-sphere $C_1 \cup A \cup C_2$ is called the *admissible parallelity region between* C_1 and C_2. The above notion defines an equivalence relation on the set of all discs $C \subset D_\Omega$ with $\partial C \subset \Delta_\Omega$. In particular, the set of all squares for $\mathcal{B}_i$ falls into a (small) collection of admissible parallelity classes. Define

$$\Sigma(\mathcal{B}_i) := \left\{ \begin{array}{l} \text{the union of } A(\mathcal{B}_i) \text{ with the admissible parallelity regions} \\ \text{between all pairs of admissibly parallel squares for } \mathcal{B}_i. \end{array} \right.$$

Recall that γ_i, γ_{i+1} lie in layer-surfaces from Ψ, and let p be the smallest index with the property that Ψ_p contains these layer-surfaces. Let $\varphi_p : M_p \to N$ be the immersion from section 27.3. For convenience, denote the pullback $\varphi_p^{-1} \Sigma(\mathcal{B}_i)$ by $\Sigma(\mathcal{B}_i)$ again. Consider the regular neighborhood $U(\Sigma(\mathcal{B}_i))$ in M_p. Then clearly $U(\Sigma(\mathcal{B}_i))$ is an admissible product I-bundle in $(M_p, \underline{\underline{m}}_p)$. Let $E(\Sigma(\mathcal{B}_i))$ be the union of all those components from $(M_p - U(\Sigma(\mathcal{B}_i)))^-$ which are 3-balls. Define

$$X(\mathcal{B}_i) := U(\Sigma(\mathcal{B}_i)) \cup E(\Sigma(\mathcal{B}_i)).$$

Then $X(\mathcal{B}_i)$ is still an admissible product I-bundle in $(M_p, \underline{\underline{m}}_p)$. Denote the image $\varphi_m(X(\mathcal{B}_i))$ by $X(\mathcal{B}_i)$ again. We say $X(\mathcal{B}_i)$ is *essential* or *inessential* if it is essential or inessential in $(M_p, \underline{\underline{m}}_p)$.

Note that $X(\mathcal{B}_i)$ cannot be the I-bundle over a disc, for γ_i is supposed to be essential in the layer-surface containing it. Hence it is an essential, product I-bundle. Now, Ψ comes from a great hierarchy (it is supposed to be a great Haken 2-complex). Hence and using [Joh 1, prop. 24.1], it follows that $X(\mathcal{B}_i)$ is a product I-bundle whose lids lie in a single layer-surface from Ψ and which meets this layer-surface from two sides (unless each $X(\mathcal{B}_i)$ lies in the regular neighborhood of an annular face of $(M_p, \underline{\underline{m}}_p)$ in which case the proposition follows immediately). Thus there is a layer-surface, S, from Ψ such that $\gamma_1, ..., \gamma_m \subset S$ (see fig. 30.1).

In this situation we can apply thm. 29.14. To make the connection, consider the two Haken graphs

$$\theta := \theta_S \text{ and } \theta' := \theta'_S \subset S$$

induced by Ψ from the two sides of S (see section 27.5 for the exact definition). Then θ, θ' are two Haken-graphs in S which are in general position. As above define

$$\Sigma(\mathcal{B}) := \left\{ \begin{array}{l} \text{the union of } A(\mathcal{B}) \text{ with the admissible parallelity regions} \\ \text{between all pairs of admissibly parallel squares for } \mathcal{B}. \end{array} \right.$$

Note that $\Sigma(\mathcal{B})$ is not just the union of all $\Sigma(\mathcal{B}_i)$, for there may well be squares for different bundles from $\mathcal{B}$ which are admissibly parallel. Note further that the terminology from section 29 allows yet another description of $\Sigma(\mathcal{B})$. Indeed, define $\bar{\gamma}_n := \gamma_1 \cup \gamma_2 \cup ... \cup \gamma_n$. Then, of course, there is a minimal and essential train track in (S, θ) which carries $\bar{\gamma}_n$. Let $\sigma(\bar{\gamma}_n)$ be the web of $\bar{\gamma}_n$ w.r.t. the previous train-track (see 29.3). Then $\Sigma(\mathcal{B}) = \sigma(\mathcal{B}) \times [0,1]$. The map, which maps start-points to end-point of fibers of the product I-bundle $\sigma(\mathcal{B}) \times [0,1]$, defines a web-extension $g : \sigma(\bar{\gamma}_n) \hookrightarrow S$ for $\bar{\gamma} := \gamma_1 \cup ... \cup \gamma_{n+1}$. Thus we are in the situation of thm. 29.14. Let a, b, c be the constants given in thm. 29.14. Then there is an index q, $1 \leq q \leq a$, and a collection $\mathcal{A} \subset \mathcal{A}(\bar{\gamma}_q)$ such that (1) and (2) of thm. 29.14 holds. In particular, $gU(\sigma^+(\bar{\gamma}_q, \mathcal{A})) = U(\sigma^+(\bar{\gamma}_q, \mathcal{A}))$ mod isotopy. Let F be the union of $U(\sigma^+(\bar{\gamma}_q, \mathcal{A}))$ with all disc-components from $(S - U(\sigma^+(\bar{\gamma}_q, \mathcal{A})))^-$. Then the following holds:

30.4. Lemma. *At least one component of $F \subset S$ is an incompressible surface in S which is different from a disc.*

Proof of lemma. Let B be a band for $\sigma(\bar{\gamma}_q, \mathcal{A})$, i.e., a component of $\sigma(\bar{\gamma}_q) - \bar{\gamma}_q$. Let $s \subset \bar{\gamma}_q$ be a side of B. Let $B_1, ..., B_s$ be all the squares into which B is split by θ. Let C_i be the closure of that component of $\sigma(\bar{\gamma}_q, \mathcal{A}) - \theta$ which contains B_i, and let l_i be a component of $(\partial C_i - \theta)^-$.

We claim there are indices $i \neq j$ with $C_i \cap C_j \neq \emptyset$. Assume the converse. Then, in particular, $l_i \neq l_j$, for all $i \neq j$. Consider the union $l = l_1 \cup ... \cup l_s$. This is a collection of arcs in $\bar{\sigma}_q - int(\bar{\sigma}_q, \mathcal{A})$. But according to thm. 29.14, we have $\#(\sigma(\bar{\gamma}_q, \mathcal{A}) - \bar{\gamma}_q) \leq \#(\sigma(\bar{\gamma}_q) - \bar{\gamma}_q)$, a collection $\mathcal{P} \subset \bar{\gamma}_q' := \bar{\gamma}_q - int(\sigma(\bar{\gamma}_q, \mathcal{A}))$ of points with $\partial \bar{\gamma}_q' \subset \mathcal{P}$, $\#\mathcal{P} \leq c$ and, for every arc $l \subset \bar{\gamma}_q$ with $l \cap \mathcal{P} = \partial l$, an index i, $1 \leq i \leq b$, with $m(g^i l)$. It therefore follows that there are constants j, c', depending on a, b, c alone, such that $m(g^j s) \leq c'$. Since $\#(\sigma(\bar{\gamma}_q, \mathcal{A}) - \bar{\gamma}_q) \leq \#(\sigma(\bar{\gamma}_q) - \bar{\gamma}_q)$, it then follows that $m(g^t \bar{\gamma}_q) \leq c''$, where t and c'' are constants depending on a, b, c alone. A moments reflection shows that this, however, contradicts our hypothesis that $size(\mathcal{B})$ is large. Thus the claim is established.

By the above claim, we have that $B \cap C_i$ contains the squares B_i and B_j. Thus the midpoints of B_i and B_j can be joined by an arc $k \subset C_i$ and an arc $l \subset B$. Consider the simple closed curve $k \cup l$. The lemma follows if $k \cup l$ is not contractible in S. Thus assume the converse. Then $k \cup l$ bounds a disc C in S. Recall that $\theta \subset S$ is a Haken graph which meets the interior of C. It follows that there is at least one proper arc in C which lies in a layer-arc of θ. This proper arc separates a 2-faced disc from C. But this is impossible since γ_i is supposed to be essential in S. Thus our assumption is wrong, and the lemma is established.

We can now finish the proof as follows. First, denote by $F \times 0$ the pullback of F under φ_m in one of the lids of $X(\mathcal{B})$. More precisely, let $F \times 0$ be the 2-manifold in the regular neighborhood $U(\sigma(\mathcal{B}) \times 0)$ (in ∂M_m) with $\varphi_m(F \times 0) = F$. By construction of g and the properties of F, it follows that $F \times I$ can be admissibly isotoped in $(M_m, \underline{m}_m)$ so that afterwards $\varphi_m(F \times 0) = \varphi_m(F \times 1)$. Then the image $\varphi_m(F \times [0, \overline{1]})$ is a Stallings fibration in N with incompressible boundary, i.e., an essential Stallings fibration. Thus we have found the desired Stallings fibration containing $\mathcal{B}$.

This completes the proof of the proposition. $\Diamond$

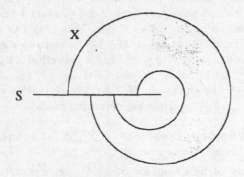

(Figure 30.1.)

We use prop. 30.3 to establish a crucial estimate regarding the baselength of bundle-paths for induced Haken graphs. For this we first introduce a modification of the complexity $base(..)$ from section 28.18. Given a Haken graph Δ and a bundle-path $\mathcal{B}$ for Δ, set

$$\ell base(\mathcal{B}) := max \{ \, b(\mathcal{B}, r) \mid r \text{ is layer-arc of } \Delta \, \}.$$

Note that the modification consists in the mere fact that r is no longer allowed to be a bound face (the letter 'ℓ' stands for 'layer-arc'). Given this notation we can further define

$$\ell base(\Delta, \alpha) := max \{ \, \ell base(\mathcal{B}) \mid \mathcal{B} \text{ is a bundle-path in } \Delta$$
$$\text{with } width(\mathcal{B}) > \alpha \, \}.$$

As an easy consequence of prop. 30.5 we get a universal constant $\alpha = \alpha(N, \Psi)$ with the property that $\ell base(\Delta, \alpha) \le \alpha$, for all useful Heegaard 2-complexes in a Haken 3-manifold N without non-trivial, essential Stallings fibrations.

30.5. Proposition. *Let N be a Haken 3-manifold and let $\Psi \subset N$ be a great and useful Haken 2-complex. Then there is a constant $\alpha = \alpha(N, \Psi) \in \mathbf{N}$ with* *the following property: If $\Omega \subset N$ is a useful Heegaard 2-complex w.r.t. Ψ and if $\mathcal{B}$ is a bundle-path for Δ_Ω with width$(\mathcal{B}) > \alpha$ and $\ell base(\mathcal{B}) > \alpha$, then $\mathcal{B}$ is contained in an essential Stallings fibration in N.*

Proof. Let $\mathcal{B} = \mathcal{B}_1 \cup ... \cup \mathcal{B}_n$ be a bundle-path in Δ_Ω with $\ell base(\mathcal{B})$ and $width(\mathcal{B})$ large (i.e., as large as we want). Then, in particular, $\mathcal{B}$ has a large size and so, by prop. 30.3, the conclusion of prop. 30.5 follows whence we show that $\mathcal{B}$ is essential. Thus assume the converse.

Now, $\ell base(\mathcal{B})$ is large and so, by definition, there is a layer-arc r from Δ_Ω which contains the end-point of a large number of layer-arcs from $supp(\mathcal{B})$. For sake of simplicity suppose that r itself contains no end of any bundle from $\mathcal{B}$ but that it contains an end-point of *every* layer-arc from $supp(\mathcal{B})$ containing such an end (the argument for the general case is similar and left to the kind reader). We also note that r contains only one, and so exactly one, end-point of the above layer-arcs from $supp(\mathcal{B})$ since Δ_Ω is useful (Ω is useful).

Since $\mathcal{B}$ is supposed to be inessential, there is, by definition, at least one layer-arc from $\mathcal{B}$ which is inessential in Ψ. Let $\mathcal{B}_i$ be the bundle from $\mathcal{B}$ which contains this particular layer-arc, and let γ_i, γ_{i+1} be the two layer-arcs from Δ_Ω joined by $\mathcal{B}_i$. Then γ_i and γ_{i+1} are layer-arcs from $supp(\mathcal{B})$ which both have exactly one end-point in the layer-arc r (see our assumption above). Let $A \subset D$ be the unique square characterized by the requirement that $\mathcal{B}_i \subset A$ and that $r, \gamma_i, \gamma_{i+1}$ and $\mathcal{B}_i$ each contain exactly one face from A. Let A_j, $1 \leq j \leq p$, be the collection of all squares into which A is split by $\mathcal{B}_i$. Let the indices be chosen so that $k_j := A_j \cap A_{j+1} \neq \emptyset$, for all $1 \leq j \leq p - 1$, and that $k_p := A_p \cap r \neq \emptyset$. Then all k_j, $1 \leq j \leq p - 1$, are *proper* arcs in layer-surfaces S_j from Ψ. By our choice of $\mathcal{B}_i$, there is at least one index s, $1 \leq s \leq p - 1$, such that k_s is inessential in S_s. Thus, by definition, k_s separates a 2- or 3-faced disc B_s from S_s.

We claim that, for every index $j, 1 \leq j \leq p - 1$, the arc k_j is inessential in its layer-surface S_j. To see this let m be the smallest index such that $\partial A_s \subset \Psi_m$. Let $\varphi_m : M_m \to N$ be the immersion from section 27.3. Now, the union of a square with a 2- or 3-faced disc is again a 2- or 3-faced disc. It therefore follows that the union of $\varphi_m^{-1} A_{s+1}$ with one of the two components of $\varphi_m^{-1} B_s$ is a 2- or 3-faced disc. After a small general position isotopy, this is in fact a proper 2- or 3-faced disc, say A'_{s+1}, in $(M_m, \underline{\underline{m}}_m)$. At this point recall from section 27.3 that $(M_m, \underline{\underline{m}}_m)$ is a 3-manifold with useful boundary-pattern (Ψ is a great Haken 2-complex). Thus, by the very definition of useful boundary-patterns [Joh 1], it follows that $\partial A'_{s+1}$ bounds a disc $A^*_{s+1} \subset \partial M_m$ such that $\mathcal{G}_m \cap A^*_{s+1}$ is the cone over $\mathcal{G}_m \cap \partial A^*_{s+1}$ with the midpoint of A^*_{s+1} as cone-point. Here $\mathcal{G}_m \subset \partial M_m$ denotes the graph given by the union of all boundary components of all faces of $(M_m, \underline{\underline{m}}_m)$. It follows that k_{s-1} and k_{s+1} must be inessential, too. The same with k_{s-2} and k_{s+2}, etc. This establishes the claim.

Assume k_s separates a 2-faced disc from S_s. Then k_{p-1} separates a 2-faced disc from S_{p-1}. Now, consider the last arc k_p. Notice that k_p is *not* a proper arc in the layer-surface S_p (i.e., the layer-surface containing the base r). But, applying the above argument to k_{p-1}, it follows that k_p is the face of a 2-faced disc which does not meet ∂S_p and whose other face lies in the boundary of another layer-surface. In other words, the graph $\Gamma_{S_p} \subset S_p$ induced by Ω is not essential w.r.t. the induced Haken graph θ_{S_p} or θ'_{S_p} (for definitions see 27.22, 27.5 and def. 27.10). Hence Ω is not essential w.r.t. Ψ. But this contradicts our hypothesis that Ω is useful.

Assume k_s separates a 3-faced disc from S_s. Then, by what we have just seen, every arc k_j, $1 \leq j \leq p-1$, separates a 3-faced disc from S_j. One vertex of these 3-faced discs is a vertex of Ψ. Moreover, it is easily verified that all these vertices must be different (use that $A(\mathcal{B}_i) \cap r \neq \emptyset$ and that Ω is useful). Hence p is smaller than the total number of vertices of Ψ. This, however, contradicts the fact that $width(\mathcal{B})$ is large.

Thus, in any case, we get a contradiction. Hence our assumption that $\mathcal{B}$ is inessential in Ψ is wrong. This finishes the proof of the proposition. $\Diamond$

30.6. *Inessential Bundle-Paths.*

We turn our attention to bundle-paths which are inessential in induced Haken graphs of Heegaard 2-complexes (one easily shows that these bundle paths are also inessential in Ψ but not vice versa). We wish to show that those bundle-paths can have a small width only. To do this we will have to introduce a new complexity for Heegaard 2-complexes. This will be the "whirl-degree" introduced next.

30.7. *Whirls.* We need the notion of a "whirl" for graphs. To introduce this notion in its proper generality let θ be a Haken graph in an orientable surface S (which will later be a layer-surface). A *basin* w.r.t. θ is a pair (A, A') of annuli in S such that

(1) $A' \subset int(A)$ and $\partial A \cup \partial A'$ is in essential position w.r.t. θ (see 30.1 for definition),

(2) $int(A')$ contains at least one closed layer-curve from θ, and

(3) $\theta \cap (A - A')^-$ is a non-empty system of arcs joining opposite boundary curves.

Given a basin (A, A'), let U_1, U_2 be the two components from $(A - A')^-$, and fix a product structure $S^1 \times I$ for U_i, $i = 1, 2$, such that $\theta \cap U_i$ are fibers. Let $p_i : U_i = S^1 \times I \to S^1$ be the projection. For the following it is tacitly being understood that every arc in U_i, joining the two components of U_i, is oriented from $\partial U_i \cap \partial A$ towards $\partial U_i \cap A'$.

Given a finite graph $\Gamma \subset S$, we say (A, A') is a basin for Γ w.r.t. θ if $\Gamma \cap U_i$, $i = 1, 2$, is a system of arcs, l, joining the two boundary curves of U_i

such that $p_i(\partial l)$ is a single point. We say (A, A', Γ) is a *whirl for* Γ *w.r.t.* θ if (A, A') is a basin for Γ w.r.t. θ. Define its *degree* to be

$$deg_\theta(A, A', \Gamma) := \min\{\ i(l_1, k_1),\ -i(l_2, k_2)\ \},$$

where k_i, l_i, $i = 1, 2$, is a component of $\theta \cap U_i$, $\Gamma \cap U_i$, respectively, and where $i(k_i, l_i)$ is the algebraic intersection number of k_i and l_i (recall our convention regarding orientations). Clearly, the definition of $deg_\theta(A, A', \Gamma)$ is independent of the choices of k_i and l_i. Moreover, define

$$whirl_\theta(\Gamma, S) := \max\{\ deg_\theta(A, A', \Gamma)\ \},$$

where the maximum is taken over all whirls (A, A', Γ_Ω) in S. Now, given a Heegaard 2-complex $\Omega \subset N$ in general position w.r.t. a Haken 2-complex $\Psi \subset N$ and given a layer-surface S from Ψ, set $\qquad$ •

$$whirl(\Omega, S) := \max\{\ whirl_{\theta_S}(\Gamma_S, S),\ whirl_{\theta'_S}(\Gamma_S, S)\ \},$$

where $\theta_S, \theta'_S \subset S$ are the Haken graphs induced by Ψ and where $\Gamma_S := \Omega \cap S$. Finally, define

$$whirl(\Omega, \Psi) := \max\{\ whirl(\Omega, S)\ |\ S \text{ is a layer-surface from } \Psi\ \}.$$

We call $whirl(\Omega, \Psi)$ the *whirl-degree of* Ω (see the picture on the right of fig. 30.1 for an example of a large whirl-degree).

The next proposition tells us that the whirl-degree can be limited.

30.8. Proposition. *Let* N *be a Haken 3-manifold and let* $\Psi \subset N$ *be a great and useful Haken 2-complex. Then there is a constant* $\alpha = \alpha(N, \Psi) \in \mathbf{N}$ *with the following property. Every Heegaard 2-complex* $\Omega \subset N$ *can be isotoped so that afterwards it is useful and* $whirl(\Omega, \Psi) \le \alpha$.

Proof. The proof is by induction on the surface-filtration $\{\Psi_i\}_{1 \le i \le n}$ of Ψ. The induction beginning is trivial.

For the induction step suppose Ω is isotoped so that it is in useful position w.r.t. Ψ_{i-1} and that $whirl(\Omega, \Psi_{i-1})$ is small. Consider the layer-surface $S := (\Psi_i - \Psi_{i-1})^-$. By the argument from the proof of prop 27.26, we may suppose Ω is isotoped, using an isotopy which fixes $\Omega \cap \Psi_{i-1}$, so that afterwards Ω is in useful position w.r.t. Ψ_i. In particular, $whirl(\Omega, \Psi_{i-1})$ is unchanged. W.l.o.g. we assume $whirl(\Omega, \Psi)$ is large (otherwise we are done). Then, by definition, there is whirl (A, A', Γ_S), $A, A' \subset S$, w.r.t. $\theta = \theta_S$, say, such that $deg_\theta(A, A', \Gamma_S)$ is large. Of course, there is a homeomorphism $h : S \to S$, $i = 1, 2$, with support in $(A - A')^-$ such that $deg_\theta(A, A', h\Gamma_S)$ is small (smaller than 10 say) and that $h(\Gamma_S)$ is still essential w.r.t. θ_S and θ'_S. Since A is an annulus, there is in fact such a homeomorphism which is isotopic to the identity through an isotopy α_t, $t \in I$, in S. But any such isotopy can be extended to an ambient isotopy of N which fixes $U(\Psi_{i-1})$. Using this extended isotopy, we see that Ω can be isotoped so that afterwards Ω is useful w.r.t. Ψ_i and that $whirl(\Omega, \Psi_{i-1})$ as well as $deg_\theta(A, A', \Omega \cap S)$ is small.

To finish the argument observe that the number of basins (A, A') w.r.t. θ_S is small (mod isotopy preserving θ_S). To see this recall A is an annulus which contains at least one closed layer-curve from θ_S. Thus it remains to show that the number of annuli A as above, containing a fixed layer curve, is small. Since the number of all vertices of θ_S is small, it remains to show that the number of all annuli A as above, containing a given layer-curve and a given collection of vertices from θ_S, is small. But this follows easily from the fact that ∂A is essential w.r.t. θ_S. Hence the claim is established. The same for basins w.r.t. θ'_S. Carrying out the above reduction-operation for all the finitely many basins successively, we have eventually isotoped Ω so that Ω is useful w.r.t. Ψ_i and that $whirl(\Omega, \Psi_i)$ is small. The induction step is therefore complete.

This finishes the proof of the proposition. $\Diamond$

30.9. *Width of inessential bundle-paths.* Combining prop. 30.8 with the next proposition, we get a universal upper bound for the complexity $bundle(\Delta_\Omega)$ for induced Haken graphs.

30.10. Proposition. *Let N be a Haken 3-manifold and let Ψ be a great and useful Haken 2-complex. Then there is a known arithmetic function depending on N and Ψ alone such that $bundle(\Delta_\Omega) \leq \varphi(\alpha)$, for all useful Heegaard 2-complex $\Omega \subset N$ with $whirl(\Omega, \Psi) \leq \alpha$.*

Proof. Assume the converse. Then there is a useful Heegaard 2-complex $\Omega \subset N$ such that $whirl(\Omega, \Psi) \leq \alpha$ and that there is an inessential bundle-path $\mathcal{B} = \mathcal{B}_1 \cup ... \cup \mathcal{B}_p$ in Δ_Ω with arbitrarily large width. Recall from section 28.19 the definition of inessential bundle-paths in Haken graphs. Recall, in particular, that the outer ends of $\mathcal{B}$ are contained in a single layer-arc r from Δ_Ω. For simplicity we assume that every end of every bundle $\mathcal{B}_j$, $1 \leq j \leq p$, lies in a layer-arc of Δ_Ω which has an end-point in r (the proof in the more general case is similar). In fact, the above layer-arcs have exactly one end-point in r (recall Δ_Ω is useful since Ω is). Let S be the layer-surface from Ψ containing r.

Given a bundle $\mathcal{B}_j$, $1 \leq j \leq p$, note that every disc $C \subset D$ with $C \cap (r \cup \mathcal{B}_j \cup supp(\mathcal{B}_j)) = \partial C$ is either a square or a triangle (triangles occur only for the bundles $\mathcal{B}_1, \mathcal{B}_p$ meeting r). Denote by $A(\mathcal{B}_j)$ the union of all these discs and set

$$A = A(\mathcal{B}) := \bigcup_j A(\mathcal{B}_j).$$

Note that A is a 2-faced disc. We next construct an immersion $g : A \to S$ with $g|A \cap r = id|A \cap r$ (see fig. 30.1).

Begin with the triangle $C_0 \subset D$ with $C_0 \cap (r \cup \mathcal{B}_1 \cup supp(\mathcal{B}_1)) = \partial C_0$. Since Ψ is a great and useful Haken 2-complex, it follows that ∂C_0 bounds a disc, C'_0, in Ψ contained in the union of three layer-surfaces. So $C_0 \cap r$ is the face of a 3-faced sliding-disc, C^*_0, for the graph $\Gamma_S := \Omega \cap S$. Since N is irreducible, it follows that C_0 is parallel to C'_0. Thus C_0 can be isotoped into C_0^*, using an isotopy which is constant on $r \cap C_0$ and which keeps the other faces of C_0 in their respective layer-surfaces of Ψ. Extend this isotopy to an isotopy of Ω which is constant outside a regular neighborhood of C_0 in $\mathcal{D}_\Omega$. Then $C_0 \subset S$. But then the square $C_1 \subset A(\mathcal{B}_1)$ for $\mathcal{B}_1$, adjacent to C_0, is turned into a 3-faced disc. Thus we can repeat the above construction for this disc, and so on. In the end all of $A(\mathcal{B}_1)$ is being pushed into S. But then the cell in $A(\mathcal{B}_2)$ meeting r is turned into a 3-faced disc, and we can apply the previous procedure to $A(\mathcal{B}_2)$, and so on. In this way we push all the discs $A(\mathcal{B}_1), A(\mathcal{B}_2), ..., A(\mathcal{B}_p)$ successively into S. In the end $A(\mathcal{B})$ is pushed into S and the map which maps every point from $A(\mathcal{B})$ to its corresponding point in S is the desired immersion $g : A \to S$.

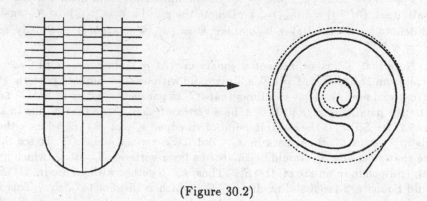

(Figure 30.2)

Let B be the regular neighborhood of $g(A)$ in S. Then B is an annulus, for Ψ has only a small number of vertices but A contains a large number of vertices from Δ_Ω. Clearly, B contains a whirl for $\Gamma_S \subset S$ with a large degree. But this contradicts the hypothesis that the whirl-degree $whirl(\Omega, \Psi)$ is small. Thus our assumption at the beginning of this proof is wrong, and so the proposition is established. $\diamondsuit$

30.11. *Squares.*

The next proposition tells us that almost all squares for a bundle contain only a small number of layer-arcs (recall from section 28.24 the definition of $square(\Delta, .)$ for Haken graphs).

30.12. Proposition. *Let N be a Haken 3-manifold and let $\Psi \subset N$ be a great and useful Haken 2-complex. Then there is a constant $\alpha = \alpha(N, \Psi) \in \mathbf{N}$ such that $square(\Delta_\Omega, \alpha) \leq \alpha$, for every Heegaard 2-complex $\Omega \subset N$ in useful position w.r.t. Ψ.*

Proof. Let $\{\Psi_i\}_{1 \leq i \leq n}$ be the surface-filtration of Ψ. Let B be an arbitrary bundle in Δ_Ω disjoint to $\partial \mathcal{D}_\Omega$. We are asked to show that there is only a small number of squares for $\mathcal{B}$ which contain a large number of layer-arcs from Δ_Ω. Assume the converse.

Recall from 30.1 the definition of essential bundles in Ψ. Since Ω is useful, it follows that a bundle in Δ_Ω is inessential in Ψ if at least one of its layer-arcs is inessential in Ψ. Moreover, the argument given in the proof of prop. 30.5 shows that the arcs from a bundle are either all essential, or all inessential.

Case 1. $\mathcal{B}$ is inessential.

Let $C \subset A(\mathcal{B})$ be a square for $\mathcal{B}$, i.e., $C \cap (\mathcal{B} \cup supp(\mathcal{B})) = \partial C$. Let m, $1 \le m \le n$, be the smallest number such that $\partial C \subset \Psi_m$. Then w.l.o.g. $C \cap \Psi_m = \partial C$. Let $\varphi_m : M_m \to N$ be the immersion from 27.3. For ease of notation set $(M, \underline{m}) = (M_m, \underline{m}_m)$. Denote the pull-back $\varphi_m^{-1}(C)$ by C again and denote by $\bar{\Psi}$ the Haken 2-complex $\bar{\Psi} := (\varphi_m^{-1}\Psi \cap int(M))^-$. Finally, set $\Delta_C := \bar{\Psi} \cap C$.

Note that C is an inessential square in $(M, \underline{m})$ since we are in Case 1. Recall from 27.3 that $(M, \underline{m})$ is a 3-manifold with useful boundary pattern (Ψ is great and useful). Thus it follows that C is parallel to a disc in ∂M. Let E be the parallelity region. Let x be a vertex of Δ_C. Note that x lies in an edge of $\bar{\Psi}$. So x is the starting point of an edge, k_x, of $\bar{\Psi} \cap E$ which either ends in a vertex of $\bar{\Psi}$ or back in C. But ∂k_x cannot lie in C. To see this note that otherwise k_x would be an arc in a layer-surface, S_x, of $\bar{\Psi}$ which has both end-points in an arc of $\Omega \cap S_x$. Thus k_x together with an arc in $\Omega \cap S_x$ would bound a 2-faced sliding-disc in S_x which is disjoint to ∂S_x. This in turn is impossible since Ω is in useful, and so in essential, position w.r.t. Ψ. Thus x is joined with a vertex of Ψ. Of course, different vertices of Δ_C are joined in this way with different vertices of Ψ. Thus we have an injection from the set, $\Delta_C^{(0)}$, of all vertices of Δ_C into $\Psi^{(0)}$. This shows that the number of vertices of Δ_C is not larger than $\#\Psi^{(0)}$. But the number of vertices of Δ_C dominates the number of layer-arcs in C. To see this note that, by our choice of C, the Haken graph Δ_C has no layer-arcs which are proper and essential in C and which are disjoint to $C \cap \mathcal{B}$. Thus the total number of layer-arcs in Δ_C is small. This proves the proposition in Case 1.

Case 2. $\mathcal{B}$ is essential.

Since $\mathcal{B}$ is a bundle in Δ_Ω disjoint to $\partial \mathcal{D}_\Omega$, there are two layer-arcs, say γ_1, γ_2, from Δ_Ω such that $\mathcal{B}$ consists of layer-arcs which all join γ_1 with γ_2. Let m, $1 \le m \le n$, be the smallest number such that $\gamma_1 \cup \gamma_2 \subset \Psi_m$. Let $\varphi_m : M_m \to N$ be the immersion from 27.3. For ease of notation set again $(M, \underline{m}) = (M_m, \underline{m}_m)$. As in 28.17, let $A := A(\mathcal{B})$ denote the union of all squares for $\mathcal{B}$. Denote the pull-back $\varphi_m^{-1}(A)$ by A again and denote $\bar{\Psi} := (\varphi_m^{-1}(\Psi) \cap int(M))^-$. As in the proof of prop. 30.3, associate an essential I-bundle $X(\mathcal{B})$ to $\mathcal{B}$ ($\#\mathcal{B}$ is large, for otherwise we are done). Recall from 27.3 that $(M, \underline{m})$ is a 3-manifold with useful boundary-pattern. Let V be the characteristic submanifold of $(M, \underline{m})$. The existence of $X(\mathcal{B})$ shows that V is non-empty (see [Joh 1, prop. 10.8]). Recall also that $\bar{\Psi}$ is a great Haken 2-complex for M and that $(\partial V - \partial M)^-$ is a layer-surface from $\bar{\Psi}$ (see 27.3).

Consider the intersection of A with $(\partial V - \partial M)^-$. We claim at most two components from $A \cap (\partial V - \partial M)^-$ are arcs from $\mathcal{B}$. Assume the converse. Then there is a non-trivial component, C, from $(A - V)^-$ such that $(\partial C - \partial A)^-$ is disconnected. But V is the characteristic submanifold and so C is parallel to some square in $(\partial V - \partial M)^-$. If G is a face of $(M, \underline{m})$ which contains a component, c, of $C \cap \partial A$, then c together with an arcs in $\partial(\partial V - \partial M)^-$ bounds a 2-faced disc, D_0, in G. The image $\varphi_m(D_0)$ is a 2-faced sliding disc for $\Omega,$ in the layer-surface from $\bar{\Psi}$ containing $\varphi_m(G)$, which does not meet the boundary of that layer-surface. But Ω is essential w.r.t Ψ and so has no such sliding disc. This establishes the claim.

Thus, passing to some smaller, but still large, sub-bundle of $\mathcal{B}$ if necessary, we may suppose w.l.o.g. that $A \cap (\partial V - \partial M)^- = \emptyset$. Hence either $A \subset V$ or $A \subset M - V$.

If $A \subset V$, then recall from 27.3 that the layer-surfaces of $\bar{\Psi}$ contained in V form a disjoint union, say Q, of vertical squares. It follows that $\mathcal{B} = Q \cap A$. Let C be a square of $\mathcal{B}$. We claim $\Delta_C := (\Delta \cap int(C))^-$ is empty. To see this note that Δ_C has no vertices since Q is a disjoint union of squares. Thus Δ_C is a system of proper arcs in C. Moreover, none of these arcs can meet $(\partial C - \partial A)^-$ since Q is a disjoint union. But, by our choice of C, there is no arc from Δ_C which is parallel to a component of $(\partial C - \partial A)^-$. Finally, no arc from Δ_C has both end-points in one component of $C \cap \partial A$, for otherwise we could again construct a forbidden 2-faced sliding disc for Ω. This proves the claim.

Thus suppose $A \subset M - V$. Then we claim $X(\mathcal{B}) \subset M - V$. If not then at least one of the parallelity regions used to built $X(\mathcal{B})$ intersects V since $A \subset M - V$. But such a parallelity region is bound by squares in A and squares in layer-surfaces of $\bar{\Psi}$. So a layer-surface of $\bar{\Psi}$ would intersect $(\partial V - \partial M)^-$ transversally. But this is impossible since $(\partial V - \partial M)^-$ itself is a layer-surface. This establishes the claim.

Since V is a characteristic submanifold and since $X(\mathcal{B})$ is an essential I-bundle in the complement of V, it follows that $X(\mathcal{B})$ is the product I-bundle over an annulus (with boundary-pattern given by its boundary curves). Moreover, there is a component, say E, of $(M_m - (V \cup E(\mathcal{B}))^-$ which meets both V and $X(\mathcal{B})$. More precisely, E is also a product I-bundle over an annulus. Now let $C \subset A$ be a square for $\mathcal{B}$ and denote $\Delta_C := \Delta \cap \bar{\Psi}$. Then, as in Case 1, it is now easy to show that every vertex of Δ_C can be joined by an arc in some edge of $\bar{\Psi}$ with some vertex of $\bar{\Psi}$. In fact, this defines again an injection $\Delta_C^{(0)} \hookrightarrow \Psi^{(0)}$. So the number of vertices of Δ in A is not larger than the number of all vertices of Ψ. But the number of vertices of Δ_C in C dominates the number of all layer-arcs in C. This proves the proposition in Case 2.

Hence the proof of the proposition is finished. ◇

30.13. *α-Dominated Heegaard 2-Complexes.*

For later reference, we now compress the results of this section into a single statement. To do this we first define:

30.14. Definition. *A Heegaard 2-complex* $\Omega \subset N$ *in general position to a Haken 2-complex* $\Psi \subset N$ *is called α-dominated,* $\alpha \in \mathbf{N}$, *w.r.t.* Ψ *if*

$$\ell base(\Delta_\Omega, \alpha), \quad whirl(\Omega, \Psi), \quad square(\Delta_\Omega, \alpha)$$

are all smaller than α.

We can now combine the results from this section to the following, more compact statement.

30.15. Theorem. *Let* N *be a Haken 3-manifold without non-trivial, essential Stallings fibrations and let* $\Psi \subset N$ *be a great and useful Haken 2-complex. Then there is a constant* $\alpha = \alpha(N, \Psi) \in \mathbf{N}$ *with the property that every Heegaard 2-complex* $\Omega \subset N$ *can be isotoped into a useful and α-dominated Heegaard 2-complex w.r.t.* Ψ.

Proof. This is a direct consequence of prop. 30.5, 30.8 and 30.12, provided N has no essential Stallings fibration at all. In the other case it follows after a straightforward additional adjustment of Ω near the boundary of N. ◇

We will show in the next section that useful and α-dominated Heegaard 2-complexes are closely related to Heegaard 2-complexes whose induced Haken graph is useful and α-dominated.

31. Finiteness for Heegaard Graphs.

In this section we establish the Finiteness Theorem for Heegaard graphs in Haken 3-manifolds without non-trivial Stallings fibrations. So far we have only considered isotopies for Heegaard graphs. Now we include slides in our consideration as well. Given a Haken 3-manifold N with a fixed Haken 2-complex $\Psi \subset N$, we define the *combinatorial length* of a Heegaard graph in N to be the smallest number of its points of intersection with Ψ modulo isotopies *and* slides. There is certainly a great variety of isotopies and slides for Heegaard graphs, and, using them, one can easily increase the intersection of Heegaard graphs with Ψ. But we will see that this is essentially the only way for creating large intersections, i.e., large intersections can always be reduced by using isotopies and slides alone. More precisely, we will show that the maximal combinatorial length grows polynomially, i.e., that there is a polynomial $p[x] \in \mathbf{Z}[x]$ associated to (N, Ψ) such that the combinatorial length of *every* Heegaard graph $\Gamma \subset N$ is smaller than $p(|\chi_0\Gamma|)$. This is the Finiteness Theorem for Heegaard graphs (see thm. 31.6). To prove it we need the help of almost all results from this chapter. As a first application of the Finiteness Theorem we show that the Tietze spectrum has polynomial growth (see thm. 31.9).

31.1. *Special Slides.*

The idea for the Finiteness Theorem is to look at the effect of isotopies and slides not only on Γ but on an entire Heegaard 2-complex Ω with $\Gamma_\Omega = \Gamma$. We have already seen how to put Heegaard 2-complexes into useful etc. positions by using certain locally defined isotopies. We now use Ω to introduce a certain type of slides with two important properties: firstly, they have only a small effect on Ω, and, secondly, they generate the collection of all possible slides in a rather efficient way. These slides are called "special slides". The Finiteness Theorem will be proven through the use of special slides and the locally defined isotopies defined before.

Let N be a Haken 3-manifold. Let $\Omega \subset N$ be a Heegaard 2-complex and let e be any edge from its Heegaard graph Γ_Ω. Given e, denote by $\underline{d}(e)$ the collection of all faces of $(\mathcal{D}_\Omega, \underline{d}_\Omega)$ corresponding to the edge e (we view $\partial \mathcal{D}_\Omega$ as an edge path in $\Gamma_\Omega \cup \partial N$). Let d be a face from $\underline{d}(e)$, i.e., a face contained in e. Then a *sliding-disc for d in* Ω is a 2-, 3- or 4-faced disc $A \subset \mathcal{D}_\Omega$ with one face equal to $(\partial A - \partial \mathcal{D}_\Omega)^-$ and the other faces in bound or free faces of $(\mathcal{D}_\Omega, \underline{d}(e))$ such that

(1) $A \cap d$ is non-empty and connected, and
(2) $A \cap \bigcup \underline{d}(e)$ is non-singular in Γ_Ω.

(recall that we view $\partial \mathcal{D}_\Omega$ as an edge-path in $\Gamma_\Omega \cup \partial N$). Every such sliding-disc A gives rise to a slide for Γ_Ω. Indeed, since $A \cap d$ is an arc in Γ_Ω, it follows that

$$\Gamma'_\Omega = \Gamma'_\Omega(A) := (\Gamma_\Omega - (A \cap d))^- \cup (\partial A - \partial \mathcal{D}_\Omega)^-$$

is a new Heegaard graph in N slide-equivalent to Γ. We say Γ'_Ω is obtained from Γ_Ω by a *special slide* (of the edge e across the sliding-disc A).

But A gives also rise to a slide for Ω. To see this note that the special slide across A is the result of a certain isotopy, namely the isotopy α_t in A which pushes the arc $d(A) := A \cap d$ across A and into $(\partial A - \partial \mathcal{D}_\Omega)^-$. Now, this isotopy extends to an ambient isotopy of $(N - U(\Gamma(A)))^-$, where $\Gamma(A) := (\Gamma_\Omega - (A \cap d))^-$. Denote this isotopy by α_t again. Collapsing $U(\Gamma(A))$ to $\Gamma(A)$, we obtain from

$$U(\Gamma(A)) \cup \alpha_1(\mathcal{D}_\Omega) \cup \alpha_1(d(A))$$

a new Heegaard 2-complex $\Omega' = \Omega'(A)$. We say that Ω' is obtained from Ω by a *special slide* (of the edge e across the sliding-disc A). The following picture depicts the effect of a special-slide on Ω (or rather: its meridian-system $\mathcal{D}_\Omega$):

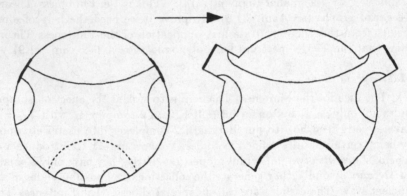

(Figure 31.1.)

31.2. *Finiteness for Heegaard Graphs.*

We begin with an important special case of the Finiteness Theorem. In fact, both the proof and the formulation of the next proposition are important for us. Indeed, the proof shows how to locate special slides and how to deal with the geometric obstruction to their existence. To formulate the result we define $\Gamma(e) := (\Gamma - e)^-$ and $N(e) := (N - \Gamma(e))^-$, for every edge e of a Heegaard graph $\Gamma \subset N$.

31.3. Proposition. *Let N be a Haken 3-manifold without non-trivial, essential Stallings fibrations (see 30.1). Let $\Psi \subset N$ be a great and useful Haken 2-complex. Let $\Gamma \subset N$ be a Heegaard graph and let $e \subset N$ be an edge of Γ. Let $\Omega \subset N$ be a Heegaard 2-complex with $\Gamma = \Gamma_\Omega$ and $\#\underline{d}_\Omega$ minimal. Suppose the induced Haken graph $\Delta_\Omega \subset \mathcal{D}_\Omega$ is useful and α-dominated, $\alpha \in \mathbf{N}$.*

Then there is a Heegaard string or Heegaard wand $e' \subset N(e)$ (see 19.15) which is slide-equivalent in $N(e)$ to e and for which $\#(e' \cap \Psi)$ resp. $\#(\text{ring}(e') \cap \Psi)$ is smaller than some constant $\varphi(\alpha)$ depending on N, Ψ and α alone.

Proof. We divide the proof in several steps. In each step we may or may not find the desired (special) slides. If we do, then we stop. If we do not, then we take the next step and probe deeper into the situation at hand. For ease of notation, we abbreviate

$$\Gamma = \Gamma_\Omega, \quad (\mathcal{D}, \underline{d}) = (\mathcal{D}_\Omega, \underline{d}_\Omega) \quad \text{and} \quad \Delta = \Delta_\Omega.$$

Step 1. By hypothesis, the induced Haken graph Δ is useful and α-dominated. Thus it satisfies the hypothesis of thm. 28.36. It therefore follows from this very theorem that e has at least one α-companion (strictly speaking e has a $\varphi(\alpha)$-companion, but for simplicity we write α for $\varphi(\alpha)$). By def. 28.2, this means that there is a face $d \in \underline{d}(e)$ and an arc $\ell(d) \subset \partial D$ such that

(1) $dist_\Delta(\partial \ell(d)) \le \alpha$, and

(2) $\ell(d)$ contains d, but no other face from $\underline{d}(e)$.

In particular, there is an arc $c(d) \subset \mathcal{D}$ with $\partial c(d) = \partial \ell(d)$ and $length_\Delta(c(d)) = dist_\Delta(\partial \ell(d)) \le \alpha$. In the terminology of def. 28.2, such an arc is an α-companion of d. Note that there is a unique disc $A(c(d)) \subset \mathcal{D}$ associated to $c(d)$ with $d \subset A(c(d))$ and $(\partial A(c(d)) - \partial \mathcal{D})^- = c(d)$.

Suppose $A(c(d))$ is a sliding-disc for d (in the sense of 31.1). Then $A(c(d)) \cap \bigcup \underline{d}(e) = d \subset A(c(d))$. So there is special slide of e (across $A(c(d))$) which pushes e into an arc (namely $c(d)$) which in turn intersects Ψ in a small number of points. Hence, in this case, the proof of the proposition is finished.

Thus we suppose that $A(c(d))$ is not a sliding-disc. In fact, we suppose that there is no α-companion c' for d whatsoever whose associated disc $A(c')$ is a sliding-disc. In particular, $dist_\Delta(\partial d)$ is large.

Step 2. Since $A(c(d))$ is not a sliding-disc, it must meet other faces from $\underline{d}(e)$. Thus one or two end-points of $c(d) = (\partial A(c(d)) - \partial \mathcal{D})^-$ lie in faces from $\underline{d}(e)$. To fix ideas suppose both end-points lie in faces from $\underline{d}(e)$ (the argument in the other case is the same). Let $d_1, d_2 \in \underline{d}(e)$ be the two faces containing the end-points of $c(d)$. Then $A(c(d)) \cap \underline{d}(e)) \subset d_1 \cup d \cup d_2$ (see fig. 31.2).

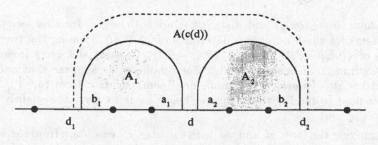

(Figure 31.2.)

Fix the boundary-pattern of $A(c(d))$ whose bound faces are given by the intersection of $A(c(d))$ with the bound faces from $(\mathcal{D}, \underline{d}(e))$. Then, by assumption, $A(c(d))$ is a 6-gon. Since $dist_\Delta(\partial d)$ is large, it follows from prop. 28.35 the existence of a system $\mathcal{P}(d)$ of at most five bands in $A(c(d))$ satisfying the conclusion of prop. 28.35. All these bands have a large width and join d with some other face of $A(c(d))$. Hence, in particular, none of them can actually end in $c(d)$ since $length_\Delta(c(d))$ is small. So, in fact, $\mathcal{P}(d)$ has at most four bands. Let $P_i \in \mathcal{P}(d)$, $i = 1, 2$, be the band joining d with d_i. Then P_1 or P_2 is non-empty since d has no α-companion whose associated disc is a sliding-disc. To fix ideas we suppose that both P_1 as well as P_2 are non-empty (the proof in the other case is similar, in fact simpler). Let $A_i \subset A(c(d))$, $i = 1, 2$, be the unique disc with $P_i \subset A_i$, $A_i \cap c(d) = \emptyset$ and $(\partial A_i - \partial \mathcal{D})^-$ a component of $(\partial P_i - \partial \mathcal{D})^-$.

Suppose A_1, say, is a sliding disc (in which case the induced orientations of d and d_1, coming from some fixed orientation of e, must both point in the same direction). Then a special slide of $A_1 \cap d$ across A_1 turns e into a Heegaard wand e' such that $ring(e')$ intersects Ψ in a small number of points. In this case the proof of the proposition is again finished.

Thus we suppose neither A_1 nor A_2 is a sliding-disc.

Step 3. This is the hardest step. By hypothesis, A_i, $i = 1, 2$, is a square with one face, say a_i, in d and the opposite face, say b_i, in another face d_i from $\underline{d}(e)$. To make progress note that every large collection of bundle-paths is contained in (the regular-neighbourhood of) a small collection of bundle-paths. It follows (cf. the proof of 28.33 and 28.35) that there is a small collection (pencil) of pairwise disjoint bands (squares) $A_{i1}, ..., A_{in} \subset A_i$, $i = 1, 2$, with the following properties:

(1) one face a_{ij} of A_{ij}, $1 \le j \le n$, lies in a_i and the opposite face b_{ij} in another bound or free face of A_i contained in $\partial \mathcal{D}$,

(2) there is a bundle-path $\mathcal{B}_{ij} \subset A_{ij}$, joining a_{ij} with b_{ij}, such that $width(\mathcal{B}_{ij})$ is large and $U(\mathcal{B}_{ij}) = A_{ij}$ (see section 28.16) and

(3) $length_\Delta(\partial A_{ij} - \partial \mathcal{D})^-$ is small and $dist_\Delta(\partial c)$ is small, for every component c from $(a_i - \bigcup_j A_{ij})^-$.

In particular, $dist_\Delta(\partial a_{ij})$ and $dist_\Delta(\partial b_{ij})$ are both large. Moreover, $supp(\mathcal{B}_{ij})$ is a system of a small number of layer-arcs of $\Delta \cap A_{ij}$, joining the two components of $(\partial A_{ij} - \partial \mathcal{D})^-$ (recall $(\partial A_i - \partial \mathcal{D})^-$ is short and every inessential bundle-path in Δ has small width). For simplicity, we suppose that not only $dist_\Delta(c)$ but also $length_\Delta(c)$ is small, for all components c from $(a_i - \bigcup_j A_{ij})^-$ (it is not hard to modify the argument below so as to cover the slightly more general case, too).

Identifying the faces d and d_i with the edge e, we obtain from the square A_i a surface $A_i' \subset N(e)$ which is homeomorphic to either an annulus or a Möbius band (fig. 31.3 illustrates the first possibility). Using this identification, we now view both a_i and b_i as arcs in the edge e. If A'_i is an annulus, then either $a_i \subset b_i$ or $b_i \subset a_i$. If A'_i is a Möbius band, then a_i and b_i must have a non-trivial intersection (we are in Case 3). In fact, the arc $a_i \cap b_i$ is always long. But, in contrast, the next lemma claims that in general none of the arcs $a_{ij} \cap b_{ij}$, $1 \leq j \leq n$, is long.

31.4. Lemma. *There is a constant* $\beta = \beta(N, \Psi) \in \mathbf{N}$ *such that, for every* $i = 1, 2$ *and every* $j = 1, ..., n$, *either* $dist_\Delta(\partial a_{ij} \cap b_{ij}) \leq \beta$ *or* $\mathcal{B}_{ij}$ *is a bundle of arcs.*

Proof of lemma. Assume the converse. Then there is a bundle-path $\mathcal{B}_{ij}$ in A_i (different from a bundle) whose ends have a long intersection. This bundle-path defines a (closed) bundle-path $\mathcal{B}'_{ij}$ in A'_i. It follows that every (proper) essential arc in A'_i intersects $\mathcal{B}'_{ij} \cup supp(\mathcal{B}'_{ij})$ in a large number of points since $width(\mathcal{B}_{ij})$ is large. Hence the lift $\tilde{\mathcal{B}}'_{ij}$ of this bundle-path to the universal cover $\tilde{A} \to A$, $A := A'_i$, has arbitrary large size. We realize this covering map as the restriction of the fiber-projection $U(A) = A \times I \to A$, to an embedding $\tilde{A} \hookrightarrow U(A) = A \times I \subset N$. In other words we realize $\tilde{\mathcal{B}}'_{ij}$ as a bundle-path in some (open) disc in N. So, by prop. 30.3, it cannot be essential (unless it wraps around a boundary-parallel torus in N in which case we easily remove the bundle-path by an isotopy constant outside the regular neighborhood of the boundary). W.l.o.g. all bundles from $\mathcal{B}'_{ij}$ are inessential in the terminology of section 30.1 (the proof in the other case is similar). Hence $C = A(\mathcal{B}'_{ij}) \subset A'$ cannot be a Möbius band. It follows that it is a compressible, non-proper annulus in N. Intuitively, it is like a long tube in N.

Since C is compressible, there are two disjoint discs $B_1, B_2 \subset N$ with $\partial C = \partial B_1 \cup \partial B_2$. In fact, it is easy to see that these discs can be chosen so that the Haken graphs $B_i \cap \Psi$, $i = 1, 2$, have a small number of layer-arcs. The union of $B_1 \cup C \cup B_2$ is a 2-sphere which may or may not be singular (it is singular if the interior of one of the discs B_1, B_2 meets the annulus C). In any case, there is an I-bundle, say E, over the disc (i.e, a 3-ball) and an immersion $f_C : E \to N$ which maps the lids of E homeomorphically to B_1 and B_2, respectively, and which maps the frontier of E homeomorpically to C. Consider the pull-back

$$\Phi_C := (f_C^{-1}(\Psi) \cap int(E))^-$$

Then Φ_C is a Haken 2-complex in E which contains a Haken sub-complex of depth 2 (recall $\mathcal{B}_{ij}$ is not a bundle). For the following we may assume that Ψ_C itself has depth 2, i.e., Ψ_C has a surface-filtration with layer-surfaces of index 1 and 2 but no other indices. Moreover, all layer-surfaces are discs. We may view the layer-discs of Φ_C of layer-index 1 as vertical squares in E and all other layer-discs as horizontal. Now, the total number of layer-discs of layer-index 1 is small since the bundle-path $\mathcal{B}_{ij}$ has a small length. For sake of simplicity, we suppose there is only one layer-disc of layer-index 1, say B (the argument in the other case is similar). Let S be the layer-surface from Ψ

which contains $f(B)$. Then, intuitively, $f_C(B)$ is a long band in S which intersects the 1-skeleton of Ψ in a large number of components, and $f_C(E)$ is a long solid pipe in the neighborhood of $f_C(B)$. Figure 31.3 gives two views of this situation. We have encountered a similar situation in the proof of prop. 15.7. We next adapt the argument given there to our present context.

Suppose $f_C : E \to N$ is an embedding (if not, the argument is similar). Then we write E for $f_C E$, etc. If $\Gamma \cap int(E) = \emptyset$, then there is a compression disc $D_0 \subset E$ for C whose interior does not meet Γ at all and whose boundary meets Γ in a single point. Now recall C comes from the disc $A_i \subset \mathcal{D}_\Omega$. It follows that $D_0 \cap A_i$ is an arc. Then, splitting $\mathcal{D}_\Omega$ along D_0 and straightening the resulting disc-system, we get from $\mathcal{D}_\Omega$ a new meridian-system which has strictly fewer faces than $\mathcal{D}_\Omega$. This, however, is a contradiction to our hypothesis that (2) of the proposition does not hold. If $\Gamma \cap int(E)$ meets a small number of layer-discs from Φ_C, then there is clearly again a compression disc for C in E which does not meet Γ, and we get a contradiction as before. Thus Γ meets a large number of layer-discs from Φ_C. But the number of edges of Γ is small. So there must be at least one edge, say e', of Γ which meets a large number of layer-discs from Φ_C. In particular, $\#(e' \cap \Psi)$ is large. Replacing e by e', we have a reduction since the annulus, C', associated to the edge e' intersects the compression disc for C in E in a smaller circle than C. Thus, repeating the process if necessary, we may suppose that $f_C : E \to N$ is an embedding such that $f_C(int(E))$ does not meet Γ. But this has already been excluded. The lemma is thereby established.

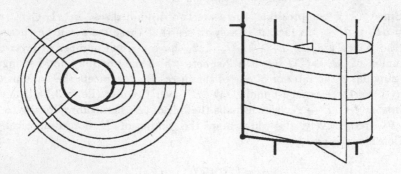

(Figure 31.3.)

Suppose A_1' is a Möbius band. Since $a_1 \cap b_1$ is long, there must be a bundle-path $\mathcal{B}_{ij}$ whose ends intersect in a long arc. But, by the previous lemma this is impossible if $\mathcal{B}_{ij}$ is not a bundle. In fact the argument from this lemma excludes also the case that $\mathcal{B}_{ij}$ is a bundle, i.e., an arc-system. Thus A_1' and A_2' are both annuli.

Since A_1' is an annulus, we have either $a_1 \subset b_1$ or $b_1 \subset a_1$. To fix ideas assume $b_1 \subset a_1$. Let y_1 be the end-point of b_1 *not* in ∂e. Now, recall that, for every point in ∂a_{ij}, there is a Δ-short arc in A_{ij} which joins this point with a face from A_{ij}, different from a_{ij} and contained in $\partial \mathcal{D}$. Using lemma 31.4, it is not hard to see that such arcs can be put together to form an arc w in A'_1, which joins the point y_1 with $\partial N(e)$ and which intersects Ψ in a small number of points. Now, a_1 and $a'_1 := (\partial A_1 - \partial \mathcal{D})^- \cup w$ are arcs in A'_1 which join the two boundary components of A'_1. It follows that a_1 and a'_1 are isotopic in A'_1. The same for A'_2. It follows that these isotopies are special slides which push e in A'_1 and A'_2 into an arc which intersects Ψ in a small number of points.

This finishes the proof of the proposition. $\Diamond$

31.5. We are now ready to prove the Finiteness Theorem. Let N be a Haken 3-manifold and let $\Psi \subset N$ be a useful Haken 2-complex. Given a Heegaard graph $\Gamma \subset N$, define the (*combinatorial*) *length of* Γ (or: Ψ-*length of* Γ) to be

$$length[\Gamma] = length_\Psi[\Gamma] := \min \#(\Gamma' \cap \Psi),$$

where the minimum is taken over all Heegaard graphs Γ' isotopic and slide-equivalent to Γ. Given this notation, we are going to prove:

31.6. Finiteness-Theorem. *Let N be a Haken 3-manifold with or without boundary and without non-trivial, essential Stallings fibrations. Let $\Psi \subset N$ be a great and useful Haken 2-complex. Then there is a polynomial $p_N[x] \in \mathbf{Z}[x]$ such that*

$$length_\Psi[\Gamma] \leq p_N(|\chi_0\Gamma|),$$

for every Heegaard graph $\Gamma \subset N$.

Proof. Let Γ' be any given Heegaard graph in N. Then we prove the theorem by showing that Γ' can be slid into a Heegaard graph each of whose edges intersect Ψ in a limited number of points (limited by some constant depending on N and Ψ alone). As usual, we do this by induction on the surface-filtration $\{\Psi_i\}_{1 \leq i \leq n}$. The induction beginning is trivial. For the induction step let $\Omega' \subset N$ be a Heegaard 2-complex with $\Gamma_{\Omega'} = \Gamma'$. Suppose this Heegaard 2-complex is chosen so that $\#\underline{d}_{\Omega'}$ is as small as possible. Moreover, according to thm. 30.15, we may suppose that Ω' is isotoped so that the above holds *and* that, in addition, Ω' is useful and α-dominated (where $\alpha = \alpha(N, \Psi)$ is the universal constant given by thm. 30.15). Then the Haken graph $\Delta' = \Omega' \cap \Psi$ is useful but not necessarily α-dominated (since we do not have an estimate for $comb(\Delta')$ and $base(\Delta')$).

Given Ω', we assume as our induction hypothesis that Ω' is slide-equivalent, via special slides (as in 31.3), to some Heegaard 2-complex Ω such that $\#(e \cap \Psi_i) \leq \varphi_i(\alpha)$, for every edge from Γ_Ω (where φ_i is a function independent of our choice of Ω').

At this point recall from 31.1 that a special slide does not really change Ω' that much. Indeed, the effect of a special slide (of an edge e of $\Gamma_{\Omega'}$) on a disc from $\mathcal{D}_{\Omega'}$ is that it subtracts some neighborhood of some of its faces (lieing in e) and attaches, at the same time, a copy of that very neighborhood to all other faces lieing in e (see fig. 31.1). In particular, there is a system of arcs (all copies of e) in the meridean-system $\mathcal{D}_\Omega$ of the resulting Heegaard 2-complex Ω which decomposes $\mathcal{D}_\Omega$ into discs (all of which are copies of parts of the original disc from $\mathcal{D}_{\Omega'}$). In the following we will refer to this decomposition as the *decomposition associated to the special slide*. The discs of this decomposition are the *pieces* of the decomposition. For every meridean-disc, there is one large piece and a number of smaller, additional pieces. The pieces intersect Δ_Ω in Haken graphs - the *pieces of* Δ_Ω. Note that the small pieces of Δ_Ω are relatively simple (they are basically just bundle-paths or unions of bundle-paths if we use the slides from 31.3). Note also that we can iterate the above construction. In this way we get a similar decomposition for sequences of special slides as well.

In order to continue note that the pieces of $\Delta := \Delta_\Omega$ have similar properties than $\Delta_{\Omega'}$. In particular, they are all useful. But we do not know whether they are α-dominated since we do not yet have an estimate for $comb()$ and $base()$. In this context, the next lemma is relevant.

31.7. Lemma *There is a universal constant $\beta = \beta(N, \Psi)$ such that $comb(\Delta_{i+1})$ $\leq \beta$ (where $\Delta_{i+1} = \Omega \cap \Psi_{i+1}$).*

Proof. Assume the converse. Then, by definition, there is top-free comb $\mathcal{C}$ in Δ_{i+1}^+ such that $width(\mathcal{C})$ is large. In particular, $\mathcal{C}$ is a Haken subgraph from Δ_i^+. Moreover, by the definition of combs (see section 28.22) there are two disjoint arcs $l_1, l_2 \subset \partial \mathcal{D}'$ such that $\mathcal{C}$ is a collection of pseudo-arcs in Δ_i joining l_1 with l_2. Now, $width(\mathcal{C})$ is supposed to be large. Thus, passing to some sub-collection if necessary, we may suppose $\mathcal{C}$ is a collection of *pairwise disjoint* pseudo-arcs (recall $depth(\Delta_i)$ is small). Moreover, every one of those pseudo-arcs consist of at most 2^d, $d := depth(\Delta_\Omega)$, layer-arcs from Δ_Ω (see lemma 28.10), i.e., a small number of layer-arcs.

For every layer-surface, S, from Ψ_i, set

$$S' := (S - U(\Gamma_\Omega))^-.$$

A proper arc in S' is called *inessential* (in S') if it separates a disc $B \subset S'$ such that $B \cap \partial S'$ is disjoint to $\partial \Psi_i$ and contains no vertex of Ψ_i. Two proper arcs $k, k' \subset S'$ are *parallel* (in S') if there is a disc $A \subset S'$ with $(\partial A - \partial S')^- = k \cup k'$ such that $A \cap \partial S'$ does not meet $\partial \Psi_i$ or any vertex of Ψ_i. Clearly, the number of parallelity classes of essential arcs in S' is small since $\#(\Gamma \cap S)$ is small.

Now, let k be a layer-arc from our comb $\mathcal{C}$. Then k lies in a (unique) layer-surface $S(k)$ from Ψ_i. More precisely, k is a proper arc in $S'(k) = (S(k) - U(\Gamma_\Omega))^-$. So it may or may not be essential.

Suppose k is inessential. Then, by definition, k separates a disc $B \subset S'(k)$ such that $B \cap \partial S'(k)$ is disjoint to $\partial \Psi_i$ and contains no vertex of Ψ_i. Hence $B \cap \partial S'(k)$ is an arc which is contained either in ∂N, or in $U(\Gamma_\Omega)$, or in an edge of Ψ_i.

If $B \cap \partial S'(k) \subset \partial N$, then k separates a disc $B' \subset D_\Omega$ with $B' \cap \partial D_\Omega \subset$ ∂N, for otherwise we could reduce the number of faces of D_Ω by cutting along B. W.l.o.g. $B \cap B' = k$. Then the union $B \cup B'$ separates a 3-ball E from N since N is ∂-incompressible. Let α_t, $t \in I$, be an ambient isotopy of N which is constant outside of the regular neighborhood of E and which pushes B' across E. Using α_t we can remove k. This operation changes $\mathcal{C}$ into another comb of pairwise disjoint pseudo-arcs. But the total number of pseudo-arcs remains the same.

If $B \cap \partial S'(k) \subset U(k)$, then, splitting D_Ω along B, we can reduce the number of faces of D_Ω. But this is excluded by our hypothesis on Ω.

If $B \cap \partial S'(k)$ is contained in an edge of Ψ, then let α_t, $t \in I$, be an ambient isotopy of N which is constant outside of the regular neighborhood of B in N and which pushes D_Ω across B and out of B. This operation changes $\mathcal{C}$ into another comb of pairwise disjoint pseudo-arcs. But the total number of pseudo-arcs remains the same. Fig. 31.4 illustrates the effect of this operation.

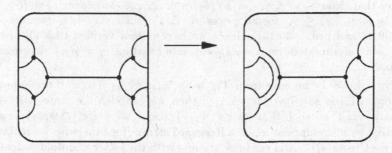

(Figure 31.4.)

Thus, altogether, we may suppose that all layer-arcs from $\mathcal{C}$ are essential in S'. Recall from above that the number of parallelity classes of essential arcs in the surface S' is small. Thus, by a simple counting argument, it follows the existence of an embedding $g : \tau_0 \times [0, p] \hookrightarrow N$, p large, such that $g(\tau_0 \times i)$ is a pseudo-arc from $\mathcal{C}$, for every integer i, $0 \le i \le p$, and that $g(k \times [0, p])$ is a parallelity region in $S'(k)$, for every layer-arc, k, from τ_0. Let $l \subset \tau_0$ be an arc joining the two ends of τ_0. Then $\bigcup_i l \times i$ is a system of pairwise disjoint arcs in D_Ω. So, in particular, there is a disc (square) $C \subset D_\Omega$ with $(\partial C - \partial D_\Omega)^- \subset \bigcup_i l \times i$ and such that $C \cap \mathcal{D}_\Omega$ consists of exactly two components joined by the arcs from $\bigcup_i l \times i$. Note that both C and $g(l \times [0, p])$ are contained in the handlebody $M(\Gamma_\Omega) := (N - U(\Gamma_\Omega))^-$. Set $X := U(C \cup g(l \times [0, p]))$, where the regular neighborhood is taken in $M(\Gamma_\Omega)$. Then X is an I-bundle whose lids are contained in $\partial M(\Gamma_\Omega)$. C and $g(l \times [0, p])$ are two vertical squares in X. Thus, using lemma 7.16, we find a homeomorphism $h : M(\Gamma_\Omega) \to M(\Gamma_\Omega)$ which is constant on $(M(\Gamma_\Omega) - X)^-$ and for which $\#(C \cap g(l \times [0, p]))$ is small (see the proof of 21.8). Set $\mathcal{D}'_\Omega := h(\mathcal{D}_\Omega)$. Then $\mathcal{D}'_\Omega$ is a meridian-system of Γ_Ω which is in general position to Ψ and which has strictly fewer faces than D_Ω. It is easily seen that this contradicts our minimality condition on $\#\underline{d}_\Omega$. Hence p cannot be large, and so the lemma is established.

According to the previous lemma, we have that $comb(\Delta_{i+1})$ is small. Using this result, it is now easy to verify that all pieces of Δ_{i+1} (i.e., the intersection of Δ_{i+1} with all pieces of the associated decomposition) are γ-dominated, for some appropriate constant $\gamma = \gamma(\alpha)$. For this we have to verify that, for each piece of Δ_{i+1}, all the complexities mentioned in def. 28.25 are smaller than some a $\gamma = \gamma(\alpha)$. This is certainly the case for $depth(..)$. Furthermore, $comb(..) \leq \gamma$, for all pieces of Δ_{i+1}, since $comb(\Delta_{i+1})$ is bounded (31.7) and since no bound face from $\mathcal{D}_\Omega$ contains the entire boundary of a layer-arc from $\Delta_{\Omega'}$ ($\Delta_{\Omega'}$ is useful). We also have that $\ell base(.., \gamma) \leq \gamma$, for all pieces of Δ_{i+1}, since Ω' is supposed to be α-dominated. Even more, the argument from prop. 30.5 shows that $\ell base(\Delta_{i+1}, \gamma) \leq \gamma$. It is easy to see that this implies that $base(.., \gamma) \leq \gamma$, for all pieces of Δ_{i+1}. Moreover, $bundle(..) \leq \gamma$ and $square(.., \gamma) \leq \gamma$, for all pieces of Δ_{i+1} since Ω' is α-dominated (cf. def. 30.14 and prop. 30.10). Hence we have indeed verified that all pieces of Δ_{i+1} are γ-dominated, for some appropriate constant $\gamma = \gamma(\alpha)$ depending on α alone.

Now, let e be an edge from Γ_Ω with $\#(e \cap \Psi_{i+1}$ large. If the hypothesis of prop. 31.3 is satisfied for Δ_{i+1}, then we can slide e into a Heegaard string or wand e' such that $\#(e' \cap \Psi_{i+1})$ resp. $\#(ring(e') \cap \Psi_{i+1})$ is small. Actually, we may suppose e' is a Heegaard string (for otherwise we replace N by $(N - U(ring(s)))^-$ and continue arguing with the latter manifold instead). In this case e' is obtained from e by a special slide as given in 31.3 (and w.l.o.g. across an outermost bundle-path). So everything goes well as long as Δ_{i+1} is useful and α-dominated. The situation is different though if Δ_{i+1} is not useful or α-dominated. Then it may well be that Δ_{i+1} has long bundle-paths with large width and in this case the above method does no longer work. However, by what we have seen above, all pieces of Δ_{i+1} are useful and α-dominated. So it follows from prop. 28.35 that any long bundle-path in Δ_{i+1} with large width is pieced together from short and essential bundle-paths (useful and α-dominated Haken graphs have no bundle-paths of large size). Now, according to prop. 30.3, any such long bundle-path must be inessential. Intuitively, it is curled-up in the direction of its length. So it is not hard to see that it can be shortened quite effectively to a short bundle-path by an isotopy (which does not change Ω' drastically). Taking into account the special nature of the special slides from 31.3, we then see again (as in the proof of prop. 28.36) that short companions (in Δ_{i+1}) for long edges of Γ_Ω do exist. As in the proof of prop. 31.3, these short companions give rise to special slides for every long edge. So we can apply the above reduction-process again. Applying this process over and over again, we get a sequence $\Omega_1, \Omega_2, \ldots$ of Heegaard 2-complexes whose associated Heegaard graphs Γ_{Ω_i} have decreasing intersections with Ψ_{i+1}. So the process must stop. But it only stops if every edge of the resulting Heegaard graph intersects Ψ_{i+1} in a limited number of points. This finishes the induction step.

The Finiteness Theorem is therefore established. $\Diamond$

31.8. *Growth of the Tietze Spectrum.*

As an application of the Finiteness Theorem we now show that the Tietze spectrum has polynomial growth. See the introduction to this chapter for the definition of the Tietze spectrum and recall from 25.2 the definitions of the distance, $d(\Gamma_1, \Gamma_2)$, between Heegaard graphs $\Gamma_1, \Gamma_2 \subset N$.

31.9. Theorem. *Let N be a Haken 3-manifold with or without boundary and without non-trivial, essential Stallings fibrations. Then there is a polynomial function $q_N : \mathbf{N} \times \mathbf{N} \to \mathbf{N}$ such that*

$$d(\Gamma_1, \Gamma_2) \leq q_N(\chi_0 \Gamma_1, \chi_0 \Gamma_2).$$

Proof. Fix a great and useful Haken 2-complex $\Psi \subset N$, and recall that $N - \Psi$ consists of open 3-balls. Moreover, recall that the closure of those open 3-balls is not only a 3-ball, but carries the boundary-pattern induced by the cell-structure of the Haken complex Ψ. Now, fix a Heegaard 2-complex $\Omega_0 \subset N$. For every open 3-ball, Q_i, of $N - \Psi$, fix one point $x_i \in Q_i - \Omega_0$. Moreover, let P_i denote the collection of all midpoints of all components into which Ω_0 splits the faces of the boundary-pattern of Q_i^-. Finally, let θ_i be the cone over P_i with cone-point x_i, and set

$$\beta_0 := \sum_i \#(\theta_i \cap \Omega_0),$$

where the sum is taken over all open balls from $N - \Psi$.

Now, let Γ be any Heegaard graph in N with $length_\Psi(\Gamma) = \#(\Gamma \cap \Psi)$. Set $E_i := (Q_i^- - U(\partial Q_i))^-$. Then E_i is a 3-ball in Q_i and $\Gamma \cap (Q_i^- - E_i)^-$ is a system of straight arcs. Contract E_i into a small ball contained in the regular neighborhood, $U(x_i)$, of the point $x_i \in Q_i$, using an ambient isotopy of Q_i^- which is constant on ∂Q_i^-. This isotopy pushes Γ so that afterwards $\Gamma \cap (Q_i^- - U(x_i))^-$ are straight arcs, i.e., the complexity of the embedding type of $\Gamma - \Psi$ is now concentrated in the neighborhood of the point x_i. In particular, it easily follows that

$$\#(\Gamma \cap \Omega_0) \leq \beta_0 \cdot \#(\Gamma \cap \Psi) = \beta_0 \cdot length(\Gamma).$$

Recall from def. 25.3, that the winding number, $w(\Gamma_1, \Gamma_2)$, of two Heegaard graphs $\Gamma_1, \Gamma_2 \subset N$ is defined as follows. First set $d_{\Gamma_1}(\Gamma_2) := \min \#(\Gamma_2' \cap \mathcal{D})$, where the minimum is taken over all meridian-systems $\mathcal{D}$ in $(N - U(\Gamma_1))^-$ and all Heegaard graphs Γ_2' in N slide-equivalent to Γ_2, and then define

$$w([\Gamma_1], [\Gamma_2]) := \min \{ d_{\Gamma_1}(\Gamma_2), d_{\Gamma_2}(\Gamma_1) \}.$$

By this very definition and by the Finiteness Theorem (thm. 31.6), it follows that

$$w(\Gamma, \Gamma_{\Omega_0}) \le \#(\Gamma \cap \Omega_0) \le \beta_0 \cdot length(\Gamma) \le \beta_0 \cdot p_N |\chi_0 \Gamma|,$$

for all Heegaard graphs in Γ in N (where $p_N(x) \in \mathbf{Z}[x]$ is some polynomial). Now, $d(\Gamma_1, \Gamma_2) \le d(\Gamma_1, \Gamma_0) + d(\Gamma_2, \Gamma_0)$. So thm. 31.9 follows from cor. 25.13 (see also remark 25.14) which relates the distance to the winding number of Heegaard graphs. $\Diamond$

32. Rigidity for Heegaard Graphs.

In this section we prove the Rigidity Theorem for Heegaard graphs.

We already know from the Finiteness Theorem (thm. 31.6) that the combinatorial lengths of Heegaard graphs in a given Haken 3-manifold N, without non-trivial, essential Stallings fibrations, are universally bounded from above. In this section we show that this puts a severe limitation on the set of Heegaard graphs. In fact, as it turns out, there can be only finitely many Heegaard graphs of given genus and given length (mod isotopies and slides) in any such 3-manifold. This is our Rigidity Theorem for Heegaard graphs (see 32.17). We deduce this theorem from the Finiteness Theorem (thm. 31.6), the Unknotting Theorem (thm. 24.5) and the general Handle Addition Lemma (prop. 23.16) without reference to any other previous result in this book. Actually, this is not all good news, for some proofs of this section would be easier if one could still work with useful Heegaard 2-complexes (unfortunately, the slides used for the Finiteness Theorem destroy this nice property).

Throughout this section N is a Haken 3-manifold and $\Psi \subset N$ is a great and useful Haken 2-complex.

32.1. *Definition of Small p-Reductions.*

Let $\Gamma \subset N$ be an irreducible Heegaard graph. By the Finiteness Theorem, we may suppose that $length(\Gamma) = length_\Psi(\Gamma)$ is a small number. As before we call an expression, such as $length_\Psi(\Gamma)$ etc., *small* if it is smaller than a known constant which may depend on Ψ, N etc. but not on the embedding type of Γ. We say the expression is *large* if it can be taken to be arbitrarily large. Again this convention is only made for convenience. Its purpose is to avoid explicit estimates as long as possible (such explicit numerical estimates can always be provided on request but in most cases they are tedious to state and in the end they only confuse the matter). Now, to make progress towards the Rigidity Theorem we begin by first refining the above length-function for Heegaard graphs somewhat.

Let $\{\Psi_i\}_{1 \leq i \leq n}$ be the surface-filtration of Ψ, and set

$$\ell_i(\Gamma) := \#(\Gamma \cap (\Psi_i - \Psi_{i-1})^-), \quad 1 \leq i \leq n.$$

Using this notation, define

$$\ell(\Gamma) = \ell_\Psi(\Gamma) := (\ell_1(\Gamma), ..., \ell_n(\Gamma))$$

with respect to the lexicographical order. We would like to use the refined length-function, $\ell(\Gamma)$, as our new complexity for Heegaard graphs. E.g., we would like to take minima with respect to $\ell(\Gamma)$ rather than $\#(\Gamma \cap \Psi)$. But *a priori* a Heegaard graph, Γ, with small $\#(\Gamma \cap \Psi)$ small may have other Heegaard graphs $\Gamma' \subset N$, slide-equivalent to Γ, with $\ell(\Gamma') < \ell(\Gamma)$ and $\#(\Gamma \cap \Psi)$ large. Thus we may run into troubles with $\#(\Gamma \cap \Psi)$ if we are not careful. To avoid damage, we will only allow "small p-reductions".

We say a Heegaard graph $\Gamma \subset N$ has a *small p-reduction* if there is a Heegaard graph $\Gamma' \subset N$ such that the following holds:

(1) Γ' is obtained from Γ by (ambient) isotopies and slides,

(2) $\ell_q(\Gamma') = \ell_q(\Gamma)$, for all $1 \leq q \leq p - 1$,

(3) $\ell_p(\Gamma') < \ell_p(\Gamma)$, and

(4) $\ell_r(\Gamma')$ differs from $\ell_r(\Gamma)$, by some small number, for all $r \geq p + 1$.

In this case we also say that Γ' is obtained from Γ by some small p-reductions. In the following we will take minima w.r.t. small p-reductions only. Much of the work of this section goes into showing that all operations on Heegaard graphs, relevant for our purpose, can indeed be taken to be small p-reductions.

32.2. *Existence of Small p-Reductions.*

Heegaard graphs with small, or even minimal, intersections with Ψ may still have small p-reductions. In this section we discuss a couple of situations in which small p-reductions come up quite naturally. E.g., this is the case whenever there is a disc in N, disjoint to Γ and with boundary in a layer-surface from Ψ, underneath of which there is some part of Γ. Criteria such as this one will be important for us. We will prove a general form of it in prop. 32.4. Its proof is based on a certain observation concerning i-faced discs, $1 \leq i \leq 3$, for Ψ. We begin with establishing this observation.

Let $(A, \underline{a})$ be an i-faced disc, $1 \leq i \leq 3$. An embedding $f : A \hookrightarrow N$ with $f^{-1}\Psi = \partial A$ is called an *i-faced disc for* Ψ if f maps every face of $(A, \underline{a})$ into a layer-surface of Ψ (but different faces into different layer-surfaces). For simplicity, we often write A for the image $f(A)$. Moreover, given an i-faced disc A, we set

$$B(A) := \text{the unique disc in } \Psi \text{ with } \partial B(A) = \partial A$$
$$\text{and contained in } i \text{ layer-surfaces.}$$
$$E(A) := \text{the unique 3-ball in } N \text{ bound by } A \cup B(A).$$

Existence and uniqueness of $B(A)$ is an easy consequence of the fact that Ψ is a great Haken 2-complex, and existence of $E(A)$ follows from the fact that $A \cup B(A)$ is a 2-sphere and the irreducible 3-manifold N. Keep also in mind that the faces of A are allowed to intersect the 1-skeleton, $\Psi^{(1)}$, of Ψ in a large number of points. So, in particular, the graph $\Psi^{(1)} \cap B(A)$ may well have a large number of edges.

32.3. Lemma. *Let $\Gamma \subset N$ be an irreducible Heegaard graph with $\#(\Gamma \cap \Psi)$ small. Let $f : A \hookrightarrow N$ be an i-faced disc, $1 \leq i \leq 3$, for Ψ with $f(A) \cap \Gamma$ connected (i.e., empty or a single point). Then there is an ambient isotopy α_t, $t \in [0, 1]$, such that*

(1) α_t is constant outside of $int(E(A))$,

(2) $\#(\alpha_1\Gamma \cap \Psi) - \#(\Gamma \cap \Psi)$ is small,

*and there is an isotopy $f_t : A \hookrightarrow N, \ t \in [0,1]$ of $f = f_0$ with $(f_t)^{-1}(\Psi) = \partial A$
such that*

(3) $\#(f_t(\alpha_1 A) \cap \alpha_1\Gamma) = \#(\alpha_1 A \cap \alpha_1\Gamma) = \#(A \cap \Gamma)$, for all $t \in [0,1]$,

(4) $f_t(A)$ is an i-faced disc for Ψ, for every $t \in [0,1]$, and

(5) $\Psi^{(1)} \cap B(f_1 A)$ is the cone over $\Psi^{(1)} \cap \partial B(f_1 A)$.

Proof. Consider the graph $\tau := B(A) \cap \bigcup_j \partial S_j$, where $\bigcup_j S_j$ is the union of all
layer-surfaces from Ψ containing a face of A. This graph is already the cone
over $\partial B(A) \cap \bigcup_j \partial S_j$ since Ψ is a great Haken 2-complex. Thus the lemma
follows without problems if, for every component B of $(B(A) - U(\tau))^-$, either
$B \cap \Gamma$ is empty or the number $\#(B \cap \Psi^{(1)})$ is small (recall $\Psi \cap int(E(A)) = \emptyset$
since $\Psi \cap int(A) = \emptyset$ and since Ψ has no layer-surfaces which are 1-, 2- or
3-faced discs).

To continue let B be a component of $(B(A) - U(\tau))^-$ such that $B \cap \Gamma \neq$
$\emptyset$ and that $\#(B \cap \Psi^{(1)})$ is large. We next describe a process for reducing
$\#(B \cap \Psi^{(1)})$ to some small number - using only isotopies as allowed by lemma
32.3. First let S be the layer-surface from Ψ containing B, and let $\theta_S, \theta'_S \subset S$
be the two Haken graphs induced by Ψ from the two sides of S (see section
27.5). W.l.o.g. let the notation be chosen so that $\theta'_S \cap B = \emptyset$ and that
$\theta_S \cap B \neq \emptyset$. Since the number of vertices of Ψ is small, it follows that $\theta_S \cap B$
has only a small number of components different from an arc and only a small
number of components meeting $U(\tau)$. Hence almost all components of $\theta_S \cap B$
are arcs whose boundaries lie in $B \cap \partial B(A)$. Let $\mathcal{K}$ be the collection of all of
those arcs. W.l.o.g. all arcs from $\mathcal{K}$ are essential in $(B - U(\Gamma))^-$, for we can
remove all other arcs by an isotopy of A not meeting Γ. Now, the number of
all parallelity classes of essential arcs from $\mathcal{K}$ is small since $\#(\Gamma \cap \Psi)$, and so
$\#(B \cap \Gamma)$, is small. Thus we simply suppose that all arcs from $\mathcal{K}$ are pairwise
parallel (the other case is similar). Let $k_1, k_2, ..., k_n$ be all arcs from $\mathcal{K}$, and
let $B_i \subset B, \ 1 \leq i \leq n$, be one of the discs into which B is split by k_i. Let
the choices be made so that (1) $B_n \subset B_{n-1} \subset ... \subset B_1$ and that (2) no disc
B_i meets $U(\tau)$. Recall that every arc from $\mathcal{K}$ lies in an edge of θ_S. So
there must be two small indices, say $i(1), i(2)$, so that $k_{i(1)}$ and $k_{i(2)}$ lie
in the same edge of θ_S. Since every edge from θ_S lies in the boundary of
some layer-surface from Ψ, we may suppose that the above holds and that,
in addition, $k_{i(1)}$ and $k_{i(2)}$ lie in a single boundary component, r, of some
layer-surface from Ψ. Moreover, we may suppose that there is a component,
say $r_{i(1)}$, of $(r - k_{i(1)} \cup k_{i(2)})^-$ for which $r_{i(1)} \cap B = \partial r_{i(1)}$ and which does
not meet θ'_S (recall $r \cap \theta'_S$ consists of a small number of points). Let $s_0 \subset B$
be an arc with $\partial s_0 = \partial r_{i(1)}$. Then the union $s_0 \cup r_{i(1)}$ is a simple closed
curve which does not meet θ'_S. Thus it can be pushed into $N - \Psi$. Hence
it is contractible in N (recall $N - \Psi$ consists of open 3-balls) and so in S

(recall S is incompressible). It follows that $r_{i(1)}$ must be an inessential arc in $(S - B(A))^-$. Let $C_{i(1)}$ be the disc in $(S - B(A))^-$ separated by $r_{i(1)}$. W.l.o.g. we may suppose $C_{i(1)} \cap B(A) \subset B$. Moreover, we may suppose that $B_{i(2)} \cap \partial A \subset C_{i(1)}$ (if $C_{i(1)}$ does not contain $B_{i(2)} \cap \partial A$, then there is a small index $i(2)$, $i(2) > i(1)$, such that $C_{i(2)} \cap C_{i(1)} = \emptyset$ and w.l.o.g. there are only a small number of such pairwise disjoint discs since $\Gamma \cap \Psi$ is small and since we can push ∂A across discs not meeting Γ).

Consider the disc $G := B_{i(1)} \cup C_{i(1)}$. Note that $a := (int(G) \cap \partial A)^-$ is a single arc and that $l := \theta_S \cap G$ is a system of arcs in G. Actually, l consists of a small number of arcs since $(G \cap \partial A) - a$ contains only a small number of end-points of $\mathcal{K}$ (recall $i(1) - i(2)$ is small). For the following we assume l is a single arc (the argument in the other case is similar). Then l splits G into two discs. Let G_1 be one of them. Now, push Γ out of G_1. Of course, this can be achieved while changing $\#(\Gamma' \cap \Psi) - \#(\Gamma \cap \Psi)$ by some small number only. Since we have $\Gamma \cap G_1$ out of the way, we can also push a out of G_1 *without meeting* Γ. Now a meets no longer l. Altogether, it follows that, by using isotopies allowed by lemma 32.3 and by increasing $\#(\Gamma \cap \Psi)$ by some small number only, we can make the number of components of $B(A) \cap \Psi^{(1)}$ is small. This finishes the proof of the lemma. $\Diamond$

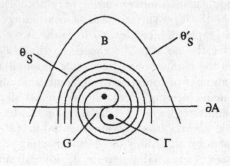

(Figure 32.1.)

Observe that, as an easy consequence of the previous lemma, we get the existence of a small p-reductions for all those Heegaard graphs Γ which allow an i-faced disc, A, for Ψ with $\Gamma \cap A = \emptyset$ and $\Gamma \cap E(A) \neq \emptyset$. But the same lemma can also be enhanced to give an even better criterion:

32.4. Proposition. *Let* $\Psi, \Gamma \subset N$ *be given as before. Suppose* $\#(\Gamma \cap \Psi)$ *is small. Let* S *be a layer-surface of* Ψ *of index* p, $1 \leq p \leq n$, *and let* $A \subset N$ *be a disc with* $A \cap S = \partial A$. *Suppose* $A \cap \Gamma$ *is connected and* $\#(\Gamma \cap A) < \#(\Gamma \cap B(A))$. *Then* Γ *has a small p-reduction.*

Proof. First some notation. W.l.o.g. A is in general position w.r.t. Γ and Ψ. Set

$$\Delta := \Delta(A) := (\Psi \cap int(A))^-.$$

Then Δ is a Haken graph in A. For every $x \in \Delta \cap \partial A$, let $k(x)$ denote the layer-arc from Δ with end-point x. Notice that $k(x)$ is a proper arc in some layer-surface, say $S(x)$, from Ψ. Let $i(x)$ be the index of the layer-surface $S(x)$, i.e., $S(x) := (\Psi_{i(x)} - \Psi_{i(x)-1})^-$. Set

$$\ell_j(A) := \text{the total number of all points } x \in \Delta \cap \partial A \text{ with } i(x) = j.$$

Using this notation, define

$$\ell(A) := (\ell_1(A),\ \ell_2(A),\ ...,\ \ell_n(A))$$

with respect to the lexicographical order. Let A be chosen so that it satisfies the hypothesis of the proposition and that, in addition, $\ell(A)$ is as small as possible. The proof of the proposition is by induction on the complexity $\ell(A)$.

If $\ell(A) = (0, ..., 0)$, then $\Delta \cap \partial A = \emptyset$ and so $A \cap \Psi = \partial A$. Hence, applying lemma 32.3, we may suppose A and Γ are isotoped so that afterwards $B(A) \cap (\theta_S \cup \theta'_S) = \emptyset$, where $\theta_S, \theta'_S \subset S$ denote the two Haken graphs induced by Ψ (see section 27.5). In this case it follows that an isotopy, which pushes $E(A)$ towards $B(A)$ and slightly beyond, defines a small p-reduction, where p is the index of the layer-surface containing ∂A. This establishes the induction beginning. The remainder of this proof is concerned with the induction step.

To carry out the induction step, we have to analyze the Haken graph $\Delta(A)$. Actually, we only need a certain subset of layer-arcs of this Haken graph. We define this subset recursively as follows. First let $\mathcal{K}_1$ be the collection of all those layer-arcs from Δ which have both end-points in ∂A. Given $\mathcal{K}_i$, let $\mathcal{K}_{i+1}$ be the collection of all layer-arcs, k, with $k \cap (\mathcal{K}_i \cup \partial A) = \partial k$ which join $\mathcal{K}_i$ with ∂A. Finally, set

$$\mathcal{K} := \bigcup_{1 \le i \le n} \mathcal{K}_i.$$

Note that $\mathcal{K}$ consists of "inessential layer-arcs". Here a layer-arc, k, from Δ is called *inessential* if it is a face of a 2- or 3-faced disc whose other faces lie in ∂A or in a layer-arc from Δ. The claim follows by a straightforward induction, using the fact that every proper arc in an i-faced disc, $1 \le i \le 3$, is inessential.

Our starting point is now the trivial observation that $\Delta \cap \partial A$ is empty if and only if $\mathcal{K}_1$ is empty. Due to this observation we will attempt to remove $\mathcal{K}_1$. This, however, will require to remove the higher collections $\mathcal{K}_i$, $i \ge 2$, first. The idea will be to start with the top-collection and work ourselves downwards.

First some preparation. Given an arbitrary arc k from $\mathcal{K}$, let $S(k)$ be the layer-surface from Ψ containing it. Then k is a proper but inessential arc in $S(k)$ (as usual the boundary-pattern of $S(k)$ is the one induced by Ψ, i.e., its bound faces lie in other layer-surfaces from Ψ but adjacent bound faces lie in different layer-surfaces). Indeed, these statements follow easily from the fact that Ψ is a great Haken 2-complex. Thus k separates a unique 2-

resp. 3-faced disc $C(k)$ from $S(k)$. We say an arc $k \in \mathcal{K}$ is *outermost* if $C(k) \cap \mathcal{K} = k$.

32.5. Lemma. *The number of all outermost arcs from $\mathcal{K}$ is small.*

Proof of lemma. Assume the converse. Let $\mathcal{O} \subset \mathcal{K}$ be a large collection of outermost arcs. Since the number of layer-surfaces from Ψ is small, we may suppose w.l.o.g. that $\mathcal{O}$ lies in one of them. Since layer-surfaces have only a small number of faces, we may suppose that $\partial \mathcal{O}$ is contained in one of those. In particular, every $k \in \mathcal{O}$ separates a *2-faced* disc $C(k)$ from $S(k)$. All these 2-faced discs are pairwise disjoint since the arcs from $\mathcal{O}$ are outermost. But $\#(\Gamma \cap \Psi)$ is supposed to be small. Thus and since $\mathcal{O}$ is assumed to be large, it follows that there is at least one $k \in \mathcal{O}$ such that $C(k) \cap \Gamma = \emptyset$. Moreover, observe that all arcs from $\mathcal{O}$ must be proper in A. So k splits A into two discs. Let $A(k)$ be one of them. Note that $A(k) \cap C(k) = k$ since k is outermost. Hence both $A(k) \cup C(k)$ and $(A - A(k))^- \cup C(k)$ are discs in N with boundaries in S. Thus, after a small general position isotopy, at least one of them, say A', is a disc which satisfies the hypothesis of the proposition. But it is easily verified that $\ell(A') < \ell(A)$. This in turn is impossible by our minimality condition of A. Hence our assumption was wrong, and the lemma is established.

Intuitively, the arcs from $\mathcal{K}$ cluster around a small collection of arcs. Thus each $\mathcal{K}$ is the union of a small number of "clusters". To be more precise, let k, k' be two arcs from $\mathcal{K}$. Then we say $k' \in \mathcal{K}$ is *parallel to* k (or: k is *parallel to* k') if (1) $C(k) \subset C(k')$, if (2) every component of $(C(k') - C(k))^- \cap \partial S(k)$ is contained in a face of $S(k)$, and if (3) every arc from $(C(k') - C(k))^- \cap \mathcal{K}$ joins the two components of $(C(k') - C(k)) \cap S(k)$. Parallelity is an equivalence relation, and parallelity classes are called *clusters in* $\mathcal{K}$. Of course, $\mathcal{K}$ is the union of its clusters. In fact, it must be the union of a small number of clusters since, by lemma 32.5, the number of all outermost arcs is small. Every cluster is contained in a single $\mathcal{K}_i \subset \mathcal{K}$, for some unique index i, and also contained in a single layer-surface from Ψ. The *index* of a cluster is defined to be the index of the layer-surface containing it. Let

$$\ell_j(\mathcal{K}(A)) := \text{the number of all clusters in } \mathcal{K} \text{ with index } j.$$

Using this notation, define the *cluster-complexity* of $\mathcal{K}$ to be

$$\ell(\mathcal{K}(A)) := (\ell_1(\mathcal{K}(A)), \ell_2(\mathcal{K}(A)), ..., \ell_n(\mathcal{K}(A)))$$

with respect to the lexicographical order. We are now ready to formulate the crucial step in the proof of the proposition.

32.6. Lemma. *Suppose Δ is not trivial. Then there is an ambient isotopy α_t, $t \in [0,1]$, of N which is constant on $A \cup U(B(A))$ and there is an isotopy $\beta_t : A \hookrightarrow N$, $t \in [0,1]$, of the inclusion $A \subset N$ such that the following holds:*

(1) $\#(\alpha_1\Gamma \cap \Psi) - \#(\Gamma \cap \Psi)$ *is small,*

(2) $\beta_t A \cap S = \partial\beta_t A$ *and* $\#(\beta_t A \cap \alpha_1\Gamma) = \#(A \cap \Gamma)$, *for all* $t \in [0, 1]$, *and*

(3) $(\, \ell(\beta_1 A),\ \ell(\mathcal{K}(\beta_1 A))\,) < (\, \ell(A),\ \ell(\mathcal{K}(A))\,)$ *w.r.t. the lexicographical order.*

Proof. To prove this lemma we have to take a closer look at clusters. It turns out that it is convenient to distinguish between "singular" and "non-singular" clusters. We explain this notation first. Let $\mathcal{L} = l_1 \cup ... \cup l_p$ be a cluster in $\mathcal{K}_1$, say. Let the indices be chosen so that l_p is the (unique) outermost arc in this cluster. Recall every l_i, $1 \leq i \leq p$, is proper in A and so separates A into two discs. Let $A_i := A(l_i)$ be one of them. We say $\mathcal{L}$ is *non-singular* if the discs A_i can be chosen so that the above holds and that, in addition, $A_1, ..., A_p \subset A$ are pairwise disjoint. Otherwise it is *singular*. In a similar way we distinguish between singular and non-singular clusters in higher $\mathcal{K}_i$, $i \geq 2$.

Case 1. All clusters are non-singular.

Let $\mathcal{L} = l_1 \cup ... \cup l_p$ be a cluster in $\mathcal{K}_1$ (and let the indices be chosen as before). Since $\mathcal{L}$ is non-singular (we are in Case 1), there is a disc $A(\mathcal{L}) \subset A$ such that $A(\mathcal{L}) \cap C(l_1) = l_1$ and that $\partial(A(\mathcal{L}) \cup C(l_1))$ bounds a disc $B(\mathcal{L}) \subset S$. Actually, the disc $A(\mathcal{L})$ is unique unless $\mathcal{L} = l_1$. Set $C(\mathcal{L}) := C(l_1)$ and let $E(\mathcal{L}) \subset N$ be the (unique) 3-ball with $\partial E(\mathcal{L}) = \partial(A(\mathcal{L}) \cup B(\mathcal{L}) \cup C(\mathcal{L}))$. The following picture illustrates the way $B(\mathcal{L})$ may intersect the disc $B(A) \subset S$ bounded by ∂A.

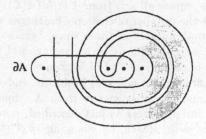

(Figure 32.2.)

Suppose first that $E(\mathcal{L}) \cap E(\mathcal{L}') = \emptyset$, for all clusters in $\mathcal{K}$ different from $\mathcal{L}$. Then $A(\mathcal{L}) \cap (\Delta(A) \cup \partial A) = \partial A(\mathcal{L})$, and so $A(\mathcal{L}) \cap \Psi = \partial A(\mathcal{L})$. Hence lemma 32.3 applies. Thus, using isotopies as in that lemma if necessary, we may suppose that Γ and A are isotoped so that afterwards ∂A bounds a disc $B(A) \subset \Psi$ for which $\Psi^{(1)} \cap B(A)$ is the cone on $\Psi^{(1)} \cap \partial B(A)$. But then let γ_t, $t \in [0, 1]$, be an ambient isotopy of the regular neighborhood $U(B(\mathcal{L}))$ in S which is constant outside of $U(B(\mathcal{L}))$ and which pushes $B(\mathcal{L}) \cap \partial A$ out of $B(\mathcal{L})$. Let β_t be the linear extension of this isotopy to an ambient isotopy of N which is constant outside of the regular neighborhood of $B(\mathcal{L})$

in N. Then, by construction, $\#(\beta_t A \cap \Gamma) = \#(A \cap \Gamma)$, for all $t \in [0,1]$, and $\#(\beta_1 \Gamma \cap \Psi) - \#(\Gamma \cap \Psi)$ is small. Moreover, it is easily verified that $\ell(\beta_1 A) \leq \ell(A)$ and $\ell(\mathcal{K}(\beta_1 A)) < \ell(\mathcal{K}(A))$. Thus $(\ell(\beta_1 A), \ell(\mathcal{K}(\beta_1 A))) < (\ell(A), \ell(\mathcal{K}(A)))$, and the lemma follows.

Suppose next that there are other clusters $\mathcal{L}' \subset \mathcal{K}$ with $E(\mathcal{L}') \cap E(\mathcal{L}) \neq \emptyset$. Then $\mathcal{K} \cap int(A(\mathcal{L}))$ need no longer be empty. Thus we have to remove all arcs from $\mathcal{K} \cap int(A(\mathcal{L}))$ first, before we can apply the process described above. Of course, we have to do this in such a way that $\#(\Gamma \cap \Psi)$ is increased by some small number only while the number of clusters is not increased at all. Here is the process for achieving these goals. First notice that the layer-arcs from $\mathcal{K}$ are either proper in A or not. Let k be first a proper layer-arc from $\mathcal{K} \cap A(\mathcal{L})$. Let $\mathcal{L}'$ be the cluster containing k. Then $A(\mathcal{L}) \cap E(\mathcal{L}')$ is a system of pairwise disjoint discs since we are in Case 1. Every component, B, from $A(\mathcal{L}) \cap E(\mathcal{L}')$ is a disc which splits $E(\mathcal{L}')$ into two 3-balls. Let $E(\mathcal{L}', B)$ be that 3-ball of them which contains the outermost arc from $\mathcal{L}'$. In this notation there is exactly one component, B_0, from $A(\mathcal{L}) \cap E(\mathcal{L}')$ with $A(\mathcal{L}) \cap E(\mathcal{L}') = A(\mathcal{L}) \cap E(\mathcal{L}', B_0)$. W.l.o.g. we may suppose k is chosen so that $k = (\partial B_0 - \partial A)^-$. Moreover, we may suppose k is chosen so that, in addition, B_0 contains no proper arc from $\mathcal{K}$ other than k. Suppose for a moment also that B_0 contains no non-proper arcs from $\mathcal{K}$ either. Then we are in the special situation discussed above. So we may suppose Γ is deformed, by an isotopy which increases $\#(\Gamma \cap \Psi)$ by some small number only, so that afterwards $E(\mathcal{L}', B_0) \cap \Gamma = \emptyset$. Then, pushing the disc $B_0 \subset A$ into $C(k)$ and a little further, we remove all arcs from $A(\mathcal{L}) \cap E(\mathcal{L}', B)$. Applying this process several times, we remove all arcs from $\mathcal{K} \cap int(A(\mathcal{L}))$. Of course, none of these processes increases the number of clusters. Furthermore, we have to apply these processes only a small number of times since, because of lemma 32.5, the total number of clusters is small. Hence, by increasing $\#(\Gamma \cap \Psi)$ by some small number and the number of clusters not at all, we can remove all proper arcs from $A(\mathcal{L}) \cap \mathcal{K}$. Non-proper arcs create no extra problems. Indeed, it is not hard to see that the number of non-proper layer-arcs from $\mathcal{K}$ must be small since we are in Case 1. Thus a similar process as just described, removes all non-proper arcs from $A(\mathcal{L}) \cap \mathcal{K}$ as well. Hence, by changing $\#(\Gamma \cap \Psi)$ by some small number and the number of clusters not at all, we may suppose $\mathcal{K} \cap A(\mathcal{L}) = k_1$. Then we are in the special situation discussed before and $(\ell(A), \ell(\mathcal{K}(A)))$ can be reduce by the argument used there.

So, altogether, the lemma is established in Case 1.

Case 2. $\mathcal{K}$ has singular clusters.

We distinguish between two types of singular clusters: those with and those without "folds". To explain this notation, let $\mathcal{L} = l_1 \cup \ldots l_p$ be a singular cluster in $\mathcal{K}_1$. Then $\partial \mathcal{L} \subset S$, and let S' be the layer-surface from Ψ with $\mathcal{L} \subset S'$. Let the discs $C(l_i) \subset S'$, $1 \leq i \leq p$, be given as before. In particular, suppose $C(l_{i+1}) \subset C(l_i)$. Set $D_i := (C(l_i) - C(l_{i+1}))^-$, $1 \leq i \leq p-1$. Then D_i is a square with two faces in A and the other faces in S (recall $\partial \mathcal{L}$ is supposed to lie in S). Moreover, $(A \cup S) \cap D_i = \partial D_i$. Since A is a disc and since S is incompressible, it follows that ∂D_i bounds at least one disc in $A \cup S$. More precisely, D_i lies either in $E(A)$ or in $(N - E(A))^-$. But S is not the 2-sphere since N is irreducible and Ψ is a great Haken 2-complex. Hence it follows that D_i bounds two discs in $A \cup S$ if $A \subset E(A)$ but only one disc if $A \subset (N - E(A))^-$. Let $D_i' \subset A \cup S$ be the unique disc with $\partial D_i' = \partial D_i$ if $D_i \subset (N - E(A))^-$. We say D_i is a *fold* if

(1) $D_i \subset (N - E(A))^-$ and $D_i' \cap A$ is connected, or

(2) $D_i \subset E(A)$ and $D_{i-1}' \cap D_{i+1}' \neq \emptyset$.

We say $\mathcal{L}$ is *folded* if it has a fold. Note that a folded cluster may have several folds. Fig. 32.3 shows the arcs $C(l_1) \cap S$ of a singular cluster $\mathcal{L}$. The picture on the left has no folds whatsoever and the picture on the right has lots of folds. The figure also illustrates how large singular clusters can occur in the presence of Heegaard graphs which intersect Ψ in a small number of points only (indicated by the dots in the picture).

Suppose $\mathcal{L}$ is a large cluster which is not folded. Then $D_i' \cap S$ is connected, for every square D_i with $D_i \subset (N - E(A))^-$. For every square D_i with $D_i \subset E(A)$, let $D_i' \subset \partial E(A) \subset A \cup S$ be the disc with $A \cap D_i' = A \cap (D_{i-1}' \cup D_{i+1}')$. Finally, set $B_i = D_i' \cap S$. Since $\#(\Gamma \cap S)$ is small, only a small number of discs B_i are pairwise disjoint. Thus there must be a square $D_i \subset E(A)$ which is parallel to a large number of squares D_j. Actually, it is not hard to see that there must be a square D_i with $D_i \cap \Gamma = \emptyset$ and $\partial D_p \subset D_i'$ (recall l_p is outermost). Thus $D_i \cap \Gamma \neq \emptyset$ and $D_i \cap \Gamma = \emptyset$. Set $A' := D_i \cup (A \cap D_i')$. Then, after a small general position isotopy, A' is a disc which satisfies the hypothesis of the proposition and such that $\ell(A') < \ell(A)$, and we are done.

Suppose $\mathcal{L}$ is a folded cluster. If $\mathcal{L}$ has only a small number of folds, then there is large sub-cluster with no folds and the lemma follows as before. So suppose $\mathcal{L}$ has a large number of folds. Two folds D_i, D_{i+2} are called *adjacent*. A fold-cluster is a maximal collection of adjacent folds $D_i, D_{i+2}, ..., D_{i+2m}$. Every fold-cluster gives rise to a product I-bundle not meeting Γ with lids in S. Since χS as well as $\#(\Gamma \cap \Psi)$ is small, there is only a small number of such I-bundles and the intersection of $C(l_1)$ with these I-bundles has only a small number of components. But every such I-bundle has a unique completion to an I-bundle over the disc (e.g., the I-bundle joining the two shaded discs in the picture on the right of fig. 32.3). Hence, by the trick used for lemma 32.3, increasing first $\#(\Gamma \cap \Psi)$ by some small number if necessary, we can isotope $C(l_1)$ in these I-bundles without meeting Γ so that afterwards it intersects A in these I-bundles in a small number of arcs only. This reduces the number of folds to a small number. After that we are in the situation discussed before.

This proves the lemma.

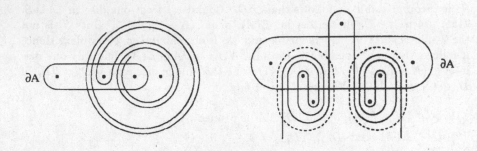

(Figure 32.3.)

By the previous lemma we can isotope Γ into Γ' and A in $N - \Gamma'$ into A' such that $\#(\Gamma' \cap \Psi) - \#(\Gamma \cap \Psi)$ is small and that $\ell(A') \leq \ell(A)$ and $\ell(\mathcal{K}(A')) < \ell(\mathcal{K}(A))$. But $\mathcal{K}(A)$ consists of a small number of clusters and the same with $\mathcal{K}(A')$, etc. Thus, after a small number of steps, the above process has to reduce $\ell(A)$. Then the proposition follows from the induction hypothesis. $\Diamond$

32.7. Corollary. *Let $\mathcal{D}$ be a system of meridian-discs for Γ. Suppose that $\mathcal{D}$ is chosen so that $\Delta := \mathcal{D} \cap \Psi$ is a Haken graph whose number of layer-curves which are proper in $\mathcal{D}$ is as small as possible. Then Γ has a small p-reduction, provided Δ has a closed curve as layer-curve.*

Proof. Let k be a layer-curve of Δ which is closed. Since $\mathcal{D}$ consists of discs, there is a disc $A \subset \mathcal{D}$ with $\partial A = k$. Let k be chosen so that, in addition, A contains no other closed layer-curve from Δ. Then all layer-arcs from Δ contained in A have strictly higher index than k. Thus there must be a layer-surface S from Ψ with $A \cap S = \partial A$. Of course, $\Gamma \cap A = \emptyset$. Moreover, because of our minimal choice of $\mathcal{D}$, we have $\Gamma \cap B(A) \neq \emptyset$. Hence A satisfies the hypothesis of prop. 32.4. Thus the corollary follows from that proposition. $\Diamond$

32.8. *Edge-Paths and Trees.*

Before we proceed we have to establish a certain technical result concerning edge-paths in finite graphs which we need in the proof of the next proposition. To formulate it let Λ be a finite graph and let $w : I \to \Lambda$ be an edge-path. Then, by definition, $w^{-1}\Lambda^{(0)}$ is a finite set of points. The points from $w^{-1}\Lambda^{(0)}$ are the *vertices*, and the closures of the components of $I - w^{-1}\Lambda$ are the *edges* of the edge-path w. The restriction of w to every one of its edges is an embedding but adjacent edges of the edge-path may well lie in the same edge of Λ, i.e., the edge-path may back-track. Note that the image $w(I)$ is a subgraph of Λ. Two vertices of w are *equivalent* if they are mapped under w to the same vertex of Λ. A pair, x, y, of vertices of w is *redundant* if $w(x) = w(y)$

and if the edge-path $w|t(x, y)$ is contractible (rel end-points) in Λ, where $t(x, y) \subset I$ is the arc with $t(x, y) \cap (x \cup y) = \partial t(x, y)$.

32.9. Lemma. *Let* $w : I \to \Lambda$ *be an edge-path. Suppose* $w(\partial I)$ *is a single vertex. Then* $w(I) \subset \Lambda$ *is a tree or there is a pair of equivalent but non-redundant vertices of* w.

Proof. The proof is by induction on the length of w, i.e., on the number of edges of w. If w has only one or two edges, then the proposition is certainly true. This establishes the induction beginning. For the induction step let w have more than two edges, and let x be that vertex of Λ equal to the end-points of w. We distinguish between two cases.

Case 1. $w^{-1}(x) = \partial I$.

Let d_1, d_2 be the two edges of w which contain end-points of w. Then $d_1 \cap d_2 = \emptyset$ since w has more than two edges.

If $w(d_1) \neq w(d_2)$, then of course w is not contractible in Λ rel end-points (we are in Case 1). Thus the end-points of w form an equivalent but non-redundant pair of vertices.

If $w(d_1) = w(d_2)$, then consider the arc $w' := w|(I - d_1 \cup d_2)^-$. Then w' is an edge-path with similar properties than w but the number of edges of w' differs from the length of w by two. Thus, by induction, w' satisfies the conclusion of the lemma. But then it easily follows that also w must satisfy the conclusion of the lemma.

Case 2. $w^{-1}(x) \neq \partial I$.

In this case the pre-image $w^{-1}(x)$ is a finite collection of points. Let $I_1, I_2, ..., I_m, m \geq 1$, be the collection of intervals into which I is split by $w^{-1}(x)$. W.l.o.g. let the indices be chosen so that the above holds and that, in addition, $I_i \cap I_{i+1} \neq \emptyset$, $1 \leq i \leq m-1$. Set $w_i := w|I_i$. Then w_i, $1 \leq i \leq m$, is an edge-path in Λ which satisfies the hypothesis of the lemma. If, for some $1 \leq i \leq m$, w_i has a pair of equivalent but non-redundant vertices, then this is certainly true for w as well, and we are done. Thus suppose the converse. Since we are in Case 2, every edge-path w_i has strictly fewer edges than w. Thus, by induction $w_i(I_i)$ is a tree in Λ, for every $1 \leq i \leq m$. For every $1 \leq p \leq m$, set $\Lambda_p := \bigcup_{1 \leq i \leq p} w_i(I_i)$. If Λ_m is a tree, then we are done since $w(I) = \Lambda_m$. So suppose the converse. Then Λ_1 is a tree but not Λ_m. Hence there is an index p, $1 \leq p < m$, such that Λ_p is a tree but not Λ_{p+1}. It follows that there must be at least one edge, e, of Λ_{p+1} such that exactly one end-point, say z, of e lies in Λ_p. Consider $w^{-1}(z)$. Then, by construction, $I_1 \cup ... \cup I_p$ as well as I_{p+1} contains a point of $w^{-1}(z)$. Let $t \subset I$ be an arc with $t \cap w^{-1}(z) = \partial t$ and which joins a point from $(I_1 \cup ... \cup I_p) \cap w^{-1}(z)$ with a point from $I_{p+1} \cap w^{-1}(z)$. Then $w|t$ is a closed edge path in Λ and the two edges of this path which have z as an end-point lie in different edges of Λ. It follows that $w|t$ cannot be contracted (rel end-points) in Λ. Hence, by definition, ∂t is a pair of equivalent but non-redundant vertices of w.

This finishes the proof of the lemma. $\Diamond$

32.10. *Equivalent Vertices and Small p-Reductions.*

Let $(\mathcal{D}, \underline{d})$ be a system of meridian-discs for Γ, where as usual the boundary-pattern is induced by the edges of Γ. Let

$$f := \text{the union of all free faces of } (\mathcal{D}, \underline{d}),$$

i.e., $f = \mathcal{D} \cap \partial N$. We view the components of $(\partial \mathcal{D} - f)^-$ as edge-paths, where the *edges* are given by the bound faces faces of $(\mathcal{D}, \underline{d})$ and where the *vertices* are given by the end-points of those faces. Recall every vertex of $(\partial \mathcal{D} - f)^-$ lies in a vertex of Γ. Every edge of $(\partial \mathcal{D} - f)^-$ lies in an edge of Γ and joins the two end-points of that edge. Thus every component of $(\partial \mathcal{D} - f)^-$ is also an edge-path in Γ. But note that in Γ this edge-path may be (and usual is) singular. In fact, even adjacent edges of that edge-path may lie in the same edge of Γ. As before, we say two vertices in $(\partial \mathcal{D} - f)^-$ are *equivalent* if they lie in the same vertex of Γ. Two vertices x, y in a component, w_0, of $(\partial \mathcal{D} - f)^-$ are called *redundant* if they are equivalent and if the edge-path $w(x, y) \subset w_0$ joining them is contractible in Γ rel end-points.

Set $\Delta := \mathcal{D} \cap \Psi$. Then Δ is a Haken graph in $\mathcal{D}$. By cor. 32.7, we may suppose that all layer-curves of Δ are arcs. Of course, Δ is non-empty. Otherwise the Heegaard 2-complex $\Omega_\Gamma = \Gamma \cup \mathcal{D}$ would lie in $(N - U(\Psi))^-$, and so in an open 3-ball. It would follow that N itself must be either a 3-ball or a 3-sphere. Every non-empty Haken graph with arc-filtration, and so also Δ, has at least one outermost disc. Here we say a disc $D_1 \subset \mathcal{D}$ is *outermost for* Δ if $(\partial D_1 - \partial \mathcal{D})^-$ is a layer-arc of Δ and if D_1 contains no other layer-arc of Δ which is proper in $\mathcal{D}$.

If D_1 is outermost for Δ, it follows that every cell of Δ in D_1, i.e., every disc $C \subset D_1$ with $C \cap (\Delta \cup \partial \mathcal{D}) = \partial C$, intersects $\partial \mathcal{D}$ in one component only. It turns out that this is a very important property for us. Indeed, this property and the existence result of the next proposition are the crucial ingredients for the proof of the existence of normal curves in Heegaard graphs (see prop. 32.15).

32.11. Proposition. *Let N be a Haken 3-manifold and let $\Psi \subset N$ be a great Haken 2-complex. Suppose $\Gamma \subset N$ is an irreducible Heegaard graph with $\#(\Gamma \cap \Psi)$ small. Let $\mathcal{D}$ be a meridian-system for Γ and let $D_1 \subset \mathcal{D}$ be an outermost disc.*

Then Γ has a small p-reduction, provided the following holds, for every component, w, of $((D_1 \cap \partial \mathcal{D}) - f)^-$:

(1) every pair of equivalent vertices in w is redundant in w, and

(2) ∂w is equivalent and redundant if $\partial w \subset f$.

Proof. Set $r := D_1 \cap \partial \mathcal{D}$. Then we distinguish between two cases:

Case 1. $r \cap f = \emptyset$.

In this case r is entirely contained in Γ.

Suppose there is no equivalent pair of vertices in r at all. Then r is a non-singular arc in Γ and $U(D_1)$ is a 3-ball. In particular, $\#(r \cap \Psi)$ is a small number (dominated by $\#(\Gamma \cap \Psi)$). Hence there is a Heegaard graph, Γ', slide-equivalent to Γ, with $(\Gamma' - U(r))^- = (\Gamma - U(r))^-$, $r \subset \Gamma'$ and such that r contains at most one vertex of Γ'. Let S be the layer-surface from Ψ which contains the layer-arc $(\partial D_1 - \partial \mathcal{D})^-$. Then S splits the 3-ball $U(D_1)$ into two 3-balls. Let E be that one of them which contains D_1. Set $A := (\partial E - S)^-$. Then A is a disc with $A \cap S = \partial A$ and such that $\Gamma' \cap A$ is connected and $\#(\Gamma' \cap A) < \#(\Gamma' \cap (S \cap U(D_1)))$. Therefore prop. 32.4 applies. So there is a Heegaard graph Γ'' obtained from Γ' by some small small p-reduction. It remains to show that Γ'' is also obtained from Γ by some small p-reduction. Note that r intersects only layer-surfaces from Ψ which have strictly higher index than S since all layer-arcs from Δ in D_1 must have strictly higher index than $k = (\partial D_1 - \partial \mathcal{D})^-$ (D_1 is chosen to be outermost). Thus $\ell_i(\Gamma') = \ell_i(\Gamma)$, for all $1 \leq i \leq p$. Moreover, $\#(\Gamma' \cap \Psi)$ differs from $\#(\Gamma \cap \Psi)$ by some small number. It therefore follows that Γ'' is also obtained from Γ by some small p-reduction.

Suppose r contains at least one pair of equivalent vertices. By hypothesis, every pair of equivalent vertices in r is redundant. Thus there is a non-trivial collection, $\mathcal{R} = \mathcal{R}(D_1)$, of pairwise disjoint, non-singular edge-paths in r such that

(1) the end-points of every component of $\mathcal{R}$ form a pair of equivalent vertices, and

(2) $\mathcal{R}$ is maximal, i.e., $\mathcal{R}$ equals every collection, $\mathcal{R}'$ as in (1) with $\mathcal{R} \subset \mathcal{R}'$.

Every component of $\mathcal{R}$ and every component of $(r - \mathcal{R})^-$ forms a tree in Γ. Moreover, the union of all those trees is a tree. Otherwise we can construct an equivalent pair of vertices of r which are not redundant (see proof of lemma 32.9). Thus r forms a tree in Γ. Now let r^* be the complement in r of the two faces of $(\mathcal{D}, \underline{d})$ containing the end-points of r. Let $T(r^*) \subset \Gamma$ be the tree formed by r^* and let x, y be the two vertices of $T = T(r^*)$ containing end-points of r^*.

Now take first the regular neighborhood, $U(x)$, of x in N and then the regular neighborhood, $U(T)$ of T in $(N - U(x))^-$. Then $U(T)$ is a 3-ball since T is a tree. $B := U(x) \cap U(T)$ is a disc and so is $A := (\partial U(T) - B)^-$, for $\partial U(T)$ is a 2-sphere. Since $U(T)$ is a 3-ball, there is an ambient isotopy α_t, $t \in [0,1]$, of N which is constant outside of the regular neighborhood of $U(T)$ in N and which pushes A rel ∂A across $U(T)$ and into B. $D_1 \cap \partial(U(x) \cup U(T))$ is an arc, r', in D next to r^*. Note that, by construction of T, r' does not meet $(\partial D_1 - \partial \mathcal{D})^-$. By definition of α_t, it follows that $\alpha_1 r' \subset \partial U(x_1)$. In general, $\alpha_1 r$ may well be a rather complicated arc in

$\partial U(x)$. To make this statement more specific, select a system, $\mathcal{C}$, of proper discs in $U(x)$ such that

(1) the component from $\partial \mathcal{C}$ are pairwise parallel in $\partial U(x)$,

(2) the intersection of every component of $\mathcal{C}$ with B is connected, and

(3) every component of $\partial U(x) - \partial \mathcal{C}$ contains exactly one point from $\Gamma \cap \partial U(x)$.

Since B is a disc, there is an isotopy, β_t, $t \in [0,1]$, in A which is constant on ∂B and which pushes $\mathcal{D} \cap B$ so that afterwards $\alpha_1 r$ intersects every arc from $B \cap \partial \mathcal{C}$ in at most one point. Finally, let g_t, $t \in [0,1]$, be the contraction which contracts $U(x)$ to an arc in $U(x)$, joining the two points from $(\Gamma - T) \cap \partial U(x)$, and which contradicts every disc from $\mathcal{C}$ to its midpoint. Define $\Gamma' = g_1 \beta_1 \alpha_1 \Gamma$ and let $\mathcal{D}'$ be the result of $\mathcal{D}$ under the above processes (see fig. 32.4).

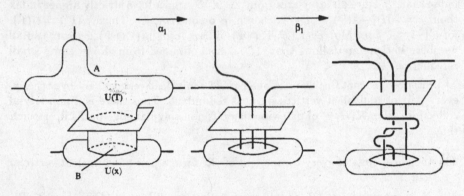

(Figure 32.4.)

Here are some properties of Γ'. First, Γ' is certainly slide-equivalent to Γ. Moreover, $\#(\Gamma' \cap \Psi) - \#(\Gamma \cap \Psi)$ is small. Also recall that the above operation does not affect $(\partial D_1 - \partial \mathcal{D})^-$. Finally, note that the layer-arcs from Δ, contained in D_1, have strictly larger indices then the layer-arc $(\partial D_1 - \partial \mathcal{D})^-$. It follows that the tree $T = T(r^*)$ does not meet the layer-surface from Ψ containing $(\partial D_1 - \partial \mathcal{D})^-$. So, if p is the index of the layer-surface containing $(\partial D_1 - \partial \mathcal{D})^-$, we have $\ell_i(\Gamma') = \ell_i(\Gamma)$, for all $1 \le i \le p$.

Now the above process straightens the edge-path r. Hence the process changes, Γ, $\mathcal{D}$, D_1 into Γ', $\mathcal{D}'$, D_1' so that $D_1' \cap \partial \mathcal{D}$ forms a non-singular edge-path in Γ. Moreover, $\#(\Gamma \cap \Psi)$ is changed by some small number only and $\ell_i(\Gamma)$ is unchanged, for all $1 \le i \le p$ (where p denotes the index of the layer-surface containing $(\partial D_1 - \partial \mathcal{D})^-$). We are therefore in the situation discussed before. Because of the above properties, the proposition follows by the argument given before.

Thus, altogether, the proposition is established in Case 1.

Case 2. $r \cap f \neq \emptyset$.

Applying a similar process as in Case 1, we may suppose that no component of $(r-f)^-$ contains a pair of equivalent vertices of $(\mathcal{D}, \underline{d})$. Thus, by hypothesis, there is no component, w, of $(r-f)^-$ with $\partial w \subset f$. Thus $f_0 = f \cap r$ is connected, and so it is an arc in $(\partial N - U(\Gamma))^-$. We now need the following lemma.

32.12. Lemma. *W.l.o.g.* f_0 *contains only a small number of end-points of* Δ.

Proof of lemma. Assume the converse. Isotope D_1, by an isotopy which fixes $D_1 \cap \partial N$ and which keeps $(\partial D_1 - \partial \mathcal{D})^-$ in its layer-surface, so that afterwards there is no 2-faced cell of Δ, i.e, there is no layer-arc from Δ which has both end-points in one other layer-arc.

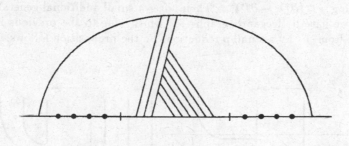

(Figure 32.5.)

Suppose there is a large collection of layer-arcs from Δ which have both end-points in f_0. These layer-arcs are inessential arcs in the layer-surfaces containing them. Let s be one of them. Then s separates a disc, $C(s)$, from the layer-surface containing it and a disc, $A(s)$, from D_1 not meeting $(\partial D_1 - \partial \mathcal{D})^-$. Moreover, $\partial(A(s) \cup C(s))$ bounds a disc $B(s) \subset \partial N$ since N is ∂-irreducible. Only a small number of discs $B(s)$ can be pairwise disjoint, for otherwise one of them contains no end-point of Γ and we get a reduction by pushing $\mathcal{D}$ across that disc. So there is a large number of inessential arcs in some layer-surface which are pairwise parallel. At least one of the corresponding parallelity regions does not meet Γ since $\#(\Gamma \cap \Psi)$ is small. Thus, cutting $\mathcal{D}$ along this parallelity region, we get a reduction. This is a contradiction. In a similar way we treat inessential arcs with only one end-point in f_0. Hence we may suppose that the number of all inessential layer-arcs with end-point in f_0 is small, i.e., almost all layer-arcs meeting f_0 are essential.

Since the depth of Δ is small, it follows that the essential layer-arcs with end-points in f_0 are the union of a small number of bundles in Δ. Thus there is at least one large bundle which joins a layer-surface S with ∂N. As usual, every large bundle of essential layer-arcs gives rise to an I-bundle containing it. This time it is a product I-bundle, X, which joins S with ∂N. But the complexity of S is minimal (see the definition of great Haken 2-complexes). Hence, in particular, X must be the I-bundle over the disc, for otherwise $(S - X)^- \cup (\partial X - S \cup \partial N)^-$ would be a non-separating surface of smaller complexity than S. Hence f_0 can be isotoped so that $\#(f_0 \cap \Psi)$ is small and $\#(\Gamma \cap \Psi)$ is changed by some small number only. This establishes the lemma.

To finish the proof we suppose $f \cap \partial(\partial D_1 - \partial \mathcal{D})^- = \emptyset$ (the argument in the other case is similar). Then notice that the arc $f_0 = r \cap f$ is either *recurrent* or not, i.e, it joins a component of $U(\Gamma) \cap \partial N$ either with itself, or with some other component of $U(\Gamma) \cap \partial N$.

If f_0 is a non-recurrent arc (see fig. 32.6 on the left), then note that every component from $(r - f_0)^-$ is an arc in Γ. Let s be one of these two arcs. Pushing vertices in s towards $s \cap \partial N$ if necessary, we may suppose s contains no vertex of Γ at all. Then slide s across the disc D_1 and into $(\partial D - \partial \mathcal{D})^-$, while fixing $s \cap (\partial D_1 - \partial \mathcal{D})^-$. Then, after a small additional general position isotopy, we have a Heegaard graph Γ' which, due to the previous lemma, is obtained from Γ by a small p-reduction. So the proposition follows.

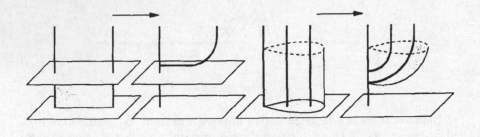

(Figure 32.6.)

If f_0 is recurrent (see fig. 32.6 on the right), then identify the two components from $(r - f)^-$. This turns the disc D_1 into an annulus which joins the layer-surface, S, containing $(\partial D_1 - \partial \mathcal{D})^-$ with ∂N. But the complexity of S is minimal (see again the definition of great Haken 2-complexes). So this annulus has to be compressible. It follows, in particular, that f_0 separates a disc from $(\partial N - B)^-$, where B denotes that component from $U(\Gamma) \cap \partial N$ which contains ∂f_0. It follows from the previous lemma, that we can push f_0 across this disc while increasing $\#(\Gamma \cap \Psi)$ by some small number. Doing this, we remove f_0, i.e., we have that $r \cap f = \emptyset$. Thus we are in Case 1, and we finish the proof by the argument given there.

Thus in any case the proposition follows. $\diamondsuit$

32.13. *Normal Curves for Heegaard Graphs.*

A *normal curve in* N (or better: in (N, Ψ)) is a non-singular curve $k \subset N$, $k \cap \partial N = \partial k$, such that $k - \Psi$ is connected and that k is in general position to the 1-skeleton, $\Psi^{(1)}$, of Ψ. Two normal curves in (N, Ψ) are *equivalent* if there is an ambient isotopy of N which preserves Ψ and which pushes one normal curve to the other. The *length* of a normal curve, k, in N is defined to be

$$length(k) := \#(k \cap \Psi^{(1)}).$$

Note that this length is an invariant for equivalence classes of normal curves. We denote

$$\mathcal{K}(N, n) = \mathcal{K}(N, \Psi, n) = \left\{ \begin{array}{l} \text{the set of all equivalence classes of normal} \\ \text{curves in } (N, \Psi) \text{ with length } n. \end{array} \right.$$

32.14. Lemma. *Let N be a Haken 3-manifold and let $\Psi \subset N$ be a Haken 2-complex. Then, for every $n \geq 1$, the set $\mathcal{K}(N, n) = \mathcal{K}(N, \Psi, n)$ is finite and constructable.*

Proof. Ψ is a 2-dim CW-complex. Let Λ be the dual graph of this complex. Then every curve in $\mathcal{K}(N, n)$ is equivalent to a curve which is the union of an edge-path in Λ of length n and an unknotted arc $k \subset N$, $k \cap \Psi = \partial k$, which joins a pair of vertices of Λ. The number of all unknotted arcs k, $k \cap \Psi = \partial k$, which join pairs of vertices of Λ is not larger than the number of all pairs of cells of Ψ. Moreover, there are clearly only a finite number of edge-paths in Λ of length n and the set of all of them can be constructed. Hence $\mathcal{K}(N, n)$ is finite and constructable. $\Diamond$

We next show the existence of short normal curves in extremal Heegaard graphs. Here a Heegaard graph Γ is *extremal* (or: *in extremal position*) if it has no small p-transformations. Note that, for every Heegaard graph in N with small intersection with Ψ, we can only apply small p-transformations a small number of times. Thus every such Heegaard graph contains an extremal Heegaard graph in its slide-equivalence class. To formulate the next result recall from section 24 the notion of Heegaard graphs which are unknotted w.r.t. a system of 3-balls. Recall $(N - U(\Psi))^-$ consists of 3-balls. Then we say an irreducible Heegaard graph $\Gamma \subset N$ is *unknotted* if it is unknotted w.r.t. $(N - U(\Psi))^-$.

32.15. Proposition. *Let $\Gamma \subset N$ be an irreducible Heegaard graph which is extremal and unknotted. Then there is an ambient isotopy α_t, $t \in [0, 1]$, such that $\alpha_1 \Gamma$ contains a normal curve with small length.*

Proof. Let $\mathcal{D}$ be a meridian-system for Γ. Set $\Delta := \mathcal{D} \cap \Psi$. Then Δ is a Haken graph and, by cor. 32.7, we may suppose all layer-curves of Δ are arcs. By an argument given before, the Haken graph Δ is not empty. Thus there is an outermost disc, i.e., a disc, D_1, such that $(\partial D_1 - \partial \mathcal{D})^-$ is a layer-arc from $\Delta = \mathcal{D} \cap \Psi$ and that D_1 contains no other layer-arc from Δ which is proper in $\mathcal{D}$. Let $r := D_1 \cap \partial\mathcal{D}$. Now Γ is extremal. Thus, by prop. 32.11, we may distinguish between the following two cases:

Case 1. There is a pair, x, y, of equivalent but non-redundant vertices of $(\mathcal{D}, \underline{d})$, in some component of $(r - f)^-$.

Let l be the arc in r which joins x and y. W.l.o.g. we may suppose that all equivalent vertices in l are redundant. Then, by the argument from prop. 32.11, we may suppose l contains no pair of equivalent vertices at all. Hence, in particular, l is a non-singular edge-path in Γ. Let C be the union of all cells, C_i, from Δ with $C_i \cap l \neq \emptyset$ and $C_i \cap \partial l = \emptyset$ (recall a *cell* is a disc, C_i, with $C_i \cap (\Delta \cap \partial\mathcal{D}) = \partial C_i$).

Suppose $(\partial C - \partial\mathcal{D})^-$ contains a small number of vertices of Δ. Consider the union $l' := (l - C)^- \cup (\partial C - \partial\mathcal{D})^-$. Notice that the arc $(\partial C - \partial\mathcal{D})^-$ is a non-singular arc in Ψ whose Ψ-length is small and that $t := (l - C)^-$ is a single arc in Γ (since the end-points of l are equivalent). Moreover, by construction of C, we have that $t \cap \Psi = \partial t$. So t is is an unknotted arc in $(N - \Psi)^-$ since Γ is an unknotted Heegaard graph. Thus, by definition, l' is a normal curve with small Ψ-length. The proposition then follows since l and l' are isotopic across the disc C. The next picture illustrates a typical but somewhat simplified situation (the dots represent vertices of $(\mathcal{D}, \underline{d})$ and the non-solid dots represent the two equivalent end-points of l).

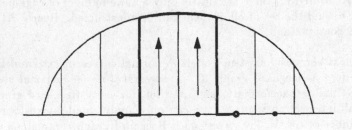

(Figure 32.7.)

Suppose $(\partial C - \partial\mathcal{D})^-$ does not contain a small number of vertices of Δ (i.e, there is no auxiliary upper bound for this number, independent of the embedding type of Γ). Recall l is a non-singular arc in Γ. Thus l contains only a small number of end-points of Δ. Thus only a small number of cells of Δ meets l. For every cell let the boundary-pattern be chosen so that the faces lie in layer-arcs from Δ and such that no two adjacent faces lie in the same layer-arc.

Suppose every cell in C has only a small number of faces. Then, by assumption, there must be a cell, C_0, in C and a face c of C which contains a large number of vertices of Δ. Then c lies in a disc in some layer-surface of S. Let θ_S and θ'_S be the two Haken graphs in S induced by Ψ. Then, more precisely, c is a proper arc in some cell, B, of θ_S, say. Recall that the Haken graph $(\theta'_S \cap int(B))^-$ has only a small number of edges since Ψ has only a small number of vertices. Thus, as usual, the arc c can be straightened in B (rel end-points) so that afterwards it intersects θ'_S in a small number of points only. In fact, isotoping Γ first, by an isotopy which increases $\#(\Gamma \cap \Psi)$ by some small number, the arc can be straightened in the complement of $\Gamma \cap B$. Carrying out this straightening-process for the faces of all cells, C_i, in C, we may suppose every face of these cells contains only as small number of end-points of Δ. Hence we are in the situation discussed before and the proposition follows.

Suppose there is a cell in C which has a large number of faces. Since D_1 is supposed to be outermost, it follows that this is only possible if there is a large number of layer-arcs of Δ which have both end-points in $(\partial D_1 - \partial D)^-$. So there is a large number of cells for Δ in C which are i-faced discs, $1 \le i \le 3$, for Ψ. But an application of lemma 32.3 shows that all these cells can be removed by an isotopy which increases $\#(\Gamma \cap \Psi)$ by some small number alone. Hence w.l.o.g. we may suppose that there is no cell in C which has a large number of faces. Thus we are back in the situation discussed above and the proposition follows.

Hence, altogether, the proposition follows in Case 1.

Case 2. There is a component, r_0, of $(r-f)^-$ whose boundary lies in f and is either not equivalent of not redundant.

We first verify that w.l.o.g. every free face of $\mathcal{D}$, i.e, every component of f, contains at least one end-point of Δ. Indeed, if there is a component f_0 of f which is disjoint to Δ, then either ∂f_0 lies in one end-point of Γ or f_0 joins two different end-points of Γ. In the latter case we remove f_0 by sliding an end-point of f_0 along f_0 and over the other end-point of f_0. This is a sliding-operation for Γ which reduces the number of its end-points without changing its intersection with Ψ. If both end-points of f_0 lie in one end-point of Γ, then f_0 is a simple closed curve. But f_0 bounds a disc in ∂N since $f_0 \cap \Psi = \emptyset$. Pushing f_0 across this disc we get again a reduction. Carrying out these operations in advance, we may suppose that every free face of $\mathcal{D}$ meets Δ.

If r_0 contains a pair of equivalent but non-redundant vertices, then we are in Case 1. Thus suppose the converse. Then, as in Case 1, we can arrange things so that r_0 contains only a small number of end-points of Δ. Moreover, note that every free face meeting r_0 must contain an end-point of Δ (for otherwise we reduce the boundary-points of Γ). Let $l \subset r$ be the arc with $r_0 \subset l$, $\partial l \subset \Delta$ and $\#(l \cap \Delta)$ minimal. Then there is an arc $l' \subset \Delta$ with $\partial l' = \partial l$. In fact, we may suppose, as in Case 1, that l' contains only a small number of vertices of Δ. Then, by definition, l' is a normal arc with small Ψ-length. The proposition then follows since l' is isotopic to l, by an isotopy which keeps end-points in ∂N and which pushes l' across the disc separated from $\overset{\cdot}{D_1}$ by l'. Fig. 32.8 illustrates a simplified version of the construction.

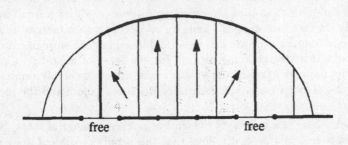

(Figure 32.8.)

This finishes the proof of the proposition. $\diamond$

32.16. *The Rigidity Theorem.*

We are now finally ready to prove our Rigidity Theorem for Heegaard graphs. Given a 3-manifold N and given an integer $n \in \mathbf{N}$, denote by $\mathcal{H}(N, n)$ the set of all slide-equivalence classes of Heegaard graphs $\Gamma \subset N$ with relative Euler characteristic $|\chi_0 \Gamma| = n$. Recall that the relative Euler characteristic is an invariant of the slide-equivalence class. Recall further that $\Gamma \leftrightarrow \partial U(\Gamma \cup \partial N) - \partial N$ induces a one-to-one correspondence between slide-equivalence classes of Heegaard graphs and isotopy classes of Heegaard surfaces.

32.17. Rigidity Theorem. *Let N be a Haken 3-manifold with or without boundary and without non-trivial, essential Stallings fibration. Then the set $\mathcal{H}(N, n)$ is finite.*

Proof. The proof is by induction on the relative Euler characteristic of Heegaard graphs. The induction beginning is trivial. For the induction step suppose $\mathcal{H}(M, m)$ is finite, for all Haken 3-manifolds M and all integers m, $m < n$. Suppose also that, for every Haken 3-manifold M, there is a great Haken 2-complex $\Psi(M) \subset M$ fixed in advance. Given $\Psi(M) \subset M$ and given $m \in \mathbf{N}$, recall from 32.13 that $\mathcal{K}(M, m)$ denotes the set of all normal curves in M with $\Psi(M)$-length smaller than m.

Now, let N be any Haken 3-manifold. We are asked to show that $\mathcal{H}(N, n)$ is finite. To do this consider a slide-equivalence class $[\Gamma]$ from $\mathcal{H}(N, n)$. Recall from 32.13 that $[\Gamma]$ contains an extremal Heegaard graph, say Γ (i.e., Γ allows no small p-transformation). Moreover, by the Unknotting Theorem (thm. 24.5), there is a Heegaard graph $\Gamma' \subset N$, slide-equivalent to Γ, such that $\Gamma' \cap \Psi(N) \subset \Gamma \cap \Psi(N)$ and that Γ' is unknotted w.r.t. to $(N - U(\Psi(N)))^-$. Since $\Gamma' \cap \Psi(N) \subset \Gamma \cap \Psi(N)$, we may suppose w.l.o.g. that Γ itself is chosen to be extremal *and* unknotted. By prop. 32.15, there is a constant $m(n)$ (which depends on n and $\Psi(N)$, but not on Γ) and an ambient isotopy α_t, $t \in [0,1]$, of N so that $\alpha_1 \Gamma$ contains a normal curve $k = k(\Gamma)$. Let

$$\varphi : \mathcal{H}(N,n) \longrightarrow \mathcal{K}(N,m(n))$$

be the map defined by the assignment $[\Gamma] \mapsto k(\Gamma)$. Since, by lemma 32.14, the set $\mathcal{K}(N,m(n))$ is finite, it remains to show that φ has finite point pre-images. This is the content of the next lemma.

32.18. Lemma. *For every* $k \in \mathcal{K}(N,m(n))$, *the pre-image of* $\varphi^{-1}(k)$ *is finite.*
Proof. Define

$$N(k) := (N - U(k))^- \quad \text{and} \quad \Gamma(k) := \Gamma \cap N(k).$$

Observe that $N(k)$ is a 3-manifold with non-empty boundary and that $\Gamma(k)$ is a Heegaard graph in $N(k)$.

We claim $N(k)$ is irreducible. Assume the converse. Then there is an essential 2-sphere, say S, in $N(k)$. By Haken's 2-sphere theorem [Ha 2] (or prop. 23.16), there is an essential 2-sphere which intersects $U(\Gamma(k))$ in a single disc. But N is irreducible and so S bounds a 3-ball in N. Thus we get a contradiction to the hypothesis that Γ is irreducible (see prop. 23.23).

We also claim $N(k)$ can be taken to be ∂-irreducible. If not, either $N(k)$ is a handlebody (in which case N a one-relator 3-manifold), or there is a system $\mathcal{D} \subset N(k)$ of essential discs such that $N^*(k) := (N(k) - U(\mathcal{D}))^-$ is a Haken 3-manifold ($N(k)$ is irreducible). Now, by the general Handle Addition Lemma (prop. 23.16), there is a Heegaard graph $\Gamma^*(k)$ in $N(k)$ which is slide-equivalent to $\Gamma(k)$ in $N(k)$ and so that $\mathcal{D} \cap \Gamma^*(k) = \emptyset$. Thus $\Gamma^*(k)$ is a Heegaard graph of $N^*(k)$. Since $\mathcal{D}$ is a system of discs, it is easily verified that every edge slide of $\Gamma^*(k)$ in $N^*(k)$ can be realized by some edge slide of $\Gamma(k)$ in $N(k)$. Thus, replacing $N(k)$ by $N^*(k)$ if necessary, we may suppose w.l.o.g. that $N(k)$ is a Haken 3-manifold (we no longer care about the intersection of Γ with $\Psi(N)$).

We finally claim that w.l.o.g. $|\chi_0\Gamma(k)| < |\chi_0(\Gamma)|$. If not, then $\chi_0\Gamma(k) = \chi_0(\Gamma)$ and k is a closed curve. In particular, the number of boundary components of $N(k)$ is strictly bigger than that of N. Since $U(k)$ contains at least one vertex of Γ, it follows that, after repeating the previous step at most $\#\Gamma^{(0)}$-times, the (absolute value of the) relative Euler characteristic is in fact diminished. This establishes the claim.

Altogether, we have show that $N(k)$ is a Haken 3-manifold and that $|\chi_0\Gamma(k)| < |\chi_0\Gamma|$. Moreover, slides and isotopies for $\Gamma(k)$ in $N(k)$ extend to similar operations for Γ in N. It therefore follows that $\varphi^{-1}(k) \approx \mathcal{H}(N(k),m)$, for some $m \leq n-1$. This finishes the proof of the lemma since $\mathcal{H}(N(k),m)$, $m \leq n-1$, is finite, by induction. $\Diamond$

The proof of the Rigidity Theorem is now complete. $\Diamond$

32.19. Remark. Note that the method described in the previous proof can be turned into an algorithm for bringing any irreducible Heegaard graph (and not just one of its sub-curves) into one of a finite set of "normal positions", using slides and isotopies only.

32.20. Corollary. *Let N be a Haken 3-manifold without non-trivial, essential Stallings fibration. Then, for every constant $n \in \mathbf{N}$, a finite set $\mathcal{G}(N, n)$ of irreducible Heegaard graphs in N can be constructed which consists of irreducible Heegaard graphs $\Gamma \subset N$ with $|\chi_0\Gamma| = n$ and contains all of them mod slides.*

Proof. Fix an integer $n \in \mathbf{N}$ and fix a great Haken 2-complex $\Psi \subset N$. Then, according to the Finiteness Theorem (thm. 31.6), the Ψ-lengths of irreducible Heegaard graphs $\Gamma \subset N$ are bounded by some known function in $\chi_0\Gamma$ (where the function depends on N and Ψ alone). In other words, the Ψ-length of every irreducible Heegaard graph $\Gamma \subset N$ with $|\chi_0\Gamma| = n$ is small. Hence, by proposition 32.15, every irreducible Heegaard graph $\Gamma \subset N$ with $|\chi_0\Gamma| = n$ contains (mod slides) a normal curve of length $m(n)$ (here $m : \mathbf{N} \to \mathbf{N}$ is some known function depending on N and Ψ alone). Now, construct the set $\mathcal{K}(N, m(n))$ of all normal curves $k \subset N$ of length $m(n)$ (by lemma 32.14 this is a constructable set). For every $k \in \mathcal{K}(N, m(n))$, construct the set $\mathcal{G}(N, n; k)$ which consists of all irreducible Heegaard graphs $\Gamma \subset N$ with $|\chi_0\Gamma| = n$ and $k \subset \Gamma$ and which contains all of those Heegaard graphs. By induction, this set is constructable (note that we can apply induction; indeed, check that we have a reduction since we basically have to construct Heegaard graphs $\Gamma' \subset M := (N - U(k))^-$ such that either $|\chi\Gamma'| < |\chi\Gamma|$ or $\#\partial M > \#\partial N$). Hence, altogether, the union $\mathcal{G}(N, n) := \bigcup_k \mathcal{G}(N, n; k)$, taken over all $k \in \mathcal{K}(N, m(n))$, is the desired finite and constructable set of Heegaard graphs. $\Diamond$

32.21. Remark. Notice that in a strict sense we have not yet *constructed* the set, $\mathcal{H}(N, n)$, of Heegaard surfaces (i.e., equivalence classes of Heegaard graphs). But we do have constructed a finite set $\mathcal{G}(N, n)$ which contains $\mathcal{H}(N, n)$. To get from $\mathcal{G}(N, n)$ to the set $\mathcal{H}(N, n)$ we would still have to decide which one of the Heegaard graphs in $\mathcal{G}(N, n)$ are equivalent under isotopies and slides. However, there is no algorithm known for this problem in general. But recall from thm. 26.27 that there is such a method for Heegaard strings. Thus the construction problem for Heegaard strings in Haken 3-manifolds (without non-trivial, essential Stallings fibrations) is completely solved.

32.22. Corollary. *The Heegaard genus is computable for every given Haken 3-manifold without non-trivial, essential Stallings fibration.*

Proof. Let N be a Haken 3-manifold without non-trivial, essential Stallings fibration and let m be any number. Then, according to cor. 32.20, we can construct the set $\mathcal{G}(N, m)$ of that corollary. In particular, we can find out whether or not $\mathcal{G}(N, n)$ is empty. So, in particular, we can determine the number n with $\mathcal{G}(N, n) \neq \emptyset$ and $\mathcal{G}(N, i) = \emptyset$, for all $0 \leq i \leq n - 1$. But this number is of course the Heegaard genus of N, and so the proof is finished. $\Diamond$

33. The Reidemeister-Zieschang Problem.

This book began with handlebodies and grew into a theory of handlebodies in Haken 3-manifolds. In order to close the circle we here return to handlebodies. Specifically, we apply the Rigidity Theorem to questions concerning curves on the boundary of handlebodies.

33.1. *Equivalence Classes of Curves.*

Given a curve, k, in the boundary, ∂M, of a handlebody, M, let us consider the sets

$$C_M^*(k) := \{\, l \mid l \text{ is a non-singular and essential curve in } \partial M$$
$$\text{homotopic in } M \text{ to } k \,\},$$
$$\mathcal{D}_M^*(k) := \{\, l \mid l \text{ is a non-singular and essential curve in } \partial M$$
$$\text{with } \pi(M, l) = \pi(M, k) \,\}.$$

Recall from section 9 that $\pi(M, k)$ denotes the fundamental group $\pi_1 M^+(k)$. Moreover, let " $\sim$ " denote the equivalence relation on the set of all curves in ∂M given by setting

$$l \sim l' \; :\Leftrightarrow \; \text{there is a homeomorphism } h : M \to M \text{ with } h(l) = l',$$

and define
$$C_M(k) := C_M^*(k)/\sim \quad \text{and} \quad \mathcal{D}_M(k) := \mathcal{D}_M^*(k)/\sim.$$

33.2. Reidemeister-Zieschang Problem. *Given a 3-manifold M, determine $C_M(k)$ and $\mathcal{D}_M(k)$.*

Let us first consider $C_M(k)$. It turns out that, for *Haken 3-manifolds*, M, the Reidemeister-Zieschang problem has a satisfactory solution for $C_M(k)$. Indeed, for every curve $l \subset \partial M$, homotopic to k, there is a map $f : S^1 \times I \to M$ with $f \mid S^1 \times 0 = k$ and $f \mid S^1 \times 1 = l$. If this map is inessential, then it can be deformed into ∂M. If it is essential, it can be deformed into the characteristic submanifold, V, of M [Joh 1, thm. 12.5]. Thus l can be deformed in ∂M either to k or into a component of $V \cap \partial M$. But V consists of I-bundle and Seifert fiber spaces and so it follows from [Joh 1, prop. 5.10], that every essential singular annulus in V can be deformed into a fiber-preserving one. The following estimate is now an easy consequence:

33.3. Proposition. *Let M be a Haken 3-manifold and let V be its characteristic submanifold. Then, for every curve $k \subset \partial M$,*

$$\# \, C_M(k) \leq \sum_R (1 + \#(V \cap R)) \leq \sum_R (3 \cdot genus(R) - 2),$$

where the sum is taken over all boundary-components, R, of ∂M. $\diamond$

Remark. Recall $\#(..)$ denotes the number of components.

Haken 3-manifolds are ∂-irreducible, and in that respect handlebodies form the other extreme case in that they are all totally ∂-reducible. Since in this case the characteristic submanifold is no longer available, the class of handlebodies has to be handled differently. We expect the result to be different as well. For instance, there now are non-trivial curves k in ∂M which are contractible in M. In fact, there are infinitely many curves on ∂M which are pairwise homotopic in M but not in ∂M.

If k is contractible, then the Reidemeister-Zieschang Problem has again a simple solution. Indeed, in this case we know from Dehn's lemma [Pa 2,St,Joh 9] that k bounds a discs. So, by the classification of discs in handlebodies modulo homeomorphisms, we have

33.4. Proposition. *Let M be a handlebody and let $k \subset \partial M$ be a simple closed curve which is contractible in M but not in ∂M. Then $\#\mathcal{C}_M(k) \leq genus(\partial M)$.* ◊

(In [Zie 1], this problem was handled without the use of Dehn's lemma. In [Zie 2] it has been further shown that $\#\mathcal{C}_M(k) = 1$, for certain curve-systems, k, on the boundary of genus 2 handlebodies; the case of a solid torus, M, being trivial).

The other extreme case occurs when $\partial M - k$ is incompressible in M. In this case we have the following result.

33.5. Proposition. *Let M be a handlebody and let $k \subset \partial M$ be a simple closed curve such that (M, k) is a simple relative handlebody. Then $\#\mathcal{C}_M(k) = 1$.*
Proof. Let $l \subset \partial M$ be a curve-system which is homotopic to k. Then there is a homotopy equivalence $f : M \to M$ with $f(k) = l$. But (M, k) and (M, l) may be considered as simple 3-manifolds with boundary-patterns. So, by [Joh 1, thm. 24.4], f can be deformed (rel k) into a homeomorphism. ◊

We finish this section with a result concerning $\mathcal{D}_M(k)$. Its proof is a combination of the Rigidity Theorem (thm. 31.22) and the classification of exotic homotopy equivalences [Joh 1].

33.6. Proposition. *Let M be a handlebody with $genus(\partial M) \geq 3$ and let $k \subset \partial M$ be a simple closed curve. Suppose $\partial M - k$ is incompressible in M and $M^+(k)$ is atoroidal. Then $\mathcal{D}_M(k)$ is finite.*
Proof. Let $k_1, k_2, \ldots$ be a sequence of simple closed, pairwise non-homeomorphic curves in ∂M with $\pi(M, k_i) \cong \pi(M, k_i)$. We have to show that this sequence has to be finite.

Set $M^+ := M^+(k)$. Then, by the handle-addition lemma (prop. 15.16), M^+ is irreducible and ∂-irreducible, and so it is a Haken 3-manifold. Moreover, by hypothesis, M^+ is atoroidal and ∂M^+ has at least one boundary component which is not a torus. Thus M^+ cannot contain an essential Stallings fibration. So the Rigidity Theorem (thm. 31.22) applies to M^+.

We claim $M^+(k_i)$, $i \geq 1$, is irreducible and ∂-irreducible. If not, then, by prop. 15.16 again, there is a disc $D \subset M$ with $D \cap \partial M = \partial D \subset \partial M - k_i$ and such that ∂D does not bound a disc in ∂M. If D is separating, then it separates a handlebody from $M^+(k)$. Thus in any case there is a non-separating proper disc in $M^+(k)$. But this is impossible since $\pi_1(M^{(}k_i) = \pi(M, k_i) \cong \pi(M, k) = \pi_1 M^+(k)$ has no non-trivial free-product decomposition.

By what we have just seen $M^+(k_i)$ is a Haken 3-manifold. Hence there is a homotopy equivalence $f_i : M^+(k_i) \rightarrow M^+$ since $\pi_1 M^+(k_i) \cong \pi_1 M^+$. It follows that $M^+(k_i)$ must be atoroidal since M^+ is.

Now, by [Joh 1, cor. 29.3], the homotopy class of any Haken 3-manifold such as M^+ contains only *finitely many* homeomorphism classes of Haken 3-manifolds. Thus, passing to a sub-sequence if necessary, we may suppose the above sequence, $k_1, k_2, ...$, has been chosen so that all 3-manifolds $M^+(k_i)$ are homeomorphic to M^+. In this case, fix a homeomorphism $h_i : M^+(k_i) \rightarrow M^+$, for every $i \geq 1$. Now let $\Gamma'_i \subset M^+(k_i)$ be the Heegaard string which is the co-core of the 2-handle from $M^+(k_i)$. Set $\Gamma_i := h_i(\Gamma'_i)$. Then $\{\Gamma_1, \Gamma_2, ...\}$ is a set of Heegaard strings in M^+. By the Rigidity Theorem, this set is finite modulo ambient isotopy. Thus the set $\{k_1, k_2, ...\}$ is finite modulo handlebody homeomorphisms. $\Diamond$

It is easy to check whether $\partial M - k$ is incompressible but, in general, it is hard to check whether $M^+(k)$ is atoroidal. However, it is straightforward to check whether (M, k) is simple (see prop. 7.3). We therefore may note the following corollary to the previous proposition.

33.7. Corollary. *Let M be a handlebody with $genus(\partial M) \geq 3$. Let k be a simple closed and separating curve in ∂M such that (M, k) is a simple relative handlebody. Then $\mathcal{D}(k)$ is finite.*

Proof. According to the previous proposition, it remains to show that $M^*(k)$ is a simple 3-manifold. Suppose the converse. Then there is an essential annulus or torus, T, in $M^*(k)$ which is not ∂-parallel. Since k is separating, it follows from prop. 15.7, that there is a surface $T^* \subset M$ obtained from T by annulus-modifications. Since T is a torus, it follows that T^* is a system of incompressible annuli. But (M, k) is simple and so all these annuli are ∂-parallel. By definition of annulus-modifications, it then follows that T is ∂-parallel. But this contradicts our choice of T. $\Diamond$

Similar results are possible for curve-systems (and not just curves) as well, but then we have to take into account the possibility of slides and the fact that dual Heegaard graphs to curve systems need no longer be minimal. Having established the finiteness of $\mathcal{D}_M(k)$ for handlebodies, M, it would of course be interesting to know more about the actual numbers $\#\mathcal{D}_M(k)$. The methods in this book and [Joh 1] could in principal be turned into an algorithm for calculating $\#\mathcal{D}_M(k)$ but the result would be a process which is so tedious that it appears to be completely useless from a practical point of view.

APPENDIX: COMPUTING SURFACES

Various parts of this book have been devoted to algorithmic problems for decision problems, etc.; mainly because of their theoretical interest. In this appendix we now wish to address the more practical question of implementing the algorithms discussed before in terms of computer programs. The goal of this section is to indicate that many of the geometric problems discussed in this book can indeed be translated into symbolic forms in which they are accessible for computers. In particular, this appendix should show, in terms of some concrete examples, the beneficial interaction of relative handlebodies, Heegaard diagrams and computers.

34.1. *Non-Separating Surfaces.*

A typical algorithmic problem, discussed in this book, is to find a minimal object in a given set of objects with respect to some given complexity-function. In general, however, these sets are infinite and only abstractly "given" by means of some property.

Take for instance the computation of the *Thurston norm*. Any torsion-free element α in the first homology $H_1(N; \mathbf{Z})$ of a 3-manifold N, determines, via Poincaré duality, a set of proper (homologous) surfaces in $(N, \partial N)$. This subset of the set of all surfaces in N is therefore abstractly determined by the property that there dual is equal to α, but this set is not given (and usually infinite) in a constructive way. Nevertheless, we are asked to search for a surface with minimal absolute Euler characteristic. At this point the computational version of the Poincaré map, as introduced in section 9, may be of help, in that it *constructs* at least one surface (whose dual is α). The absolute Euler characteristic of this surface can therefore be computed and gives already an upper bound for the Thurston norm. So we then only have to consider the set of all normal surfaces whose absolute Euler characteristic is smaller or equal to this bound. But we know that (modulo Dehn twists along annuli and tori - which incidentally do not change the Euler characteristic) the latter set is not only finite, but constructible. The actual construction in turn is a tedious process and asks for a mechanical procedure carried out by a machine. This procedure should not only construct surfaces, but should also determine their topological properties automatically. Here is were computers should come into play.

Let us now focus our discussion by considering again Thurston's example as introduced in sections 21 and 22. Recall that this example is a one-relator 3-manifold $N = M^+(k)$. Its fundamental group has the presentation

$$< g_1, g_2, g_3 \mid g_1^{-1} g_2 g_3^{-1} g_1 g_2 g_3^2 g_2 g_1^{-1} > .$$

The Heegaard diagram of this manifold is given in fig. 22.1. It gives the relative handlebody, (M, k), from fig. 33.1. This relative handlebody is simple (and so is the associated 1-relator 3-manifold, $M^+(k)$), i.e., it is ∂-irreducible and contains no strongly essential annuli. It follows that the homeomorphism type of $M^+(k)$ is given by the isomorphism type of its fundamental group [Joh 1]. If, moreover, l is another curve in ∂M homotopic in M to k with

$M^+(l) \cong M^+(k)$, then (M, k) is homeomorphic to (M, l) (see prop. 32.5). Thus (unless the one-relator presentation of N is non-unique), the curve k is the only simple closed curve (mod handlebody homeomorphisms) representing the relator of the above presentation. In order to encode this curve, and so the relative handlebody (M, k), in some symbolic form accessible to computers, we translate it into the string

$$-0.0 + 1.2 - 2.1 + 0.2 + 1.1 + 2.2 + 2.0 + 1.0 - 0.1.$$

Here $\pm i.j$ stands for the j-th intersection of the relator curve(s) with the i-th meridian-disc in the positive resp. negative direction (w.r.t. some fixed choices of the indices and orientations of the meridian-discs).

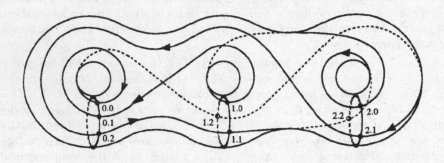

(Figure 34.1.)

It is not hard to write a computer program which calculates for strings such as the one above the basis for the null space of the Seifert matrix. It turns out that for the example at hand this null-space is two-dimensional and generated by the following basis vectors

$$[\,3\ 1\ 0\,] \text{ and } [\,1\ 0\ 1\,]$$

Given a linear combination of this vectors produces a solution vector for the Seifert matrix. For example, the linear combination $1 * [\,3\ 1\ 0\,] + 0 * [\,1\ 0\ 1\,]$ produces

$$2\ 2\ 0\ 1\ 2\ 2\ 2; 3\ 1\ 0.$$

This solution vector then represents an admissible surface S in (M, k). The entries on the right of the semicolon indicate the numbers of copies of the respective meridian-discs while the entries on the left describe how these copies are piped together (they are the x-values in the process from 14.5) to form the surface in question. In fact, the homology class of this surface is the image of a homology class, given by $[\,3\ 1\ 0\,]$, under the Poincaré map (see section

9). To produce such a surface one has to fix one of the several possible piping schemes. If one is not only interested in one example of an essential surface but in constructing all of them, then one has to take into account all the other piping schemes as well. But given one piping scheme it is usually not very difficult to produce all other piping schemes. This is due to some basic properties concerning normally oriented arc-systems in discs. To describe this property, recall that a piping scheme stands for a system $\mathcal{K}$ of arcs in a disc(s) D ($=$ 2-handles of $M^+(k)$) with $\mathcal{K} \cap \partial D = \mathcal{K}$. A selection of normal vectors for every arc from $\mathcal{K}$ is called a *normal orientation* for $\mathcal{K}$. The relevance of this notion lies in the fact that any normal orientation of the surface S induces a normal orientation of $\mathcal{K}$. Given a normally oriented arc-system and an arc l in D with $l \cap \mathcal{K} = \partial l$, we define $\mathcal{K}_l := (\mathcal{K} - U(l)) \cup l' \cup l''$, where l', l'' are the two copies of l in $\partial U(l)$. Note that $\mathcal{K}_l$ is again an arc-system iff l joins different components from $\mathcal{K}$ and that $\mathcal{K}_l$ carries an induced normal orientation iff l meets $\mathcal{K}$ from the same side (given by the normal orientation). In the latter case we say that $\mathcal{K}_l$ is obtained from $\mathcal{K}$ by an *oriented saddle move*. The relevance of this move is due to the following observation:

34.2. Proposition. *Let* $\mathcal{K}$ *and* $\mathcal{L}$ *be two arc-systems in* D *with* $\partial \mathcal{K} = \partial \mathcal{L}$ *and such that the normal orientations coincide at* $\partial \mathcal{K}$. *Then* $\mathcal{L}$ *can be obtained from* $\mathcal{K}$ *by a finite sequence of oriented saddle moves.*

Proof. Let l be some outermost arc from $\mathcal{L}$, i.e. an arc which separates a disc D_0 from D with $D_0 \cap \mathcal{L} = l$. Then w.l.o.g. $l \cap \mathcal{K} = \partial l$ and l is isotopic (fixing end-points) to an arc k either from $\mathcal{K}$ or $\mathcal{K}_l$. Moreover, $\mathcal{K}_l$ is obtained from $\mathcal{K}$ by an oriented saddle move, and so w.l.o.g. we may assume that k belongs to $\mathcal{K}$. Thus, replacing $\mathcal{K}$, $\mathcal{L}$ by $\mathcal{K} - k$, $\mathcal{L} - l$, the proposition follows by induction. $\Diamond$

The previous result tells us that all piping schemes can be obtained from some given one by orientable saddle moves alone. The point being that the number of possible orientable saddle moves is often not very big, and sometimes even non-existence (but in most cases much smaller then the total number of arc-systems in D supported by $\partial \mathcal{K}$). This may be helpful in calculating Thurston norms.

34.3. *Separating Surfaces.*

If a 3-manifold has no non-separating essential surfaces, it still can have separating ones. The same is true for relative handlebodies. In fact, a completed n-relator 3-manifold $M^*(k)$ contains a non-separating and essential surface if and only if its underlying relative handlebody (M, k) does. The same needs not be true for separating surfaces. Indeed, it is our goal to show that Poincaré's homology sphere provides us with a concrete counterexample.

To be more specific, recall from [He, p. 19] that Poincaré's homology sphere is (in our terminology) the completed 2-relator 3-manifold $N = M^*(k)$ corresponding to the relative handlebody, (M, k), shown in fig. 33.2. The fundamental group of N has the presentation

$$\mathcal{P} = <\, g_1, g_2 \mid g_1 g_2^{-1} g_1^{-1} g_2^{-1} g_1 g_2, \ g_1 g_2^{-1} g_1^{-1} g_2^2 g_1^{-1} g_2^{-1} \,> .$$

Geometrically, the two relator curves k_1 and k_2 of k have the reading

$$k_1 = +0.1 - 1.2 - 0.3 - 1.6 + 0.5 + 1.4 \text{ and } k_2 = +0.0 - 1.3 - 0.4 + 1.0 + 1.1 - 0.2 - 1.6.$$

Using the algorithm from 7.3, it is easily verified that (M, k) is simple. It is not full, but it is easily seen that the argument from 9.13 applies to this special relative handlebody as well, and so in the following it will be treated as if it were a full relative handlebody.

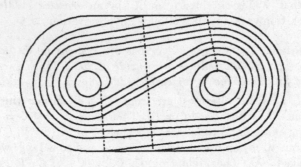

(Figure 34.2.)

To find essential surfaces in (M, k) we have to calculate the positive solution set of the Haken matrix (section 13). Again it is not hard to write a computer program for doing this. The problematic part is that such a program usually takes a very long time. Fortunately, this problem can be overcome for the example at hand.

Indeed, the Haken matrix associated to (M, k_1) is given by

$$A_{(M,k_1)} = \begin{bmatrix}
1 & 1 & 0 & 0 & 0 & 0 & 0 & 0 & 0 & 0 & 0 & 0 & -1 & 0 \\
0 & 1 & 1 & 0 & 0 & 0 & 0 & 0 & 0 & 0 & 0 & 0 & 0 & -1 \\
1 & 0 & 1 & -1 & -1 & 0 & 0 & 0 & 0 & 0 & 0 & 0 & 0 & 0 \\
0 & 0 & 0 & 0 & 1 & 1 & 0 & 0 & 0 & 0 & 0 & 0 & -1 & 0 \\
0 & 0 & 0 & 1 & 0 & 1 & -1 & -1 & 0 & 0 & 0 & 0 & 0 & 0 \\
0 & 0 & 0 & 0 & 0 & 0 & 0 & 1 & 1 & 0 & 0 & 0 & 0 & -1 \\
0 & 0 & 0 & 0 & 0 & 0 & 1 & 0 & 1 & -1 & -1 & 0 & 0 & 0 \\
0 & 0 & 0 & 0 & 0 & 0 & 0 & 0 & 0 & 0 & 1 & 1 & -1 & 0 \\
0 & 0 & 0 & 0 & 0 & 0 & 0 & 0 & 0 & 1 & 0 & 1 & 0 & -1
\end{bmatrix}$$

and similarly with (M, k_2). All positive solution vectors of $A_{(M,k_1)}x = 0$ are linear combinations (with positive integers as coefficients) of the following linear independent solutions

$x1$	$y1$	$z1$	$x2$	$y2$	$z2$	$x3$	$y3$	$z3$	$x4$	$y4$	$z4$	g_1	g_2
[0	0	2	2	0	0	1	1	1	2	0	0	0	2]
[0	1	0	0	0	1	0	1	0	0	0	1	1	1].
[0	1	0	0	0	1	1	0	1	1	1	0	1	1]
[0	2	2	0	2	0	0	0	4	3	1	1	2	4]
[2	0	0	1	1	1	2	0	0	0	2	0	2	0]

Only the x-values and the values for g_1, g_2 are relevant since the y- and z-values are then completely determined. The relevant basis vectors may therefore be written

		$x1$	$x2$	$x3$	$x4$	g_1	g_2
b_1	$:=$ [0	2	1	2	0	2]
b_2	$:=$ [0	0	0	0	1	1]
b_3	$:=$ [0	0	1	1	1	1].
b_4	$:=$ [0	0	0	3	2	4]
b_5	$:=$ [2	1	2	0	2	0]

In the same way we obtain a basis for the positive solutions of $A_{(M,k_2)}x = 0$:

		$x1$	$x2$	$x3$	$x4$	$x5$	g_1	g_2
c_1	$:=$ [0	1	0	0	1	0	1]
c_2	$:=$ [0	1	1	1	1	0	1]
c_3	$:=$ [0	0	0	0	1	2	2].
c_4	$:=$ [0	0	2	3	2	2	2]
c_5	$:=$ [0	0	0	1	2	2	4]
c_6	$:=$ [2	1	2	2	0	2	0]

Now, any positive solution of the spectral equation $A_{(M,k)}x = 0$ for (M, k) can be obtained by writing together the x-values of two solutions $b = \sum \beta_i b_i$ and $c = \sum \gamma_i c_i$ with the properties that $b[5] = c[6]$ and $b[6] = c[7]$. But note that all g_1-values for the vectors c_i, $1 \le i \le 6$ are even, but some g_1-entries of the b_i's are odd. This gives a restriction. It follows that the following set is a basis for the positive solutions of the spectral equation.

$$\begin{cases} [\,0\,2\,1\,1\,2;\ 0\,2\,] & c_1 + c_2 \\ [\,0\,2\,1\,2\ ;\ 0\,2\,] & b_1 \end{cases} \qquad \begin{cases} [\,0\,0\,0\,0\,1;\ 2\,2\,] & c_3 \\ [\,0\,0\,1\,1\ ;\ 2\,2\,] & b_2 + b_3 \end{cases}$$

$$\begin{cases} [\,0\,0\,2\,3\,2;\ 2\,2\,] & c_4 \\ [\,0\,0\,1\,1\ ;\ 2\,2\,] & b_2 + b_3 \end{cases} \qquad \begin{cases} [\,0\,0\,0\,0\,1;\ 2\,2\,] & c_3 \\ [\,0\,0\,0\,0\ ;\ 2\,2\,] & b_2 + b_2 \end{cases}$$

$$\begin{cases} [\,0\,0\,2\,3\,2\,; & 2\,2\,] & c_4 \\ [\,0\,0\,0\,0\ \ ; & 2\,2\,] & b_2 + b_2 \end{cases} \qquad \begin{cases} [\,0\,0\,0\,0\,1\,; & 2\,2\,] & c_3 \\ [\,0\,0\,2\,2\ \ ; & 2\,2\,] & b_3 + b_3 \end{cases}$$

$$\begin{cases} [\,0\,0\,2\,3\,2\,; & 2\,2\,] & c4 \\ [\,0\,0\,2\,2\ \ ; & 2\,2\,] & b_3 + b_3 \end{cases} \qquad \begin{cases} [\,0\,0\,0\,1\,2\,; & 2\,4\,] & c_5 \\ [\,0\,0\,0\,3\ \ ; & 2\,4\,] & b_4 \end{cases}$$

$$\begin{cases} [\,2\,1\,2\,2\,0\,; & 2\,0\,] & c_6 \\ [\,2\,1\,2\,0\ \ ; & 2\,0\,] & b_5 \end{cases} \qquad \begin{cases} [\,0\,2\,0\,0\,2\,; & 0\,2\,] & c_1 + c_1 \\ [\,0\,2\,1\,2\ \ ; & 0\,2\,] & b_1 \end{cases}$$

$$\begin{cases} [\,0\,2\,2\,2\,2\,; & 0\,2\,] & c_2 + c_2 \\ [\,0\,2\,1\,2\ \ ; & 0\,2\,] & b_1 \end{cases}$$

Explicitly, the fundamental solutions of the spectral equations are therefore given by

$$f_1 = [\,0\,2\,1\,2, 0\,2\,1\,1\,2; 0\,2\,], \qquad f_2 = [\,0\,0\,1\,1, 0\,0\,0\,0\,1; 2\,2\,],$$

$$f_3 = [\,0\,0\,1\,1, 0\,0\,2\,3\,2; 2\,2\,], \qquad f_4 = [\,0\,0\,0\,0, 0\,0\,0\,0\,1; 2\,2\,],$$

$$f_5 = [\,0\,0\,0\,0, 0\,0\,2\,3\,2; 2\,2\,], \qquad f_6 = [\,0\,0\,2\,2, 0\,0\,0\,0\,1; 2\,2\,],$$

$$f_7 = [\,0\,0\,2\,2, 0\,0\,2\,3\,2; 2\,2\,], \qquad f_8 = [\,0\,0\,0\,3, 0\,0\,0\,1\,2; 2\,4\,],$$

$$f_9 = [\,2\,1\,2\,0, 2\,1\,2\,2\,0; 2\,0\,], \qquad f_{10} = [\,0\,2\,1\,2, 0\,2\,0\,0\,2; 0\,2\,],$$

$$f_{11} = [\,0\,2\,1\,2, 0\,2\,2\,2\,2; 0\,2\,].$$

These 11 fundamental solution vectors are linear independent and therefore also form a basis of the null space of the Haken matrix $A_{(M,k)}$ since, by prop. 13.14, the dimension of this null space is at most 11.

According to prop. 13.16, every strongly essential surface in the relative handlebody (M, k) is of the form S_f^+, where f is a linear combination of the fundamental solutions, f_i, $1 \le i \le 9$ ((M, k) is not full, but this statement holds nevertheless - see above remark). In particular, we see that every such surface is obtained, by piping and the +-construction, from a disc system in (M, k) which consists of *even* numbers of copies of meridian-discs of (M, k). It therefore follows that every strongly essential surface in (M, k) is separating - well in line with the fact that the first homology of $M^*(k)$ is trivial. (Here we like to emphasize the fact that a solution vector always determines a separating 2-manifold if *all* its latter entries are even, providing us with a simple criterion for the separability property of surfaces, or rather 2-manifolds.)

According to section 13, there is a set of $\le 2^{11} = 2048$ basic solutions associated to the above set of 11 fundamental solutions. This set is given by all solution vectors of the form

$$\sum\nolimits_{1 \le i \le 11} \epsilon_i f_i, \quad \text{with } \epsilon_i = 0 \text{ or } 1.$$

By cor. 13.23, (M, k) contains a strongly essential surface if and only if there is a surface which is equivalent to S_b^+ for some basic solution b and which is not totally compressible and ∂-compressible. The latter property is comparatively easy to test. This will be discussed next.

To fix ideas, let us consider the fundamental solution f_1. (M, k) is ∂-irreducible, and note that admissible surfaces in ∂-irreducible handlebodies are totally compressible if and only if all its boundary curves are contractible (since all closed surfaces in handlebodies are totally compressible). Therefore it is interesting to calculate these boundary curves. But these curves represent conjugacy classes in the fundamental group of M. This group in turn is a free group $\mathcal{F}_n = <g_1, g_2, ..., g_n>$, $n = $genus$(M)$, and so ∂S_{f_1} gives rise to a list of words in $\mathcal{F}_g$. In the case of S_{f_1}, we obtain the following list:

```
-2 -1 -2 1 2 1 -2
-2 -1 -2 1 2 0 -2 -1 2 -1 -2 0 1 2 1 -2 1 2 0
-1 0 1 0 0
-1 0 0 1 0
1 0 -1 0
```

Here $-a$ represents the generator g_a^{-1} and 0 means that the boundary curve just touches a meridian curve somewhere without intersecting it there. We see that three, but not all, words of the above list cancel to the trivial word. In particular, it follows that S_{f_1} is not totally compressible. The Euler characteristic $\chi S_{f_1}^+$ is given by the formula

$$\#(\text{meridian} - \text{discs}) - \tfrac{1}{2}\sum_i d_i \cdot \alpha(a_i) + \#(\text{trivial boundary words}).$$

Here d_i is the number of copies of the i-the meridian-disc, and $\alpha(i)$ is the total number of g_i's occurring in all the relators of (M, k). For the example at hand, we have

$$d_1 = 0, \ d_2 = 2, \ \alpha(1) = 6 \ \text{and} \ \alpha(2) = 7.$$

Thus, altogether, we have

$$\chi(S_{f_1}^+) = 2 - 7 + 3 = -2.$$

We further know that $S_{f_1}^+$ has exactly two boundary curves and so $S_{f_1}^+$ is the twice punctured torus.

In order to see whether $S_{f_1}^+$ is strongly essential in (M, k), we now apply the algorithm from section 12. For this, we consider the Heegaard diagram $\mathcal{D}$ associated to (M, k).

In order to construct the circle-pattern $C_{f_1} \in \mathcal{C}(\mathcal{D})$ representing the surface $S_{f_1}^+$, we proceed as follows. First we translate the solution vector into the piping information given as follows:

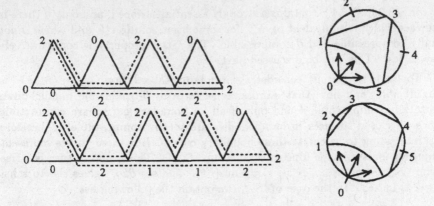

(Figure 34.3.)

From this we can easily produce a marking as well as the circle-pattern for S_{f_1}. Performing obvious ∂-compressions and forgetting trivial circles, we obtain the circle-pattern below for $S_{f_1}^+$:

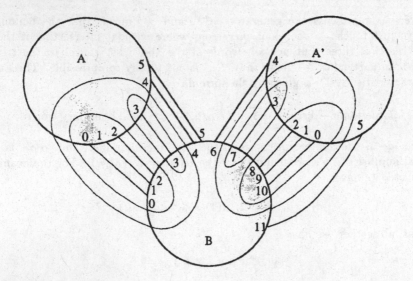

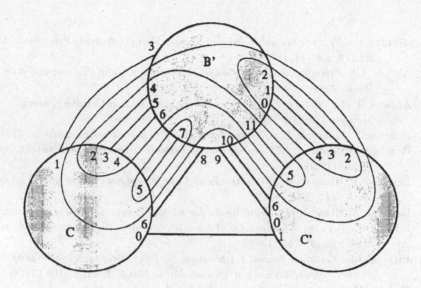

(Figure 34.4.)

Compressing along compression regions, as described in section 12, we now see that we cannot obtain any $(\partial\text{-})$compression disc, and so $S_{f_1}^{+}$ is really a *strongly essential*, twice punctured torus in (M, k).

Thus (M, k) contains an essential surface although $M^{*}(k)$ does not. As optimists, we take this as an indication that relative handlebodies contain some additional information about 3-manifolds which may be useful for a deeper study of 3-manifolds.

REFERENCES

[Al]   Alexander, J.W.: *On the subdivision of 3-space by a polyhedron*. Proc. Nat. Acad.
        Sci. 10, 6-8 (1924).

[AhBe] Ahlfors, L.V.-Bers, L.: *Riemann's mapping theorem for variable metrics*. Ann. of
        Math. 72, 385 - 404 (1960)

[Ap]   Apostol, T.M.: *Modular Functions and Dirichlet Series in Number Theory*.
        Springer (1976)

[BS]   Behnke, H. - Sommer, F.: *Theorie der analytischen Funktionen*. Springer (1972)

[BP]   Benedetti, R.-Petronio, C.: *Lectures on Hyperbolic Geometry*. Universitext, Springer
        Verlag (1991)

[BH 1] Bestvina, M.-Handel, M.: *Train tracks and automorphisms of free groups*. Ann. of
        Math. 135, 1-51 (1992)

[BH 2] Bestvina, M.-Handel, M.: *Train tracks for surface homeomorphisms*. Preprint

[Bin]   Bing, R.H.: *An alternative proof that 3-manifolds can be triangulated*. Ann. of
        Math. 69, 37 - 65 (1959)

[BGM]   Birman, J.S.- Gonzáles-Acuna, F.- Montesinos, J.M.: *Heegaard Splittings of
        Prime 3-Manifolds are not Unique*. Mich. Math. J. 23, 97-103 (1976)

[BZ 1] Boileau, M.-Zieschang, H.: *Genre de Heegaard d'une variété de
        dimension 3 et generateurs de son groupe fondamental*. C.R. Acad. Sc.
        Paris, t.296, série I, 925 - 928 (1983)

[BRZ]   Boileau, M.-Rost, M.-Zieschang, H.: *Décompositions de Heegaard des
        extérieurs des noeuds torique et des variétés de Seifert associées*. C.R.
        Acad. Sc. Paris, t.302, série I, n^0 18, 661 - 664 (1986)

[Bo]   Bonahon, F.: *Difféotopies des espaces lenticulaires*. Topology 22, 305 - 314 (1983)

[BO 1] Bonahon, F.-Otal, J.-P.: *Scindement de Heegaard des espaces lenticulaires*. C.R.
        Acad. Sc. Paris, t. 294, série I, 585 - 587 (1982)

[BO 2] Bonahon, F.-Otal, J.-P.: *Scindements de Heegaard des espaces lenticulaires*. Ann.
        Sc. Ec. Norm. Sup. (4) 16, 451 - 466 (1983)

[Br]   Brown, K.: *Presentations for groups acting on simply-connected complexes*. J. Pure
        Appl. Alg. 32, 1-10 (1984)

[Can]   Cannon, J.W.: *The combinatorial structure of cocompact discrete hyperbolic groups*.
        Geometriae Dedicata 16, 123-14 (1984)

[CB]   Casson, A.J. - Bleiler, S.A.: *Automorphisms of Surfaces after Nielsen and Thurston*
        London Math. Soc. Student Text 9 (1988)

[CG 1] Casson, A.J. - McA.Gordon, C.: *Reducing Heegaard splittings*. Topology Appl. 27,
        275 - 283 (1987)

[CG 2] Casson, A.J. - McA. Gordon, C.: *Manifolds with irreducible Heegaard splittings
        of arbitrary high genus* (unpublished)

[Chi]   Chillingworth, D.: *Simple closed curves on surfaces*. Bull. Lond. Math. Soc. 1,
        310 - 314 (1969)

[Cr]   Crowell, R.H.: *Genus of alternating link types* Ann. of Math. 69, 258 - 275 (1959)

[CF]   Crowell, R.H.- Fox, R.H.: *Introduction to Knot Theory*. Springer Verlag (1963)

[Cu]   Culler, M.: *Lifting representations to covering groups.* Adv. Math. 59, 64-70 (1986)

[CGLS]   Culler, M.-McA.Gordon, C.-Luecke, J.-Shalen. P.B..: *Dehn surgeries on knots.*
         Ann. of Math. 125, 237 - 300 (1987)

[De]   Dehn, M.: *Die Gruppe der Abbildungsklassen.* Acta math. 69, 135 - 206 (1938)

[Eng]   Engmann, R.: *Nicht-homöomorphe Heegaard-Zerlegungen vom Geschlecht 2 der
         zusammenhängenden Summe zweier Linsenräume.* Abh. Math. Sem. Univ.
         Hamburg 35, 33 - 38 (1970)

[FL]   Fathi, A.-Laudenbach, F.: *Difféomorphismes pseudo-Anosov et décomposition de
         Heegaard.* C.R. Acad. Sc. Paris, t. 291, série A, 423 - 425 (1980)

[FLP]   Fathi, A. - Laudenbach, F. - Poenarue, V.: *Travaux de Thurston sur les surfaces.*
         Astérisque 66 - 67, Soc. Math. de France, 66-67 (1979).

[FO]   Floyd, W.J. - Oertel, U.: *Incompressible surfaces via branched surfaces.* Topology 23,
         117-125 (1984).

[FHS]   Freedman, M.-Hass, J.-Scott, P.: *Least area incompressible surfaces in
         3-manifolds.* Invent. math. 71, 609-642 (1983)

[Fro]   Frohmann, C.: *The topological uniqueness of triply periodic minimal
         surfaces in R^3.* J. of Diff. Geom. 31, 277-283 (1990)

[Fu]   Funcke, K.: *Nicht frei äquivalente Darstellungen von Knotengruppen
         mit einer definierenden Relation.* Math. Z. 141, 205-217 (1975)

[Ga]   Gabai, D.: *Foliations and the topology of 3-manifolds. III.* J. Diff. Geometry 26,
         479-536 (1987)

[Ha 1]   Haken, W.: *Theorie der Normalflächen.* Acta math. 105, 245-375 (1961)

[Ha 2]   Haken, W.: *Über das Homöomorphieproblem der 3-Mannigfaltigkeiten I.* Math. Z. 80
         89-120 (1962).

[Ha 3]   Haken, W.: *Some results on surfaces in 3-manifolds.*
         Studies in Modern Topology, No. 5, pp. 39-98, Math. Assoc. Amer. (1968)
         (distributed by Prentice-Hall, Englewood Cliffs, N.J.)

[Ha 4]   Haken, W.: (unpublished)

[Ham]   Hamilton, A.J.S.: *The triangulation of 3-manifolds.* Quart. J. Math. Oxford Ser. 27,
         63 - 70 (1976)

[Han]   Hansen, V.L.: *Braids and coverings.* LMS student text 18 (1989)

[Has]   Hass, J.: *Genus two Heegaard splittings.* Proc. A.M.S. 114, 565 - 570 (1992)

[Hee 1] Heegaard, P.: *Forstudier til en topologisk teori för de algebraiske Sammenhäeng.*
         (Dissertation Univ. of Copenhagen (1898); published by det Nordiske
         Forlag Ernst Bojesen, Copenhagen (1898)).

[Hee 2] Heegaard, P.: *Sur l'Analysis situs.* Bull. Soc. Math, France 44, 161-242 (1916)
         (translation of [Hee 1])

[He]   Hemion, G.: *On the classification of homeomorphisms of 2-manifolds and the
         classification of 3-manifolds.* Acta math. 142, 123-155 (1979)

[Hem]   Hempel, J.: *3-manifolds.* Ann. of Math. Studies 86, Princeton U. Press (1976)

[Ja 0]   Jaco, W.: *Lectures on three-manifold topology.* CBMS regional conf. ser. math.
         43 (1977)

[Ja 1]   Jaco, W.: *Adding a 2-handle to a 3-manifold: An application to property R.* Proc.
         A.M.S. 92, 288 - 292 (1984)

[JO]   Jaco, W.-Oertel, U.: *An algorithm to decide if a 3-manifold is a Haken 3-manifold.*
         Topology 23, 195 - 209 (1984)

References

[JR]   Jaco, W.-Rubinstein, J.H.: *PL equivariant surgery and invariant decompositions of*
           3-manifolds. Adv. Math. 73, 149 - 191 (1989)
[JS]   Jaco, W.-Shalen, P.B.: *Seifert fibre spaces in 3-manifolds.* Memoirs of the Amer.
           Math. Soc. 21, No. 220 (1979).
[Joh 1]   Johannson, K.: *Homotopy equivalences of 3-manifolds with boundaries.* Springer
           LNM 761 (1978).
[Joh 2]   Johannson, K.: *On the mapping class group of simple 3-manifolds.* in: Topology
           of Low-Dimensional Manifolds, Sussex 1977, R. Fenn ed., Springer LNM,
           48 - 66 (1979)
[Joh 3]   Johannson, K.: *On surfaces in one-relator 3-manifolds.* London Math. Soc. Lecture
           Note Ser. 112, (ed. by D.B.A. Epstein),
           157-192 (1986).
[Joh 4]   Johannson, K.: *Classification problems in low-dimensional topology.* Geometric and
           algebraic topology, Banach Center Publ. 18, 37 - 59 (1986)
[Joh 5]   Johannson, K.: *On surfaces and Heegaard surfaces.* Trans. A.M.S. 325, 573 - 591
           (1991)
[Joh 6]   Johannson, K.: *On Heegaard graphs in surfaces.* Preprint (1991)
[Joh 7]   Johannson, K.: *On the Reidemeister-Singer theorem.* Preprint (1991)
[Joh 8]   Johannson, K.: *Heegaard surfaces in Haken 3-manifolds.* Bull. (new series) A.M.S. 2
           91-98 (1990)
[Joh 9]   Johannson, K.: *On the Loop- and Sphere Theorem.* Preprint (1993)
[KM]   Kalliongis, J.-Miller, A.: *Equivalence and strong equivalence of actions*
           on handlebodies. Trans. A.M.S. 308, 721 - 745 (1988)
[Ka]   Katok, S.: *Fuchsian groups.* Chicago Lecture Notes in Math. (1992)
[Ke 1]   Kerkhoff, S.P.: *The Nielsen realization problem.* Ann. of Math. 117, 235 - 265 (1983)
[Kne]   Kneser, H.: *Geschlossene Flächen und dreidimensionale Mannigfaltigkeiten.* Jahresb.
           d. Deut. Math. Verein., 38, 248-260 (1929)
[Knu]   Knuth. D.: *The Art of Computer Programming I-III.* Addison-Wesley (1968)
[Ko 1]   Kobayashi, T.: *Non-separating essential tori in 3-manifolds.* J. Math. Soc. Japan 36
           11 - 22 (1984)
[Ko 2]   Kobayashi, T.: *Heegaard genera and torus decompositions of Haken manifolds.*
           Preprint (1985)
[Ko 3]   Kobayashi, T.: *Torus decomposition with incompressible tori intersecting a*
           Heegaard surface in essential loops. Preprint (1985)
[Ko 4]   Kobayashi, T.: *Structure of full Haken 3-manifolds.* Osaka J. Math. 24, 173 - 215
           (1973)
[Ko 5]   Kobayashi, T.: *Heights of simple loops and pseudo-Anosov homeomorphisms.*
           Contemporary Math. 78, 327 - 338 (1988)
[Ko 6]   Kobayashi, T.: *A construction of 3-manifolds whose homeomorphism*
           classes of Heegaard splittings have polynomial growth. Preprint (1991)
[Ko 7]   Kobayashi, T.: *A criterion for detecting inequivalent tunnels for a knot.* Math.
           Proc. Camb. Phil. Soc. 107, 483 - 491 (1990)
[Kr]   Kramer, R.: *The twist group of an orientable cube-with-two-handles is not*
           finitely generated. Preprint
[Li 1]   Lickorish, L.B.R.: *A representation of orientable combinatorial 3-manifolds.* Ann.
           of Math. 76 (3), 531 - 540 (1962)

[Lo]   Long, D.D.: *On pseudo-Anosov maps which extend over two handlebodies.* Proc.
         Edinburgh Math. Soc. (2) 33, 181-190 (1990)

[LM]   Lustig, M.-Moriah,Y.: *Nielsen equivalence in Fuchsian groups and Seifert fibre
         spaces.* Topology 30, 191 - 204 (1991).

[LS]   Lyndon, R.C.-Schupp, P.E.: *Combinatorial Group Theory.* Springer Verlag (1970)

[MKS]   Magnus, W.-Karras, A.-Solitar, D.: *Combinatorial Group Theory.* Interscience
         (1966)

[MY 1]   Meeks, W.H.,III-Yau, S.T..: *The classical Plateau problem and the topology
         of three-dimensional manifolds.* Topology 21, 409 - 440 (1982)

[MY 2]   Meeks, W.H.,III-Yau, S.T.: *Topology of three-dimensional manifolds and the
         embedding problems in minimal surface theory.* Ann. of Math. 112,
         441 - 484 (1980)

[McC 1] McCullough, D.: *Twist groups of compact 3-manifolds.* Topology 24, 461 - 474
         (1985)

[McC 2] McCullough, D.: *Virtually geometrically finite mapping class groups of
         3-manifolds.* J. Diff. Geom. 33, 1-65 (1991)

[MMZ]   McCullough, D.-Miller, A.-Zimmermann, H.: *Group actions on handlebodies.* Proc.
         London Math. Soc. (3) 59, 373 - 416 (1989)

[McCP]   McCool, J.-Pietrowski, A.: *On free products with amalgamation of two infinite
         cyclic groups.* J. Algebra 18, 377 - 383 (1971)

[Moi 1] Moise, E.E.: *The triangulation theorem and Hauptvermutung.* Ann. of Math. 56,
         96 - 114 (1952)

[Moi 2] Moise, E.E.: *Geometric Topology in Dimensions 2 and 3.* Graduate Texts in
         Mathematics, Springer Verlag (1977)

[Mor]   Moriah, Y.: *Heegaard splittings of Seifert fibre spaces.* Invent. math. 91,
         465-481 (1988)

[Mos]   Mostow, G. D.: *Quasi-conformal mappings in n-space and the rigidity of
         hyperbolic space forms* Publ. IHES 34, 53 - 104 (1968) (see also [Th 1]
         and [BP])

[Nie 1] Nielsen, J.: *Abbildungsklassen endlicher Ordnung.* Acta math 75, 23 - 115 (1943)
         (for correction see: [Zieschang, H.: *On decompositions of discontinuous groups
         of the plane.* Math. Z. 151, 165 - 188 (1976)

[Nie 2] Nielsen, J.: *Surface transformation classes of algebraically finite type.* Math. fys.
         Meddelelser Kgl. Danske Vidensk. Selsk. XXI, 2 (1944)

[Och]   Ochiai, M.: *On Haken's theorem and its extensions.* Osaka J. Math. 20, 461 - 488
         (1983)

[Ot]   Otal, J.-P.: *Sur les scindements de Heegaard de la sphère S^3.* Toplogy 30,
         249 - 257 (1991)

[Pa 1]   Papakyriakopoulos, C.D.: *On Dehn's lemma and the asphericity of knots.* Ann. of
         Math. 66, 1-26 (1957)

[Pa 2]   Papakyriakopoulos, C.D.: *On solid tori.* Proc. London Math. Soc. VII, 281-299
         (1957)

[Pa 3]   Papakyriakopoulos, C.D..: *Some problems on 3-dimensional manifolds.* BAMS 64,
         317 - 335 (1958)

[Pr]   Przytycki, J.H.: *Incompressibility of surfaces after Dehn surgery.* Michigan Math.
         J. 30, 289 - 308 (1983)

[Rei 1]   Reidemeister, K.: *Zur dreidimensionalen Topologie*. Abh. Math. Sem. Univ.
          Hamburg 9, 189-194 (1936).

[Rei 2]   Reidemeister, K.: *Über Heegaard-Diagramme*. Abh. Math. Sem. Univ. Hamburg
          25, 140-145 (1961).

[Ru]   Rubinstein, J.H.: *Polyhedral minimal surfaces, Heegaard splittings and decision
          problems for 3-dimensional manifolds*. Preprint (1994)

[RS]   Rubinstein, J.H.- Scharlemann, M.: *Comparing Heegaard splittings of
          non-Haken manifolds*. Preprint 1994

[Scha 1] Scharlemann, M.: *Outermost forks and a theorem of Jaco*. Proc. Rochester Conf.,
          A.M.S. Contemporary Math. Series 44, 189 - 193 (1985)

[Scha 2] Scharlemann, M.: *The Thurston norm and 2-handle addition*. Proc. A.M.S. 100,
          362 - 366 (1987)

[SchT]   Scharlemann, M.- Thompson, A.: *Heegaard splittings of (surface) × I are standar*
          Preprint (1991)

[SchT2]   Scharlemann, M.- Thompson, A., *Thin position and Heegaard splittings of the
          3-sphere*. J. Diff. Geom. 39, 343-357 (1994)

[Schu]   Schubert. H.: *Bestimmung der Primfaktorzerlegung von Verkettungen*. Math. Z. 76,
          116 - 148 (1961)

[Schul]   Schultens, J.: *The classification of Heegaard splittings for (compact orientable
          surface) × S^1*. Proc. London. Math. Soc. (3) 67, 425-448 (1993)

[Sei]   Seifert, H.: *Topologie dreidimensionaler gefaserter Räume*. Acta math. 60,
          147 - 238 (1932)

[ST]   Seifert, H. - Threlfall, W.: *Lehrbuch der Topologie*. Teubner (1934)

[Sh]   Shalen, P.B.: *A "piecewise linear" proof of the triangulation theorem for 3-manifold*
          Adv. in Math. 52, 34 - 80 (1984)

[Si]   Singer, J.: *Three-dimensional manifolds and their Heegaard-diagrams*. Trans. Amer.
          Math. Soc. 35, 88-111 (1933).

[Sta]   Stallings, J.: *On the loop theorem*. Ann. of Math. 72, 12 - 19 (1960)

[Sta 2]   Stallings, J.: *On fibering certain 3-manifolds*. Topology of 3-manifolds, Prentice Hall
          95 - 100 (1962)

[Th 1]   Thurston, W.P.: *Geometry and topology of 3-manifolds*. Princeton Notes, Preprint
          (1979)

[Th 2]   Thurston, W.P.: *Three dimensional manifolds, Kleinian groups and
          hyperbolic geometry*. Bulletin A.M.S. (New Series) 6 (3), 357 - 381 (1982)

[Th 3]   Thurston, W.P.: *A norm for the homology of 3-manifolds*. Memoirs of the A.M.S. 5!
          100 - 130 (1986)

[Th 4]   Thurston, W.P.: *Hyperbolic structures on 3-manifolds, I: Deformation of
          acylindrical manifolds*. Ann. of Math. 124, 203 - 246 (1986) (see also
          Preprint)

[Tu 1]   Tukia, P.: *Quasiconformal groups of compact type and the Nielsen
          realization problem*. Preprint

[VKF]   Volodin, I.A.-Kuznetsov, A.T.-Fomenko, A.T.: *The problem of discriminating
          algorithmically the standard three-dimensional sphere*. Russian Math.
          Surveys 2972-168 (1974).

[Wa 1]   Waldhausen, F.: *Eine Klasse von 3-dimensionalen Mannigfaltigkeiten I,II*. Invent.
          Math. 3, 308-333; ibid. 4, 87-117 (1967).

[Wa 2] Waldhausen, F.: *On irreducible 3-manifolds which are sufficiently large.* Ann. of Math. 87, 56 - 88 (1968)

[Wa 3] Waldhausen, F.: *Heegaard-Zerlegungen der 3-Sphäre.* Topology 7, 195-203 (1968).

[Wa 4] Waldhausen, F.: *Recent results in the theory of 3-manifolds.* Proc. of Symp. in Pure Math. 32, 21 - 38 (1978)

[Wa 5] Waldhausen, F.: *Some problems on 3-manifolds.* Proc. of Symp. in Pure Math. 32, 313-322 (1978)

[Wh 1] Whitehead, J.H.C.: *On certain sets of elements in a free group.* Proc. London Math. Soc. (2) 41, 48-56 (1936); Collected Works Vol II, Pergamon Press, New York, 69-77 (1962)

[Wh 2] Whitehead, J.H.C.: *On equivalent sets of elements in a free group.* Ann. of Math. 37, 782-800 (1936); Collected Works Vol. II, Pergamon Press, New York, 79-97 (1962)

[Zie 1] Zieschang, H.: *Über einfache Kurven auf Vollbrezeln.* Abh. Math. Sem. Univ. Hamburg 25, 231-250 (1962).

[Zie 2] Zieschang, H.: *Classification of simple systems of paths on a solid pretzel of genus 2.* Soviet Math. 4, 1460-1463 (1963) Transl. of Doklady Acad. Sci. USSR 152, 841-844 (1963)

[Zie 3] Zieschang, H.: *On simple systems of paths on complete pretzels.* Amer. Math. Soc. Transl. (2), 127-137 (1970)

[Zie 4] Zieschang, H.: *Finite groups of mapping classes of surfaces.* Springer LNM 875 (1981)

[ZVC] Zieschang, H.-Vogt, E.-Coldewey, H.-D.: *Surfaces and planar discontinuous groups.* Springer LNM 835 (1980)

INDEX